Ian Johnston, Niki Papavramidou

Galen on the Pulses

Medicine in the Medieval Mediterranean

Edited by
Alain Touwaide

Volume 10

Ian Johnston, Niki Papavramidou

Galen on the Pulses

Medico-historical Analysis, Textual Tradition, Translation

DE GRUYTER

ISBN 978-3-11-061161-8
e-ISBN (PDF) 978-3-11-061267-7
e-ISBN (EPUB) 978-3-11-061188-5
ISSN 2569-314X

Library of Congress Control Number: 2023911827

Bibliographic information published by the Deutsche Nationalbibliothek
The Deutsche Nationalbibliothek lists this publication in the Deutsche Nationalbibliografie; detailed bibliographic data are available on the Internet at http://dnb.dnb.de.

Cover illustration: Redrawings of two enamels on the Holy Crown of Ungary, Byzantium, ca. 1070 A.D.
Printing and binding: CPI books GmbH, Leck

www.degruyter.com

Preface

This project has had a very long gestation followed by a protracted parturition. It began in 2004 when Niki, then in residence at the US National Library of Medicine in Bethesda, MD, unearthed two boxes of typed translations of a number of Galen's works done by the Harvard anatomist, Robert Montraville Green (1880–1955). These translations were prepared between 1950 and 1953. When Green died in 1955, only one of his translations, that of Galen's *Hygiene*, had been published. The initial plan of the present book was to publish Green's translations of Galen's four treatises on the arterial pulses which had not previously been translated into any modern language, nor had any critical editions of the Greek texts been prepared. The project was taken on by Alain Touwaide, editor of the series *Medicine in the Medieval Mediterranean* at that time being published by Ashgate Variorum.

The original idea was to leave Green's original translations intact but supplement them with explanatory footnotes focussing particularly on a number of the possible variant readings to be found in Greek manuscripts collated by Niki. I was asked to write an introduction to the work directed largely at the medical aspects of Galen's treatises, which I did. In doing this, it became obvious that Kühn's edition was in need of revision and Green's translations were rudimentary and unsuitable for publication as they were—rough initial drafts only.

A certain momentum had, however, been gained and Alain was keen to see the works published in translation. They are very detailed and complex descriptions of all aspects of sphygmology and were to become a cornerstone of clinical practice in Western and Arabian medicine for many centuries. In addition to establishing the theoretical foundation and methodology of pulse diagnosis definitively, they are a mine of information on the historical development of ideas on the nature and clinical relevance of the arterial pulses from the very beginnings of Western medicine, and well worth publishing on these grounds alone.

Alas, the project stalled. It seemed better to reshape it and to revise Green's translations to make them suitable for publication. I agreed to do this but quickly realised that the translations needed to be redone completely. I therefore discarded Green's rough drafts altogether and started the translations again from scratch. The center of gravity of the book moved accordingly from Green's manuscript to a new translation and a focus on Galen's medical system.

In the meantime, Ashgate had been absorbed into Routledge and, shortly afterwards, the series *Medicine in the Medieval Mediterranean* under Alain's editorship was transferred to De Gruyter, who agreed to honour existing contracts. So work on the project continued. The introduction was completely redone, and I added three short Galenic treatises on the pulses—*Pulses for Beginners, On the Use of the Pulses* and *Synopsis of the Pulses*—together with the sections on the pulses in the pseudo-Galenic *Medical Definitions*, so making the present volume a comprehensive collection of Galen's writings on the subject.

https://doi.org/10.1515/9783110612677-001

What it was hoped would be the final version was submitted in 2018, but in fact, following readers' reviews, a further revision of the introductory material was needed. This was completed in early 2019 and the 'final' final manuscript submitted in August 2019. Then COVID struck, affecting 'not less than everything.' The project was one of a myriad endeavours that was halted but now, despite the ongoing problems with COVID, there were signs of life again.

In preparing this work, Niki was responsible for gathering and collating the manuscripts of the four major treatises and providing footnotes where there were significant textual variations. She also read through some parts of the translation. I was responsible for the translation itself and also for the medical-historical introduction. The section on the textual tradition has been a collaborative effort on the basis of Niki's initial collection of material and collation of the manuscripts.

The hope is that all the efforts, including those of parents, colleagues, librarians, and friends, have resulted in a useful contribution to the knowledge of the great physician's writings.

Ian Johnston, Cloudy Bay, April 2023

Contents

Acknowledgements

Substantial contributions have been made by several people to whom profound thanks are due. First and foremost, Alain Touwaide must be recognised as the driving force who kept this project alive when it had become almost moribund—indeed with a barely palpable pulse! In particular, his help with the treatment of the manuscripts was invaluable. Three "medical" people contributed significantly. Susie Collis undertook the very considerable task of reading the whole manuscript, looking particularly for passages which lacked clarity or were confusing so they could be improved upon, while retaining fidelity to Galen's often convoluted prose. My daughter, Justine, and a friend, Shiva Roy, a practising cardiologist, read the medical introduction and offered critical and helpful suggestions. Elaine Hawkins, a retired medical secretary, typed out the whole manuscript. Tapes were hand-delivered to Elaine via a walk through the bush separating our somewhat isolated dwellings. This provided a pleasant and regular social component to the whole process. The final word of gratitude must go to Alain. Without his particular combination of attention to detail, scholarship and tenacity this project would never have got across the line. Certainly, he went above and beyond the ordinary duties of an editor—thank you Alain!

Ian Johnston, Cloudy Bay, April 2023

First and foremost, I wish to thank Elizabeth Fee (1946–2018). When in residence at the National Library of Medicine (2003–2004), Elizabeth entrusted me with Green's manuscripts and urged me to publish them, besides providing guidance and unfailing moral support with admirable love. I am particularly grateful to Alain Touwaide and Emanuela Appetiti who gave me access to their splendid and exceptional library on the history of Greek medicine in 2007, including copies of, and documentation on, numerous Greek manuscripts with medical contents. They hosted me at the Smithonian Institution, offering me a truly unique opportunity to do research in close collaboration on a great many medico-historical topics which we discussed together. I thank all the libraries which provided copies of their manuscripts before the era of massive digitization, particularly Julia Zakhariadi who contacted the State Historical Museum of Moscow for me and arranged for a reproduction of the 464 Moscow codex. Last, but far from least, I cannot say enough how grateful I am to my parents, Spiros and Giota Papavramidis, for their support during my research stay in the USA and throughout my life, my husband Panos Kotsalis, and our two daughters, Giota and Kalliopi, for their patience, encouragement, and endless love.

Niki Papavramidou, Thessaloniki, April 2023

https://doi.org/10.1515/9783110612677-002

List of Schemas and Tables

https://doi.org/10.1515/9783110612677-003

Part One: **Medico-historical Analysis and Textual Tradition**

I Introduction

Recorded recognition of the arterial pulse as a significant component of human physiology and medicine starts with the following statement from the so-called Ebers papyrus dating from 1550 B.C. ca.[1]

> There are vessels in him at every part of the body. As far as these [vessels] are concerned: if with reference to them any *swnw*-physician, and priest of Sekhmet, any magician gives both hands, his fingers on the head, on the back of the head, on the hands, on the place of the heart, on both arms, on both legs then he is measuring for the heart. For its [the heart's] vessels [lead] to every part of his body. It is the case: it [the heart] speaks in front, in the vessels in every part of the body.

After a hiatus of a thousand years or so there is further evidence of attention to the pulse in all three of the major enduring medical traditions, dating from the *Hippocratic Corpus* and Aristotle (5th and 4th cent. B.C.) in Western medicine, Bian Que (late 5th—4th cent. B.C.) in Chinese medicine, and Sage Kanada (also known as Kashyapa) (between the 6th and 2nd cent. B.C.) in Ayurvedic medicine. These were the beginnings of a continuous and increasing development of sphygmology to the present day.

In Western medicine, in the history of which Galen (A.D. 129—ca. 216) and his revered predecessor Hippocrates (ca. 460—between 375 and 351 B.C.) stand preeminent, the history of ideas and uses of the arterial pulses may be divided into three stages. First, the period from Hippocrates to Galen, some 600 years, saw the progressive clarification of the anatomical basis of the pulses, establishment of the theoretical foundations of sphygmology, and the increasing practical application of this component of the art. This culminated in the formulation of a detailed account of the nature of the pulses and their role in medical practice, as documented by Galen. Second, the period from Galen's death (after A.D. 216) to William Harvey (1578–1657) and his ground-breaking work on the circulation in 1628, saw Galenic sphygmology, documented in the treatises translated in the present work, dominate the teaching and practice of medicine in the West. It extended from the medical encyclopaedists of the centuries immediately following Galen, through the flourishing of Arabic medicine from the 8th to the 12th century. The 13th to 16th centuries in Europe saw the beginnings of university teaching based on the use and, later, rediscovery of the classical texts from Antiquity, and their translation into Latin. In medicine this particularly applied to Galen's works which were widely used for both teaching and practice. Third, the period from Harvey to the present day saw the gradual attrition of Galenic medicine generally, and his sphygmology in particular, with a move of the latter to an increasingly quantitative and instrument-based approach and away from the simple palpation of the pulse at the patient's bedside or in the doctor's office.

1 Cited from von Staden 1989: 9.

https://doi.org/10.1515/9783110612677-004

Galen's substantial body of work on sphygmology constitutes a watershed between the first and second periods. In summary, Galen's writings on the pulses accomplished three things:

1. they provided a comprehensive survey of work done on the subject in the Graeco-Roman world prior to and including his time;
2. they established a clear theoretical and practical basis for the use of the pulses in medicine;
3. they remained the authoritative statement on the subject, a foundational component of medical practice, until the start of the 20th century, Harvey's fundamental revision of the concept of the circulation of the blood in the 16th century notwithstanding.

There are seven genuine Galenic treatises on the pulses included in the early-19th century edition of Galen's Greek text (*Galeni Opera Omnia*, 20 volumes in 22 tomes, 1821–1833) by the German physician and historian of medicine Karl Gottlob Kühn (1754–1840).

The aim of the present work is to present complete English translations of these seven works prefaced by translations of the entries on the pulses in the pseudo-Galenic *Definitiones medicae.*

Such a work provides in English as full an account of Galen's views on the arterial pulses and their relevance to clinical practice as the extant material allows. It also provides useful information on earlier and near-contemporary writings on the subject, almost none of which have survived.

Only one of Galen's seven treatises has a critical edition—that of *De usu pulsuum* prepared by David John Furley (1922–2010) and James Sterling Wilkie (1906–1982). Apart from this, all the translations here are made from Kühn's Greek text with some reference to Kühn's Latin and several of the manuscripts listed in this *Introduction.* It was not our intention to provide a critical edition of any of the treatises. It was simply to provide the first English translation of these works in full based largely on Kühn's edition.

In the first part of this medico-historical analysis of pulse theories, we shall start with an overview of the development of ideas on the nature and uses of the pulses in Western medicine up to the present time in the following seven sections:

1. Galen's predecessors and contemporaries
2. Galen's writings on the arterial pulses;
3. Galen's views on the pulses;
4. the pulse from Galen to Harvey;
5. William Harvey;
6. from Harvey to the 20th century;
7. modern views on the pulses.

This will be followed by a short section (8) on the pulses in the other two major traditions mentioned above (Chinese and Ayurvedic medicine) with particular attention to correspondences with Galen's views.

Following this, the first part deals with the textual tradition of Galen's writings on pulses with the following two sections:

1. the manuscripts and their linkages (stemma codicum);
2. an attempt toward a textual history of the Galenic corpus sphygmologicum.

II Medico-historical Analysis

1 Galen's Predecessors and Contemporaries

There was clearly a substantial body of thinking and writing on the arterial pulse prior to Galen, material which Galen utilised, responded to, and criticized, particularly in the four major treatises, but also in his other writings on the pulse and elsewhere. Unfortunately, much of the pre-Galenic written material, including a number of works specifically devoted to the pulse, has been lost. Further, our knowledge of what this material was, and of the observations and concepts contained in it, depends heavily on what Galen himself has documented. In the four major treatises, 25 individuals are mentioned specifically in relation to the pulse, together with the Dogmatic and Empiric schools generally. Also mentioned are several philosophers, most notably Plato (427–347 B.C.), Aristotle (384 – 322 B.C.) and Chrysippus (ca. 280 – 207 B.C.), referred to in regard to more general matters, which make up the frequent and often substantial digressions that characterise these four works.

We must begin with Hippocrates (ca. 460-between 375 and 351 B.C.) and Aristotle (384 – 322 B.C.). The general consensus, based on available material, is that Hippocrates and other contributors to the *Hippocratic Corpus* were aware of the phenomenon of the pulse, at least in certain arteries, but did not make use of the arterial pulse as a diagnostic indicator. Thus, in two editions of the Hippocratic writings, William H. S. Jones (1867–1939), in the first instance, writes: "... in all the Hippocratic collection no attention is paid to the pulse" and "It is a fact that no use is made of this knowledge (of the pulse) in any treatise of the collection, but we must not infer from this that Hippocratic writers were ignorant of pulses. We can only infer that they were ignorant of their medical importance."[2] In the second instance, Geoffrey E. R. Lloyd states: "Although the phenomena of pulsation, throbbing and palpitation are referred to by Hippocratic writers, the value of the pulse in diagnosis was not appreciated until after the date of most of the Hippocratic treatises."[3] There is mention of pulsation in the Hippocratic *Nutriment* (XLVIII), but it is linked with the veins, while there are two references to pulsation of the heart in the late treatise on the heart.[4] What Galen himself says, early in the first of the four treatises (*De pulsuum differentiis*) is as follows:

> Hippocrates, then, was the first of all those we know to write the name of the pulse and he does not seem to have been ignorant of the art relating to it. He did not elaborate further on this part of the art nor did he assign a name to every movement of arteries.[5]

2 Jones 1923: xx and xxxii.

3 Lloyd 1978: 31.

4 Lloyd 1978: 348, 350.

5 Galen, *De pulsuum differentiis* I.2 (VIII.497K).

https://doi.org/10.1515/9783110612677-005

Aristotle, who had a profound and wide-ranging interest in biological matters, refers to the pulse on several occasions, as below:

> In connection with the heart there are three phenomena, which, though apparently of the same nature, are really not so, namely palpitation, pulsation, and respiration ... The beating of the heart, which, as can be seen, goes on continuously, is similar to the throbbing of an abscess ... In the heart the beating is produced by the heat expanding the fluid, of which the food furnishes a constant supply. It occurs when the fluid rises to the outer wall of the heart, and it goes on continuously; for there is a constant flow of the fluid that goes to constitute the blood, it being in the heart that the blood is first created. That this is so we can perceive in the initial stages of generation, for the heart can be seen to contain blood before the veins become distinct. This explains why pulsation in youth exceeds that in older people, for in the young the exhalation is more abundant. All the veins pulsate, and do so simultaneously with each other, owing to their connection with the heart. The heart always beats, and hence they also beat continuously and simultaneously with each other and with it. Palpitation, then, is the recoil of the heart against the compression due to cold; and pulsation is the volatilization of the heated fluid.[6]

> The blood in the blood-vessels (*phlebes*) pulsates in all animals altogether at the same time.[7]

> Whereas there are three motions belonging to the breath in the windpipe—respiration, pulsation, and a third which introduces and assimilates the nutriment—we must define how and where and for what purpose each takes place ... The pulse is something peculiar and distinct from the other motions and in some respects may be seen to be contingent, assuming that when there is an excess of warmth in a fluid, that fluid which is evaporated must set up a pulsation owing to the air being intercepted in the interior, and pulsation must arise in the originating part and in the earliest stage, since it is inborn in the earliest parts. For it arises firstly and in the greatest degree in the heart, and thence extends to the other parts ... That the pulse has no connection with the respiration is shown by the following indication—whether one breathes quickly or regularly, violently or gently, the pulse remains the same and unchanged, but it becomes irregular and spasmodic owing to certain bodily affections and in consequence of fear, hope, and anguish affecting the soul. Next we ought to consider whether the pulse also occurs in the arteries and with the same rhythm and regularity. This does not appear to be so in the case of the parts widely separated, and, as has been noted, it seems to serve no purpose whatsoever.[8]

It is said that the differentiation between arteries and veins was due to Praxagoras (late 4th / early 3rd cent. B.C.), although this is a complex question.[9] It may also be said that Praxagoras initiated the method of palpating the pulse for diagnostic purposes, at least in Western medicine. In terms of overall circulatory physiology, Praxagoras held that the arteries contained only air, whilst blood was present in the veins. The key features of his pulse theory, pieced together from fragmentary evidence only, are the following:

1. the pulsation of the arteries does not depend on the heart in that they have their own innate capacity to pulsate;

---

6 Aristotle, *De juventute et senectute* 479b15–480a15 (Engl. transl. Ross 1984: 761–762).

7 Aristotle, *Historia animalium* 521a7–8 (ed. and Engl. transl. Peck 1979: 220–221).

8 Aristotle, *De spiritu* 482b14–483a8 (Engl. transl. Dobson 1984: 767–768).

9 On Praxagoras' sphygmology, see Lewis 2017.

2. the arteries are filled with *pneuma* because they pulsate (dilate)—i. e., the 'bellows' model;
3. the physician can make inferences about the qualities of the humors from the pulse, even though the humors are not present in the arteries;
4. the pulse is an indicator of events taking place in the veins by virtue of "bubbles" arising in the course of normal digestion.

It is clear, then, that Praxagoras is of particular importance in the history of ideas about the pulse, not least because of his influence on his most famous pupil, Herophilus of Chalcedon (335–280 B.C.). Fritz Steckerl (1911–1988) writes:

> Praxagoras' system of the pulse was probably a very elaborate one since he considered the pulse as such an important diagnostic tool. Praxagoras must have advanced the study of the pulse in a considerable way. His pupil, Herophilus, wrote a book on the pulse. The interest of his teacher was certainly stimulating. Once this method of diagnosis had become 'modern', there was probably no limit to the subtleties which were considered to be significant for the condition of the patient ... Observation of the pulse, together with the examination of the urine, probably served him as a first indication of the specific cause of a given disease.[10]

In his writings on the pulses, Galen takes issue with Praxagoras' theory over the following points:

1. there must be clarity and consistency in terminology, specifically regarding the terms σφυγμός, σπασμός, παλμός, and τρόμος. It is not clear from available evidence whether Praxagoras considered the distinction between these four types of movement of an artery as one of magnitude only, or as that between movements κατὰ φύσιν and παρὰ φύσιν (*in accord with* and *contrary to nature*, respectively). Galen makes the point, already made by Herophilus, that σφυγμός (pulse) must be restricted to the normal movement of arteries and is the only such movement;
2. the arteries have their own innate capacity to pulsate—Galen takes the view that this capacity comes from the heart;
3. one cannot take a position on what is contained in the arteries on the basis of palpation of the pulse—this in relation to the view that the arteries contain air/*pneuma* alone, as Praxagoras and Erasistratus (ca. 304–250 B.C.) believed.

Herophilus substantially advanced the study of the arterial pulse and was the author of the first known Greek book on the subject—*On Pulses*—, which was written in the 3rd century B.C. Our knowledge of Herophilus' contribution to sphygmology, analysed by Heinrich von Staden,[11] is heavily dependent on Galen. Thus, of the 48 testimonia provided by von Staden, 35 are from Galen, predominantly from the four long treatises translated in the present work. The remaining 13 are from a miscellaneous group in-

10 Steckerl 1958: 26–27.

11 von Staden 1989: 262–287 and 322–360.

cluding Marcellinus (2nd cent. A.D. [?]) (4 testimonia), Rufus of Ephesus (1st cent. A.D.) (2 testimonia) and Pliny (A.D. 23/24–79) (2 testimonia). For purposes of clarity and brevity, the main elements of Herophilus' pulse theory may be enumerated as follows:

1. the identity of the four key features to be evaluated in examining the arterial pulse —size/magnitude (μέγεθος), rapidity/speed (τάχος), strength (σφοδρότης), and rhythm (ῥυθμός). There is some debate on whether he also made distinctions in terms of regularity and irregularity (τάξις and ἀταξία), and equality/evenness (ὁμαλότης) and inequality/unevenness (ἀνωμαλία), and possibly also fullness (πλῆθος/πληρότης);
2. the use of musical and of metrical analogies in relation to rhythm;
3. the identification of different pulse types in relation to the stages of life—infancy, childhood and adolescence, prime, post-prime and old age;
4. the use of a portable water-clock (clepsydra), calibrated according to age and temperament, to provide a quantitative evaluation of the pulse rate;
5. the use of names for specific types of pulse—gazelle-like, ant-like etc.

In terms of cardiovascular function, Herophilus recognised the connection of the arterial system with the heart and believed the former to contain both blood and *pneuma*. The arteries, he thought, had no innate capacity to pulsate, but received this capacity from the heart. He identified the two phases of the pulse as systole (contraction) and diastole (dilatation), the former being an active function (ἐνέργεια) of the arteries and the latter a return (ἐπάνοδος) to their natural condition. Upon dilatation, the arteries were thought to draw in what filled them and then to expel this by their contraction. He considered that the contraction and dilatation of all arteries occurred simultaneously.

Although Galen does take issue with Herophilus and his followers on a number of points with regard to the pulse, he is, on balance, favourably disposed towards him as the following quotation indicates:

> Herophilus ... is a man who is known by everybody to have surpassed the great majority of the ancients, not only in breadth of knowledge but in intellect, and to have advanced the art of medicine in many ways; as, for instance, by his *logos* of the pulsation of 'veins',[12] which one needs more now and finds more useful than any other *logos*, for deriving benefit therefrom, while those before overlooked it and neglected to investigate it. We find, however, that this Herophilus concedes no small importance to experience ...[13]

12 von Staden 1989: 325 (from whom this passage is quoted here) adds the following: "The original must have read ἀρτηριῶν (arteries) since Herophilus attributed pulsation only to the arteries, not to the veins."

13 von Staden 1989: 325.

Among the individuals mentioned by Galen in the four major treatises on the pulses there are eleven who might be regarded as Herophileans.[14] These are, in probable chronological order:

- Bacchius (3rd cent. B.C.);
- Zenon (2nd cent. B.C.);
- Chrysermus (1st cent. B.C.);
- Hegetor (1st cent. B.C.);
- Heracleides of Erythrae (1st cent. B.C.);
- Alexander Philalethes (ca. 50 B.C. to A.D. 25);
- Heracleides of Tarentum (last half of 1st cent. B.C.);
- Apollonius Mus (50 B.C.-A.D. 30);
- Aristoxenus (first half of 1st cent. A.D.);
- Demosthenes Philalethes (ca. 20 B.C. to A.D. 50);
- Philonides of Sicily (1st cent. A.D.).

There are no extant writings for any of these physicians, all of whom appear to have contributed to thinking about the arterial pulses. Heinrich von Staden, who provides a summary of the different definitions of the pulse offered by seven of the Herophileans listed above, has this to say on the group as a whole:

> Herophilus' analysis of the pulse was not simply transmitted with piety and reverence from one generation to the next, like an orthodoxy that had become canonical or obligatory. Instead, each Herophilean tried to improve not only on Herophilus' definition but also on those of his own immediate predecessors and contemporaries within the school. The result is a striking example of the incessant shuffling and dissent that lurk not too far beneath the homogeneous surface suggested *prima facie* by the label 'Herophileans'.[15]

Two physicians against whom Galen directed a great deal of polemic in other works, Erasistratus and Asclepiades (1st cent. B.C.), receive very little attention in the works on the pulses. Neither man, in fact, is in any way renowned for his contributions to pulse theory—in each case their importance lies elsewhere. Two references, one for each of them, which Galen does make on the matter of arterial fullness, are, however, worth noting:

> But neither (i.e., the opinions of Praxagoras and Herophilus) satisfies Erasistratus, for he wishes the arteries to pulsate in an opposite way to the heart. The heart is filled when it is dilated due to the filling of what is evacuated whereas the arteries dilate when they are filled—they say they are filled when the *pneuma* is sent from the heart.[16]

> For if, perchance, one were to follow strictly the doctrines of Herophilus, systole (contraction) is a function of the arteries, whilst diastole (dilatation) is a return to the proper and natural state of

14 von Staden 1989: 447–448 with the collected testimonia on the Herophileans on 445–578.
15 von Staden 1989: 446–447.
16 Galen, *De pulsuum differentiis* I.2 (VIII.703K).

> their bodies. For he means that just as one sees in the case of those who are dead, the wall of the artery is open, so also it is open as far as it can be in those who are living, which is the opposite of the view of Asclepiades. For this man thinks that both the heart and the arteries are dilated when they are filled with *pneuma* since it flows into them due to the *leptomeria* which they have within themselves. On the other hand, whenever they are filled there is no longer any forward flow and the wall falls back to the natural state which it was formerly in them.[17]

Along with Herophilus, Archigenes (1st–2nd cent. A.D.) was one of the two most important contributors to the pre-Galenic literature on the pulse. A disciple of Agathinus (1st cent. A.D.), he flourished in Rome during the time of Trajan (A.D. 98–117). He has been described as an Eclectic in terms of the medical schools, with both Dogmatist and Pneumaticist tendencies. He made substantial contributions to a number of areas of medicine and surgery. Sadly, all his writings have been lost. Now we must rely on what is recorded particularly by Galen and by later writers such as Aretaeus of Cappodocia (ca. A.D. 150–190), Oribasius (ca. A.D. 325–400), Aetius of Amida (ca. A.D. 500–550), and Paul of Aegina (ca. A.D. 630–670). Regarding his book on the pulse, William Henry Broadbent (1835–1907) states: "Archigenes wrote a treatise on the pulse which is often referred to by Galen in a controversial spirit, with the effect of exciting regret that the book has been lost."[18] It is obviously a work of particular importance—indeed, authors of the entry on Archigenes in the third (1996) edition of the *Oxford Classical Dictionary* suggest that Galen's own "theory of the pulse was borrowed from that of Archigenes, while at other points Galen reacts against his teaching."[19]

Be that as it may, Archigenes is clearly the most frequently referred-to writer in the four long treatises that follow in this volume. In particular, Books II and III of the first treatise are devoted to his theories. Whilst at times Galen is complimentary towards Archigenes (as he also is towards Herophilus), he is, at other times, highly critical. For example:

> Archigenes, for it is right to mention him alone after Herophilus, when he quotes Homer as follows,
>
> For dead though he is,
> Persephone has left to him, and him alone,
> a mind to reason with.
> The rest are mere shadows flitting to and from.[20]
>
> seems to be confused not only in the names, but also much more importantly in the matters themselves. And because of this, he does not, for a moment, perceive that he is contradicting himself.[21]

17 Galen, *De pulsuum differentiis* IV.10 (VIII.747–48K).
18 Broadbent 1890: 4.
19 Rathbone and Nutton 1996.
20 See *Odysseia* X.491–495 (ed. and Engl. transl. Murray and Dimock 1995: 394).
21 Galen, *De pulsuum differentiis* II.7 (VIII.602K).

According to Galen, Archigenes identified eight basic qualities of the arterial pulse: magnitude, strength, rapidity, frequency, fullness, regularity/irregularity, equality/inequality, and rhythm, thus adding a further four qualities to those definitely associated with Herophilus. Galen immediately takes issue with Archigenes on the listing of the eight qualities and is also critical of Archigenes' understanding of classification in relation to these qualities.[22]

Apart from his criticisms regarding terminology and the method of classification, Galen takes issue with Archigenes on a number of other points of varying importance which include such aspects as the number of possible variations in magnitude, the analysis of strength, the question of fullness, prognostic issues with regard to frequency, other specifics to do with named varieties of the pulse, and the prognostic significance of various pulse variations. Obviously, without Archigenes' original work, it is impossible to determine to what degree Galen was dependent on him for his own theories and classifications, and to what extent his criticisms are justified. It is clear, however, if only from the amount of attention which Galen gives to the treatise, that it was of considerable importance.[23]

Finally, three other pre-Galenic physicians who are thought to have contributed significantly to the literature on the pulse are Rufus of Ephesus (1st–2nd cent. A.D.), Soranus (2nd cent. A.D.), and Marcellinus (2nd cent. A.D. [?]). It is noteworthy that in none of their writings on the pulse is Galen referred nor does Galen himself refer to any of them in his treatises on the pulse.

Considering Rufus, he is believed to have worked during the later part of the first century A.D. and the early part of the second century. He has, fortunately, a small corpus of writings extant. There is, however, uncertainty about the treatise on the pulse (Σύνοψις περὶ σφυγμῶν) associated with his name. According to George Sarton (1884–1956)[24], it first came to light as a 12th-century Latin translation published only in 1639, however, by René Chartier (1572–1674) in the 8th volume of his massive edition of Galen[25]. Its Greek text was published by Charles Daremberg (1817–1872) in 1846,[26] and further reproduced in the edition of Rufus' works prepared by Daremberg and posthumously published in 1879 by Charles-Emile Ruelle (1833–1912)[27]. As Charles Reginald Schiller Harris (1896–1979), best known as CRS Harris, says, Daremberg "... appears to have had considerable doubts as to its authenticity" whilst Ruelle "... accepted

22 Galen, *De pulsuum differentiis* II.4 (VIII.576–583K).

23 Galen's commentary on Archigenes' work, referred to several times in the treatises translated here, is mentioned in his *De libris propriis*. He refers there (5) to "the eight volumes of commentary and criticism of Archigenes' major work on the pulses ..." (XIX.33K). Archigenes' work, like Galen's commentary, is no longer extant.

24 Sarton 1927: 281–282.

25 Chartier 1639: 330–332, under the title *Compendium pulsuum Galeno adscriptum*.

26 Daremberg 1846.

27 Daremberg and Ruelle 1879: 219–232.

it as genuine."[28] In a brief article in 1924, Pierre Eugène Menétrier (1859–1935) considered the work to be much earlier than Rufus: "… probablement même d'un élève d'Hérophile."[29] Issues of authorship notwithstanding, Broadbent is lavish in his praise of the work: "His (i.e., Rufus') description of the different characters of the pulse leaves little to be added at the present day".[30] Perhaps the most interesting aspect of the work is the correlation of the arterial pulse with cardiac systole rather than diastole. CRS Harris' comments on this point are given below:

> The process of the pulse is explained as follows: 'The heart, when it draws in breath or *pneuma* from the lungs, first receives it in its left ventricle, then as this contracts immediately afterwards, it furnishes and supplies the *pneuma* to the arteries themselves. The arteries of the body, when they are filled with *pneuma* by the contraction of the heart, give rise to the pulse … but when they are emptied, they contract … The heart produces the pulse when it is being emptied (i.e., at systole) but the arteries when they receive the *pneuma* and are filled.' This is a description, which at first sight looks as if it might have been taken straight out of Erasistratus, so close is it to the facts. For *pneuma* read *blood*, and Rufus would perhaps have discovered the circulation, for he avoided Bacchius' fatal error of the simultaneous expansion and contraction of the heart and the arteries. 'Because heart and arteries', he continues, 'perform their double motion in the same time interval, nearly all the authorities assume that they are both filled simultaneously: I want to establish their error. Those who hold it', he concludes, 'had better learn the facts by going to see a vivisection' (ἀνατομή).[31]

Soranus is also credited with some writings on the pulse which have only survived in Latin translation and are also of questionable authenticity.[32] However, Galen makes no mention of either Rufus or Soranus in his treatises on the pulse. Two points of note in Soranus' writings are the description of pulsation in the veins, at least in certain pathological conditions, and affirmation of the Empiric view that only the "beat" (πληγή) of the pulse can actually be palpated, corresponding more or less to the modern view and contrary to the elaborate claims of Galen and other early non-Empiric writers.

The third treatise, this time extant and of agreed authenticity, is that of Marcellinus entitled Περὶ σφυγμῶν. Its author has been described as "… a Greek physician, Eclectic, but with Dogmatic tendencies (chiefly Pneumatic)."[33] Whilst this short work takes into account much of the earlier writings on the pulse (particularly that of Herophilus and Archigenes), there is no mention of Galen, nor is Marcellinus referred to by Galen. Harris sees the work as being that of a practising physician who devotes rel-

28 Harris 1973: 263.
29 Menétrier 1924: 97–98. Harris 1973: 263, describes Menétrier's thesis as 'a judgement which seems to me pure guesswork'.
30 Broadbent 1890: 4.
31 Harris 1973: 263.
32 See Harris 1973: 261. More recently, Hanson and Green 1994, who do not mention the treatise.
33 Harris 1973: 257.

atively little attention to the theoretical aspects of pulse analysis. Its author does, however, offer three possible hypotheses on the cause of the pulse as follows:[34]

1. The cause of the pulse is a vaporization of the innate heat condensing under the force of attraction and expanding under that of transpiration in diastole, the arteries moving their bodies in accordance with each of these movements.
2. The cause of the pulse is the flowing of *pneuma* dancing in the arteries in a rhythmic course in obedience to the thrust of the heart and the force of attraction along the sides of the arteries.
3. The cause of the pulse is the necessary passage of air in accordance with the movement of light substances, since *pneuma* has altogether a tendency to move outwards by being drawn through the surface and through the mouths of the arteries in thousands of passages.

In general, Marcellinus accepts Archigenes' definition of the pulse and the eight aspects of relevance in pulse examination. He particularly provides an account of the various kinds of abnormal pulse and the clinical significance of each.[35]

2 Galen's Writings on the Arterial Pulses

As mentioned at the outset, Galen occupies a very significant place in the long history of the understanding of the nature of the arterial pulse and its relevance to clinical diagnosis and prognosis in Western medicine. Not only is he the author of the earliest extant texts specifically on the pulse, but also a substantial part of his surviving *oeuvre* is on this topic.

There are, in all, seven treatises on the pulse accepted as genuine in Kühn's twenty-two-tomes edition of Galen's works. These total 1085 pages, amounting to approximately one complete Kühn volume. The four long treatises included in the present translation account for 82.6% of this material. What they represent is a comprehensive review of the art (science?) of sphygmology in Western medicine towards the end of the second century A.D. and an unmatched repository of information on the views of important earlier writers on the subject whose works are no longer extant—in particular, Herophilus and Archigenes.

Galen himself has this to say about his own writings on the pulse (and the other works included under prognosis) in *On My Own Books:*

> First are the three on critical days (*De diebus decretoriis*, IX.769–941K) and second, in addition to these, three on crises (*De crisibus*, IX.550–768K), then the works on the pulses within which the first concerns the *differentiae* of these, second other such things about the diagnosis, and a

34 For the Greek text and further comment on the hypotheses, see Harris 1973: 257–258. The translations are those given by Harris.

35 See particularly Lewis 2015.

> third besides these, an equal number about their causes and a fourth about prognosis from them. There are sixteen in all and, apart from them, there is that written about the use of the pulses for beginners. Some questioned why, in this, the pulse specific to fevers was not spoken of. My answer to them is that the subject is greater than is appropriate for beginners since it requires a very considerable enquiry. But it was, at least, said at the beginning of this book that there are two opinions among doctors: the one of those who think the *systole* of arteries is also perceptible, and the other of those who say it is imperceptible. I think it is easier for the beginner to practise first on the grounds that the *systole* is imperceptible. On this basis, I am persuaded that there is no specific sign of fever in pulses. On the other basis, which I have written of in the large work of sixteen books, the sign has been spoken of. But it is those who have not learned from teachers and those who, as the saying goes, are guided by books, who seek such things. For me, as was said, the teaching of beginners is different from that for those who have been thoroughly taught all things in their entirety. There is, then, one other book of mine in which I prepared a synopsis of the sixteen books and, apart from all these, there are eight books on Archigenes' work on the pulses and also that on the use of the pulse. One might also place the three-part work on difficulty in breathing (*De difficultate respirationis*) in this part of the craft.[36]

The seven treatises specifically on the pulse accepted as genuine Galenic works, and included in Kühn, are the following:

1. the four long treatises, each subdivided into four books, on four distinct aspects of the pulses:
   - differences/*differentia* (*differentia*),
   - recognition/diagnosis (*dignoscentia*),
   - causes/causation (*causa*),
   - prognosis/prognostication (*praesagitio*);
2. two synoptic works, one specifically for beginners and one a later, systematic synopsis of the four long treatises;
3. a short work on the use/need (χρεία) of the pulses.

The seven genuine works are the following with the references to their edition in Kühn's *Galeni Opera omnia:*

1. *De pulsuum differentiis, libri IV*—Kühn VIII, pp. 493–765.
2. *De dignoscendis pulsibus, libri IV*—Kühn VIII, pp. 766–961.
2. *De causis pulsuum, libri IV*—Kühn IX, pp.1–204.
3. *De praesagitione ex pulsibus, libri IV*—Kühn IX, pp. 205–430.
4. *De pulsibus ad tirones*—Kühn VIII, pp. 452–492; English translation by Singer.[37]
5. *Synopsis librorum suorum sedecim de pulsibus*—Kühn IX, pp. 431–549 (pp. 533–549 in Latin only).
6. *De usu pulsuum*—Kühn V, pp.149–180; English translation by Furley and Wilkie.[38]

---

36 Galen, *De libris propriis* 5 (XIX.32–33K).

37 Singer 1997: 325–344.

38 Furley and Wilkie: 185–228.

An eighth work, *De pulsibus ad Antonium* is also included in Kühn (XIX.629–642) but not in the present translations. Excerpts from a ninth work (*Definitiones medicae*, XIX.346–462K), although spurious, are included in the translations.[39] Following, a brief outline of the content and intent of each of the works is given.

2.1 *On the Differences of the Pulses* (4 Books)

Book I is in essence the key to understanding Galen's sphygmology. After two brief opening sections offering some historical observations and a digression on one of Galen's favourite topics—the need to concentrate on matters rather than names—, the remaining 28 sections are concerned with features of the pulse that can be distinguished by palpation and their variations. These are:

- the changes in the three dimensions of the artery (length, breadth, height) with each pulsation;
- the consistency of the arterial wall;
- the fullness or otherwise of the arterial lumen;
- the times of the components of each pulsation—dilatation, external pause, contraction, internal pause.

On the basis of the possible combinations of the dimensional variations, 27 different pulses can, in theory, exist. These are tabulated. Equality/inequality and regularity/ irregularity of pulses are considered, both in relation to a single pulse and in relation to a series of pulses. Six different pulse rhythms are identified, and a number of specific pulses are described, all on the basis of the palpable variations listed.

Books II and III are essentially a critique of other views of the features of the pulses, with particular attention being given to Archigenes. In Book IV on the definition of the pulse, Galen offers several variations of his own preferred definition followed by detailed consideration of the views of eighteen of his predecessors.

The key issues addressed by Galen in this treatise are the list of identifiable characteristics of the pulse, regardless of whether or not the contraction (systole) is perceptible, and what causes the pulses. He defines the arterial pulse as a movement (dilatation and contraction) of the arteries and the arterial component of the heart (left ventricle). On the question of what causes arterial pulsation, he sees it as an extension of the inherent capacity of the heart to pulsate which is transmitted through the arteries. This is related, of course, to the question of whether the arteries dilate because they are filled passively (as in a wineskin, for example) or are filled because they dilate actively (as in a bellows). Galen takes the latter view—that is, the arteries sequentially

39 Also listed in Galen's *De libris propriis* 5, is his commentary on Archigenes' treatise on the pulses. Of this, he writes: "Outside of all these is the eight-part book on Archigenes' treatise on the pulses, having both explanation and use." (XIX.33K). See also note 23 above.

dilate and contract (the dilatation being the active component), drawing material (i.e., blood and pneuma) into themselves (i.e., the bellows model). The material is then distributed throughout the body. Although not explicitly addressed in this work, contraction would seem to be viewed by Galen as a passive restoration following the active dilatation.

These issues are further considered in his other treatises, but to reiterate, the features of the arterial pulse which may be observed by the physician, and on which a classification may be based, are the size and strength of the arterial dilatation, its rate and rhythm, spatial and temporal variations in the sequence, and the physical characteristics of the artery. Although Galen's subdivisions and classification in this treatise may seem overly theoretical and unlikely to lend themselves readily to practical application, the treatise is undoubtedly a rich source of information on the views of Galen's predecessors and contemporaries.

2.2 *On the Diagnosis of the Pulses* (4 Books)

In Book I, Galen deals with fundamental issues, some theoretical and some practical. Basically, the term "pulse" refers, in his usage, to the sequences of dilatation and contraction and their intervening pauses palpable in the peripheral arteries. The essential factors involved in the arterial pulse are:
1. the body of the artery, which is long, hollow and round, has a perpetual twofold motion (i.e. dilatation and contraction) in three dimensions;
2. what is within the artery, whether the interior is "empty" or "full";
3. whether the tunic (wall) of the artery is hard or soft.

The art of pulse palpation and recognition of the many variations is one that takes a long time to learn, according to Galen. Two issues pertaining to this art which he raises here (and elsewhere) are, first, the importance of the amount and state of the tissue intervening between the palpated artery and the palpating fingers, and second, whether the contraction phase is, in fact, palpable.

In Book II, the important points to be identified in the midst of much that is superfluous, are that Galen does recognise the desirability of establishing norms, factoring in age, season, place, nature, κρᾶσις (temperament) and "all other such things". He concludes, however, that it is not possible to establish an average. He does not, in these texts, pursue the practice of timing the pulse initiated by Herophilus. He does note that all the arteries move synchronously with one another and with the heart, an observation not actually borne out by modern studies (vide infra). What he does do in this book is to identify and enumerate the key elements of the pulse as follows:
1. Degree of dilatation—large, moderate, small.
2. Movement of dilatation—rapid, moderate, slow.
3. Pause—frequent, infrequent.
4. Tension—strong, weak.

5. Body of artery—hard, soft.

Because there is no precise quantitation and, as he claims, average values cannot be established, the assessment of these factors is necessarily subjective to a degree.

In Book III, Galen, after a further digression, considers some practical aspects of the technique of pulse palpation, and also deals with the issue of rhythm, which he defines as the ratio of the time of dilatation to the time of contraction. Since each of these times can, according to Galen, be identified as either rapid, moderate, or slow, there are nine possible combinations. Whether all these variations are detectable, given the actual difficulty of detecting contraction, is another matter.

In Book IV of this second treatise, Galen concentrates on the "full" pulse with reference to the Atomists, and to Archigenes and Agathinus. He raises the question of whether a "full" pulse relates to the quality of the arterial wall, the quantity of substance in the artery (which is, in Galen's view, a mixture of humor or spirit—either pure and ethereal or cloudy and vaporous—and blood) or is a function of the quality of bodily heat. Thus, there are three kinds of "full" pulse:

- that due to the condition of the artery;
- that due to the substances in the arterial lumen;
- that due to the functioning of the intrinsic spirit.

2.3 *On the Causes of the Pulses* (4 Books)

In Book I, Galen sets out his tripartite division of causes which he uses, albeit variably and to a variable extent, in his various considerations of causation. This is the threefold division into συνεκτικόν (containing), προηγούμενον (internal antecedent) and προκαταρκτικόν (external antecedent) causes.[40]

The συνεκτικά causes are the causes actually responsible for the generation of the pulse. Within this group of causes are the determinants of the perpetual motion of the heart and arteries. On these, Galen says:

> So then, the heart appears to move continuously with all the arteries in the so-called pulsatile movement while the animal is alive but, when it dies, it is not possible to still see any pulsation of the arteries nor in relation to the heart itself. From this it is clear that there is some cause by which they are moved up to this point but it is difficult to discover what this is. For they say it is the innate heat, or the tonus, or the specific nature of the κρᾶσις, or the whole constitution of the body, or the πνεῦμα alone, or some of these, or also all of them at the same time. Some now propose a incorporeal capacity which uses the specific organs of movement spoken of, either all, or some, or one. This cause, then, which produces the pulses, whatever it may be, and even if we don't know its essential substance, we called a capacity (faculty, power, potency—δύναμις) for the ability

40 Johnston 2006: 33–35 and 102–120.

> to create pulses just as, I think, we are accustomed to call every other capacity from its ability to do what it does.[41]

The other two categories of cause, προηγούμενον (internal antecedent) and προκαταρκτικόν (external antecedent), are causes of changes of the pulses but not of their creation. The former are causative factors arising within the organism while the latter are external factors which influence the pulses. In general terms, the use of the pulse is for maintaining warmth and cooling where each is appropriate, for elimination of what is "smoky", and for contributing to the production of animal spirit.

The remaining three books are on specific subjects. Book II deals with the different kinds of pulses and what causes them. Apart from discussing inequalities of the pulse, rhythm changes, variations in the dimensions of the pulse, and arterial distortion (due to masses adjacent to the artery in question) in a general way, Galen covers the various named pulses, identified by terms such as dicrotic, vermicular, formicant, bounding, and undulant, some of which remain in use today. In the final two books of this treatise Galen considers factors that change the pulses. In Book III he deals with "natural" causes, listing such things as exercise, bathing, sleep and waking. In Book IV, the "non-natural" causes are considered.[42] In this last group, Galen lists 23 individual diseases or groups of diseases.

2.4 *On Prognosis from the Pulses* (4 Books)

In Book I, Galen makes quite explicit what he means by the term "prognosis": it must be understood in a broad sense rather than in the specific sense which characterises modern medical usage. What Galen is talking about in this treatise is not only what is expected to happen, but what has happened and what is happening. In fact, in Galen's view the pulse gives important information on the first two by helping to identify the prevailing condition, while it is this identification that allows prediction of future events.

Book I is largely given over to general considerations. In defining his aims, they are to set out what sorts of pulse occur in particular diseases, and what diseases are indicated by a particular pulse insofar as these aims are distinguishable. In relation to the previous treatise, the pulse at any particular moment gives information on the present state of the συνεκτικά causes and on the effects of whatever προηγούμενα and προκαταρκτικά causes are operative. Two other general points in the first book are the need to take into account the nature of the tissue surrounding the artery being palpated (still

41 Galen, *De causis pulsuum* II.2 (IX.4–5K). Also, *De naturalibus facultatibus* I.4 (II.9–10K).
42 See Bylebyl 1979: 482–485.

relevant today), and that pulselessness never occurs, which is, of course, generally true although rare exceptions are now known.[43]

In Book III, Galen considers particularly four pairs of terms (fast/slow, large/small, and strong/weak in relation to the pulse, and hard/soft in relation to the artery), and what these features signify when detected. He also makes the point that, if the heart has a normal *krasis*, the pulse is strong whereas, if the cardiac κρᾶσις is defective, the pulse is weak. Various specific pulses, those named in the previous book, are individually considered. In the third book, Galen deals with the pulse in various δυσκρασίαι. Three key points which he makes are:

- that a δυσκρασία doesn't affect the pulses unless it involves the heart and arteries directly;
- that a δυσκρασία of the heart spreads to the arteries;
- that a δυσκρασία of the arteries does not affect the heart.

In general, a hot δυσκρασία produces large, rapid and frequent pulses, whilst a cold δυσκρασία produces small, slow and infrequent pulses. Galen also considers here the relationship between fever and the pulse. In the fourth book, he lists diseases on a regional basis and indicates what kind of pulse is likely to occur in each case.

2.5 *Pulses for Beginners*

This is a short, clear text of obvious didactic value. A single book divided into twelve sections, it is notable for the absence of the prolixity and complexity that mark the four major texts. Its practical focus is evident at the outset where Galen identifies the merits of the radial artery for pulse evaluation—the relative lack of overlying tissue, the straightness of the artery at the point of palpation, and the fact that palpation can be carried out without the need for the patient to remove any clothes. Galen lists the five components to be evaluated during palpation of the pulse. The primary considerations are the four pairs of qualities: whether the pulse is rapid or slow, its vigour or feebleness, whether the individual beats are long or short, and the hardness or softness of the artery. To these are added a fifth, the interval between the impacts (beats), which is, in effect, the interval between successive dilatations, since Galen acknowledges here that both contraction (systole) and the two pauses (between diastole and systole and between systole and diastole) may be impalpable, at least for the beginner. Further, these four events may occur evenly (equally) or unevenly (unequally) and any variations may be regular or irregular. Setting aside differences in the understanding of what underlies the several perceptible components, this summary of the palpable events and their types of variation is as applicable today as it was in Galen's time. The outline of the palpation of the pulse, which occupies the first eight sections of

43 For example, Takayasu's Disease, on which see *Hurst's The Heart* 1990: 2165–2166.

the treatise is, Galen avers, enough for the beginner. In fact, it might reasonably be claimed that it is enough for the expert, since it contains the practically applicable substance of the first major treatise shorn of the numerous digressions and general prolixity, and the interesting, but unnecessary historical information.

The remainder of the treatise (sections 9–12) is about changes in the pulse, and equates with the other major treatises, in particular, Books 3 and 4 of *On the Causes of the Pulses.* The first basic point is that almost all changes within the body are reflected in changes in the pulses, which are, of course, what the doctor is endeavouring to analyse for purposes of diagnosis and prognosis. Galen uses a somewhat different, and more practical, classification of changes in this treatise, a classification which he also uses elsewhere: κατὰ φύσιν (according to nature), παρὰ φύσιν (contrary to nature), and neither κατὰ nor παρὰ φύσιν.[44] He prefaces his consideration of the constituents of these three categories with the important remark that ideally the analysis should be based on a knowledge of the normal or average, but that this is impossible to establish. In summary, the changes κατὰ φύσιν, dealt with first, include sex, age, place, season, sleep/wakefulness, temperament (κρᾶσις) and bodily habitus. Those that are neither κατὰ nor παρὰ φύσιν, dealt with second, are listed as exercise, hot and cold baths, and eating and drinking. The causes of change in the παρὰ φύσιν category, dealt with third, are, according to Galen, essentially innumerable both in overall quantity and in kind. He does, however, offer a broad subdivision into those that dissolve and dissipate the vital capacity (faculty) and those that compress and burden the vital capacity (faculty). The last, and by far the largest, section of the treatise[45] deals with some of the specific factors which can produce a pulse that is παρὰ φύσιν. These are, essentially, emotional changes, pain, and the largest subgroup—various diseases. This short and highly practical treatise is dedicated to Teuthras.[46] The only other writer on the pulse who is mentioned is Archigenes, and he only once.

2.6 *Synopsis of the Sixteen Books on the Pulses*

This, a single book, summarises the content of the four major treatises sequentially, reducing them to a little over 10 % of their combined length. On this work, Vivian Nutton writes:

> Twenty-five years later (i.e. after the four treatises), around A.D.195, Galen returned to the theme. He admitted that the 16 books were prolix, hard to assimilate, and, one might add, expensive to

44 On these terms see Johnston 2006: 48–49.

45 Section 12 (VIII.473–492K).

46 Presumably one of Galen's students. He is also addressed in *De venae sectione adversus Erasistratum* 1 (I.193K).

> have copied. So, for the benefit of those who might otherwise be tempted to leave the subject alone, he produced his own summary in a single book (in length, the equivalent of two).[47]

The *Synopsis* comprises 34 sections and, after a short introductory section, broadly follows the order of the four major treatises. The final six sections in Kühn are in Latin only.[48]

2.7 *On the Use of the Pulses*

Here, Galen addresses the fundamental question indicated by the title and focusses on the analogy between the regularly repeated sequences of pulses and respirations. In examining this matter, Galen necessarily considers the function (or use) of the arteries more generally, and utilises data gathered from experiments on ligating arteries. On the basis of these experiments, he calls into question the assumed close analogy between the two actions or functions (if the pulse can be called a function) in that respiration is essential for life whereas individual arteries can be ligated, and their pulses thus obliterated, without even the area of supply suffering any apparent damage. The essential nature of the heart and the arteries is summarised at the end of section 4 of the treatise as follows:

> For the power in the body of the heart, by which it dilates and contracts, flows down through their walls to all the arteries, and in this way dilates and contracts them, just as the heart itself does. Thus, as the heart is dilating, it draws to itself what is adjacent to its openings, whereas it expels when contracting. In this way also, the arteries, when they dilate, draw in from all directions, whereas when they contract, they expel in every direction.[49]

In the following section (5), Galen presents his concept of the way in which arteries function in the exchange of material both with other areas of the body and with the outside world. In short, the arteries are in a two-way communication with virtually everything:

1. back and forward exchange with the veins;
2. vent-like spaces in the arterial tunics allow the arteries to draw in from and eliminate into surrounding tissues;
3. arteries are able to eliminate and draw in through the skin;
4. there is even a two-way exchange with the heart despite the cardiac valves, particularly identified by Erasistratus and well-known to Galen.

---

47 Nutton 1995: 69.

48 For a reconstruction of the Greek of these sections, which was used in the translation in this volume, see Cortés Gabaudan and Martínez Manzano 2013.

49 Galen, *De usu pulsuum* 4 (V.164K).

In this book Galen also makes clear his position on whether the arteries are filled because they expand (bellows) or expand because they are filled (wineskin)—he believes it is the former. The pulse itself is produced by vital power from the heart and has the purpose of maintaining the natural heat of the body which it cools during expansion and purges through contraction. One key difference between the pulse and respiration is, in Galen's view, that the former is dependent on the vital power and the latter on the psyche.

The two other texts of relevance are the *De pulsibus ad Antonium* and the *Definitiones medicae*, both regarded as spurious. The first is a short work (14 pages in Kühn's edition), not included in the translations here, which sets out to answer four clearly articulated questions:

1. what a pulse is;
2. why it is called a pulse;
3. what its use is;
4. what an artery is.

The definition of the pulse which is given is as follows:

> A pulse is a movement of heart or artery which is effected through systole and diastole for the cooling of the innate heat and the excretion of smoky superfluities.[50]

The second work, *Medical Definitions*, gives detailed consideration to the pulses in two separate places[51]: definitions 110 and 111[52] offer several different definitions of the pulse, whilst the definitions 205–206 and 208–233[53] provide a series of descriptions of the different types of pulse and other matters, including a consideration of rhythm. Both groups of definitions are included in the translations here.

The primary definition reads:

> A pulse is a natural διαστολή (dilatation) and συστολή (contraction) of heart and arteries. There are two parts of the pulse, διαστολή and συστολή. Alternatively, a pulse is a movement, by dilatation and contraction, of heart and arteries, and brain and meninges which is natural and involuntary.[54]

---

50 The questions are posed on XIX.629K and the definition given on 629–630.

51 The study here and the translation are based on Kühn's edition. The new edition published by Jutta Kollesch in the *Corpus Medicorum Graecorum* in 2023, could not be taken into consideration. However, since the two editions number the definitions in different ways, the correspondence between their numbering has been included. Jutta Kollesch's edition is identified as CMG. The number of the entries according to her edition is followed (between parentheses) by the page number in her edition.

52 *Definitiones medicae* 110–111 (XIX.375–377K) = 107 CMG (45–46) and 36CMG (190–192).

53 *Definitiones medicae* 205–206 (XIX.402–403K) = 220–221 CMG (84), and 208–233 ((XIX.404–412K) = 223–256 CMG (86–96).

54 *Definitiones medicae* 110 (XIX.375–376K) = 107 CMG (45–46).

In the definition of rhythm, Herophilus and Bacchius are mentioned,[55] and the terms εὔρυθμος, ἄρρυθμος, κακόρυθμος, παράρυθμος, ἑτερόρυθμος ἔκρυθμος are considered.[56] Heinrich von Staden raises the question as to whether some of this material may be due to later Herophileans.[57]

3 Galen's Views on the Pulses

Galen's views on the pulses can be summarized in eleven major points as follows:

1. The heart and the arteries throughout the body maintain a continuous series of pulses while the animal lives, distributing blood and πνεῦμα to the various parts. The pulsations of the heart and arteries occur simultaneously but are to an extent independent functions/actions. The power (capacity, δύναμις) responsible for the pulsation of the arteries is separate from that of the heart but does arise from the heart and is distributed from the heart through the walls of the arteries.

2. The single arterial pulse consists of four components: dilatation (diastole, expansion, distension), an external pause, contraction (systole), and an internal pause. Arguably, each is able to be independently recognised, although this is questionable with regard to the beginning of diastole and the end of systole, and even perhaps to the systole itself.

3. Filling and emptying of the arteries are active rather than passive phenomena.

4. The pulsation of the arteries is attributable to three *synektic* causes:
- the pulsatile power/capacity referred to in point one above;
- physical needs which include control of bodily temperature and maintenance of the innate heat; generation of physical and psychical *pneuma* (dilatation); elimination of waste materials (contraction);
- the state and tonus of the arterial wall.

5. Variations in the pulses are attributable to variations in one or more of these factors, which are in turn due to προηγούμενα (internal antecedent) and προκαταρκτικά (external antecedent) causes. The variations are reflected in the parameters listed in point six below.

55 *Definitiones medicae* 220 (XIX.408–409K) = 240 CMG (90).
56 *Definitiones medicae* 221–222 (XIX.409–410K) = 242–243 CMG (92).
57 von Staden 1989: 513–514.

6. Clinical examination of the pulse is a critical component of the assessment of health and disease and involves the evaluation of *single pulses* and *series of pulses*. In the list below the parameters which apply predominantly to a *single pulse* are listed separately from those that apply predominantly to a *series of pulses*, although it must be recognised that there is some overlap. For the *single pulse*, there are seven basic parameters, as follows:
- magnitude: large (μέγας), small (μικρός), intermediate;
- rapidity: rapid (ταχύς), slow (βραδύς), intermediate;
- strength: strong (σφοδρός), weak (ἀμυδρός), intermediate;
- hardness: hard (σκληρός), soft (μαλακός), intermediate;
- moistness: moist (ὑγρός), dry (αὐχμηρός), intermediate;
- fullness: full (πλήρης), empty (κενός), intermediate;
- temperature: hot (θερμός), cold (ψυχρός), intermediate;

For a series of pulses, there are four basic parameters, as follows:
- frequency: frequent (πυκνός), infrequent (ἀραιός);
- equality/evenness: equal (ὁμαλός), unequal (ἀνώμαλος);
- regularity: regular (τάξις), irregular (ἀταξία);
- rhythm: there are six terms, which are transliterated below, with the addition of a translation/explanation for the last four:
  eurhythmic/εὔρυθμος;
  arrhythmic/ἄρρυθμος;
  kakorhythmic/κακόρυθμος (bad rhythm);
  pararhythmic/παράρυθμος (out of tune, irregular);
  heterorhythmic/ἑτερρόρυθμος (different or false rhythm—that is, inappropriate for age);
  ekrhythmic/ἔκρυθμος (out of rhythm, irregular)

7. A normal pulse is one in which all the parameters are in a median state. Reference is made to what may be called a "theoretical median state" or, more valuable clinically, the assessment of the pulse in a particular person when he/she is demonstrably healthy. The latter may not of course be feasible in many instances.

8. Evaluation of the pulses can give information on what might be called normal variations, the effects of natural and non-natural factors, and specifically among the latter, disease states/conditions.

9. A number of specifically identifiable abnormal pulses are described by Galen and given names (some of which antedated Galen) that persisted for many centuries. These are listed in the *Glossary of Pulse Terms* contained in this volume.

10. The non-disease states which Galen identifies as causing pulse changes are as follows:[58] gender, hot or cold nature, thin bodily state, age/stage of life, seasons, places, pregnancy, sleep, waking from sleep, well-fleshed bodily state, κρᾶσις of the body, exercise, hot/cold baths, food, wine, psychical affections (anger, joy, grief, fear), and pain.

11. The disease states which Galen identifies as causing pulse changes are as follows:[59] Inflammation, pleuritis/pleurisy, empyema, marasmus, consumption (φθίσις), peripneumonia, lethargy/lethargic attacks, phrenitis, κατοχή/καταφορά, catalepsy, convulsions, paralysis, epilepsy, apoplexy, synanche, orthopnoea, hysterical choking (πνίξ), stomachical affections, dropsies (ascites, tympanitic ascites, anasarca), elephantiasis and jaundice.

4 The Pulse from Galen to Harvey

The accepted view, one not without substance, is that knowledge of the cardiovascular system and ideas on the arterial pulse did not change significantly during the period between Galen and Harvey, that is, from about A.D. 200 to the early 17th century. Broadbent, for example, in his outline of the history of ideas on the pulse, covers this period of over 1,400 years in three brief paragraphs, all of which are devoted to discoveries in the 16th century that contributed to Harvey's definitive formulation.[60] Certainly, the evidence suggests that Galen's ideas on the pulse, and on much else, remained dominant over this period. The five points supporting this view are the following:

1. Galen's writings on the pulses are frequently referred to and quoted during this period;
2. the four long treatises on the pulses and the *Synopsis* of these are included in the sixteen "books" of the *Alexandrian Summaries*;[61]
3. the chapter on the pulse in Paul of Aegina's important work, dating from around A.D. 630, consists entirely of a summary of Galen's views;
4. the inclusion of the work *De pulsibus* attributed to Philaretus of Byzantium in the *Articella*, which was a central component of Western medical teaching from the 13th to 16th centuries. This is in essence a summary of Galen's views;[62]

58 These are listed in the order in which they appear in *De pulsibus ad tirones* and *De causis pulsuum*.
59 These are also listed in the order in which they appear in *De pulsibus ad tirones* and *De causis pulsuum*. A summary of these conditions and the pulses associated with them is given in Glossary 3.
60 Broadbent 1890: 11–12.
61 For a list of the works included in the Alexandrian Summaries, see Johnston and Horsley 2011: lii-liv.
62 Philareus was a Byzantine physician during the later part of the first millennium. Doubt clouds his dates, and even his name in that he has been confused with Philagrius and with Theophilus Protospatharius. Be that as it may, his work *De pulsibus* was, as mentioned, a component of the *Articella* and is still extant. The work itself bears a close, but not clearly defined, relationship to the pseudo-Galenic *De pulsibus ad Antonium* included in Kühn (XIX.629–642K). On Philaretus and his work, see Pithis 1983; Nutton and Savvidis 2007, and Touwaide 2008.

5. Harvey himself makes a number of references to Galen in his own work.

This is not to say, however, that there were no significant writings on the pulse during this period—there were, and both from those writing in Greek or Latin, and later from those writing in Syriac and Arabic.

In Byzantium, the first two physicians who contributed to the literature on the pulse during this period in chronological order were Alexander of Tralles (ca. A.D. 525–605), whose works were translated into Syriac, Hebrew, Arabic and Latin, and John of Alexandria (7th cent.), who is thought to have prepared an epitome of Galen's four treatises on the pulse. There is also the work attributed to Philaretus, *De pulsibus*, from the later part of the first millennium.[63]

Most important for this period is Paul of Aegina who has already been mentioned. His chapter on the pulses is taken directly from Galen.[64] He gives a definition of the pulse as "... a movement of the heart and arteries, taking place by diastole (dilatation) and systole (contraction)" and identifies a twofold purpose for the pulse—during diastole cold air enters to ventilate and resuscitate the animal spirits and therefore the form of the vital spirit (πνεῦμα), and during systole, there is evacuation of fuliginous superfluities. He lists the ten features of the pulse for the physician to observe:
1. time of motion—which may be rapid, slow or moderate;
2. extent of diastole—which occurs in three dimensions and may be broad or narrow, long or short, high or low;
3. tonus or vital strength—which may be strong, weak or moderate;
4. consistency of the body of the artery—which may be hard, soft or moderate;
5. contents of the arterial lumen—which may be full, empty or moderate;
6. quality of heat in the artery;
7. time of rest—which may be dense, rare or moderate (there are two "rests"—between diastole and systole and vice versa);
8. rhythm—which is the ratio of the time of motion to the time of rest (or the time of one motion and rest to that of another motion and rest). Three types of arrhythmia were identified: a slight departure from a regular rhythm, a moderate departure, and no detectable rhythm at all;
9. equality or inequality—an equal pulse is one in which successive beats are the same in all characteristics and an unequal pulse is one in which there is variation in one or more of the characteristics. The various terms such as *dorcadissens* (bounding), *dicrotic*, vermicular (worm-like), or *formicant* (ant-like), were applied to unequal pulses;
10. regular or irregular—these terms are applied to unequal pulses. Paul writes:

63 See note 62 above.

64 Paulus Aegineta II.11 (ed. Heiberg 1921: 81–93; Engl. transl.: Adams 1849: 202–221).

> The unequal pulses being divided into those which are alike as to periods, and those which are wholly unlike, the regular and irregular are formed according to each of these divisions; from the equality of periods, the regular is formed, and from the entire inequality, the irregular.

To these, some add the position of the artery and the times of expansion. Paul also follows Galen's division of the types of things that change the pulses into three groups—κατὰ φύσιν (related to such factors as age, season, place, κρᾶσις), παρὰ φύσιν (related to δυσκρασία, putrefaction of humors, and diseases generally), and non-κατὰ non-παρὰ φύσιν (related to such things as the temperature of the immediate environment, baths, food, wine, heating medications etc.). It can be seen, then, how closely Paul, as representative of the medical writers in the first several centuries after Galen, adheres to the latter's concepts regarding the arterial pulse.

In the Arabic World, when it comes to the major figures of Arabic medicine, particularly from the 10th to the 14th century, they too were heavily dependent on Galen. One major difference from their Greek predecessors is that they were aware of, and took into account, both Chinese and Ayurvedic medical writings. Thus, Rashīd al-Dīn Faḍlullāh Hamadānī (1247–1318) prepared an encyclopaedia of Chinese medicine which included material from the early Chinese writings on the pulse. The following two quotations are from 'Ali ibn al-'Abbas al-Majusi (d. late 10th cent.) and ibn Sina (980–1037), known in the West as Haly Abbas and Avicenna, respectively. The first highlights the similarity to Galen and the second provides an interesting insight into the diagnostic range of pulse analysis:

> The pulse is a messenger that does not lie and a mute announcer that tells of secret things by its movement. ... The heart and the arteries move in one motion, in one style, and at one time, which means that the movement of each is equal to that of the other and does not differ from it. Therefore, we can have the condition of the heart and its movements from the conditions of arteries and the pulse. The condition of the pulse differs a great deal according to the differences in the moving power, the differences in the instinctive heat and the differences in the condition of the arteries and what they contain of blood and spirit. ... Early physicians classified these differences in ten categories:
>
> 1. indications taken from the amount of relaxation;
> 2. indications taken from the time of movement;
> 3. indications taken from the amount of power;
> 4. indications taken from the consistency of the artery;
> 5. indications taken from the contents of the artery;
> 6. indications taken from the volume of the artery;
> 7. indications taken from the time of diastole;
> 8. indications taken from the times of movements and intervals;
> 9. indications taken from the special amount;
> 10. indications taken from the number of beats.[65]

65 See Amber and Babey-Brooke 1993: 82, with n7 for the source of the quote from al-Majusi (Haly Abbas).

> And hereby it is possible to arrive at the identity of the beloved person, if the patient will not reveal it, such knowledge offering one means of treatment. The device whereby this may be affected is that many names should be mentioned and repeated while the finger is retained on the pulse, and when it becomes very irregular and almost ceases, one should then repeat the process. I have tried this method repeatedly and have discovered the name of the beloved. Then, in like manner, mention the streets, dwellings, arts, crafts, families and countries, joining each one with the name of the beloved, and all the time feeling the pulse, so that when it alters on the mention of any one thing several times, you will infer from this all particulars about the beloved as regards name, appearance and occupation.[66]

In a 2011 article on the history of the arterial pulse, Nima Ghasemzadeh and Abamard Maziar Zafari have this to say on Avicenna in relation to the pulse:

> Avicenna came close to a general understanding of the various arrhythmias based on the characteristics of the pulse. He described different pulses being observed in atrial and ventricular arrhythmias. He categorized the arterial pulse more thoroughly than his predecessors. Moreover, he explained three factors—vital power (capacity, *dunamis*), resistance and elasticity—as being important in making the size of the pulse beat. This was the first time that concepts like resistance and elasticity were defined in a physiological manner. In describing different pulse rhythms, Avicenna also compared pulse rhythms to musical rhythms ... he described several types of pulses in different age groups and in both genders. Likewise, he related different pulses to different physiological and pathological states ... Based on Avicenna's observations and descriptions of the arterial pulse, it is conceivable that he understood the pulse as a wave rather than an impact generated by a cylindrical tube, a basic concept that is the foundation for the studies on the pulse wave in modern medicine.[67]

A number of developments occurred during the first six centuries of the second millennium, which, for circulatory physiology, paved the way for Harvey's momentous discovery and a reevaluation of Galenic concepts of the pulse. First, in general terms, was the translation of many important Greek and Arabic works into Latin, an endeavour associated particularly with Constantine the African (ca. 1020-before 1098/99) and the school of Salerno. At Monte Cassino, Constantine devoted his activity to the translation of Arabic treatises, including many Arabic versions of Galen. Vivian Nutton writes:

> The importance of Constantine's translations cannot be overestimated. They put the Latin-speaking world in touch with the tradition of Hippocratic learning, promoted by Galen and extended by the Arabs.[68]

Other translators of importance were Gerard of Cremona (fl. 1150–1185), Burgundio of Pisa (1110–1193), and Niccolo da Reggio (fl. 1315–1348). Coupled with this ongoing translation movement was the development of universities and hospitals. These several fac-

66 For this and other examples, see Browne 1921: 84–89.
67 Ghasemzadeh and Maziar Zafari 2011: 4.
68 Nutton 1995b: 141.

tors contributed substantially to a revision of Galen's anatomy, which had been dominant for over a thousand years.

Regarding the Medieval West, brief notes on significant medieval physicians regarding the physiology of the circulation follow.

Berengario da Carpi (1470–1530), on the basis of many dissections, showed that the *rete mirabile* (important in Galen's concept of vascular physiology) does not exist in the human brain.[69]

Michael Servetus (1511–1553) had a varied, and indeed chequered career, combining religion, law and medicine, which came to an untimely end when he was burned at the stake in Geneva, along with copies of his work, *Christianismi Restitutio*, deemed heretical. Cecilia C. Mettler (1909–1943) writes:

> In the fifth book of this text, of which only two complete copies are extant, he discusses the nature of the Holy Spirit and gives an account of the pulmonary circulation, stating that the blood passed into the heart after it was mixed with air in the lungs. 'It is also not simply air,' he wrote, 'but air mixed with blood that is sent from the lungs to the heart through the arterial vein; therefore, the mixture is made in the lungs. The bright colour is given to the sanguine spirit by the lungs, not by the heart'.[70]

Andreas Vesalius (1515–1564) made the most significant impact as regards the revision of Galenic anatomy generally in *De humani corporis fabrica* (*On the Fabric of the Human Body*) published in 1543. Although he certainly refined Galen's ideas on the heart, among other things casting doubt on the supposed permeability of the interventricular septum, he did not overthrow Galen's *two-systems theory*—veins originating from the liver and arteries from the heart, both functioning essentially independently and centrifugally.

Realdo Columbo (1516–1559) was Vesalius' prosector in Padua and later professor of anatomy in Rome. He is credited with the discovery of the pulmonary circulation, although this was partly understood by Galen and more completely by Servetus. Cecilia Mettler has this quote from him:

> I for my part hold quite a different view, namely, that this vein-like artery was made to carry blood mixed with air from the lungs to the left ventricle of the heart. And this is not only most probable but is actually the case; for if you examine not only dead bodies but also living animals, you will find the artery in all instances filled with blood, which by no manner of means would be the case if it were constructed to carry air forsooth and vapours. Wherefore I cannot wonder enough at those anatomists who have not observed a matter so clear and of such importance, eminent though they wish to be considered and indeed are considered by many of their fellows.[71]

---

69 On this issue, see Rocca 2003: 201–238 and 249–254.
70 Mettler 1947: 45.
71 Mettler 1947: 46.

Andrea Caesalpino (1519–1603) was a professor of medicine at Pisa who later became physician to Pope Clement VIII, and lecturer of medicine in Rome at the university, La Sapienza. Cecilia Mettler writes:

> Caesalpinus is frequently regarded by the members of the Italian historical school as the legitimate discoverer of the circulation. In his text, *Quaestionum peripateticarum* (Florence, 1571), he discussed the theoretics of the systemic and pulmonary circulations, mentioning that, in systole, the heart sends blood into the aorta, and in diastole, receives it back from the vena cava. His theory has been rejected by British historical writers, because it was not supported by any experimental evidence.[72]

Girolamo Fabrici d'Acquapendente, also known as Girolamo Fabrizio or Hieronymus Fabricius ab Aquapendente (1537–1619), wrote a treatise on the valves in the venous system—*De venarum ostioles*—published in Padua in 1603, although he failed to identify their true purpose, being misled by the long-standing concept of centrifugal blood flow in the veins. He did, however, have a significant influence on his English student, William Harvey.

5 William Harvey

The culmination of the various studies referred to in the previous section, and others not mentioned there, was the publication of Harvey's *De motu cordis* in 1628 (Frankfurt).[73] It was this above all that revolutionized the understanding of cardiovascular physiology. Two excerpts that have a particular bearing on the concept of the arterial pulse are as follows:

> From these facts it is manifest, in opposition to commonly received opinions, that the diastole of the arteries corresponds with the time of the heart's systole; and that the arteries are filled and distended by the blood forced into them by the contraction of the ventricles; the arteries, therefore, are distended because they are filled like sacs or bladders and are not filled because they expand like bellows. It is in virtue of one and the same cause, therefore, that all the arteries of the body pulsate, viz. the contraction of the left ventricle; in the same way as the pulmonary artery pulsates by the contraction of the right ventricle ... Whence it appears that wherever the motion of the blood through the arteries is impeded ... there do the remote divisions of the arteries beat less forcibly, seeing that the pulse of the arteries is nothing more than the impulse or shock of the blood in these vessels.[74]

Also, from Harvey's reply to the criticism by the French physician and anatomist Jean Riolan (1577/80–1657):

---

72 Mettler 1947: 50.

73 Full title of the work: *Exercitatio anatomica de motu cordis et sanguinis in animalibus.*

74 Harvey 1628/1906: 29–30.

> The authority of Galen is of such weight with all, that I have seen several hesitate greatly with that experiment before them, in which the artery is tied upon a tube placed within its cavity; and by which it is proposed to prove that the arterial pulse is produced by a power communicated from the heart through the coats of the arteries, and not from the shock of the blood contained within them; and thence, that the arteries dilate as bellows, and are not filled as sacs. This experiment is spoken of by Vesalius, the celebrated anatomist; but neither Vesalius nor Galen says that he had tried the experiment, which, however, I did. Vesalius only prescribes, and Galen advises it, to those anxious to discover the truth, and for their better assurance, not thinking of the difficulties that attend its performance, nor of its futility when done; for indeed, although executed with the greatest skill, it supplies nothing in support of the opinion which maintains that the coats of the vessel are the cause of the pulse; it much rather proclaims that it is owing to the impulse of the blood.[75]

The appearance of Harvey's work marked the beginning of the end for Galen's pulse lore which had endured for almost fifteen hundred years. But the beginning was a long way from the end of the end. The new understanding of how the pulse arose and what, at least in broad outline, it represented did not mark the transition from using the pulse as a general indicator for a wide variety of states of health and disease to its more restricted role as a simple, initial clinical sign of cardiac function.

6 From Harvey to the Twentieth Century

Speaking of the 18th century, Roy Porter (1946–2002) writes:

> In general, physicians rested content with the traditional diagnostic uses of the 'five senses'; they would feel the pulse, sniff for indications of decay, taste urine, listen for breathing irregularities, and observe skin colour. This time-honoured approach was almost exclusively qualitative. Thus, what standardly counted in pulse lore was not the number of beats per minute but their strength, firmness, rhythm, and 'feel'.[76]

Further, Broadbent refers to the "idea of the critical pulses of Solano [Francisco Solano de Luque, 1684–1738] and Nihell [Jacob/James Nihell, 1705–1759], by which the occurrence of epistaxis, purgation, etc., was foreseen" and the "organic pulses of Bordeu [Théophile de Bordeu, 1722–1776] and Fouquet [Henri Fouquet, 1727–1806], a special variety of pulse for disease in each organ." He concludes: "These fantastic descriptions had no reference to the physiology of the circulation and might more easily have been written before its discovery."[77]

Three specific developments, of varying importance, which contributed to the end of the general use of pulse analysis in Western medicine were the following:

1. introduction of the pulse watch by Sir John Floyer (1649–1734) in 1707, which signaled a quantitative turn;

---

75 Harvey 1649/1906: 134–135.
76 Porter 1995: 403.
77 Broadbent 1890: 12.

2. the first measurement of arterial blood pressure (in a horse) by Stephen Hales (1677–1761) in 1733, and the extension of this by Jean Léonard Marie Poiseuille (1797–1869) and others which culminated in the introduction of a clinically reliable sphygmomanometer by Scipione Riva Rocci (1863–1937) in 1896, and Nikolai Sergeyevich Korotkov (1874–1920) in 1905;
3. the development of the sphygmograph by Frederick Akbar Mahomed (1849–84) and Étienne Jules Marey (1830–1904), which opened the way for a detailed objective and quantitative study of the arterial pulse.

Several important books devoted specifically to the pulse appeared in the later part of the 19th century and the first part of the 20th century. These were the works of Jules Marey (*La circulation du sang à l'état physiologique et dans les maladies*, 1881), Charles Ozanam (1824–1890) (*Le pouls*, 1886), William H. Broadbent (*The Pulse*, 1890), and James MacKenzie (*The Study of the Pulse*, 1902). Of note is the fact that the sphygmograph did not find a place in day-to-day clinical practice, whereas today the sphygmomanometer is on every doctor's desk.

In summary, then, while Harvey's ground-breaking work was done early in the 17th century, it was not until the 20th century that the pulse became primarily linked to analysis of the cardiovascular system (at least in Western medicine) and ceased to occupy the important general role it had hitherto held in the diagnosis and prognosis of the whole range of diseases. It is noteworthy too that the qualitative analysis of the pulse as the indicator of various bodily states and disease conditions has not been supplanted by the quantitative to the same extent in the Chinese and Ayurvedic medical traditions.

7 Modern Western Views

In modern Western medicine,[78] the diagnostic value of palpation of the arterial pulse (particularly a peripheral arterial pulse such as the radial) is seen as much more restricted and, it must be said, less important, than in earlier times and in other medical traditions in modern times. Primarily, palpation of the pulse provides information about the health and function of the cardiovascular system, although secondarily, it does give some information about non-cardiovascular diseases and health more generally. First, the information gained is usually and necessarily supplemented by other evaluations, particularly blood pressure measurement and electrocardiography (ECG), but also by a wide variety of more complicated, and often invasive, investigations. Second, pulse evaluation is supplemented by other, more definitive investigations directed at the suspected condition. As an example of the downgrading of the impor-

78 An excellent and concise account of the arterial pulse in modern Western medicine can be found in Moran 1990.

tance of palpation of the pulse, in a standard modern cardiology text—John Willis Hurst, *The Heart*—only six of the 2476 pages are devoted to the arterial pulse. There has, however, been one major monograph devoted specifically to the arterial pulse[79] in which the authors offer the following view of the pulse in the chapter entitled 'What is the Pulse', this making an interesting comparison with the definitions of Galen and other early writers:

> In general terms, the arterial pulse is any periodic fluctuation that is caused by the heart and occurs at the same frequency as the heartbeat. Cardiac contraction ejects surges of blood into the ascending aorta during systole. These in turn lead to flow pulsations, pressure pulsations, and diameter pulsations in arteries throughout the body. Any of these can be described as a pulse, as can the pressure, diameter, and flow fluctuations in the veins leading blood back to the heart.[80]

Of the three wave (pulsation) types referred to, it is the pressure waves that are detected on palpation. Left ventricular systolic ejection creates a pressure wave in the ascending aorta which is propagated to arteries throughout the body. The detection of the wave will depend, *inter alia*, on the pressure of the applied finger(s) and the mobility of the palpated artery. Diameter waves are created by the pressure changes with an almost identical contour but the actual changes in diameter are small—in a child's aorta approximately 10 % of the mean diameter, in a child's radial artery approximately 5 % of the mean diameter and in older people only 1–2 % of the mean diameter. These figures are of interest in relation to Galen's elaborate scheme of classification of potentially palpable three-dimensional changes in arterial size. Flow waves occur due to the longitudinal movement of blood that follows systolic ejection.

One important point the authors make is that the amplitude of the palpable pulse not only depends on the amplitude of the pressure pulse, but also on the size of the palpated artery as well as the degree of deformity of the Meissner's (pressure-sensitive) corpuscles and the number of corpuscles activated in the palpating fingers. Thus, a pulse of apparently low amplitude may be a pulse of actual low-pressure amplitude in a normal artery or a pulse of normal pressure amplitude in a narrow artery. The same type of distinction applies to a pulse of apparently high amplitude.[81]

In effect, modern Western medicine, in clinical examination of the pulse, assesses four principal features: rate and rhythm, contour and volume, the members of both the first and second pairs being closely connected. An analysis of the four features follows.

1. rate and rhythm: under normal circumstances, the initiating electrical impulse for the heartbeat arises in the sino-atrial (SA) node, spreads through the walls of the

79 O'Rourke, Kelly, and Avolio 1993.
80 O'Rourke, Kelly, and Avolio 1993: 15.
81 O'Rourke, Kelly, and Avolio 1993: 19.

atria which contract, activating the atrio-ventricular (AV) node which, in turn, distributes branches via the bundle of His to the ventricular walls.

The recognition of this system dates from the early 20th century. When the mechanism is functioning normally the heart beats at a rate between 55 and 100 beats per minute (BPM) with a regular rhythm apart from some recurring variation related to respiration. Alterations of rate can occur with or without preservation of a regular rhythm. Thus, sinus bradycardia (rate < 55BPM) can occur in a resting state in a young adult in good physical condition, for example, and with pathology of the conducting system such as sick sinus syndrome, secondary AV block or tertiary AV block with junctional escape or ventricular escape. It can, however, also occur in severe infections.[82] Sinus tachycardia (rate > 100BPM) with a regular rhythm can also occur in a normal cardiovascular system—for example, during exercise, anxiety, and pregnancy—and in various non-cardiovascular diseases such as hyperthyroidism, anaemia and fever. It can also occur in primary cardiac electrical disorders such as atrial flutter with regular AV block and paroxysmal supraventricular tachycardia (PSVT).[83] It should also be mentioned that the normal rate with an apparently normal rhythm can occur in some pathological conditions such as atrial flutter with a 3:1 or 4:1 AV block.

Abnormalities of rhythm can occur with or without an abnormal rate and these abnormalities can themselves be regular (regularly irregular) or irregular (irregularly irregular). Examples of the former are *coupled rhythm* (bigeminy, trigeminy) and *grouped beats with dropped beats*, characteristic of some infectious diseases including rheumatic fever, but also of organic heart disease. Typical of an irregular rhythm occurring with a normal or abnormal rate are multiple ectopic beats (from an atrial or ventricular ectopic focus), and atrial fibrillation. In the latter, the atria do not electrically conduct in a synchronous fashion, so there is a disorderly arrival of stimuli at the AV node with a variable minority of these stimuli being transmitted to the ventricles. Atrial fibrillation is characteristic of organic heart disease but can also occur in other conditions such as acute infections and hyperthyroidism. In atrial flutter, where regular impulses are generated in the atria but at a very high rate, the actual rhythm can be regular, if there is a consistent degree of heart block at the AV node but may be quite irregular if the degree of block is variable. As with atrial flutter, fibrillation can occur with both organic heart disease and with other conditions. In summary, then, pulse rate can be slower or faster than normal without signifying any pathology (e.g. a low resting rate in fit young adults or an increased rate in exercise, anxiety and pregnancy) but also can be normal in the presence of pathology, either cardiovascular or otherwise. Variations in rhythm (apart from sinus arrhythmia) do signify some abnormality but may be essentially benign. In modern practice, disorders of rate and rhythm require at least ECG analysis for proper evaluation. Furthermore,

---

82 Wood 1967: 252–54

83 Wood 1967: 250–252.

the rapidly expanding subject of cardiac electrophysiology involves invasive assessment and management of rhythm disorders.

2. contour and volume: the arterial pressure wave begins with aortic valve opening following the onset of left ventricular ejection. It shows first a swift upward (*anacrotic*) stroke when the left ventricular stroke volume enters the aorta faster than run-off can occur. Peak systolic pressure coincides with the end of the upstroke. There is then a downward movement of the pulse pressure wave as run-off exceeds ejection with an incisura (*dicrotic notch*) occurring approximately at the end of systole which is related to a transient reversal of flow from the central arteries towards the ventricle just prior to aortic valve closure. There is then a small, secondary positive wave attributable to the elastic recoil of the aorta and aortic valve as well as reflected waves from the distal arteries. It is generally agreed that the dicrotic wave is not palpable, so what one palpates is the sharp upstroke and the more gradual downstroke to the trough with the peak of the upstroke representing the systolic blood pressure and the depth of the trough representing the diastolic blood pressure.

The changes in contour of the wave are as follows:

- the twice-peaking pulses—*bisferiens* with two palpable waves in systole and the dicrotic with one palpable wave in systole and one in diastole—which signify, in the first instance, a hyperkinetic state, aortic valve disease or hypertrophic cardiomyopathy, and, in the second instance, impaired left ventricular performance or low cardiac output, as in dilated cardiomyopathy or cardiac tamponade;
- a plateau or tardus pulse in which there is a slow upstroke and delayed peak characteristic of aortic stenosis.

The changes in volume are as follows:

- absent pulse(s) found in the rare Takayasu's disease in which there is a granulomatous arteritis involving the proximal aorta and its major branches, so the peripheral pulses may be variably reduced or absent;
- *pulsus alternans* where pulse waves alternate between greater and lesser volume characteristic of impaired left ventricular function;
- *pulsus bigeminus* where secondary premature beats will have a lesser volume;
- a bounding or collapsing (water-hammer) pulse characterised by vigorous contraction with an increased arterial pressure and reduced peripheral resistance (rapid run-off) typical of normal subjects with a hyperkinetic circulation, or conditions such as aortic regurgitation, patent *ductus arteriosus* or arteriovenous fistula;
- *pulsus paradoxus* where there is a marked decrease in pulse amplitude during normal inspiration. This is characteristic of pericardial tamponade and obstructive airways disease.

So, in summary, comparing broadly Galen's concept of the pulse and what it indicates clinically with the modern view, the following five points may be made:

1. the palpable arterial pulse is a result of injection of blood into the arterial system by left ventricular systole which causes a largely passive expansion/dilatation of the arteries, cardiac ventricular contraction being coincident with arterial dilatation—that is, the wineskin model as opposed to the bellows model favoured by Galen;
2. the palpated pulse is primarily a pressure wave with expansion followed by relaxation. The nature of this wave is essentially dependent on cardiac function; there is no independent power (δύναμις) of the arteries producing their expansion/dilatation/pulsation, as Galen thought. Nonetheless, the nature of the pulse wave is influenced by the physical state of the arterial wall, as Galen recognised, and by neural and humoral mechanisms;
3. while in clinical terms, the pulse examination in modern medicine focusses primarily on the heart and the arteries themselves, the pulse remains an indicator of natural states (age, gender etc.), non-natural states (exercise etc.) and unnatural states (that is, disease states generally), as Galen described;
4. there are various characteristic pulses, as listed above which have a partial correspondence to Galen's list of named pulses;
5. in other medical traditions (Chinese, Ayurvedic) the qualitative evaluation of the pulse, with numerous distinctive forms being observed, still has a significant place in current practice.

To conclude, Galen's copious writings on the arterial pulses are important for two main reasons. First, since they are the earliest extant account of the theory and practice of sphygmology in Western medicine and give a comprehensive analysis of previous ideas on the subject (particularly those of Herophilus/Herophileans and Archigenes), they are of great value in determining the historical development of sphygmology up to Galen's time. Second, they established the theory and practice of sphygmology that formed a critical component of diagnosis and prognosis in both Western and Arabic medical practice from the 2nd century A.D. until the progressive revision of his ideas on the circulation in general and the arterial pulse in particular which began in the later part of the first half of the second millennium and continued over the next several hundred years.

8 Early Chinese and Indian Writings on the Pulse

The purpose of this brief interpolated section is to draw attention to three parallels and a single divergence that are apparent when comparing the development of ideas on the arterial pulses and their clinical relevance in the three main enduring medical traditions—Ayurvedic, Chinese and Western. The three parallels are:

1. contemporaneity of the initial recognition of the pulse—6th to 5th centuries B.C.;

2. formalization of a detailed system of qualitative analysis and classification of the kinds of pulses and the use of this system clinically for diagnosis and prognosis—2nd century A.D.;
3. the enduring use of these systems of qualitative analysis in all three traditions, running parallel for almost two millennia (3rd to 19th centuries).

The divergence is, of course, the movement away from the qualitative and towards the quantitative in assessing cardiovascular function generally, and the concomitant downgrading of simple palpation of the pulse in Western medicine in particular, a movement which really gathered pace at the end of the 19th century. This development, while having a considerable impact on the other traditions, has not however, led to such a displacement of the enduring methods in the other two traditions, which will be briefly considered below.

About Chinese medicine, the following short quotation from the *Records of the Grand Historian* (*Shiji* 史記), a work dealing with the several centuries prior to the Western Han dynasty (206 B.C.–6 A.D.), is an anecdote about the famed, and possibly historic, physician Qin Yueren 秦越人 (Bian Que 扁鵲) who is said to have practiced in the late 6th or early 5th century B.C.:[84]

> Bian Que, looking up and sighing, replied 'your ideas of medicine are no better than viewing the sky through a narrow tube or reading a piece of writing through a narrow crack. In my practice of medicine, I need not even feel the pulse, look at the complexion of the patient, listen to him, or visually examine his physical condition, in order to say where the disease is located'.

The Chinese system of pulse analysis, and its use in diagnosis and prognosis, is set out in the earliest extant medical text, the *Huangdi Neijing Suwen* 皇帝內徑素問 (*The Yellow Emperor's Inner Classic—Simple Questions*) and in the *Nan Jing* 難經 (*Classic of Difficulties*) from around the same date (possibly the 1st cent. B.C.). However, what might be termed the culmination of the system is to be found in a work devoted specifically to the pulse, the *Mai Jing* 脈經 or *Pulse Classic* attributed to Wang Shuhe 王叔和 from the 2nd-3rd century A.D. If Galen's system of pulse analysis seems complex to modern Western readers, it must be viewed as simplicity itself when compared with the Chinese system. In essence, the pulse was felt at three points along the line of the radial artery (*cun* 寸, *guan* 關 and *che*尺 from distal to proximal) separately in both wrists. There were thus six points, and each was associated with specific structures or regions—for example, the *cun* pulse on the left was predominantly associated with the heart. The pulse beats were evaluated for volume, strength and regularity. There were also thought to be important and detectable seasonal variations in the pulse. One notable feature of the Chinese analysis of the pulse was the attempt to quantitate the rate and establish the range of normal. This was, however, done by relating the rate of

84 See Sivin 2000: 45.

the pulse to another variable, respiration, which is, of course, influenced by some of the factors which also influence the pulse. The following excerpt from the *Huangdi Neijing Suwen* (*The Yellow Emperor's Classic of Medicine*) sets out the method of determining whether the rate was normal or abnormal:

> Huang Di asked: "What kind of pulse do you find in a normal person?"
> Qi Bo answered: "The pulse of a normal, healthy individual will beat twice with each inhalation and twice with each exhalation. With one complete breath, there are four beats. Occasionally, it is normal to detect five beats per breath, depending on the patient's lung capacity. When palpating the pulse, one should feel it in reference to the patient's breath, if the patient is normal. However, if the pulse is more than five beats per patient's breath, this is abnormal. In this case, the physician should examine the pulse in reference to his own breathing."[85]

In the Preface to the *Pulse Classic*, written around the same time as Galen wrote his treatises on the pulses, Wang Shuhe said:

> The mechanisms of the pulse are fine and subtle, and the pulse images are difficult to differentiate. The bowstring and the tight, the floating and the scallion-stalk confusingly resemble one another. They may be readily distinguished at heart (i.e. their verbal definitions may have been memorized), but it is difficult for the fingers to distinguish them. If a deep pulse is taken as a hidden one, the formula and the treatment will never be in the right line. If a moderate pulse is taken as a slow one, crisis may crop up instantly. In addition, there are cases where several different kinds of pulse images appear all at once or several different categories of disease may exhibit the same kind of pulse.[86]

In a recent textbook on pulse diagnosis in traditional Chinese medicine (TCM), Zheng-hong Lin lists the standard 28 pulse types under six categories as follows:[87]

1. *Floating Pulse Category:* floating pulse, surging pulse, soggy pulse, scattered pulse, scallion-stalk pulse, drumskin pulse;
2. *Sunken Pulse Category:* sunken pulse, hidden pulse, firm pulse, weak pulse;
3. *Slow Pulse Category:* slow pulse, moderate pulse, rough pulse, bound pulse;
4. *Rapid Pulse Category:* rapid pulse, racing pulse, skipping pulse, stirred pulse;
5. *Vacuous Pulse Category:* vacuous pulse, faint pulse, fine pulse, intermittent pulse, short pulse;
6. *Rapid Pulse Category:* replete pulse, long pulse, slippery pulse, string-like pulse, tight pulse.

About Ayurvedic Medicine, in a recent article on the history of sphygmology, the authors refer to Sage Kanada (also known as Kashyapa) (between the 6th and 2nd cent. B.C.), an ancient Indian physician, alchemist and philosopher, who described a variety of pulses during different physiological and pathological states. According to his theory,

85 Ni 1995: 71.
86 Yang Shou-zheng, *The Pulse Classic.* Preface to the Engl. transl. by Wang Shuhe 1997: xi.
87 Lin 2004:32–34.

each pulse had three stages, abnormality in any of which reflects diseases of the three main humors of the human body: *bayu/vata* (air), *pitta* (bile) and *kaph/kapha* (phlegm).[88]

Subsequently, Caraka, a physician from Kashmir who is thought to have practiced in the 2nd century A.D., compiled a compendium of medical practice in pulse diagnosis. Several of the early Ayurvedic texts remain available, not only in the original Sanskrit, but also in English translation, and remain relevant to the teaching of Ayurvedic medicine.[89] Key features of pulse diagnosis in this tradition, a number of which are also to be found in Chinese medicine, are briefly enumerated below:

1. all diseases can be detected by analysis of the pulse;
2. the pulse should be examined in the morning, prior to breakfast, by a doctor who is attentive, free of disease himself, and calm;
3. the patient should be quiet and free of external and internal factors which might influence the pulse—for example, exercise, bathing, eating, hunger, thirst, and fatigue;
4. the radial pulse is examined with three fingers of the right hand with the left hand being used to support the arm. The right radial pulse is examined in the male and the left in the female;
5. the movements of various animals and birds are used to describe different types of pulses.

In summary, the recognition of the true nature of the circulation by Harvey, supported by the discovery of capillaries by Marcello Malphighi (1628 – 1694), which was reported in 1661, sounded the death-knell of Galenic sphygmology. But the demise was protracted, as was the supplanting of Galen's system of medicine generally. In the other two traditions, no such momentous discovery occurred, and although Western methods have now spread widely to regions where these traditions were (and are), they have by no means supplanted them. In the pulse, the methods of Galen are now entirely obsolete in the West whereas in Chinese traditional and Ayurvedic medicine, the methods of Wang Shuhe and Caraka respectively remain relevant.[90]

88 Ghasemzadeh and Maziar Zafari 2011, with reference to Kanada's original text.

89 For bibliographical details of these works and their translations, see Amber and Babey-Brooks 1993: 73 – 75.

90 The whole history of Chinese medicine, including the influence of Western medicine, is well covered in the work by Wong and Lien-the 1936, esp. 57 – 66 on the pulse. Interestingly, on the topic of the pulse, they refer to *Clavis Medica ad Chinarum Doctrinam de Pulsibus* published in 1686 as a posthumous work by the Polish Jesuit missionary in China, Michael Boym (1612 – 1659), that should have been part of a larger work entitled *Medicus Sinicus.*

III Textual Tradition

The translations in this volume have been made on the basis of the available printed versions of Galen's Greek text. Examination of the transmission of these works through the centuries, from the most ancient traces and witnesses to the editions followed here, allow for assessing how much these editions reflect Galen's text. The following is a synthesis of the current state of research on the topic.

1 Manuscripts and Stemma codicum

The Greek manuscripts of our texts (except *De usu pulsuum*) have been inventoried in the early 20th century in the catalogue edited by the German historian of classical philosophy Hermann Diels (1848–1922) and published in 1906–1908.[91] Data about manuscripts below include location and shelfmark, updated as necessary on the basis of Touwaide's *A Census of Greek Medical Manuscripts*, with the major codicological characteristics (date or century[92] and copyist when known [signature or identification on the basis of the writing]), followed by the identification of the texts on the basis of a consultation of the manuscripts for the four major treatises, and on the basis of images (microfilms) and available catalogues and literature for the others. The list below includes references to modern catalogues and relevant studies.

1.1 The four major treatises (*De pulsuum differentiis*; *De dignoscendis pulsibus*; *De causis pulsuum*; *De praesagitione ex pulsibus*).

"The Great Sphygmic" (Ἡ Μεγάλη Σφυγμική) as these treatises are often called in the manuscripts, are contained in the following manuscripts (all codices are identified by a capital letter that follows their shelfmark, between parentheses; for three manuscripts we have added a second identifier made of a capital letter followed by an Arabic number; this second identifier follows the first):[93]

91 Diels 1905, with a supplement in Diels 1908, and a revised edition in Touwaide 2021, tome 1.

92 The period of the manuscripts below does not reproduce Diels' data; it has been updated on the basis of current research.

93 For a list of the manuscripts, see Diels 1905 and 1908, and the revised edition by Touwaide 2021:
- *De pulsuum differentiis:* Diels 1905: 87, and Touwaide 2021: 1.128–129; Diels 1908: 33, and Touwaide 2021: 1.371;
- *De dignoscendis pulsibus:* Diels 1905: 87–88, and Touwaide 2021: 1.129; Diels 1908: 33, and Touwaide 2021: 1.371;
- *De causis pulsuum:* Diels 1905: 88, and Touwaide 2021: 1.129–130;
- *De praesagitione ex pulsibus:* Diels 1905: 88–89, and Touwaide 2021: 1.130–131.

https://doi.org/10.1515/9783110612677-006

Città del Vaticano, Biblioteca Apostolica Vaticana, *graecus* 281 **(C)** = **(Va1)**
Late 15th century, Georgios Moschos.
(ff. 1–42) *De pulsuum differentiis*; (ff. 42v–73) *De dignoscendis pulsibus*; (ff. 73v–107v) *De causis pulsuum*; (ff. 106–150) *De praesagitione ex pulsibus.*
Mercati and Franchi de' Cavalieri 1923: 384; Touwaide 2016: 291, no. [1526].

Città del Vaticano, Biblioteca Apostolica Vaticana, *graecus* 1064 **(K)**
15th century (?). The manuscript is divided in two parts: ff. 1–125 and 126–267.
(ff. 1–83v) *De pulsuum differentiis*; (ff. 84–140) *De dignoscendis pulsibus*; (ff. 141–200) *De causis pulsuum*; (ff. 201–267v) *De praesagitione ex pulsibus.*
Amati et al. 1800–1834: ff. 43v–44r; Touwaide 2016: 297, no. [1561].

Firenze, Biblioteca Medicea Laurenziana, plut. 74.18 **(A)** = **(F1)**
Late 12th century, Ioannikios with notes by Burgundio of Pisa[94].
(ff. 189–248v) *De pulsuum differentiis libri*; (ff. 249–275) *De dignoscendis pulsibus*; (ff. 275–297v), *De causis pulsuum*; (ff, 297v–322) *De praesagitione ex pulsibus.*
Bandini 1764–1770: 3.120–121; Touwaide 2016: 76, no. [0331].

Firenze, Biblioteca Medicea Laurenziana, plut. 74.28 **(B)** = **(F2)**
13th–14th century.
(ff. 1–114v) *De pulsuum differentiis*; (ff. 114v–199v) *De dignoscendis pulsibus*; (ff. 199v–294v) *De causis pulsuum*; (ff. 294v–399) *De praesagitione ex pulsibus.*
Bandini 1764–1770: 3.130–131; Touwaide 2016: 77, no. [0341].

Firenze, Biblioteca Medicea Laurenziana, plut. 74.31
14th century.
(ff. 207v) *De pulsuum differentiis* (table).
Bandini 1764–1770: 3.130–131; Touwaide 2016: 77, no. [0344].

Some identifications of manuscripts included in Diels' lists are not correct:
- *Paris. gr.* 2137 and 2157 mentioned by Diels 1906: 87 and 88, respectively, both with a question mark, are mistakes for *Paris. gr.* 2157 (Touwaide 2016: 215, no. [1038], and 222, no. [1076]);
- *Paris. gr.* 2158 listed in Diels 1906: 88, with a question mark for *De dignoscendis pulsibus*, is a mistake, probably a confusion with *Paris. gr.* 2153 (see Touwaide 2016: 222, no. [1078]);
- *Paris. gr.* 2276, listed in Diels 1906: 87–89, with a question mark, for all treatises, is a mistake: a confusion with *Paris. gr.* either 2161 or 2167 (see Touwaide 2016: 239, no. [1221]);
- Vienna, Österreichische Nationalbibliothek, *medicus graecus* 28 listed in Diels 1905: 89 does not contain Galen, *De praesagitione ex pulsibus*, but Theophilus, *De pulsibus* (see Hunger and Kresten 1969: 76–78).

Leiden, Universiteitsbibliothek, *Vulcanianus* VUL 43 (**E**)
15th century.
(ff. 4–16v) *De pulsuum differentiis*; (ff. 17–56) *De dignoscendis pulsibus*; (ff. 57–98) *De causis pulsuum*; (ff. 98–153) *De praesagitione ex pulsibus.*
Molhuysen 1910: 16–17; Touwaide 2016: 106, 115, no. [0520].

London, British Library, *Harleianus* 5625 (**F**)
16th century.
(ff. 1–90v) *De pulsuum differentiis*; (ff. 91–156v) *De dignoscendis pulsibus*; (ff. 157–225v) *De causis pulsuum*; (ff. 226–300v) *De praesagitione ex pulsibus.*
McKendrick 1999: 138; Touwaide 2016: 119, no. [0543].

Milano, Biblioteca Ambrosiana, O 50 sup. (**N**)
Late 13th century.
(ff. 13–129v) *De pulsuum differentiis*; (ff. 129v–230) *De dignoscendis pulsibus.*[95]
Martini and Bassi 1906: 679–680; Touwaide 2016: 133, no. [0642].

Modena, Biblioteca Universitaria e Estense, α. O. 4. 12 (**G**)
16th century. This codex has been consulted or owned by the Renaissance physician Agostino Gadaldini (1515–1575).[96]
(ff. 1–86) *De pulsuum differentiis*; (ff. 86–144) *De dignoscendis pulsibus*; (ff. 144–204) *De causis pulsuum*; (ff. 204–271) *De praesagitione ex pulsibus.*
Puntoni 1896: 514; Touwaide 2016: 142, no. [0711], 143.

Moskva, Gosudarstvennyi Istoricheskii Muzei (GIM). Sinodal'naia Biblioteka Moskovskoi Patriarchii, Sinod. 51 (Vlad. 464) (**M**)
14th century, Konstantinos Sofos (partim).
(ff. 8–42) *De pulsuum differentiis*; (ff. 42–67) *De dignoscendis pulsibus*; (ff. 67–93) *De causis pulsuum*; (ff. 93–127) *De praesagitione ex pulsibus.*
Vladimir 1894: 701–703, no. 464; Touwaide 2016: 147, no. [0731], 148.

Oxford, Bodleian Library, *Laudianus Gr.* 57 (**O**)
15th century, Demetrios Damilas.
(ff. 1–75v) *De pulsuum differentiis*; (ff. 75v–128) *De dignoscendis pulsibus*; (ff. 128–183) *De causis pulsuum*; (ff. 183–244) *De praesagitione ex pulsibus.*
Coxe 1853: 538–539; Touwaide 2016: 177, no. [0848], 182.

---

94 Fortuna and Urso 2009: 144–146.
95 Diels 1906: 88, does not include it in the list of the copies of *De dignoscendis pulsibus.*
96 Petit 2007: 137, where the codex is identified as *Mutinensis gr.* 226.

Paris, Bibliothèque nationale de France, *graecus* 2153 (**H**)[97]
15th century.
(ff. 47–78v) *De pulsuum differentiis*; (ff. 79v–106) *De dignoscendis pulsibus*; (ff. 106v–130) *De causis pulsuum*; (ff. 130v–169) *De praesagitione ex pulsibus.*
Omont 1886–1888: 2.205, and 3.396; Touwaide 2016: 221, no. [1069].

Paris, Bibliothèque nationale de France, *graecus* 2161 (**I**)
Late 15th century, Alfonsos Doursos and Demetrios Moschos.
(ff. 283–353v) *De pulsuum differentiis*; (ff. 353v–406) *De dignoscendis pulsibus*; (ff. 406–460v) *De causis pulsuum*; (ff. 460v–521v) *De praesagitione ex pulsibus.*
Omont 1886–1888: 2.207; Touwaide 2016: 223, no. [1082].

Paris, Bibliothèque nationale de France, *graecus* 2167 (**J**)
16th century. It might have been the source of Aldina edition (1525) for a part at least of *De pulsuum differentiis*,[98] and other Galenic texts (*De febrium differentiis* and *De simplicium medicamentorum temperamentis ac facultatibus*).
(ff. 287–303) *De pulsuum differentiis* (Book I); (ff. 303–343) *De dignoscendis pulsibus*; (ff. 343–387) *De causis pulsuum*; (ff. 287–446v) *De praesagitione ex pulsibus.*
Omont 1886–1888: 2.208–209; Touwaide 2016: 223, no. [1088].

Venezia, Biblioteca Nazionale San Marco, *graecus* Z 287 (= 709) (**L**)
1469, Ioannes Rhosos.
(ff. 157–196) *De pulsuum differentiis*; (ff. 196v–224v) *De dignoscendis pulsibus*; (ff. 225–253v) *De causis pulsuum*; (ff. 254–287) *De praesagitione ex pulsibus.*
Mioni 1981: 410–411; Touwaide 2016: 332, no. [1695], 340.

Wolfenbüttel, Herzog August Bibliothek, *Gudianus graecus* 11 fol. (**B**)
15th/16th century, Giovanni Benardo Regazzola (ca. 1490–ca. 1552),[99] source of the Aldina edition (1525) for *De pulsuum differentiis* (partim) and *De pulsibus ad tirones.*[100]
(ff. 12–21v) *De pulsuum differentiis*, Book II.
Köhler and Milchsack 1913: 6; Touwaide 2016: 370, no. [1845]–[1846], 371.

The relationships between the manuscripts for two of our treatises (*De pulsuum differentiis* and *De dignoscendis pulsibus*) have been examined by Heinrich von Staden in his study on Herophilus.[101] von Staden collated the following five manuscripts (listed below in von Staden's order):

97 In Diels 1906: 87–88, this manuscript appears only for *De dignoscendis pulsibus*, whereas it contains all four treatises.
98 Pietrobelli 2013: 140n21.
99 Pietrobelli 2013: 141–143.
100 Pietrobelli 2013.
101 von Staden 1989: xxiii.

- *Vaticanus graecus* 281 **(A)** (= our **Va1**)
- *Florentinus* plut. 74.18 **(L)** (= our **F1**)
- *Florentinus* plut. 74.28 **(B)** (= our **F2**)
- *Leidensis Vulcanianus* 43 **(V)**
- *Londinensis Harleianus* 5625 **(H)**

von Staden limited his collations to the earliest manuscripts. Since he had not collated all of them, he thought that a comprehensive assessment was impossible. Nevertheless, he concluded that the manuscripts Vat. Gr. 281 **(A/Va1)**, Laur. plut. 74.28 **(B/F2)**, and Laur. plut. 74.18 **(L/F1)** seemed to be the most reliable and useful. Also, he thought that *Harleianus* **(H)** might be a copy of *Vaticanus* **(A/Va1)**, and *Leidensis* **(V)** a copy of Laur. plut. 74.28 **(L/F1)**. On the other hand, he stated that the two *Laurentiani* belong to the same tradition, whereas the *Vaticanus* derived from a different hyparchetype.

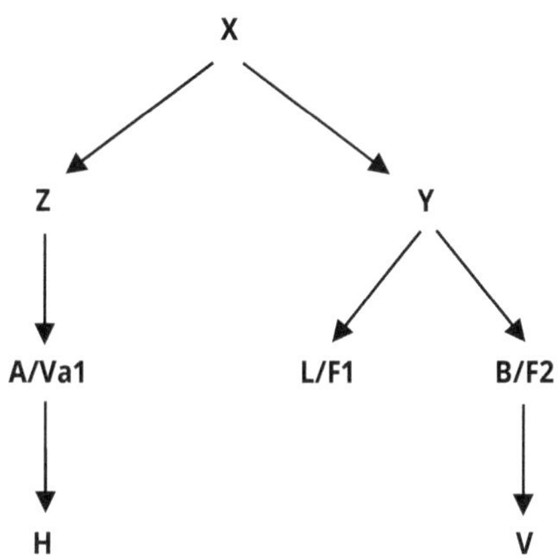

Schema 1: Stemma codicum according to von Staden 1989

Globally von Staden estimated that all these manuscripts deviate from Kühn's text only to a limited degree.

From our collation of the manuscripts, we concluded that the most reliable and useful ones are the following three:

- Firenze, Laur. plut. 74.18 **(A, F1)**
- Firenze, Laur. plut. 74.28 **(B, F2)**
- Città del Vaticano, Vat. gr. 281 **(C, Va1)**.

It seems likely that **(A/F1)** and **(B/F2)** come from the same tradition.

1.2 *De pulsibus ad tirones*

Diels' catalogue lists fourteen manuscripts.[102] The list has been revised by Garofalo,[103] and results in the following thirteen manuscripts:

Berlin, Staatsbibliothek zu Berlin (Preussischer Kulturbesitz) (SBB-PKB), Phillipps 1524
16th century.
(ff. 41–46v) *De pulsibus ad tirones.*
Studemund and Cohn 1890: 48–49; Touwaide 2016: 29, no. [0120].

Bologna, Biblioteca universitaria, 3632
15th century.
(f. 63v) *De pulsibus ad tirones* (fragment).
Olivieri and Festa 1895: 442–456; Touwaide 2016: 35, no. [0165].

Città del Vaticano, Biblioteca Apostolica Vaticana, *Palatinus Graecus* 400
16th century.
(ff. 48–60v) *De pulsibus ad tirones.*
Stevenson 1885: 257–259; Touwaide 2016: 281, no. [1491].

Firenze, Biblioteca Medicea Laurenziana, plut. 75.5[104]
12th century, Ioannikios with notes by Burgundio of Pisa.[105]
(ff. 320–324v) *De pulsibus ad tirones* (fragment: until 8.482.2 Kühn).
Bandini 1764–1770: 3.145–147; Fortuna and Urso 2009: 145, 147; Touwaide 2016: 77, no. [0348].

Leiden, Universiteitsbibliothek, *Vossianus Graecus* VGF 59[106]
Mid-15th century, Georgios Kalofrenas.[107]

102 For the manuscripts see Diels 1905: 86, and Touwaide 2021: 1.127–128; Diels 1908: 33, and Touwaide 2021:1.371. In the list that follows, we provide the same information as for the manuscripts of the four major treatises.

103 Garofalo 2010: 89–91. According to the same, 90n12, the following two manuscripts of Paris, Bibliothèque nationale de France, do not contain the text but a re-arrangement of it: *graecus* 2276 (on which, see Omont 1886–1888: 2.228; Touwaide 2016: 239, no. [1220]), and *graecus* 2830 (on which see Omont 1886–1888: 3.46; Touwaide 2016: 251, no. [1320]). The study by Bacalexi 2014, does not provide information about the Greek manuscripts of the treatise, but covers, instead, the history of the text from the time of the School of Alexandria (early 6th century) to the end of the 16th century, with a major focus on the Renaissance, particularly the Latin translations.

104 In Diels 1906: 86, this manuscript is identified as plut. 74.5. This is a mistake for plut. 75.5 (see Touwaide 2016: 75, no. [0317]).

105 Fortuna and Urso 2009: 146.

106 This manuscript is listed in Diels 1906: 87, as *Vossianus* 2168 (on its identification as VGF 59, see Touwaide 2016: 111–112, no. [0495]). It is not included in the list of Garofalo 2010.

(ff. 308–320) *De pulsibus ad tirones.*
de Meyïer 1955: 68–72; Touwaide 2016: 111–112, no. [0495], and 114, no. [-515].

Moskva, Gosudarstvennyi Istoricheskii Muzesi (GIM). Sinodal'naia Bibliotecka Moskovskoi Patriarchii, Sinod. 283 (Vlad. 466)
Manuscript made of three parts, the first of which (ff. 1–119), that contains the treatise, dates to the 15th century.[108]
(ff. 11–19v) *De pulsibus ad tirones* (fragment: to 8.849.10 Kühn).
Vladimir 1894: 703–705, no. 466; Touwaide 2016: 147, no. [0733], 149.

Paris, Bibliothèque nationale de France, grec 2207
16th century.
(f. 217) *De pulsibus ad tirones* (fragment).
Omont 1886–1888: 2.214; Touwaide 2016: 227, no. [1132].

Paris, Bibliothèque nationale de France, grec 2260
16th century.
(ff. 120v–136v) *De pulsibus ad tirones.*
Omont 1886–1888: 2.225–226; Touwaide 2016: 236, no. [1198].

Paris, Bibliothèque nationale de France, supplément grec 446
10th century.[109]
(ff. 95–105v) *De pulsibus ad tirones.*
Omont 1886–1888: 3.262 and 398; Touwaide 2016: 255, no. [1351].

Venezia, Biblioteca Nazionale Marciana, gr. Z 281 (= 581)
15th century, Ioannes Rhosos.
Whereas this manuscript has been identified as the source of the Aldina edition (1525) for the treatise by Ivan Garofalo,[110] this is not the case.[111]
(ff. 72–75v) *De pulsibus ad tirones.*
Mioni 1981: 404–405; Touwaide 2016: 331, no. [1677], 340.

Wien, Österreichische Nationalbibliothek, *medicus graecus* 52
15th century (first half).
(ff. 23–31) *De pulsibus ad tirones* (fragment: up to 8.487.1K).
Hunger and Kresten 1969: 103–105.

107 Touwaide 1981: 25; Cronier et al. 2015: 102.
108 Garofalo 2010: 90.
109 Garofalo 2010: 90.
110 Garofalo 2010: 91, 95–96, 102.
111 See Pietrobelli 2013.

Wolfenbüttel, Herzog August Bibliothek, *Gudianus graecus* 11 fol.
(ff. 1–10v) *De pulsibus ad tirones* (fragment: up to 8.490.3K).

The tradition of the text has been explored only recently.[112] It can be summarized by the following stemma codicum.

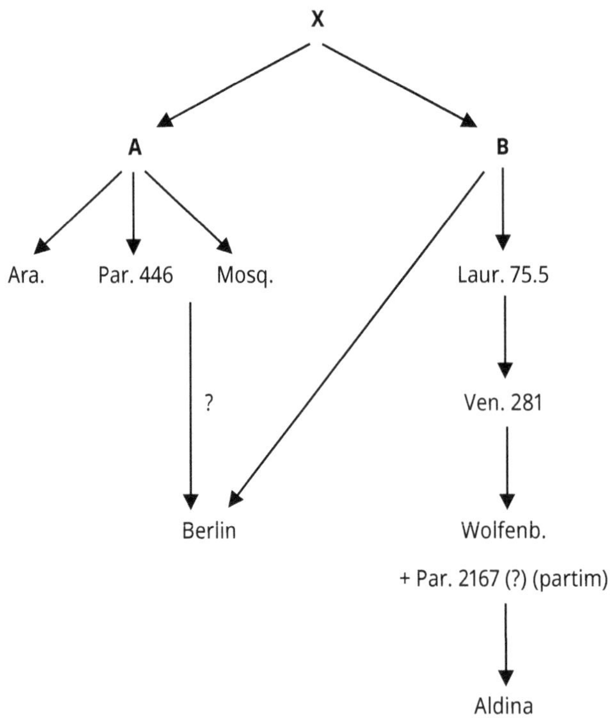

Schema 2: Stemma codicum for *De pulsibus ad tirones*

1.3 *Synopsis librorum suorum sedecim de pulsibus*

The list of fourteen manuscripts provided in Diels' catalogue duly amended is the following with only six items:[113]

112 Whereas Garofalo 1998: 391, referring (in note 36) to a critical edition then in preparation by Heinrich von Staden for the Collection des universités de France, this edition has not been yet published. A study of the textual tradition can be found in Garofalo 2010.

113 *Synopsis librorum suorum sedecim de pulsibus:* Diels 1905: 89, and Touwaide 2021: 1.131; Diels 1908: 33, and Touwaide 2021: 1.371. The manuscript Paris, Bibliothèque nationale de France, *graecus* 2332, listed in Diels, does not seem to contain our text. The manuscript listed in the catalogue as Padova, Johannes Rhodius, cannot be identified (see Touwaide 2016: 190 no. [0892]). For supplementary information on the manuscripts, see Cortés Gabaudan and Martínez Manzano 2013.

Bethesda, National Library of Medicine, Census 82[114]
16th century, Camillo Zanetti.[115]
This manuscript of 40 ff, contains only *Synopsis librorum suorum sedecim de pulsibus* interfoliated with the Latin translation by Antonio Gadaldini (below).
Tunis 1989: 6; Cortés Gabaudan and Martínez Manzano 2013: 68; Touwaide 2016: 52 no. [0221] and 34.

Firenze, Biblioteca Medicea Laurenziana, plut. 59.14
14th century.
(ff. 178–213v) *Synopsis librorum suorum sedecim de pulsibus* (fragment: up to IX.531.4K).
Bandini 1764–1770: 2.524–526; Cortés Gabaudan and Martínez Manzano 2013: 65–67.

København, Det Kongelige Bibliotek, e don. var. 14, 2°
16th century, Camillo Zanetti. This codex has been consulted or owned by the Renaissance physician Agostino Gadaldini.[116]
This manuscript of 58 folios contains only *Synopsis librorum suorum sedecim de pulsibus* (fragment: the text stops at 9.533.13K).
Schartau 1994: 453–454; Cortés Gabaudan and Martínez Manzano 2013: 67.

Milano, Biblioteca Ambrosiana, C 120 sup.
16th century, Camillo Zanetti.
This codex contains two copies of the text: ff. 69–107v, and ff. 109–147v.
Martini and Bassi 1906: 224–225; Cortés Gabaudan and Martínez Manzano 2013: 68.

Milano, Biblioteca Ambrosiana, D 293 inf.
16th century, writing similar to that of Andreas Darmarios.
This manuscript of 54 folios contains only *Synopsis librorum suorum sedecim de pulsibus* (fragment: the text is interrupted at IX.531.4K).
Martini and Bassi 1906: 1049; Cortés Gabaudan and Martínez Manzano 2013: 67.

Milano, Biblioteca Ambrosiana, D 518 inf.
16th century, Camillo Zanetti.
This manuscript of 38 folios contains only *Synopsis librorum suorum sedecim de pulsibus.*
Martini and Bassi 1906: 1069; Cortés Gabaudan and Martínez Manzano 2013: 69.

114 This is the manuscripts that was Cheltenham, Philipps 24.386 in the time of Diel's catalogue (see Touwaide 2016: 52 no. [0221]; also 34).
115 See Cortés Gabaudan and Martínez Manzano 2013: 68 (with a reference to the on-line catalogue of the library), and 70n28.
116 Petit 2007: 136.

The following manuscripts of Paris, Bibliothèque nationale de France, listed by Diels do not contain Galen's text, but later re-arrangements:[117]

- Paris, Bibliothèque nationale de France, grec 2155, ff. 325v–319;
- Paris, Bibliothèque nationale de France, grec 2229, ff. 43v–44;
- Paris, Bibliothèque nationale de France, grec 2315, ff. 393v–400;
- Paris, Bibliothèque nationale de France, grec 3035, ff. 209–219v;
- Paris, Bibliothèque nationale de France, supplement grec 446, ff. 31v–33v.

The manuscript of Paris, Bibliothèque nationale de France, suppl. gr. 629, ff. 59–65, which contains *Synopsis librorum suorum sedecim de pulsibus*, is a recent copy made by Minoide Mynas (1788–1859). Its model has not been identified. For another of its texts the model is a *Parisinus.*[118]

The tradition of the text has not been fully researched.[119] However, according to Francisco Cortés Gabaudan and Teresa Martínez Manzano, the manuscripts of Florence and Milan D 293 inf. are linked, with the latter being a copy of the former.[120] The relationships between the other manuscripts, all copied by Camillo Zanetti, have not yet been clarified.

According to the same, the longer version contained in Zanetti's manuscripts was Zanetti's work. Probably considering that the text of the *Laurentianus* plut. 59.14 (or that of another manuscript with a similar text) was incomplete, he created a supplement by extracting relevant information from other of Galen's works.[121]

1.4 *De usu pulsuum*

This text was not included in the catalogue edited by Diels. It seems to be known through two manuscripts only:[122]

Cambridge, Gonville and Caius College, 355/582[123]
15th/16th century.
(ff. 16 recto–42 recto) *De usu pulsuum.*

---

117 As early as 1907, the text in most of the Paris manuscripts has been identified by Diel's student Johannes Carolus Germanus Gossen (b. 1884) as extracts from Galen's text that do not correspond to Galen's text ad litteram (Gossen 1907: 36n1). In the same sense, see more recently Francisco Cortés Gabaudan and Teresa Martínez Manzano 2013: 65n11. To these, we add the *supplementum graecum* 446, which does not seem to contain the text of our treatise *ad litteram.*

118 See Lamagna 2010: 242n6, about the text of ff. 43–45, Stephanus Athenaeus, *De urinis.*

119 In this sense, see Cortés Gabaudan and Martínez Manzano 2013: 70.

120 Cortés Gabaudan and Martínez Manzano 2013: 69.

121 On this, Cortés Gabaudan and Martínez Manzano 2013: 70–72, with the *loca similia* from Galen's works in the column to the right of the critical editions on pp. 92–110.

122 Furley and Wilkie 1984: 187.

123 Touwaide 2016: 40, no. [0187], 42, no. [0190], and 47.

Touwaide 2016: 42 no. [0190], and 40 no. [0187].

El Escorial, Biblioteca del Real Monasterio de San Lorenzo, Φ III 11[124]
14th/15th century.
(ff. 16 recto—20 verso) *De usu pulsuum.*
de Andrés 1965: 60–64; Touwaide 2016: 65 no. [0270].

According to the study by David J. Furley and James S. Wilkie, the *Cantabrigensis* is a copy of the *Scorialensis.*

1.5 *Definitiones medicae*

The manuscripts have been listed in Diels' catalogue, with a total of over thirty codices.[125] Jutta Kollesch further revised and complemented this list.[126] Supplementary manuscripts have been brought to light:
- London, Wellcome Library MS. 289.[127]
- Paris, Bibliothèque nationale de France, supplément grec 1328.[128]
- Torino, Biblioteca Nazionale Universitaria, B. VI. 21.[129]
- Venezia, Biblioteca Nazionale Marciana, gr. Z. 521 (coll. 316).[130]
- Venezia, Biblioteca Nazionale Marciana, *appendix graeca* IV.10 (coll. 833).[131]

Except the Torino copy, all are included in the new edition of the treatises by Jutta Kollesch.[132]

---

124 Touwaide 2016: 65, no. [0270].

125 Diels 1905: 111; Diels 1908: 36, with a revised edition in Touwaide 2021: 1.159–160, and 375, respectively. For an update and corrections, see Touwaide 2016:
- 42, no. [0190] about the manuscript identified as Caius College [948] in Diels;
- 73, no. [0310] about Firenze, plut. 73.9 in Diels;
- 120, no. [0554] about London, *Regius* 12 F III;
- 221, no. [1067] about *Paris. gr.* [2151 (?)];
- 232, no. [1183] about *Paris. gr.* [2252 (?)];
- 306, no. [1606] about *Vat. gr.* 4423, which is a Latin translation of the treatise.

126 Kollesch 1973: 47–48. More recently, see Cronier 2021, which, however, does not list the manuscripts with a fragmentary text (see p. 51).

127 Touwaide 2016: 122.

128 Touwaide 2016: 261.

129 Touwaide 2016: 316.

130 Touwaide 2016: 342.

131 Cronier 2021: 48n2, 50.

132 Kollesch 2023: XIII–XIV.

The tradition of the text has been studied by Jutta Kollesch as early as 1973[133] with an update in 2023.[134]

Most of Galen's sphygmological treatises have been translated into Syriac and Arabic. With regard to Syriac, no manuscripts have been preserved or located thus far, although we know that Galen's treatises were translated into Syriac.[135] For Arabic, manuscripts can be found in libraries worldwide. Lists can be found in Diels' catalogue[136] and in more recent works.[137] Below, we present an outline of the Syriac and Arabic translation movement on the basis of available contemporary literature.

The four major treatises were also translated into Latin in the late Middle Ages and the Renaissance. This was also the case with *De pulsibus ad tirones* and *De usu pulsuum*. Their manuscripts have been inventoried as early as Diels' catalogue and, more recently, in the major contribution to the Latin translations of Galen's works compiled by Richard J. Durling (1932–1999).[138] Copies of some of these translations have subsequently been brought to light. These translations are briefly referred to in the survey of the fortunes of Galen's sphygmological work below.

Apart from *De usu pulsuum*, which was edited in a printed version only recently,[139] all the treatises under consideration here have been included in the major editions of Galen's works in Greek:[140]

– 1525 Aldina edition. It was the *editio princeps* of Galen's *opera omnia* in Greek, published by Aldo Manuzio's heir, Andrea d'Asola (ca. 1498–1557/1558).[141]
– 1538 Basel edition. Georg Agricola (1494–1555) identified many errors in the 1525 Aldina edition that needed emendation. The task was taken up by Leonard Fuchs (1501–1566), Hieronymus Gemuseus (1505–1543), and Joachim Camerarius (1500–1574), who prepared a new edition for the Swiss printer Andreas Cratander (1490–1540) and a consortium of other Basle printers. This edition appeared in 1538.

---

133 Kollesch 1973: 141–144. The conclusions of her study have been slightly revised recently in Cronier 2021.

134 Kollesch 2023: XIV–XXXIX.

135 Degen 1981.

136 See Diels 1906: 87 for *De pulsuum differentiis*; 88 for *De dignoscendis pulsibus* and *De causis pulsuum*; 89 for *De praesagitione ex pulsibus*.

137 For *De pulsibus ad tirones*, see Sezgin 1970: 81–82, no. 5, and Ullmann 1970: 44, no. 32; for the four major treatises (grouped under the general title περὶ σφυγμῶν πραγματεία), see Sezgin 1970: 91–94, no. 16, and Ullmann 1970: 43–44, no. 31; for *De usu pulsuum*, see Sezgin 1970: 104, no. 32; Ullmann 1970: 41 no. 16.

138 Durling 1961: 289–290.

139 Furley and Wilkie 1984: 185–227, with the Greek text on pp. 194–227 with facing English translation.

140 For an overview of the editions of Galen text, see Boudon-Millot 2007: CCX–CCXXVIII.

141 On this edition, see Cataldi Palau 1998: 639–640 for a bibliographic description. Its manuscript sources have been differently studied according to the texts.

- Chartier edition: this edition in nine volumes published in thirteen tomes from 1639 to 1679 in Paris, was compiled by the French physician and scholar René Chartier (1572 – 1654).[142] It is bilingual, offering the Greek text with a Latin translation.
- Kühn edition: published in Leipzig, in twenty volumes in twenty-two tomes from 1831 to 1833, by the German physician and Hellenist Karl Gottlob Kühn (1754 – 1840).[143] Its Greek text is accompanied by a Latin translation.

Below is a table of concordance of the editions of the Greek text of the seven sphygmological treatises.

Table 1: Table of concordance of the editions of the Greek text of Galen's seven sphygmological treatises

Treatises	Aldina vol. 3	Basel vol. 3	Chartier vol. 8	Kühn
De pulsuum differentiis	4v–27v	7 – 52	14 – 102	VIII.493 – 765
De dignoscendis pulsibus	28r–44v	52 – 84	103 – 166	VIII.766 – 961
De causis pulsuum	45r–62r	84 – 117	167 – 235	IX.1 – 204
De praesagitione ex pulsibus	62r–81r	117 – 154	236 – 308	IX.205 – 430
*De pulsibus ad tirones**	1r–4r	1 – 7	1 – 13	VIII.452 – 492
Synopsis librorum suorum sedecim de pulsibus	–	–	309 – 330 Latin	IX.431 – 533 Greek
				IX.533 – 549 Latin**
			330 – 332 Latin	–
De usu pulsuum	81v–83v	154 – 159		V.149 – 180

* On the editions, see Bacalexi 2014: 99.
** *Editio princeps* of the corresponding passages in Greek in Cortés Gabaudan and Martínez Manzano 2013: 92 – 110, with the editorial method pp. 89 – 91.

2 Toward a Textual History

Until the textual tradition of the Greek text of Galen's treatises under consideration here has been fully examined, it is difficult to offer a reconstruction of the fortunes of Galen's sphygmology in Byzantium. The recent collection of essays toward a synthesis of the tradition of Galen's oeuvre in Byzantium has not approached our specific topic.[144]

142 The edition of Chartier has been abundantly studied. See recently the volume of collected studies of Boudon-Millot, Cobolet, and Jouanna 2012.
143 On Kühn, see, among others Nutton 2002 and Tassinari 2019.
144 Bouras-Vallianatos and Zipser 2019.

The first firm point that we can identify is the presence of the treatises on sphygmology in the corpus of Galen's works used as textbooks in the teaching of the Alexandrian school of medicine in the 6th and 7th century.[145] From its most recent analysis, it appears that the so-called *Alexandrian Canon* included the introductory *De pulsibus ad tirones* commented on by Stephanos, Palladios, and Ioannes Grammaticus. The four major treatises, although present in the *Canon*, were not studied in their entirety; only the first book of each of them was. Remarkably enough, none of these commentaries has been preserved in Greek, all being known through translations and derivatives in both Latin and Arabic. The introductory treatise *De pulsibus ad tirones* crossed the Mediterranean from Alexandria to Ravenna. At a time that it is not exactly known but might be the period of the exarchate after the Ostrogothic kingdom in the late fifth/ early sixth century, the otherwise unknown Agnellus did comment on it in his lessons in Latin.[146]

From the sixth century onwards Galen's sphygmological works were translated into Syriac.[147] Although no manuscripts have been preserved or located thus far, information about these translations has been provided in the *Risala* by the translator Hunayn ibn Ishaq (808–873), responsible for the translation of many Greek medical and other Greek works into Syriac and Arabic.[148] The *Risala* is a history and inventory of Hunayn's translations of Greek medical and scientific literature into Syriac and Arabic by Hunayn, his collaborators and others, in the form of a letter addressed in 855 or 856 to Ali ibn Yahyā, who asked him to compile such a list.[149] It is the translation into Arabic of a list previously drawn up by Hunayn in Syriac, apparently not long before its translation. Through this letter we can identify the Syriac and Arabic translations.[150]

The well-known Syriac scholar Sergios of Reshayna (d. 536)[151] translated the four major treatises from Greek. However, apart from *De praesagitione ex pulsibus*, of which he translated the four books, he limited himself to the first book of each of the other

145 See Garofalo 2019, particularly 62, 64, 75, and 77.
146 See Palmieri 2005 for a study that includes (pp. 1–67) a critical edition of the text (on the basis of the unique manuscript currently known), with facing French translation and abundant notes of commentary (pp. 69–194).
147 On this process, see for example Le Coz 2004.
148 On Hunayn's biography and translations, see most recently Lamoreaux 2016: xii-xviii with the references to earlier bibliography.
149 The text was first edited by Bergsträsser in 1925, and, more recently, by Lamoreaux 2016.
150 On *De pulsibus ad tirones*, see Bergsträsser 1925: 6 (Arabic numerals) for the Arabic text, and 5, no. 5 for the German translation, and Lamoreaux 2016: 12–15 for the Arabic text with facing English translation. On the four major treatises, see Bergsträsser 1925: 13–15 (Arabic numerals) for the Arabic text, and 10–12, no. 16 for the German translation, and Lamoreaux 2016: 28–33, no. 18 for the Arabic text with facing English translation. On *De usu pulsuum*, see Bergsträsser 1925: 25 (Arabic numerals) for the Arabic text, and 20, no. 41 for the German translation, and Lamoreaux 2016: 56–57, no. 43 for the Arabic text with facing English translation.
151 The abundant bibliography on Sergios has been conveniently listed with a succinct biography in Bhayro and Hawley 2014: 293 and n12.

three treatises.[152] He seems to have also translated the *Synopsis de pulsibus*.[153] Sergios' incomplete work on three of the four major treatises was completed in the 9th century with the translation of Books II–IV of all three by Ayyūb al-Ruhāwī also known as Hiob von Edessa (d. 832) for the physician Djibrīl ibn Bakhtīshū (d. 827).[154] At an unknown time, *De pulsibus ad tirones* was translated by Bār Sahdā.[155] Besides translations of Galen's *De sectis ad eos qui introducuntur* and *Ars medica*,[156] this otherwise unknown scholar authored a history of the church and a refutation of the Zoroastrians.[157] The movement of translation from Greek into Syriac was completed and improved on in the 9th century by Hunayn ibn Ishaq himself. He translated the full text of all four major sphygmological treatises anew for the physician Yūhannā ibn Māsawayh (777–857), in addition to *De pulsibus ad tirones* for Salmawayh bin Būnān.[158]

As for Arabic, Hunayn translated *De pulsibus ad tirones*, and the four major treatises. For the latter, however, he started the translation, but he did only a part, leaving the rest to his nephew Hubaysh b. al-Hasan al-A'sam al-Dimasqi (9th cent.), who translated it from Hunayn's Syriac version. It is not clear whether Hunayn translated the *Synopsis de pulsibus*, as he believed it was not authentic. *De usu pulsuum* was translated by a collaborator of Hunayn, Istafan ibn Basil (9th cent.), most probably of Greek descent, also on the basis of a previous Syriac version.[159] We have no knowledge that our texts were translated again, be it by the same or later translators, directly from the Greek text, or subsequently revised as has been the case for other Greek medical works. The most significant and best-known case of this process of second-generation translations and revisions is that of Dioscorides, *De materia medica*.[160]

The importance of these Syriac and Arabic translations cannot be underestimated. The Syriac ones were made on the basis of Greek manuscripts copied at the latest in the early 6th century, that is, on manuscripts anterior by far to the oldest Greek copies that have been preserved (the oldest one dates to the 12th century). These translations have not been preserved or not located thus far. However, they have been translated into Arabic. The fact is that the translators (in both Syriac and Arabic) often used transliterations for the rendering of technical terms, thus reproducing Greek terms as such, without offering an exact translation. Such terms introduced into the Syriac versions might have passed through the Arabic translations. This might also have been the case of Syriac terms reproduced in Arabic. As a result, Greek terms written in Syriac

152 Degen 1981: 144–145, and, more recently, Muraviev 2014: 267.
153 See Degen 1981: 156 no. 105, and Lamoreaux 2016: 76–77 and n1 ad no. 70.
154 Degen 1981: 144. On Ayyūb al-Ruhāwī, see ibid. 163, where Ayyūb al-Ruhāwī is credited with 39 translations of Galenic works from Greek into Syriac.
155 Degen 1981: 144.
156 Degen 1981: 134 and 135, respectively.
157 Baumstark 1922: 135, for whom Bār Sahdā lived in the pre-Islamic period. On him, see also Degen 1981: 162.
158 Degen 1981: 144–145.
159 Furley and Wilkie 1984: 188.
160 For an overview of all the translations and their manuscripts, see Sadek 1983: 7–19.

and Arabic alphabet, and Syriac terms written in Arabic alphabet and, in both cases, duly adapted to the phonetics of the receiving languages can be found in the Arabic texts. Since the manuscripts at the origin of these translations are much anterior to the preserved Greek copies, it might be the case that the Arabic translations attest to readings better than those in the Greek codices through their Hellenisms and Syriacisms.

No translation into Latin is known before the 12th century. Interestingly, preserved translations have been made in two different places and in two different ways. In the East (Constantinople) or in Italy, the judge and translator known as Burgundio of Pisa (ca. 1110–1193),[161] translated *De pulsibus ad tirones*,[162] *De pulsuum differentiis* (only I),[163] *De causis pulsuum* (III–IV),[164] and *Synopsis de pulsibus*.[165] He sojourned in Constantinople where he had access to manuscripts offering a text of good philological quality and often superior to that of many other manuscripts.[166] Notes by Burgundio in manuscripts in which the hand of Ioannikios has been recognized, appear in such manuscripts.[167] This is the case for *De pulsibus ad tirones* and manuscript *Laurentianus* 75.5, which can be considered as the source of Burgundio.[168] It is also the case for the manuscript *Laurentianus* 74.18 of *De pulsuum differentiis*.[169] At almost the same time, or perhaps a little later, the translator identified as Marc of Toledo (fl. 1193–1216),[170] translated *De usu pulsuum* from Hunayn's Arabic version.[171] His work was typical of the Toledo school, as the local translators are known to have worked extensively on Arabic texts, including the *Quran*, which Marc rendered into Latin.[172]

In the Renaissance, Burgundio's translations were included in the volumes containing the Latin translation of all Galenic works published between 1490 and 1528.[173] Marc of Toledo's translation was not printed until the edition of Pavia dated to 1515–1516 and the Giunta edition of 1528.[174] Several other Latin translations were made afresh. Their source(s) have not necessarily been identified. The first one, in a

161 On Burgundio, see recently (with the references to earlier bibliography) Rodriguez Suarez 2016: 105–107.

162 Durling 1961: 289 no. 92; Fortuna and Urso 2009: 148; Bacalexi 2014: 100, and, more recently Urso 2019: 367 about Burgundio's manuscript.

163 Durling 1961: 290 no. 95; Fortuna and Urso 2009: 148.

164 Fortuna and Urso 2009: 148; Urso 2019: 367. Critical edition by Scimone 2021a.

165 Haskins 1924: 208 and n100, followed by Urso 2019: 367. This translation does not appear in Durling 1961.

166 For a chronology of Burgundio's translations, see Scimone 2021b.

167 See Fortuna and Urso 2009: 141–149.

168 Fortuna and Urso 2009: 146, and Garofalo 2010: 96, 101.

169 Fortuna and Urso 2009: 146 with the previous literature on this point.

170 On Marc of Toledo, see the fundamental study of d'Alverny 1989/1994 and, more recently, Burnett 2005; also, Sangrador Gil 1997: 44. On the school, see recently and for example Polloni 2018.

171 Durling 1961: 290 no. 96a; Garofalo 2010: 101, and Bacalexi 2014: 100.

172 See Martínez Gázquez 2001.

173 Bacalexi 2014: 100.

174 Durling 1961: 290, no. 96, referring to the editions cited p. 280.

chronological order, was *De usu pulsuum* by the Englishman, Thomas Linacre (ca. 1460–1524).[175] A physician, Linacre traveled to Italy. In Florence, he frequented Angelo Poliziano (1454–1494) and also the Byzantine scholar Demetrius Chalcondylas (1423–1511) who was teaching Greek. In Venice, he was in contact with the humanist Ermolao Barbaro (1454–1493) and the Ferrarese physician, Nicolao Leoniceno (1428–1524), who had an exceptional library collection and was a Galenic scholar of distinction. There, Linacre also participated in the Academy founded by the printer and publisher Aldo Manuzio (1449–1515), whose company later produced the first printed version of all Galen's works in Greek (below). He translated several other Galenic treatises.[176] The second translation was by the German scholar Gunther von Andernach (1505–1574) who translated *De pulsibus ad tirones*[177] among many other Galenic treatises.[178] The source of his translation, whether a Greek manuscript or the printed editions of 1525 and 1538 discussed below, has not been identified.

The German scholar, Herman Cruser (ca. 1510–1575),[179] translated *De pulsibus ad tirones* and the four major treatises.[180] A physician and Hellenist, in addition to a jurist and a diplomat, Cruser translated into Latin the four sphygmological treatises, and several other treatises by Galen,[181] in addition to the *Vitae et Moralia* by the Greek historian Plutarch.[182] In 1537, another Latin translation of *De pulsibus ad tirones* was published.[183] It was by Martin Grégoire (d. 1552) of Tours.[184] Grégoire also translated into Latin two of Galen's works devoted to nutrition: *De alimentorum facultatibus* and *De victu attenuante*,[185] in addition to a French translation of *De compositione medicamentorum per genera* (Books I–IV).[186]

---

175 His translation was first published in London, possibly in 1522 (Durling 1951: 253, sub 1522(?).1). It is not listed in Bacalexi 2014.

176 Durling 1961: 297 sub nomine.

177 His translation was first published in 1531, in Paris by the well-known printer Simon de Colines. See Durling 1961: 290, no., 92 (ctd) b, and 257 sub 1531.1. The same year it was also published in Basel by the famous printer and publisher Andreas Cratander. See Durling 1961: 290, no., 92 (ctd) b, and 257 sub 1531.10. On both editions see also Bacalexi 2014: 100.

178 See Durling 1961: 297 sub nomine.

179 On him, see (in chronological order): Eloy 1778: 4.738–739; *Allgemeine Deutsche Biographie* 4 (1876): 628–629 (Harles); Hirsch 1884–1888: 6.657–658 (Pagel); and, more recently: Maillard, Kecskeméti and Portalier 1995: 138.

180 All published in first edition in the same volume in Paris in 1532 by Simon de Colines. See Durling 1961: 257, no, 1532.1. Also, Bacalexi 2014: 101.

181 See Durling 1961: 296 sub nomine.

182 Lyon 1564.

183 Durling 1961: 290, no. 92 ctd, d; Bacalexi 2014: 101. This translation was published in Paris by Chrétien Wechel.

184 On him see Boutineau 1904, and, more recently, Maillard, Kecskeméti and Portalier 1995: 219. For his Galenic translation, see Durling 1961: 297 sub nomine.

185 First published in the same volume in 1538 in Paris by Chrétien Wechel (Durling 1961: 283, no. 13 c, and 260, no. 1538.3).

Antonio Gadaldini (1515 – 1575) published a Latin translation of many Galenic treatises[187] among which was *Synopsis librorum suorum sedecim de pulsibus.*[188] A physician, Gadaldini was born in Ferrara and moved to Venice, where he collaborated in the production of the Giuntini edition of Galen's works in Latin published in 1541 – 1542. He discovered multiple manuscripts and devoted his activity to improving the Galenic text of previous editions.[189] He is known to have owned or consulted two manuscripts of the *Synopsis* (above). He used the manuscript of Milan, D 518 inf., as a source for his translation, and might have drafted it in the interfoliated leaves of the manuscript of Bethesda.[190]

The Italian scholar Giovanni Battista Rasario (1517– 1578)[191] translated multiple Galenic treatises,[192] including *Synopsis librorum suorum sedecim de pulsibus.* He seems to have reproduced Gadaldini's translation and also used the same manuscript of Milan as Gadaldini.[193]

The last Renaissance Latin translation was *De pulsibus ad tirones* by the Spaniard, Ferdinando de Mena (ca. 1520 – 1585).[194] A humanist physician and a scholar, de Mena was a professor of medicine at the Alcalá university, north of Madrid, in addition to being the private physician to King Philip II (1527– 1598). Besides two other Galenic works,[195] de Mena authored a commentary on the Galenic treatise on phlebotomy,[196] and translated into Latin the novel by the Greek Heliodoros.[197]

Apart from Gadaldini's translation, the source(s) of these translations has not been researched. The fact is that, when they are not based on earlier printed editions, they might have been made on manuscripts that are now lost. If they do offer better readings than preserved manuscripts, they might be considered as a substitute of these now lost manuscripts. A case in point is Cruser's translation of our four major treatises. Interestingly enough, his Latin sometimes offers a better reading than the Greek text of

---

186 Durling 1961: 284, sub 29 c., and 269, no 1549.29, published in Paris by Michel Vascosan. On this translation see Botineau 1904: 51 – 55.

187 Durling 1961: 197 sub nomine.

188 It was in the Latin opera omnia published by the Giunta in Venice (1556, 1565, 1576 – 1577, 1586, and 1596 – 1597), the first two of which were edited by Gadaldini. See Durling 1961: 282, no 11.a. For the editions, see Durling 1961: 280, sub A12, and A15-A18. On the Giunta editions of Galen, see also Fortuna 2012b: 399 – 407.

189 On his activity, see Garofalo 2004.

190 Cortés Gabaudan and Martínez Manzano 2013: 82.

191 On him, see recently, Savino 2012.

192 See Durling 1961: 198 sub nomine.

193 See Cortés Gabaudan and Martínez Manzano 2013: 80.

194 *Claudii Galeni de pulsibus ad tirones liber, graeco in latino sermone conversus.* Alcalá de Henares: J. Brocar, 1553. This translation does not appear in Durling 1961. It is cited in Bacalexi 2014: 101.

195 *De urinis*, first published in 1553 by J. Brocar in Alcalá de Henares, 1553 (Durling 1961: 271, no. 1553.5) and in 1558 in Valencia by the same publisher (Durling 1961: 273, no. 1558.4).

196 *Commentaria in libros Galeni de sanguinis missione, & purgatione* (2ª ed.). Turín, 1587.

197 *Traducción de Heliodoro, Historia dos leales amantes*, first published in Salamanca in 1581, with several subsequent editions (Alcalá, 1587, Barcelona, 1614, among others).

the editions, suggesting that Cruser had access to one or more Greek manuscripts different from those consulted by the contemporary authors of printed editions.

The 1525 Aldina edition relied on different manuscripts, including *Parisinus graecus* 2167, and the codex of Wolfenbüttel, which has been considered for a long time a copy of the Aldina,[198] was its source as recent research has demonstrated.[199] For *De pulsibus ad tirones*, the manuscript of Venice *graecus* Z 281 has been incorrectly identified as the source.[200]

The Basel edition reproduced mostly the Aldina, and the Chartier edition did the same with the Basel edition. Nevertheless, Chartier consulted several manuscripts of the Royal Library (*Bibliotheca regia*) in Paris. For the Latin translations, he reproduced earlier texts. For our four major treatises and *De pulsibus ad tirones*, he took Cruser's translation,[201] whereas for *Synopsis librorum suorum sedecim de pulsibus*, he took that of Gadaldini, followed by a second, shorter version by Burgundio.[202]

As for Kühn's edition, its Greek text was not, as Kühn himself stated,[203] a new edition made afresh on the basis of manuscripts, but a reproduction, with revisions, of Chartier's edition.[204] Kühn nevertheless also collated some manuscripts, the Basel edition, and three copies of the Aldina with annotations by scholars.[205] He integrated original corrections by Gottfried Heinrich Schäfer (1764–1840), a classical scholar and a librarian at Leipzig University.[206] He did provide, however, the *editio princeps* of the Greek text of *Synopsis librorum suorum sedecim de pulsibus*. The Latin translations of all our treatises are those of Chartier.

The recent edition of *De usu pulsuum* by Furley and Wilkie in 1984 is based on the two extant Greek manuscripts supplemented by the Arabic translation contained in the manuscript Istanbul, Süleymaniye Kütüphanesi, Aya Sofya 3690. The Latin translations by Marc of Toledo and Thomas Linacre are also referred to.

198 See recently Garofalo 2010: 91, 93.
199 Pietrobelli 2013.
200 Garofalo 2010: 102, and Pietrobelli 2013.
201 On this, see Fortuna 2012a: 321.
202 Fortuna 2012a: 321.
203 See vol. 1, pp. XII–XVI.
204 Kühn used both Chartier's edition and the so-called *Basileensis* of 1538. For some examples of this, see recently (in chronological order of publications): Fortuna 1993:304, and 1997:43, about *De constitutione artis medicae*; Gärtner 2015: 184–185, about *De locis affectis*; Vagelpohl 2016: 61, about *In Hippocratis Epidemiarum Librum II Commentariorum I–VI*; and Perilli 2017: 81, about *Vocum Hippocratis Glossarium*
205 See Tassinari 2019: 510–511.
206 Tassinari 2019: 511.

Part Two: **Translation**

I Introduction

This brief introduction provides information about the texts that have been used as a basis for the translation, explains the method followed in the translation process, and concludes with the list of sigla used in the translations that follows.

1 Text used

As mentioned at the outset, the translations have been made from the Greek text in Kühn's bilingual (Greek and Latin) edition *Galeni Opera Omnia*, in 20 volumes and 22 tomes published between 1821 and 1833, with the four major treatises, "The Great Sphygmic" (Ἡ Μεγάλη Σφυγμική), as they are often called in the manuscripts, as follows:
- *De pulsuum differentiis:* vol. 8 (1824), pp. 493–765;
- *De dignoscendis pulsibus:* vol. 8 (1824), pp. 766–961;
- *De causis pulsuum:* vol. 9 (1825), pp. 1–204;
- *De praesagitione ex pulsibus:* vol. 9 (1825), pp. 205–430.

On the basis of the examination of the tradition of the text through the centuries, we have also consulted Greek manuscripts, particularly the following three:
- Firenze, Biblioteca Medicea Laurenziana, plut. 74.18 (**A**) = (**F1**);
- Firenze, Biblioteca Medicea Laurenziana, plut. 74.28 **(B)** = **(F2)**;
- Città del Vaticano, Biblioteca Apostolica Vaticana, *graecus* 281 (**C**) = (**Va1**).

This choice was made on the basis of the collation of the full text of the treatises in all the extant manuscripts of the work currently known, on the basis of digital images or black and white microfilms, except the following two manuscripts that were unavailable:
- Milano, Biblioteca Ambrosiana, O 50 sup.;
- Oxford, Bodleian Library, Laud. Gr. 57.

Taking into account that Kühn consulted the following four manuscripts:
- *Leidenensis Vulcanianus* 43 (**E**)
- *Venetus Marcianus graecus* 287 (**L**)
- *Parisinus graecus* 2161 (**I**)
- *Parisinus graecus* 2167 (**J**),

it is not surprising that the differences found in the manuscripts are not considerable in quantity or quality, since (**E**), which is very similar to (**J**), depends on (**A** = **F1**), and (**L**) is very similar to (**I**).

https://doi.org/10.1515/9783110612677-007

It must be noticed that Kühn's Latin text in the lower part of the pages does not always correspond to the Greek text of which it is supposed to be a translation. This is the result of an assemblage made by Chartier who reproduced earlier Latin translations. In our specific case, the translation of the four major treatises was that by Cruser.[207] The source of Cruser's translation for the sphygmological treatises has not been identified thus far. Interestingly enough, this Latin translation offers sometimes a better reading than the Greek text, suggesting that Cruser had access to one or more Greek manuscripts different from those consulted by the contemporary authors of printed editions. When appropriate, this Latin text was followed rather than the Greek on the grounds of sense. These are indicated by footnotes.

For the four short treatises, Kühn's Greek text was used for *De pulsibus ad tirones* and *Definitiones medicae*; for the other two, more or less complete reliance on Kühn was as follows:

- for *Synopsis librorum suorum sedecim de pulsibus*, we followed Kühn's text until section 28 (= 533.13: ἐρρωμένη γὰρ ἡ δύναμις). From there on, we used the reconstructed text offered in the paper by Cortés Gabaudan and Martínez Manzano. This corresponds in length and content to the Latin text published in the Kühn edition. Unfortunately, the authors did not take into consideration another manuscript held in Paris, Bibliothèque nationale de France, *graecus* 3035, which contains a rearrangement of the text.
- for *De usu pulsuum*, we used the text given by David J. Furley and James S. Wilkie. Reference was also made to Kühn's Greek text. Points of significant difference from the Furley and Wilkie's text are indicated by footnotes.

2 Method

With regard to the style of the translation, we have followed two guiding principles:

1. the translation should follow the Greek text as accurately as possible;
2. the style of the English translation should reflect the style of the original as faithfully as possible, the difficulties of Galen's somewhat convoluted prose notwithstanding.

We consider both these principles to be paramount.

Accuracy is obviously of particular importance in the translation of technical literature (*Fachprosa*). On this point, although we fall short of Niccolò da Reggio's stated goal "neither to add nor remove anything from the Greek original,"[208] we have at least aimed for a close correspondence to the Greek.

---

207 On him, see (in chronological order): Eloy 1778: 4.738–739; *Allgemeine Deutsche Biographie* 4 (1876): 628–629 (Harles); Hirsch 1884–1888: 6.657–658 (Pagel); and, more recently: Maillard, Kecskeméti and Portalier 1995: 138.

208 Nutton 1995b: 144.

On the matter of style, a relatively close correspondence is also important in giving the flavour of Galen's writing, including the contrast between the more discursive introductory and general passages, the pure clinical and factual sections, the dialogues and the case reports. He is even capable of purple patches as evidenced by his diatribe against Thessalus in the opening book of *De methodo medendi.* What follows is an excerpt from the section on Galen's *Fachprosa* and style taken from the *Introduction* to the Loeb translation of that work [209]:

> On the translation itself, the general objective is that characteristically claimed by translators: to provide a smoothly readable version in the language of the translation while remaining faithful to the language of the original—an objective much easier to state than to achieve! Moreover, in our even approaching such an aim, Galen's Greek presents certain specific and substantial problems quite apart from the problems of the text itself. In our view there are two particularly intrusive problems. The first is how to deal with Galen's complex sentence construction—what Hankinson has called his "syntactically and paratactically involved style." The second problem is how to deal with what are, in effect, technical terms of critical importance in Galen's own system of medicine, especially when these reflect structural and functional concepts now outmoded. To make matters worse, some of these terms remain in use today, albeit with a somewhat different meaning in some cases. A subsidiary and related problem is how to deal with the names of diseases and symptoms, medications and other treatment methods—that is, whether to retain the original Greek term transliterated, use the Latin version, or attempt to find the correct English equivalent.
>
> In brief, in our attempt to render Galen's elaborate Greek *Fachprosa* into modern English that may be read with relative ease (bear in mind that this is a medical textbook) we have adopted several specific strategies. First, we have been prepared to split up Galen's long and involved sentences and to change the order of the clauses where this seemed to improve clarity without unduly sacrificing fidelity to the original. Second, we have not attempted to translate every Greek particle, particularly when they can often be functionally replaced by punctuation in creating a modern English version. Third, we have given ourselves some licence in dealing with the voices, moods and tenses of the Greek verbs where not to do so would result in an English sentence that reads awkwardly. Nevertheless, despite all these 'liberties', we have attempted to provide an English translation that can be matched with the Greek without undue difficulty.

3 Sigla

Most used sigla (in the footnotes):
F1: manuscript Firenze, Biblioteca Medicea Laurenziana, plut. 74.18.
F2: manuscript Firenze, Biblioteca Medicea Laurenziana, plut. 74.28.
K and **Kühn:** Karl G. Kühn's edition of Galen, *Opera omnia*, 1821–1833, 20 volumes in 22 tomes.
All references to specific passages include, between parentheses, the number of the volume and tome, followed by the page number.

209 Johnston and Horsley 2011: cxi-cxii.

LSJ: Henry George Liddell and Robert Scott, *A Greek-English Lexicon.* Revised and augmented throughout by Henry Stuart Jones. Oxford: Clarendon Press, with several editions.
Va1: manuscript Città del Vaticano, Biblioteca Apostolica Vaticana, *graecus* 281.

The bold numbers between brackets in the text of the translation refer to the volume and page numbers in Kühn's edition (e.g., **[XIX.346]**).

II The Short Treatises

Galen's short treatises comprise the *De pulsibus ad tirones* (Pulses for Beginners), *De usu pulsuum* (On the Use of the Pulses) and *Synopsis librorum suorum sedecim de pulsibus* (Synopsis of the Pulses). To this we add the entries from *Definitiones medicae* (Medical Definitions) that deal with pulses.

1 Excerpts from *Medical Definitions* (*Definitiones medicae* XIX.346–462K)

This is the most substantial surviving work of the genre devoted to medical definitions. Although it is included in Kühn's Galen edition, it is generally considered to be spurious and the work of an unknown writer, perhaps of the Pneumatist school around the end of the 1st century A.D. On the work, see Touwaide 2008a, and, more recently, Kollesch 2023: XL-XLI. In the translation below, correspondence between the numbers of the definitions in the editions by Kühn and Kollesch are provided. Kollesch's edition is identified as CMG.

110 (= 107 and **^{I}36 CMG).** A pulse is a natural dilatation and contraction of heart and arteries. There are two components of the pulse: dilatation and contraction. Alternatively, a pulse is a movement involving dilatation **[376K]** and contraction of heart, arteries, brain and meninges (leptomeninges) that is natural and involuntary. It can also be defined in this way: a pulse is a natural and involuntary movement of the heat in heart and arteries, moving together similarly to itself and from itself in respect of the heart and arteries. Or otherwise, a pulse is [a natural movement] involving dilatation and contraction which pertains to cooling of the innate heat, increase of the physical (vital) *tonus*, and generation of the psychical *pneuma.* Or otherwise, a pulse is a truthful messenger of those things hidden in the depths, a revealer of the unseen, and evidence of those things that are obscure in harmonious movement;[210] by a prophetic stroke it announces an unseen condition.

111 (= ^{I}36 CMG). The generic movement of the pulse is endless and involuntary, existing in the dilatation and contraction of the perpetual motion which has its origin in the heart and arteries. The artery is moved in a circular (circumferential) and in a linear fashion. It is not obviously moved in a circle, while you do not clearly see it here moved in a straight line. All those things moving in a straight-line change from place to place and have a limit **[377K]** of movement. It is not, then, moved in a straight line. However, if there are two movements, and neither movement is moved, it is not then moved. But this is false and at odds with perception. Therefore, it is moved, and since it is moved, it behoves us to show in what kind of movement the artery is moved. Accordingly, we say it makes a mixed movement—in a circle and in a straight

210 Ἁρμονικός is essentially a musical term. LSJ refers to Galen's usage here as follows: "of the pulse, in harmony with physical state."

https://doi.org/10.1515/9783110612677-008

line (circumferential and linear). If the movement of the artery is mixed, as we said, a demonstration of the centre of this is a reasonable objective.

112 (= I36 CMG). I call "centre" the median point in relation to the collapsing of the dilated artery in this way; that is, the point from which it flows outward, and again from the outward point falls back again to the centre. It seems to be moved in a straight line. If, however, you wish to look at the matters themselves in relation to the movement of the artery, accept it as it is clearly described. If someone picks up a stone and throws it into water, consider the impact of the stone as the artery; the circle occurring from this makes the dilatation clear to you. In turn, **[378K]** consider the situation in which the circle of this doesn't continue indefinitely, but stops, and then, after a short time, begins to be reduced until it comes back down to that place from which the circle took its origin.

205 (= 220 CMG). Dilatation (διαστολή/*diastolē*) is a raising and, as it were, a swelling up **[403K]** of heart, arteries, brain and meninges[211] in length, breadth and depth, perceptible through the lifting up of the touch. Then there are those [who define it] in this way: Dilatation is a separation and, as it were, a swelling up of heart and arteries in length and breadth, perceptible through the raising up of the touch.

206 (= 221 CMG). Contraction (συστολή/*systolē*) is a coming together and collapsing of these things which, to perception, are not moving. It is comprehended by sound reasoning and by natural sequence.

207 (= 222 CMG). Palpitation (παλμός/*palmos*) is a rising and falling[212] of a body that is yielding and not resisting, occurring due to πνεῦμα/*pneuma*. It does not always follow continuously but occurs at certain times. There are those [who define it] in this way: Palpitation is a rising and falling of muscular bodies brought about by πνεῦμα/*pneuma*, but not always occurring. Or otherwise: Palpitation is a movement that is unwilled and involuntary when there is a lifting up and bringing down of the palpitating parts. The palpitation is a dilatation contrary to nature; it occurs in all those parts that naturally dilate. **[404K]**

208 (= 223–225 CMG). A large pulse comes about when the artery is spread in length, depth and breadth to a great degree.

A small pulse is the opposite and comes about when the artery is increased to the least extent in length, depth and breadth.

An intermediate pulse is one that is between both of these (large and small) in length and circumference, and which impacts on the surface in proportion.

209 (= 226sq CMG). A full pulse is one which stands out when it impacts on the touch, so that the actual wall of the artery seems to have become more prominent. In particular, what is within it is detected as being more full and more substantial.

---

211 Galen here refers to the meninges in the plural and doesn't differentiate the dura mater from the arachnoid and pia mater.

212 Kühn's Greek text has εὕρεσις whereas the Latin translation has *submissio*. The Latin is obviously to be preferred here. In the second definition ὕφεσις is used.

An empty pulse is one in which the circumference of the artery itself is altogether thin and bubble-like, and the content is indistinct and attenuated, so that if someone were to press down with the fingers, it would give the impression of a cavity.

An intermediate pulse is one which is in due proportion between full and empty, and which is in accord with nature. **[405K]**

210 (= 228sq CMG). A hard pulse is one in which the artery seems sinewy, as one might say, and hard, and the contained *pneuma* tense, so that it also has an "apoplectic" (forceful)[213] beat.

A soft pulse is the opposite of the hard pulse, the artery being slack and soft. The contained *pneuma* has been dispersed and the beat is more gentle.

An intermediate pulse is one between hard and soft; it is of due proportion in accord with nature.

211 (= 230sq CMG). A moist pulse is one which is soft and has a soft substance in it, and a certain moisture falls on the touch.

A dry pulse is such that the moisture underlying has been exhausted and has been mixed with a dry and airy vapour.

An intermediate pulse is one which seems neither more moist nor more dry to the touch, but is in due proportion and in accord with nature.

212 (omitted in CMG). A pulse is hot when the artery is hotter to the touch than the adjacent parts, as in a hectic fever. **[406K]**

A pulse is cold when the artery is found to be colder.

An intermediate pulse is one which has a due proportion of cold and hot.

213 (= 232–234 CMG). A strong pulse is one having a *eutonic* (vigorous) movement, making the beat forceful.

A weak pulse is one in which the *tonus* is dissipated, making the beat weak.

An intermediate pulse is one which preserves a certain proportion in each of these.

214 (= 235 CMG). A rapid pulse is one having an abbreviated dilatation and contraction, or a rapid pulse is one in which the movement of the artery occurs over a short time.

A slow pulse is one having a slow dilatation and contraction, or one in which the movement of the artery occurs over a long time.

An intermediate pulse is one having a due proportion in respect of dilatation and contraction, or is intermediate and in due proportion, the movement of the artery occurring over a moderate time. **[407K]**

215 (omitted in CMG). A frequent pulse is one in which the arterial dilatation occurs over a short time, or when the time between the dilatation and the contraction is short.

---

213 Galen uses the term ἀποπληκτικός, "paralyzed" or "suffering a sudden stroke". The Latin translation in Kühn uses the term *repercutio* which would seem appropriate to the meaning of "forceful" included in parentheses above.

An infrequent pulse is one in which the arterial dilatation occurs over a long time. Or thus: a pulse is infrequent when the time of the pause between the dilatation and the contraction takes a moderately long time.

An intermediate pulse is one which is in between frequent and infrequent, and is in proportion, or one in which the time between arterial dilatations is moderate.

216 (= 236 CMG). Regularity (order) of a pulse is a state relating to size, or strength, or rhythm, or any other *differentia.*

A pulse is regular (orderly) in which the proportion during the cycles preserves a certain order. Or otherwise: a regular pulse is one which is equal in terms of periodicity.

217 (= 237 CMG). Irregularity of a pulse is instability in the *differentiae* in the pulses. Or thus: an irregular pulse is one in which there is sometimes greater frequency in certain beats, **[408K]** and sometimes greater infrequency. Or a pulse is irregular when it preserves no order in the cycles (in terms of periodicity).

218 (= 238 CMG). Equality (evenness) of a pulse is equality in relation to certain *differentiae* of the pulses.

An equal (even) pulse is one in which equality exists in a series [of beats]; or in which all the *differentiae* of the pulses remain equal.

219 (= 239 and **246sq CMG).** Unevenness of a pulse is an inequality of pulses in certain of the sequential *differentiae* in them. Or thus: an unequal (uneven) pulse is one in which sometimes the beats occur more strongly and sometimes more weakly. Or again in this way: A pulse is unequal (uneven) when, in relation to some part, the impact is higher and lower, and more so when the artery is inflated and dilated, as if constrained in respect of amount. Unequal in strength is like some part of the artery making the rising up more forcefully, or also in the sinking down being weaker.

220 (= 240 CMG). Bacchius the Herophilean[214] said rhythm is movement **[409K]** having order in times. Zeno the Herophilean[215] said rhythm is order of the times in which the arteries distend in relation to those in which they contract. It would be better defined in this way: rhythm is a temporal proportionality of the parts concerning the pulses having a certain kind of state in relation to each other. A different definition is like this: rhythm is a state (σχέσις/*schesis*)[216] in which the times during which the arteries dilate are in proportion to those during which they contract.

221 (omitted in CMG). A pulse is *eurhythmic* which preserves the rhythm specific to the time of life, nature and κρᾶσις/*krasis*, and to each season, and to other circumstances.

A pulse is *arrhythmic* which preserves no rhythm specific to each [factor].

---

214 Bacchius of Tanagra (250 – 200 B.C.), active in Alexandria.

215 Zeno of Laodicea (250 B.C.?—A.D. 80).

216 A distinction is made here and elsewhere between two terms used to denote the bodily state—ἕξις/*hexis* and σχέσις/*schesis*—, the former indicating a stable and enduring state and the latter an unstable and transient state.

A pulse is *kakorhythmic* which preserves this rhythm badly; it is opposite to *eurhythmic.*

222 (= 242sq CMG). We say a pulse is *pararhythmic* which approaches quantitatively the stage of life in which the rhythm is seen to exist but is not altogether in that rhythm. We say *heterorhythmic* of a pulse **[410K]** that has a rhythm of another stage of life. An *ekrhythmic* pulse is one in which the rhythm departs from any stage of life or is of some other circumstance altogether.

223 (= 244 CMG). A tremulous pulse is one which yields a beat that is neither equal nor unchanging but seems to change.

224 (= 245 CMG). A non-straight (crooked, oblique)[217] pulse is one which is not brought to completion in a straight artery, on account of the position.

225 (= 248 CMG). A tapering (mouse-tailed, dying away gradually) pulse is one in which, when comparing the position, some parts of the artery are brought upward and some conversely downward. Here the distension reaches a maximum in the parts raised upward. The majority call a pulse tapering when, from the larger size of the beat that is brought about, those following in relation to the reduction are always made increasingly smaller. They are also called tapering (mouse-tailed) when, from the frequent pulse, more infrequent pulses are brought about. [A pulse] is also called tapering when, from being rapid,[218] it becomes very much slower.

226 (= 249 CMG). A dicrotic pulse is one in which the apparent **[411K]** retreat of the artery doesn't become complete, but there is a delay in the amount of withdrawal, or of the *pneuma* being pushed back, then a failure to complete the contraction that ought to happen.

227 (= 250 CMG). A cobweb-like (arachnoidal) pulse is one which is small and not settled but is moved in the manner of a spider's web, oscillating due to a slight breeze.

228 (= 251 CMG). A pulse is intermittent whenever, in the accomplished contractions and dilatations of one pulse, the artery leaves out (intermits) an interval of time. Or in this way: A pulse is intermittent in the case when not only in two but in three or still more beats, the artery leaves out one dilatation, or two, or more.

229 (= 252 CMG). A pulse is intercident when, between two beats, it moves in relation to a specific order, from which some middle beat is intercident.

230 (= 253 CMG). A pulse is called remittent in which not in two only, but in three or more beats the artery remains immobile. Or like this: Remittent is when, from a large and strong beginning, the pulse always drops down more, **[412K]** until it has the sort of size that is called dying away (tapering); when it drops down and is smaller, as when it inclines downward to become smaller, it is called remittent.

231 (= 254 CMG). A pulse is bounding (like a gazelle) when the artery seems to be expanded but is not altogether expanded, so another beat has rushed in suddenly be-

---

217 The Greek term is ἐσκολιωμένος (Latin *obliquatus*), from the verb σκολιόομαι.

218 Reading ταχέος for παχέος as in Kollesch 2023: 94. The Latin translation in Kühn has *celeri.*

fore providing an appearance of contraction, while the second beat becomes more forceful.

232 (= 255 CMG). A recurring pulse is one which departs from the amount in respect of magnitude or strength and in the case of other [beats] is then restored again to this amount.

233 (= 256 CMG). A formicant (ant-like) pulse is one which is small, empty, frequent, weak and brought down to an extreme degree in all aspects (parameters). Or in this way: A formicant (ant-like) pulse is one in which the impression occurs, when the fingers are pressed down firmly against the artery, of an ant walking around, so that the pulse itself is both small and weak.

2 *Pulses for Beginners* (*De pulsibus ad tirones* VIII.453 – 492K)

This treatise is divided into 12 sections, as follows:

1. An introductory statement that the pulses at various sites and in the heart are the same as each other, although variations in palpability and visibility exist according to the site. Examples are given.
2. Palpation reveals that arteries expand in all three dimensions—length, breadth and height. When in accord with nature all three expansions are symmetrical, when contrary to nature this may not be so. A long, broad and high pulse, when the expansions are symmetrical, is a large pulse; a short, narrow and low pulse, when the expansions are symmetrical, is a small pulse.
3. The movement of dilatation may be fast, slow or moderate; the impact of the pulse-beat may be strong, weak or moderate; the artery itself may be hard, soft or moderate.
4. The basic unit of the pulse is a beat and an interval. Novices should ignore the difficult issue of the contraction (whether it is palpable or not). A frequent pulse is characterized by a short time between beats and an infrequent pulse by the opposite.
5. Equal and unequal apply to all the aforementioned in a continuous series of pulses. Equal is the continuous, sequential equality of two or more pulses and may apply to the various parameters mentioned.
6. Sometimes a single unequal beat occurs separating a number of equal beats. If the intercident beat occurs regularly, the pulse is regular; if it occurs irregularly, it is irregular.
7. Inequality may also occur in a single pulse. Examples of inequality occurring in a single pulse beat are the bounding (gazelle-like) and the dicrotic pulses. It also occurs when the pulse begins faster and ends slower, or vice-versa.
8. Galen gives a brief account of compound inequalities, citing three examples: worm-like (vermicular), ant-like (formicant) and hectic pulses. He then summarizes the main terms as follows:
   - a large pulse is when there is a large symmetrical distension in all three dimensions;
   - a strong pulse is one with a strong palpable impact;
   - a rapid pulse is one in which the dilatation occurs over a short time;
   - a frequent pulse is one with a short interval between beats;
   - an equal (even) pulse is one in which all these parameters remain equal over a series of beats;
   - a regular pulse is one with equal periods.
9. There are three kinds of change of the pulses:
   - those in accord with nature;
   - those neither in accord with nor contrary to nature (non-natural);

- those contrary to nature.

10. The first group includes gender, a hotter nature, a thinner nature, age/stage of life, seasons, places, pregnancy, sleep, the transition from sleep to wakefulness, acquired states (e.g. thin, fat), and change in natural κρᾶσις/*krasis*.
11. The second group (non-natural causes) includes exercises, hot and cold baths, food, wine, and water.
12. In the third group, the changes contrary to nature are in general terms either dissolution and dissipation of the capacity, which produce pulses that are small, weak and frequent, or compression and weighing down of the capacity, which produces inequality in strength, size and everything else. Advanced dissolution of the capacity will give rise to ant-like (formicant) and worm-like (vermicular) pulses. This third group may occur due to psychical disturbances—anger, joy, grief, fear, pain—or somatic disturbances which include various conditions/diseases. Galen lists the following: inflammation (which generally produces a serrated pulse), pleurisy, empyema, wasting, fever, consumption, peripneumonia, lethargy, phrenitis, a mixed condition between lethargy and phrenitis, κατοχή/*katochē*, convulsions, paralysis, apoplexy and epilepsy, συνάγχη/*synanchē*, orthopnoea, hysterical choking, stomachical conditions, dropsies (ascites, tympanitic ascites, anasarca), elephantiasis, jaundice, and the taking of hellebore. He gives a summary of the pulses in all these conditions. They are considered sequentially, but in greater detail, in Books 3 and 4 of *On the Causes of Pulses.*

1. [453K] I shall speak here, my dearest Teuthras of those things concerning the pulses which are useful for beginners to know. You have the whole art of the pulses written out elsewhere.[219] All the arteries pulsate in the same way with each other and with the heart, so it is possible to draw conclusions about them all from one of them. However, it is not possible to similarly perceive the movement of them all; the perception is clearer in the fleshless parts but more indistinct in the fleshy parts. With those arteries that are covered by thick flesh, or are internal to the bones, **[454K]** or have other bodies in front of them, you would not perceive their movement in an animal in accord with nature. On the other hand, if the body is very emaciated, the movement of the artery adjacent to the spine[220] is often clearly discernible to those palpating the epigastrium, and some of those in the limbs, which were previously not apparent become perceptible. However, the arteries in the temples and the feet, those behind the ankles, those of the arms, and those on the inner aspect of the wrists have a movement that is always perceptible. Less easily perceptible than those but not imperceptible are those in the head behind the ears and those on the inner aspect of the arms.[221] And indeed, there are certain others not covered by too much flesh, but there is no other more readily available, more in keeping with decorum, or more conducive to use when it comes to palpation than those present at the wrists. These are particularly clear due to lack of flesh in the region and the absence with these of any need to strip the

---

219 These are the four major treatises translated in the present work. Teuthras, the dedicatee, was one of Galen's students. He is also mentioned in *De venae sectione adversus Erasistratus Romae degentes* 1 (XI.193K).

220 The abdominal aorta.

221 The arteries referred to are, in order: superficial temporal, dorsalis pedis, posterior tibial, radial, posterior auricular and brachial.

body, as is the case with many others, and because they are aligned straight. This is of no small benefit to precision of diagnosis. **[455K]**

2. By palpation, it is apparent to you that the artery expands in every dimension. There are three dimensions of every body—length, depth and breadth. In the case of the animal that is in accord with nature, you will find the artery extending very symmetrically, whereas in the animal that is contrary to nature, there is at one time[222] deficiency and at another time excess in one or other dimension. In this you must remember how the pulse was in the normal state; if the abnormality is found more in breadth, call it "broad"; if in length, "long"; and if in depth, "high". The opposite to these that are less than accords with nature are named by analogy, "narrow", "short" and "low". If the deviation towards abnormality is similar in all dimensions, the overall reduction is termed "small" and the overall increase "large". These are the differences of expansion in a pulse in terms of quantity.

3. Of the specific kinds of movement, there are fast and slow, the former being a kind of hurried movement while the latter **[456K]** is relaxed. And one must judge these by comparison with the normal. In the nature of the impact, strength and weakness exist—strength when the touch is impacted strongly, and weakness when it is impacted weakly. Softness and hardness are qualities of the actual covering (wall) of the artery. Softness is when the artery seems, as it were, more fleshy in relation to the impact; hardness is when it is, as it were, drier and more leathery. This sort of difference of the pulses is determined simultaneously with the movement of the artery. But it is not specific to movement, as the three previously mentioned are. Of the latter, fast and slow pertain to the quality of the movement; strong and weak pertain to the quality of the pulse-beat; and large and small to the amount of the dilatation. However, the dilatation is not without movement. The soft and hard are not in need of movement to be as they are. You will discover these four differences (*differentiae*) of the pulses in relation to the pulse-beat.

4. There is another, fifth *differentia* in relation to the **[457K]** interval between the beats. For it is a custom among doctors to name in this way the time between two beats in which the artery dilates and contracts. And I think it appropriate to train beginners at first that the contraction is imperceptible. Let me refer to the beat and then the interval. A beat arises from the moving artery impacting on the touch. An interval is the pause between two beats, in relation to which there is "frequent", "infrequent" and what is in between both, in which the pulses exist in accord with nature. And you will discover these by the amount of time. "Frequent", then, is whenever the time of pause is short; "infrequent" is when it is long. There is no difference in saying "pause", "the interval between beats" or "contraction".[223]

---

222 Following the Latin translation in Kühn : *nunc ... nunc.*

223 The beat here is the dilatation. There is the issue of whether the contraction is perceptible. If it is, there are two intervals or pauses termed external and internal.

5. There is equal (even) and unequal (uneven) in relation to all the aforementioned *differentiae*. Equal is the continued sequential equality of these—for example, whenever the magnitude of successive beats is equal, the pulse may be said to be equal in magnitude. And if the rapidity (speed, rate) is equal, it is equal in rapidity. The same also applies to strong, frequent and weak. The inequality **[458K]** is a loss of the equality in a *differentia* of the pulses which happens to exist. Thus, there will be the inequality pertaining to magnitude, that pertaining to rapidity, those pertaining to strength, weakness and frequency, and to other qualities similarly.

6. Sometimes a single unequal pulse intercides,[224] separating a certain number of equal beats; this occurs variably. Thus, a fourth unequal beat may follow three equal ones, and this always follows, or a fifth may follow four, and in every other amount similarly. Thus, often the sixth unequal beat follows five that are equal, and in turn a seventh that is not equal follows six that are equal. Therefore, the equality in such pulses is no longer preserved and because of this, the pulse is unequal, although a certain order is preserved, and through this it is regular. Because the one unequal beat always intercides after an equal number of equal pulses, it destroys the equality. This in fact preserves a certain orderliness in relation to the periods. However, if no periodicity at all is preserved, such a pulse is called irregular (ἄτακτος/*ataktos*). **[459K]**

7. Inequality also occurs in a single pulse: here the parts of the artery differ from each other in position and movement, and when comparing the movement of each single pulse in relation to itself. Thus, the inequality exists in the position of the parts, comparing up and down, forward and backward, left and right, as the artery seems to be moved from place to place. In the movement, there is quicker or slower, earlier or later, stronger or weaker, in a greater or lesser time, or always, or not at all when it moves. Inequality in relation to each part is in being clearly interrupted; among these, there is also the bounding (gazelle-like) pulse, and similarly also, in the recurring, from which there is also the double-beating (dicrotic) pulse. There is unequal speed of the movement when it begins faster and ends slower, or conversely begins slower and ends faster. Similarly, in strength and weakness, and smallness and largeness, the movement is not divided into two times of movement only, but into several—as many as may be distinguished by perception. These, then, are the simple inequalities of a single pulse.

8. The compound inequalities, in terms of the number possible, are: one with another; one mixed with one; one mixed with many; and many mixed with many. And some of these have acquired a specific name—for example, the worm-like (vermicular), the ant-like (formicant) and the hectic.

– The vermicular is when there is an impression of a worm crawling along the artery, which is stirred up like a wave, and of the whole artery not dilating at one time. If this occurs with a small dilatation, it is called "vermicular"; if it is with

224 The Greek verb is παρεμπίπτω.

a large dilatation, it is called simply wave-like (undulant). It is clear that the vermicular pulse is weak and frequent.

- When the pulse has fallen to an extreme degree and is weak, frequent and small, it is called formicant. It seems to be rapid but is not rapid.
- Hectic, like a fever, is how a pulse is named which makes no great change but remains always similar; it stays together and never resolves. The whole bodily state (ἕξις/*hexis*) is turned towards disease in such fevers and pulses.

These things seem to me to be sufficient to have said about the *differentiae* of the pulses for beginners. If someone **[461K]** should wish to pursue the matter in greater detail, I have written a whole book about the *differentiae* of the pulses.[225] It is not, therefore necessary to speak now about the full and empty pulse or about rhythms. Enough is said about these in detail in that work and the discussion is too obscure for beginners. Let us, then, recapitulate what has previously been said in summary before we go on to those things that are next.

Accordingly, a large pulse is one which arises when the artery is very distended in length, breadth and depth—long pertains only to length, broad to breadth and height to depth. A strong pulse is one which beats strongly to the touch and a soft pulse one in which the wall of the artery is soft. A rapid pulse is one in which the dilatation of the artery occurs over a short time, while a frequent pulse is when there is a short interval [between dilatations]. An equal (even) pulse is one in which the equality is sequential, and a regular pulse is when the periods are equal. A pulse in which one beat is unequal is called unequal in relation to one beat. And the opposites of these are clear—small, short, narrow, low, weak, hard, slow, infrequent, unequal and irregular. And it is not unclear that of the other opposites, there is **[462K]** the median of each, but there is no median of equal and unequal (even and uneven) or of regular and irregular (orderly and disorderly). And it is clear also that the medians of all the others are in accord with nature, but in these latter [two pairs] only the equal (even) and the regular (orderly)[226] are in accord with nature; the others are not in accord with nature—that is, the unequal (uneven) and the irregular (disorderly) are not.

9. Since the pulses are changed naturally in many ways, you will discover no cause, so to speak, which does not change them. It seems to me most important to make a threefold differentiation of the change of these, and in this way to speak specifically in turn about each. The first of these changes is that which accords with nature. The second is that which does not accord with nature but is not yet already contrary to nature. The third is contrary to nature.[227] All these come after the natural differences, for the arteries are moved differently in different natures. And it is necessary to know these primary differences, if someone is going to recognize by what cause and to

225 *De pulsuum differentiis* (VIII.493–765K).
226 Added on the grounds of sense.
227 See Galen, *Ars medica* 1–4 (I.305–318K).

what amount the pulse happens to have been changed. But someone may learn by experiencing the specific nature of each precisely. And there is need for the artery to be palpated repeatedly, particularly when a person is faultlessly healthy and at rest **[463K]** from all vigorous movement, and subsequently also in the other conditions. Since it is not possible to have gone through all these by experience, for many doctors are often called to those [patients] they have not encountered when healthy, it is best even here for the doctor to have greater technical skills than a layman. And he would have more by being provided with knowledge from many of those similarly affected. To speak in outline, there is one nature which is common to men and another to women, and some are hotter in *krasis,* and some colder. And of each of these there is a common nature, and also one common to all those who are thin and similarly to those who are fat. And it is rare in each of the previously mentioned common natures for there to be what is not like the majority. As a result, it is also rare for someone with an accurate knowledge of the common to err in what is true.

Men have a pulse that is generally much larger and similarly much stronger and a little slower and significantly more infrequent than that of women.

Those who are hotter in nature have a pulse that is much larger, more rapid and more frequent but not much stronger. **[464K]**

Those who are thinner in nature have a pulse that is larger and a lot more infrequent but not much stronger. In nature, then, they differ in this way.

Pulses change in relation to the stages of life (ages) in the following way: the pulse of the newborn child is more frequent; that of the old person is more infrequent. The pulses in all the stages in between differ to the extent that people happen to be nearer to childhood or old age. Similarly, the pulse of the child is very rapid; that of the old person is slower. Those of the other stages of life are in between. The difference in infrequency is much greater between the old person and the child compared to that in rapidity. In the difference in magnitude and strength in the stages of life, the pulse of those in their prime is very large. That of the aged is very small. That of children is in between these but slightly larger. The pulse of those in their prime is very strong; whereas in those who are old, it is very weak. That of children is in between these. The pulses are like this in the age groups (stages of life).

In relation to the seasons, in the middle of spring the pulses are very large and very strong, and at times rapid and **[465K]** moderately frequent. They are similar also in the middle of autumn. As the spring proceeds, there is a loss of size and strength, but the pulses advance in rapidity and frequency. And finally, when summer comes in, the pulses become weak, small, rapid and frequent. As autumn proceeds, there is a loss in all respects—size, strength, rapidity and frequency—so that also, as winter comes in, they are changed to smallness, weakness, slowness and infrequency. The pulses at the beginning of spring are like those in the last stages of autumn, while those in the last stages of spring are like those in the first stages of autumn. And those in the first stages of summer are like those in the last stages of summer, and those in the first stages of winter are like those in the last stages of winter. As a result, those in the middle of summer and the middle of winter stand apart equally from each and change similarly. Mid-summer is in some way the same and in some way the opposite to mid-winter, for the pulses are small and weak in each of the two. Those of summer are rapid and frequent, while those of winter are slow and infrequent. They are not as small in summer as they are in winter, but less in summer and not in this way weak in winter, as they are in summer, but less in winter. **[466K]** These, then, are the changes of the pulses according to the seasons.

Regarding places, there is a similarity to the seasons. In very hot places, the pulses are like those in the middle of summer. In very cold places, they are like those in the middle of winter. In *eukratic* places, they are like those in the middle of spring and proportionately in those places in between. And of the other states of the ambient air, the hot are like the hot seasons, the cold are like the cold seasons and the intermediate are like those in the middle of spring.

In pregnancy, the pulses become larger, more frequent and more rapid. The other features preserve an accord with nature.

Sleep should be natural, if anything is. This also changes the pulses at the beginning making them smaller, slower, more infrequent and weaker, while as sleep proceeds, the slowness and infrequency are increased, and particularly in relation to food. However, they become larger and stronger. Over time, they change in turn to weakness and smallness, but preserve slowness and infrequency. **[467K]**

The pulses of those who have undergone the change from sleep to wakefulness are in the immediate term large, strong, rapid and frequent, and have some clonus.[228] After a short time, they return to moderation.

The acquired states (σχέσις/*schesis*) of the body change the pulses similarly to the natural states. A person who is thin in nature, when he becomes well-fleshed, has a pulse in proportion to such a nature, while the well-fleshed person who becomes thin, has a pulse like those who are thin in nature. It is clear, however, that apart from the change in capacity, it is necessary to examine the difference relating to thin and well-fleshed, and similarly in the case of the other factors, so that, in relation to the one change that has occurred alone, we make the argument about this in each case. Those things said in the case of the well-fleshed it seems necessary to extend further to the case of those who are obese.

Also the κράσεις of the body which are acquired change the pulses analogously to the natural κράσεις.

10. Next, it is time to speak about the other changes occurring from the non-natural causes.[229] Exercises at **[468K]** the beginning, [and practiced] up to the point of moderation, make the pulses stronger, large, rapid and frequent. Much exercise, exceeding the capacity of the one exercising, makes the pulses small, weak, rapid and fre-

---

228 It is not entirely clear what the Greek term κλόνος signifies in relation to the pulse. LSJ lists "shaking" and "agitation" in the physical sense, citing Galen, IX.76K and Alexander Aphrodisiensis, *In Aristotelis Topicorum*, 466.25. *Clonus* in English now has the medical meaning of rapidly alternating contraction and relaxation of muscles.

229 On these, see for example García Ballester 1993.

quent in the extreme. Great excess [of exercise], such that the person can hardly still move, and only with long rests or not at all, and is very enfeebled, makes the pulses very small, weak, slow and infrequent. If he is overwhelmed to the point of dissolution of the capacity, the pulses are specific for that. I shall speak a little later about what a dissipated capacity makes the pulses like.

Hot baths, whilever they are moderate, make the pulses large, rapid, frequent and strong. Those that are immoderate make the pulses small and weak, although they are still rapid and frequent. If, however, someone were not to stop at this, they would be small, slow, infrequent and weak.

Cold baths immediately make the pulses small, weaker, slow and infrequent.[230] Afterwards, what they are made like depends on chance. At all events, there is either torpor or strength. **[469K]** In someone who is made torpid and cooled, the pulses are small, weak, slow, and infrequent; when he is warmed up and strengthened, the pulses are large and strong but moderate in rapidity and frequency.

A large amount of food, such that it weighs down the capacity, makes the pulses unequal and irregular. Archigenes says they are more rapid more than they are more frequent. Food in a moderate amount makes the pulses large, strong, rapid and frequent. Too little food, so that the person is insufficiently nourished, is not like the moderate, but creates less change and of short duration.

For the rest, wine changes the pulses similarly to foods, but differs in making the change immediately and the change from wine stops before that from food. In this, rapidity and magnitude increase more than strength and frequency. For by almost as much as moderate food provides more vigorous and lasting strength of the body, so wine increases the magnitude. **[470K]**

Water, of all the things offered, brings about a very brief change, although it does bring about a change analogous to food.

All the other things, to the extent that they effect a change, are able to heat or cool, and to such an extent also, change the movement of the arteries. In this way the pulses are changed by the so-called non-natural causes.

11. Let us speak, in order, of the changes due to causes contrary to nature, bearing in mind those things I wrote before in the general discussion on things contrary to nature.[231] Conditions of the air and amount of food, such that they burden the capacity, and excessive exercise, bathing and sleep contrary to nature, for these are excessive in amount, change those things we term 'in accord with nature' and 'non-natural' to a contrariety to nature. Those things that are contrary to nature, not only in amount but also in class, are infinite in number and because of this indeterminate. However, the art exists even in these to measure the infinite by dividing them as far as possible into classes and kinds (species). With each of the causes contrary to nature, it is not

---

230 This is the statement from *On the Causes of Pulses* on the same matter; the passage in the present work reads "small, slow, infrequent and more infrequent."

231 See the earlier reference to Galen, *Ars medica* 1–4 (I.305–318K).

unreasonable for someone to say, it is either **[471K]** as if it were dissolving and dissipating the vital (physical) capacity,[232] or as if it were compressing, or weighing it down. The capacity, then, is dissipated by the lack of food, by the *kakoētheia* (κακοήθεια)[233] of diseases, by the strength of psychical affections, by severe and long-lasting pains, and by excessive evacuations. It is weighed down by large amounts of material and by affections in the organs, like inflammations, indurations (scirrhosities), tumors, abscesses and deteriorations of various kinds.

The dissipated capacity makes the pulse small, weak and very frequent; the compressed and, as it were, weighed down capacity changes the pulses to inequality and irregularity in terms of strength, size and everything else. These inequalities are particularly characteristic of a compressed capacity, and of one that is weighed down greatly in many *differentiae*, and to a slight extent in a few. In minor harms, the large pulses are more than the small and the strong pulses more than the weak. In major harms, the opposite obtains; some whole movements are lost, and some are intercident[234] in such conditions. But when they are intercident, **[472K]** the harm is smaller; when it is destroyed, they are obviously greater. These are the common changes of every dissolution and compression. What is specific to each is taken on besides from the affecting cause.

The change of the pulses due to the lack of the dissipated capacity is, at the beginning, to weakness, smallness, rapidity and frequency; in the intermediate period, it is to weakness, smallness, slowness and infrequency; finally, it is to extreme smallness, weakness and frequency and a false impression of rapidity. This is called the ant-like (formicant) pulse. The worm-like (vermicular) pulse occurs when the capacity is already dissipated but is still holding up to a slight degree. And it differs from the formicant by not yet being brought to extreme weakness and smallness and has a clear inequality in relation to a single beat, which is manifest in the parts beginning to move earlier or later. From this, it is less slow than it is small, while sometimes it is not slow at all. Because of this also, it is least abnormal. The collapses (syncopes) occurring due to the fatal and acute fevers do not have the vermicular pulse, whereas in the **[473K]** other dissipations of the capacity, the vermicular pulses for the most part follow, and particularly in those occurring apart from fevers or with those that are very small. The formicant pulses occur in the cardiac syncopes, in choleras, in strong fluxes of the abdomen, in haemorrhages, in the menstrual flow and in all the acute depleting affections. The vermicular pulses follow in most instances, but the formicant

---

232 There is some issue with the term "vital capacity" here and in similar contexts, given its specific meaning in respiratory physiology. "Physical" is probably to be preferred.

233 This term is transliterated Greek here as elsewhere. The definition in LSJ is 'malignant character' which is perhaps misleading given the current connotation of 'malignant'; 'bad effects' might suit here.

234 The verb "to intercide" and its derivatives is rare and is not listed in some modern English dictionaries. It is however included in the 1933 edition of the *Oxford English Dictionary* with a meaning of "to cut through, to fall between," citing R. Copland, 1541, in his *Guydon's Quest. Chirurg.* referring to Galen. It is used here throughout as the translation of παρεμπίπτω.

follow in the extreme cases. Whenever these occur without fever, you will then discover the vermicular pulse more and it will be clearly apparent that it is long lasting. Such are the most common forms of the causes contrary to nature.

12. Next I shall speak of these according to kind. [The pulse] of anger is high, large, strong, rapid and frequent. [The pulse] of pleasure is large and infrequent, but not in fact different in strength. [The pulse] of grief is small, slow, weak and infrequent. The pulse of fear that is sudden and severe is rapid, clonic, irregular and unequal, while if the fear has already lasted for a long time, **[474K]** it is like the pulse of grief. In all these affections, when they last for a long time or become very severe, the pulses that follow are like those of dissipation of capacity. And furthermore, all these dissipate the capacity—those that are strong, acutely; those that are the opposite, chronically.

Pain is something that changes the pulse. Pain that is severe or in important parts changes them, as inflammation also does. Pain that is still mild and just beginning makes the pulse larger, stronger, more rapid and more frequent, whereas when it is increased and has become very severe, so as already to have damaged the vital *tonus* (tension),[235] it makes the pulse smaller, weaker, rapid and frequent. And the more the pain lasts or becomes stronger, the more it increases each of the aspects mentioned. Pain which already also dissipates the capacity effects the change to weakness, smallness, a false impression of rapidity and excessive frequency.

The pulse common to every inflammation is saw-like (serrated), as one part of the artery seems to be dilated but another does not, and this latter clearly appears to be harder. **[475K]** This pulse has some clonus and is rapid and frequent, although not always large. The specific characteristic of each is that at the beginning the pulse is larger than accords with nature, and stronger, more rapid and more frequent. However, when the inflammation increases still further, all these features are increased, and the pulse is now clearly harder and more clonic. As the inflammation reaches its peak, the pulse is more clearly harder and more clonic, but smaller than before, although not in fact weaker, unless the affection exceeds the capacity. But also, the pulse becomes more frequent and rapid. If, however, the inflammation is sufficiently chronic and is already hardened into a scirrhosity, then in addition to the features mentioned, there is thinness and hardness of the pulse. These things come about when the inflammation changes the pulse in the animal as a whole, through either the magnitude or the importance of the part in which it is situated. When the whole animal is not involved, the pulse in the inflamed part is as described. Also, each of the things mentioned either increases or decreases according to the amount of the inflammation or to the nature of the inflamed organ itself. Thus, the more sinewy parts **[476K]** make the pulse harder, more saw-like (serrated) and smaller, while the more vein-like and artery-like parts do

235 *Tonus* is retained in the Latinised form to make the distinction from *strength*. In the 1933 edition of the *Oxford English Dictionary*, it is defined as: "The condition or state of muscular tonus; the proper elasticity of the organ." Several late 19th-century references are cited applying the term to the state of the arteries.

the opposite. The greater of these is in the artery-like parts, and this pulse readily becomes unequal and irregular.

It should already be clear what the pulse would be like in inflammations involving the liver, spleen, kidneys, bladder, stomach or colon, and in the pleuritides and peripneumonias, and in short, all those parts in which a fever follows the inflammation of the part, except for those due to the nature of the symptoms that follow of necessity in them, and of those by chance accompanying features that would be able to change each pulse. And this will happen to change the pulse to such a degree when a mixed change occurs in it by reason of the inflammation, and what the nature of the place is, and what the present symptom produces. For in those with inflammations involving the brain,[236] convulsions will readily be produced, chokings in inflammations of the lung, syncopal episodes in those with inflammations of the opening of the stomach (cardiac orifice),[237] atrophies in those with inflammations involving the liver, failures of concoction (ἀπεψία) in those with inflammations involving of the stomach, and retention of urine in those with inflammations involving the kidneys. And the more sensitive parts change the pulses due to the pains, whereas **[477K]** the more insensitive parts do so in relation to the condition alone. From all these then, various kinds of changes of the pulses arise in the inflammations. That it is necessary to distinguish these changes was described comprehensively in other works.[238] What is stated now is as much as is useful for beginners.

[The pulse] of those with pleurisy is rapid, frequent and not very large. It would also seem to be strong. It is not weak but is not immediately of a strength commensurate with the affection. It is necessary to bear in mind in all cases that we need to examine the change in each of the matters, distinguishing how much it is due to the affection and how much to something else contingently, and not due to the affection. Accordingly, the pulse of those with pleurisy, since it somehow makes the artery more sinewy and harder, so as to tend towards strength, deceives the unpractised who are unable to distinguish a hard impact (beat) from a strong one. In this way too, the majority of doctors are unable to distinguish the many other differences of pulses and are quick to find fault with those things written here. Not being acquainted with the things correctly stated, they are prejudiced [against them]. But **[478K]** it is not

236 It is not clear what τὰς φρένας refers to here. The word φρήν has several meanings: (i) the diaphragm/midriff, (ii) the heart as the seat of the passions (LSJ), (iii) the mind. Present concepts of pathophysiology clearly favour the last, but the other organs referred to perhaps suggest the first. The Latin translation in Kühn has *septum transversum* which is a term still in use applied to "the mesodermal mass separating the pericardial and peritoneal cavities; it is covered with mesothelium except where intimately associated with the liver, which originally develops within it. The septum is definitively incorporated into the diaphragm as the central tendon" (*Stedman's Medical Dictionary* 2000: 1622).
237 Here and elsewhere τὸ στόμα τῆς γαστρός is taken as a reference to the cardiac orifice of the stomach, and not the pylorus as some translators have thought. This does have a specfic name—ὁ πυλωρός.
238 Presumably the four major treatises included in the present work, and particularly the third on causes.

necessary to spend time on these in the present discussion. I have written specifically about the diagnoses concerning the pulses.[239]

Therefore, I advise training the reasoning power and the touch at the same time, so as to be able to recognise the pulses in their very actions, and not to distinguish by reason alone. The starting point of the practice in the actions is the teaching through reason. And for sure it is not possible to explain the quantity of the frequency by a statement, although there is a great difference certainly between what exceeds the customary measure of the pleuritides and what falls short. Once there is a change to peripneumonia or impending collapse, there will inevitably be excesses, while the deficiences end in a lethargic attack or damage of nerves. In this way too, the kind of inequality is, as it were, saw-like, which is no less specific for those with pleurisy. When reduced, it is a sign of pleurisy that is soft and readily brought to concoction; when increased, it is a sign of a pleurisy that is severe and difficult to concoct. Such inequalities, with a weak capacity, are acutely dangerous. With a strong capacity, they will be concocted over time or changed to empyema, or a consumptive marasmus succeeds them. Therefore, **[479K]** the pulse of what is concocting gradually is devoid of every change contrary to nature. When there is a change to empyema, the pulses become specific for those with empyema. The pulses in the marasmuses are the same as those in consumptive wastings.

The pulse of those suffering from empyema is, at the beginning, like that of inflammation at its peak. For this is the beginning of the actual empyemas. Sometimes, the pulse is unequal and irregular; in all cases it is hectic. When pus is already present, the other aspects are similar, but the pulse is more equal (even). When the empyema bursts, it is weaker, broader, slower and more infrequent.

The pulse of those who are wasting doesn't change in one way. As far as possible it is necessary to distinguish these through clearly apparent differences. Those with inflammations that are not being resolved, when they are gradually fading away and dying, have pulses that are weak, more rapid, very frequent and tapering (mouse-tailed) in size in a single beat. Archigenes calls them "nodding" and "inclining",[240] wishing to make abundantly clear the brevity of the dilatation along with the **[480K]** kind of "nodding" of the limits on either side (i. e. starting and finishing). For it is not that they cease abruptly, but contract as if making an angled return of the parts on either side gradually, there being a tapering away in magnitude in relation to each of the parts. This is not in these cases alone, but also exists in the majority of those who are wasting away in any way whatsoever.

This is the case, then, in all those who waste away due to inflammation, but also in many of the others who are wasting away, if not also in those who are wasting away

239 This refers to his treatise *De dignoscendis pulsibus* VIII.766 – 961K, which is the second of the four major treatises included in the present work.

240 The terms here (ἐπινενευκώς, περινενευκώς) are derived from the verbs ἐπινεύω (to consent, nod, incline) and περινεύω (to incline, incline forward, incline to one side and then the other). For further details see the appendix on pulse terms.

due to some occult inflammations. And this would be specific for those wasting away through inflammations and would not be present in any of those wasting away otherwise. The pulse is hectic in all those wasting away, and this is very common in them. Secondly, there is a tapering inequality in the size of the dilatation. Furthermore, this also exists in the majority. Thirdly, there is frequency. And this exists without exception in all those wasting away from inflammations. But it is also present in those acutely endangered from cardiac conditions or gastric syncopes,[241] if they seek to escape the acidity by a drink of wine, and are wasted away over time, unless you were to say these people perish due to small inflammations not apparent to us. **[481K]** And some of these people have the "nodding" pulse, unless again it is those with inflammations, although someone might say the others waste away without inflammation. This is hard to envisage. These people have a hectic pulse or one that is weak and very frequent, and some of them have a "nodding" pulse. This is the second difference in the pulses of those wasting away.

The other and third difference is having an infrequent pulse. But in these at any rate the preceding fever makes it frequent, and the extreme dissipation of the capacity makes it very frequent. However, in the intervening stage, when all fevers are cooled, if those affected are not yet dead, the pulse makes the change to infrequent. This form of the wastings is characteristic of old age, and particularly when those parts related to the thorax and lungs happen to have been affected. Such people preserve the feverish hardness of the pulse, even if it is infrequent. In very few of those who are wasting away is the pulse changed to another inequality, apart from that mentioned in relation to magnitude.

The pulse of those termed *phthisical* (consumptive) is small, weak and soft, moderately rapid and hectic. **[482K]**

The pulse of those with peripneumonia is large and somewhat wave-like (undulant), and weak and soft, similar to the pulse of those with lethargy, except those which go beyond bounds in inequality in a single pulse-beat and the so-called systematic [inequality]. In relation to a single pulse-beat, the pulse sometimes becomes interrupted, as it were, turbulent and dicrotic. In the systematic inequality, there are other differences; sometimes the pulse is intermittent and sometimes intercident. Since all the fevers in the peripneumonias are acute, but some also have some lethargic states, it is in relation to that particularly whichever of these should predominate, that the amount of the frequency is discovered. When the person with peripneumonia is more feverish, the pulse is very frequent; when he is more lethargic, it is less frequent.

The pulse of those with lethargy is similar to that of those with peripneumonia in terms of magnitude, weakness and softness; it is slower than this, and weaker and less unequal, and has more dropped or extra beats (is more intermittent or intercident);

241 The term here is στόμαχος which can mean throat, oesophagus, cardiac orifice or stomach.

sometimes it is also dicrotic. However, it is always undulant[242] in the deep καταφορά (lethargic attack), about which these things are said. For **[483K]** we shall go over the pulses of the diseases that have come to completion with all the signs of these, so that we are enabled also to recognize adequately the magnitude of those things that are lacking in what is not yet complete—as much as there already is and as much as might possibly be added to them. Since we have made mention of inequalities frequently but irregularities infrequently, it is necessary to know that in general the irregularity for the most part is consequent on the inequality. It is rare to find an unequal pulse that is regular (orderly). The smaller harms create pulses that are unequal and regular, whereas greater harms create pulses that are unequal and irregular.

The pulse of those with phrenitis is small; sometimes, but very rarely, it is seen to be large and have moderate *tonus.* Also, it is hard and sinewy, very frequent and rapid. It is also somewhat wave-like (undulant). Sometimes it will also seem to you to tremble somewhat, and sometimes to be cut off spasmodically (convulsively). The pulse of those with fevers clearly and typically acquires a specific symptom in the rapidity, particularly in relation to both the limits of the dilatation, and more the external.[243] And you sometimes discover the kind of inequality in relation to the position occurring strongly in these. But also **[484K]** it will often seem to you that the whole artery, leaving its own position, is carried upward clonically, being thrown up more than dilating in a pulsatile manner. In the same way too, it withdraws back downward, being displaced downward more than upward (contracting). The great frequency of this portends imminent syncope.

There is also another affection between lethargy and phrenitis, which we must name, as it is not the same as either or common to both but is a mixture of the forms of phrenitis and those of lethargy. We shall look at this specifically, and now speak about its pulses. And so that it is not brought forward as some riddle, I shall make it clear by those features that accompany it. In most instances [the sufferers] close their eyes, are sleepy and snore. Again, for the most part, they gaze in a strained way and continue unblinking like those with κατοχή.[244] And if someone should ask them a question or force them to converse, they have difficulty answering and are unresponsive. In many instances, they speak in a distracted way, don't answer correctly and talk nonsense heedlessly. Such is the affection I now wish to make clear by recognition of the accompanying features due to the lack of a specific name. The pulses **[485K]** of this disease are rapid and frequent like in those with phrenitis, but less so. In the same way too, they have less strength than those. [In phrenitis] the pulses are broad and short, and the part related to the outward movement is not suddenly

---

242 Reading κυματώδης for κωματώδης (as in Kühn's text) on the grounds of sense (Kühn's Latin translation has *undosus*).

243 Presumably the external pause which follows dilatation.

244 It is not clear what corresponds to κατοχή (*katochē*) in modern medicine, or indeed what the disease is that is being described—see the appendix on diseases and symptoms. The pulse in this is described in what immediately follows.

cut off, but in another manner, it is as if striving to withdraw inward, speeding up the contraction while concealing the dilatation.

These are not in fact like those in relation to the dilatation of those with phrenitis —they don't have this kind of cutting off.

The pulses of those with κατοχή—for the doctors of old called these κάτοχος and κατεχόμενος while doctors of recent times term the affection catalepsy (κατάληψις)[245] —are like the others in the lethargies, as to largeness, slowness and infrequency, just as the whole affection is not far removed in kind from the other affection. The pulse of those with κατοχή is neither weak nor soft; rather, in these features they certainly differ greatly, just as also in the dissolution and swelling in the whole state (ἕξις) in those with lethargy, there is binding and keeping together in those with κατοχή. In the same way too, they differ from each other in inequality and equality. Thus, the pulse of those with κατοχή is equal (even) **[486K]** while in those with lethargy it is unequal (uneven). Archigenes says the place of the artery is found to be specifically hotter in those cases, just as in those suffering convulsions along with lethargic attacks (καταφορά).

The actual body of the artery of those convulsing seems to be contracted and constricted from all sides, but not like it has been compressed by something or confined. Nor is it altogether uneven (shivering)[246] like with a fever, and particularly in those who are symptomatic. Nor is it very difficult to distend due to hardness, such as exists over a long period of time, and particularly with certain faults or bad states of the internal organs, but is as if it is a hollow sinewy body like the intestine, or something similar, extending from both the ends. In the same way too, the movement is unequal (uneven) when the artery changes position up and down like the string of a musical instrument. For there is no impression of either dilatation or contraction; it is more like clonus—like a leaping outwards and in turn a drawing inwards. This is not an affection occurring separately, but in time often some part of it seems to be carried upward, as if shot out and carried in as though drawn by something, or one part **[487K]** is moved quickly and one part slowly.

The pulse of those who are convulsing seem to be strong and large. It is neither weak nor small, nor as strong or large as it gives the impression of being. For the pulse-beat is deceptive due to the tension seeming to be strong and due to the clonus leaping out. Wherefore there is sometimes also the impression of greater height, and this brings a kind of rough sound to the touch. And this pulse is not something hidden from those who are practised. For it is like nothing else—not the tension in each direction nor the spasmodic movement. When the movement is mixed with that of the acute lethargic attack (καταφορά), it is hard to detect. Only someone who has trained himself to recognize accurately each movement in it is also able to recognize the combination.

---

245 See the previous note.

246 The Greek term is πεφρικός the perfect active of φρίσσω (to be rough or uneven on the surface, to shiver, to stand out). The Latin translation in Kühn has *rigidus.*

The pulse of paralysis is small, weak and slow. In some cases, it is infrequent and in some cases frequent, but is irregularly intermittent.

The pulses of those with epilepsy and those with apoplexy are similar. Those things that will be said about those with epilepsy must be considered as said also about those with apoplexy, **[488K]** but extended further. Certainly, in the case where the disturbance is moderate and the affection is not yet much stronger than the nature, it is clear that there is no change to find in magnitude, strength, rapidity, frequency and hardness. Only in those convulsing is there a likelihood that the artery is extended in each direction. If, however, the affection is so strong as to weigh down the capacity, the pulse takes on a certain inequality (unevenness), the tension is strong, and the pulse becomes smaller, weaker and more infrequent. When it is greatly compressed and the capacity overthrown, this makes the pulse weak, frequent and rapid.

The pulse in those with συνάγχη (angina, sore throat)[247] has a certain tension similar to that in the convulsion but is large and undulant (wave-like), as in those with peripneumonia. And which of the two predominates greatly in this is that from which you must expect the change. For if the peripneumonia is of the kind to predominate, the συνάγχη will end in peripneumonia, whereas, if the spasmodic is, it will end in convulsion. In those who are choked strongly, the pulse becomes small and infrequent. In the late stages, it is already frequent and unequal. **[489K]**

The pulse in those with acute orthopnoea is unequal, irregular, and failing by degrees.[248] And that of the intermediate state in badness is frequent; when the patient is *in extremis*, it is forceful, slow and remittent; when death is already imminent, it is frequent and weak.

The pulse of hysterical choking[249] is lengthened spasmodically and infrequent. In those who are terminal, it is frequent, irregular and failing by degrees.

When the *stomachus* is afflicted—for let the mouth (opening, cardiac orifice) of the stomach be termed thus by us for the present, following the customary usage of the majority—this doesn't change the pulse in one way only. Rather, if it is only inflamed, it brings about a change of the kind we said occurs in an inflamed sinewy body. If, however, there is compression, biting, dissipation (lysis), vomiting, nausea, anorexia or pain, it will accord with the kind of symptom. The bitings, vomitings, nauseas, hiccups, anguishes and dissipations strongly condense the pulse, and make it small and weak, **[490K]** and in some cases moderately more rapid. Compression alone, apart from any of these things, makes the pulse infrequent, slow, small and weak. Such a compression arises from burdensome foods which have no strong potency, but are disturbing due to the actual amount alone, and some non-biting fluids that flow into it. If it is cooled by-these, then at that time there will also particularly be such a pulse. And that of those with bulimia is of this sort.

---

247 On this term, see Galen, *In Hippocratis de Acutorum Morborum Victu* 27 (XV.790K).

248 For ὑπεκλείπω LSJ has "to fail or flag by degrees", citing Theophrastus, *De igne* 55 and this passage.

249 On this term, which might be rendered 'choking in women', see Aretaeus, *De causis et signis acutorum morborum* II.11 (ed. Hude 1958: 32; Engl. transl. Adams 1856: 285).

All the conditions that change the pulse towards frequency, if they are chronic or stronger, make the pulse vermicular. Those conditions which change the pulse towards infrequency, in addition to extending the previously mentioned differences, generate by themselves such a kind of pulse, with inequality in a single pulse, so that to a great extent the body of the artery seems to have been perforated, such that it does not seem to be continuous. And the perception to palpation in the dilatation becomes like sand falling.

The pulse of those with the fluid collections (dropsies)[250] [is as follows]: in ascites, is large, frequent and somewhat hard with a certain tension. In tympanitic (ascites), **[491K]** it is larger, not weak, more rapid, frequent, somewhat hard and with a certain tension. In anasarca, it is undulant (wave-like), broader and soft.

The pulses of those with *elephantiasis* are small, weak, slow and frequent.

The pulses of those with jaundice without fever are smaller, more frequent and harder but neither weak nor rapid.

The pulses of those who have taken hellebore, a short time before the vomiting, when there is compression, are broad, infrequent, weaker, and slower. When they are vomiting and retching, the pulses are unequal and irregular. When they are already becoming better, they are regular but still unequal, although less than before. When they are coming near to an accord with nature, they are equal, greater than before and stronger. In those among them who are suffering syncope, convulsing or hiccupping, the pulses are small, weak, irregular, quicker and very frequent. In those who experience choking, they are small, weak, **[492K]** irregular and unequal; they are neither frequent nor rapid but slow still more. Also, some undulation is apparent, and breadth, and at times some small degree of tension of the artery.

3 *On the Use of the Pulses* (*De usu pulsuum* V.149–180K)

This treatise is divided into eight sections:

1. On whether the use of the pulses is the same as that of respiration—the common view, according to Galen. He considers the major difference, which is that death follows cessation of respiration, whereas loss of pulsation is tolerated, at least regionally. He compares nerves and arteries in terms of the effects of ligature, stressing the differences. Despite these differences, he defends the value of the pulses for diagnosis and prognosis. In particular, they can reveal things about the heart. The reason why the pulses and respiration are thought to have the same use is that they change in the same ways in various conditions.
2. The use of respiration is twofold:
   - preservation of innate heat;
   - nourishing the psychical *pneuma*.

250 The three conditions listed under the collective term 'dropsies', with their current meanings, are: *ascites*—accumulation of serous fluid in the peritoneal cavity; *tympanites*—swelling of the abdomen from gas in the intestinal or peritoneal cavity; *anasarka*—a generalised infiltration of oedema into subcutaneous connective tissue. The first two are distinguished by percussion of the abdomen.

He considers the effects of carotid artery ligation and attempts to explain why there is so little effect. He discusses the *rete mirabile.* He notes the difference between the heart and other parts in response to arterial ligation. If the heart is deprived of innate heat, there is a general effect on other parts of the body, although there is still some distribution of innate heat. He gives a description of the experiment of exposing the heart and removing the pericardium, particularly in relation to the temperature of the heart.

3. Galen considers the experimental evidence that heat flows to the parts, not only through the arteries, but also through the veins and other structures. He thinks there is one use for respiration and the pulses, as stated in section 1. This is preservation of the innate heat and maintenance of the psychical *pneuma.* Dilatation of the arteries draws in air-like material; contraction expels smoky superfluities generated by the burning up of humors in the whole organism.
4. Galen criticizes Archigenes' view that arteries are filled on contraction and emptied on dilatation and counters the analogy of contraction of the mouth and nostrils on inspiration. He concludes that what inspiration is to the organs of respiration, so dilatation is to the arteries, and what expiration is to respiration, so contraction is to the arteries. The dilatation and contraction of the arteries is controlled by the heart in the way Hippocrates and Herophilus supposed, and not as Erasistratus thought—i.e., flow of the capacity from the heart through the walls of the arteries.
5. Galen considers and affirms the view that the arteries draw in from all directions when they dilate and expel in every direction when they contract. He speaks of "pores" in the arterial walls and *sunanastomoses* between arteries and veins, considering experimental evidence for the latter. He criticizes two of Erasistratus's opinions:
   - there is no retrograde flow from aorta to heart;
   - the arteries are "lifeless conduits" and dilate when filled, rather than being filled when they dilate—the wine-sack versus bellows discussion.
6. This section examines the issue of whether both dilatation and contraction are to be regarded as functions/actions of the arteries. Galen considers experimental evidence and concludes that both dilatation and contraction are active functions of the arteries dependent on a physical capacity. On the other hand, respiration is dependent on psychical capacity.
7. On the difference between expiration (passive) and exhalation (active). Is there a comparable pairing in the arteries? Galen concludes there is. He gives examples of situations which require active contraction to expel superfluities, such as sleeping after eating, changes in the ambient air and psychical affections. An alternative view is that softness of the arterial walls favours contraction. However, it is probably that softness facilitates the action of the contracting power. Galen concludes there are both active and passive components to contraction.
8. A summary: the natural (innate) heat in the animal is cooled in the dilatations and purified in the contractions, and the movements are directly analogous to those of respiration. These movements are also important for the maintenance of the psychical *pneuma.*

1. [149K] What is the use[251] of the pulses? Is it the same as that of respiration, as is supposed by almost all doctors and philosophers, or is it something else besides this? For we certainly must not believe them without examination, realizing there are not a few other phenomena in opposition [to this view], and not least what I am going to speak of now. Thus, we die immediately when deprived of respiration, whereas many of the limbs can be rendered pulseless without great harm. Anyway, if you

251 The Greek term is χρεία which has a wide range of meanings. The two which variably apply in Galen's discussions of the pulse are "use" and "need".

should wish **[150K]** to cut off either of the arteries passing down through the groins to the legs with a ligature, or those through the axillae to the arms, you would immediately make all the arteries in the limbs pulseless, but you would not deprive the limbs of voluntary movement, just as you would not deprive them of sensation. If in time, however, they become numb, cold, pale and atrophic, perhaps this happens more from a sympathetic affection[252] than from the destruction of the pulses. If, on the other hand, you were to cut off the nerves with ligatures, you would immediately make the parts immobile and anaesthetic. Thus, it should also be, when the arteries are damaged like the nerves, that they should immediately harm each of the parts they previously benefitted. But the strangest thing of all is that, if you cut off the arteries in the neck with ligatures, you will do no obvious harm to the animal, even though these are the arteries that connect the heart to the brain—that is, which connect the most important organ of the physical powers to the most important organ of the psychical powers.[253] If, then, in this way the damage of the vital arteries does no clear harm to the animal, it is hardly likely that damage to the other arteries would cause harm. If they do no harm **[151K]** when affected, it should somehow be clear to everyone that they do nothing beneficial, even if the animal should happen to be in good health.

How is it then that we make our most important prognostications from the pulses? It is not, as someone might say, from very powerful causes but, as from vital signs, it is rather that someone might make a judgement about death from curved and blackened nails.[254] For whenever things follow of necessity those things that are useful to life, the danger is dependent on the injury of what is useful, while the consequent sign is inevitably of the danger. The heart, then, is a dominant organ, while the arteries grow out of it, and move in the same way it does. As a result, they are also harmed in the same way as it is and are naturally able to reveal something important about it. But it has been shown that to reveal important things is not the same as being able to produce them. Wherefore, then, has it occurred to all doctors and philosophers to attribute the same use to respiration and the pulses, for the former obviously makes a great difference to us, whereas the latter, it would seem, make no difference at all, or at all events, very little?

They seem to me, then, as **[152K]** most of them also write, to think both of them are of a common use, because they are changed in similar ways following the same causes. For when people exercise, or bathe, or are heated in any other way whatsoever, not only is the respiration seen to become more rapid, more frequent and greater, but

---

252 See, for example, *De constitutione artis medicae* 15 (I.280–282K; Engl. transl. Johnston 2016: 99), where other references are given.

253 Perfusion of the brain can be maintained by a single patent vertebral artery through the anastomotic arrangement at the base of the brain. See, however, Galen's experiment in the present work (section 2 below).

254 See, for example, Caelius Aurelianus, *De acutis et chronicis morbis* 2.32 (Engl. transl. Drabkin 1950: 247).

also the pulses change similarly, whereas when they are resting or are cooled in any other way whatsoever, not only is the respiration seen to become less frequent, slower and smaller, but also the pulses change similarly. And in the burning fevers people breathe very deeply, very rapidly and very frequently and have very large, rapid and frequent pulses. On the other hand, if after immoderate intake of foods, the pulses become greater while the respirations become smaller, this is not a problem for the argument, for the respirations are smaller than the pulses due to the restriction of the midriff,[255] but the amount by which respiration is small [is compensated for] by the amount to which it is more frequent, making up for the smallness of the respiration by the continuity **[153K]** of the function. For to the extent that Nature "enjoys" a greater deficiency of air through not being able to expand the chest very much, there is correction by the frequency. And because of this, breathing that is smaller and more frequent can be equal to that which is larger and more infrequent. We say now that a pulse is more infrequent following food, but not in relation to the frequency before the food,[256] for it is more frequent than that, but in comparison to the kind of respiration. Accordingly, if the respiration, in like manner to the pulses, follows the same causes, but does not harm similarly when lost, (and this was shown before), no less could conclusions opposite to each other be inferred—i. e. that both arise for the same use and that they do not. But this is impossible; one or other of them must be true, not both.

2. We must enquire, therefore, into what way one or other of the arguments is fallacious, making the criterion of our judgement the use of respiration, which we showed, in the discussions of this, to be twofold, as it also seemed to Hippocrates.[257] **[154K]** One use, and the more important one, is the preservation of the innate heat; the other, and the less important one, is nourishing of the psychical *pneuma*. But we said the brain is benefitted in relation to both these aspects from inspiration through the nostrils. As a result, it is not surprising, since little assistance is supplied to the brain from the heart, that there is also little harm when the arteries called carotid are cut off with ligatures. But perhaps someone may say, not even the least harm is observed, for the animal remains unharmed throughout a whole day, as we have often put to the test. And he will in fact be right. Certainly, when we were perplexed by this, we devised a way for such a phenomenon to be put to the test. First, however, I shall speak of the reasoning from which we come to this.

Since it has been shown by us in other writings that the cavities (ventricles) in the brain are full of psychical *pneuma* requiring continual nourishment, and also that this *pneuma* is used up in the voluntary movements,[258] it seemed better to compel the an-

---

255 Of the four main meanings of φρήν listed in LSJ the midriff or diaphragm is clearly indicated here.
256 See Furley and Wilkie 1984: 271n8, on the addition of πρὸς before τὸν πρὸ τῆς τροφῆς, based on the Arabic text and the sense.
257 See *De usu partium* VIII.3 (III.620 – 625K; Engl. transl. May 1968:389 – 393; see particularly 390n14). Also Furley and Wilkie 1984: 271n10.
258 *De placitis Hippocratis et Platonis* II.8 (V.279K).

imal in which the arteries were cut off by ligatures to run about. **[155K]** When it ran well for a long time but was not able to do so indefinitely, we thought to seek the reason for it running for a long time, for it should not hold out for a long time but collapse immediately, once the psychical *pneuma* was used up.[259]

But the *rete mirabile* (net-like plexus), as it is called by the followers of Herophilus,[260] seemed to be the cause of this very thing. For in that place the carotid arteries, passing to the brain, are divided into many branches before coming to the hard membrane (dura mater), interweaving in many layers like a number of nets lying one on top of another and occupying the large space called the base of the brain, it being possible for them to come immediately out of the meninges and implant themselves in the brain, which they were going to from the beginning. Thus, the remarkable plexus, which was placed there in a such a safe place by Nature, who does nothing in vain, seemed to be indicative of some major use. But since we saw the convolution of the intestines and that of the vessels growing into **[156K]** the testes for the sake of perfect concoction of the contained materials, and besides, that this existed for an abundant provision of subsequent functions, it seemed reasonable here too for Nature to have contrived such a thing to work up over a long time the material in the arteries, which is hot, thin and vaporous blood, so providing abundant nutriment for the psychical *pneuma* in the brain. And because of this, even if the brain is deprived of connection with the heart, the *rete mirabile* is sufficient for it for a long time, and particularly when the animal is at rest, as it is not then using up the psychical *pneuma* in voluntary function. So, what seemed most difficult and incomprehensible, which was the problem of the carotid arteries, doesn't seem to be a problem still.

Similar to this also, when cutting off of each of all the other arteries with a ligature, the part suffers no immediate harm. For I think it behoved us to consider here also that it is not the same **[157K]** if the actual source of the innate heat is affected, or one of the parts heated by this source. For there must always be sufficient heat [in the heart], as it moves both itself and other parts in a pulsatile fashion at the same time and heats them. These other parts, however, need only a small provision of heat for their preservation. But if you deprive the heart of the innate heat, you will cool it itself and all the other parts that were previously heated by it. However, you will not be able to deprive any of the other parts of heat completely, even if you cut off the arteries with ligatures; something will flow through the arterial walls. For the whole body also breathes and flows in itself, according to Hippocrates' theory.[261] As a result, some of the heat will be carried to every part, if not through the arteries, then through the other parts, and particularly through veins by the continuity of the participation.

---

259 See Furley and Wilkie 1984: 271n13 who have a detailed note on the deficiencies of the Greek text here and supply a more satisfactory account based on the Arabic.
260 On the rete mirabile in particular, see Harris 1973: 354 where the relevant references to Galen are given (*De usu partium* and *De placitis Hippocratis et Platonis*). See also von Staden 1989: 158–159, 179.
261 Hippocrates, *Epidemiae VI* 6 (ed. and Engl. transl. Smith, 1994: 260–261).

All such things, then, agree with the observed phenomena, and it will not seem remarkable for the source of the heat to be harmed more than the other parts, if deprived of respiration. **[158K]** For also, in the medical cupping glasses, the flame is destroyed at once, but the heat remains longer both in the contained air and in the body [of the cupping glass] itself,[262] although not nourished. In the same way too, even in houses heated by fire, the heat often manifestly remains, although the fire is extinguished.

Therefore, what is in accord with nature is not the same for the heart as it is for the other parts, for while the heart must always be seething, it is enough for the other parts not to be altogether cooled. You will understand clearly what is being said, if you expose the heart of an animal and take off its covering, which people call the pericardial tunic, without penetrating the rest of the thorax. For the animal dies very quickly if you cool the heart. If, however, you preserve the heat, it suffers nothing. You will, then, cool the heart, if you carry out the surgery in cold air and pour on cold water. You will preserve the heat for the longest time through the opposites. But if, when it is already cooled, and because of this the animal is dead, you choose to immediately cut open one or other of the ventricles, and particularly the left, and insert a finger into it, **[159K]** you will feel the large amount of heat, and that it is much more than in the other parts that are disposed in accord with nature. For the equal part of the heat which is least in the heart is very great in the other parts. Accordingly, whenever this is not preserved, that which is, as it were, the flame from the heart is dispersed, whereas that which is in the other parts remains for a long time. Respiration alone is preservative of the heat in the heart, whereas there is a twofold preservation in other parts—a pulse which stands in, as it were, for respiration, and that which flows from the source in addition. And so, for these two reasons, each of the parts gains a greater share than the heart in terms of not being quickly harmed by lack of heat. Also, because the heart needs a lot, while the other parts need little, and because there is nowhere from which it flows to the heart, whereas it flows to parts from the heart, the latter readily departs from an accord with nature, while the former do not.

3. Concerning the fact that the heat flows from the heart to the parts not only through the arteries but also through the veins and all the other structures, since **[160K]** this was stated briefly earlier, we must take it up again and must prove it, stating first the observations from which someone might infer this. It has often happened that many gladiators, generals, soldiers and hunters[263] are injured in the veins and arteries so that doctors are compelled to cut them off with ligatures. And all these are aware of the parts being colder after a short time; earlier if you cut off veins and arteries with a ligature, but later if it is only the arteries, and least if it is only the veins. It is clear from these things that some heat also comes to the parts through the veins, but

262 The body being referred to here seems to be that of the cupping-glass itself. See Furley and Wilkie 1984: 272n18.

263 There is some uncertainty regarding this list due to textual variation. “Monarchs” is included in Kühn but is probably an error and is not included in the translation. See Furley and Wilkie 1984: 273n20.

much less than comes through the arteries. If, apart from a wound, you wish to take some part of the body and bind it strongly, you will see this become immediately livid and cold, making it clear that it is deprived of the heat flowing down to it from above through all the parts.

Therefore, when this has also been shown, and it is apparent there is nothing against there **[161K]** being one use of respiration and the pulses, it is readily inferred that the pulses exist for the sake of preservation of the heat in each part. As a consequence, what is done by respiration for the heart alone, is done by the pulses for the heat in the whole animal, and what is common to both but more specific to the arteries is the concoction of the psychical *pneuma*, if at least we recall those things said a little earlier regarding the *rete mirabile* (net-like plexus). Accordingly, since it was shown in the work, *On the Use of Respiration* that there is cooling of the innate heat through inspiration,[264] while through expiration there is a kind of purification, with evacuation of the smoky superfluity, and that both these contribute to the preservation of this heat, it is clear that we shall also say in the case of the pulses, that in the dilatations some air-like substance is drawn in, while in the contractions there is expulsion of a kind of smoky superfluity from the burning up of the humors occurring in the whole organism. **[162K]**

4. And indeed I know that the followers of Archigenes, and some others before them, thought the arteries are filled in the contractions and emptied in the dilatations. They thought the contraction to be most suitable for drawing in, taking as evidence, as they say, particularly the contractions of the mouth and nostrils in the inspirations and their dilatations in the expirations, which is seen to occur in the case of those weakened in respect of this capacity. But in the case of these, this does not occur in any part of the nostrils other than the most distal and cartilaginous. However, this is also seen sometimes in those who are running quickly, or are exercising vigorously in some other way, but not in anyone else, either healthy or diseased. It is, however, necessary for the natural state, whatever it may be, to appear free from disturbance and perfect in those who are healthy more than in others. But suppose this occurs in all similarly, and suppose the nostrils and lips appear to contract in those breathing in—what belief in the enquiry arises from this? For they will certainly not say **[163K]** the lips and nostrils are analogous to the arteries, but only that there is an analogy between the terminal parts of the arteries and these structures, whereas analogous to the arteries themselves are the pathways of the *pneuma* (air), as it were, from the lips and nostrils to the heart.

If, then, they are able to show those contracting in the inspirations, it would be a better example for them. Otherwise, in addition to not being helped by it, they are brought by this to oppose themselves.[265] We shall say, just as the pharynx and the

264 See *De usu respirationis* 8 (Furley and Wilkie 1984: 253n34).

265 See Furley and Wilkie 1984: 277n26, for a discussion of the textual issues here, including consideration of the Arabic version.

lungs and the whole chest expand in the inspirations, so it is necessary for the arteries to expand in this way at the time they draw in and not at the time they expel the air (*pneuma*). But also, the pause of the arteries after the contraction is of much longer duration than that after the dilatation, just as also that before inspiration is greater than that after it—this shows that in regard to this, there is some analogy between the pulses and respiration. Anyway, what inspiration is to the organs of respiration, so dilatation is to the arteries, and what expiration is to respiration, so contraction is to the arteries. **[164K]**

The heart directs this same double and compound movement of the arteries, just as we showed in other places,[266] and was shown by many others before us. But it is not in the manner Erasistratus supposed, but as Herophilus and Hippocrates did,[267] and almost all the most notable doctors and philosophers of ancient times. For the power in the body of the heart, by which it dilates and contracts, flows down through their walls to all the arteries, and in this way dilates and contracts them, just as the heart itself does. Thus, as the heart is dilating, it draws to itself what is adjacent to its openings, whereas it expels when contracting. In this way too, the arteries, when they dilate, draw in from all directions, whereas when they contract, they expel in every direction.

5. Now I shall define for you more clearly what "from all directions" and "in every direction" mean. There are many pores in the arteries; some are openings, as it were, in the actual walls, while others penetrate into the intestines, stomach and the external skin. But also they are continuous with each other and **[165K]** with the heart by very large channels and by the arteries themselves, particularly all those with a broad lumen. They are, however, not in fact continuous with veins by large channels, but by the *synanastomoses* of these which escape the senses.[268] Consequently, even if you have good reason to doubt that these exist, you would perhaps be convinced by the other things said by the ancients, and not least through this particular observation: if someone, taking any animal whatsoever of those having large and obvious veins and arteries, such as an ox, pig, ass, horse, sheep, bear, monkey, leopard or man himself, or one of the others of a similar kind, and wound many large arteries, you will empty out all the blood of the animal through these. We have often made trial of this observation, and have always found the veins are emptied out along with the arteries. It is from [these] observations we were persuaded that the doctrine of *synanastomoses* is true.[269]

At any rate, it is through these anastomoses that arteries draw from the veins in the dilatation and squeeze out into the veins in the contractions, just as they eliminate

---

266 Galen, *De placitis Hippocratis et Platonis* VI.3 and 7 (CMG V.4.1.2, Engl. transl. De Lacy, 1978).

267 For a detailed account of the views of Erasistratus and Galen, see Harris 1973: 177–233.

268 See Furley and Wilkie 1984: 274n31, for a detailed discussion on the textual issues here and an alternative reading. Above is an attempt to translate these two problematic sentences following the Greek text as given. On the matter generally, and this passage in particular, including the term *synanastomoses*, see Harris 1973: 280.

269 On this point, see Furley and Wilkie 1984: 274–275n32.

through the openings ending in the skin **[166K]** all the vaporous and smoky superfluity and take into themselves in exchange no small part of the ambient air. And this is what is said by Hippocrates: "It is the whole body which breathes in and out."[270] In the same way, the arteries draw from the stomach and the intestines and in turn eliminate. In this way too, from the spaces surrounding, in part they draw in and in part they eliminate out through the kind of holes in all their walls. They receive much from the heart but give less. The cause is the outgrowths of the membranes, about which, since Erasistratus discourses sufficiently, it is now superfluous to write.[271] But he seems to assume that nothing at all is transferred to the heart from the arteries, except in fact through those in the lungs.

But this is not so. For perhaps also, since the animal is governed by the actual law of Nature, a little is transferred. For the membranes do not seem to me in this way to bar the passage so precisely in respect of the mouth of the great artery (aorta), that **[167K]** nothing runs back from the aorta into the heart. If it does not, but in fact if some violent circumstance has taken hold of the animal, it is essential for it to happen in this way. This too we demonstrated through other writings,[272] and there is not a great need of this in the present discussion. For if the arteries do give something to the heart also, they could in this way draw from all sides, and in turn send back. However, if not, they will draw in from all directions and send on in every direction except to the heart. This doctrine seems to me much better than the suppositions of Erasistrateans. For the body is unable to breathe or flow together with itself if the arteries draw in from all directions, but do not send out in every direction. Furthermore, the use of their function would in this way be better extended to the whole animal. For in this way every part individually will be able to be cooled and purified, being controlled by the different movements of the arteries.

However, Erasistratus assumed that **[168K]** the arteries in animals serve the action of lifeless conduits, and not of living organs. Also, through another whole book, we have shown in various ways that blood is contained in the arteries themselves, even when the animal is in accord with nature. If this is so, it is somehow clear to everyone that it is not by being filled with *pneuma* (air) sent out from the heart, as Erasistratus thought, that the arteries are dilated more; it is because they are dilated that they are filled. For if they were empty of blood, then it would perhaps be possible in a short time for the flow from the heart to come to the extremities. If, however, they have blood in them, it would in no way be possible for the speed of the movement to agree with the dilatation, due to their being filled from the heart. So, it is not that

270 See Hippocrates, *Epidemiae VI* 6 (ed. and Engl. transl. Smith 1994: 260–267), and Galen, *De usu respirationis* 5 (IV.501–511K).

271 Furley and Wilkie 1984: 275n35 have the following note on this: 'More than once Galen compliments Erasistratus for his work on the heart valves (*De placitis Hippocratis et Platonis* I.10 [V.206K] and VI.6 [V.548K]). The existence and function of the valves was first discovered about the time of Erastistratus, if not by Erasistratus himself.

272 *De usu partium* VI.16 (III.487–492K). See May 1968: 1. 319–321.

they are filled and because of this dilate, but that they are dilated and because of this are filled. It has, in fact, been often and comprehensively demonstrated by us, and by many of those before us, that the arteries themselves function, as the heart also does, contracting and dilating in turn in relation to the same **[169K]** capacity in that, which we described as originating from the heart and being transmitted through the arterial walls.

6. Those who look into whether the dilatation, or the contraction, or both must be thought of as functions of the arteries, seem to me to do better, just as I think it was reasonable to enquire, in regard to the parts of respiration, whether expiration, or inspiration, or both must be supposed to be functions. But the probabilities about those were spoken of in other places;[273] now we must look into the movement of the pulses. If therefore, as was said, respiration and pulses have arisen for the same function and involve the same demiurgical powers, it would be easy to conjecture something about the pulses from the similarity to respiration. However, since the physical (vital) capacity of the pulses is the mechanism arising from the heart, while that of respiration is, as we showed, the psychical capacity from the brain, none of those things discovered about respiration would clearly help us in the more difficult discovery of the present matters,[274] and why **[170K]** we don't see the arteries collapse in on themselves after death as the veins do. For the veins, whenever they are emptied of blood, completely collapse in on themselves, so their walls above fall on those below. On the other hand, the arteries seem to remain separated (i. e., with their lumen preserved) throughout, due obviously to the different hardness of their walls. And yet some say this very thing does occur after death, since they are solidified by the cooling and are not naturally like this. Others, having put the arteries into hot water and seeing them still expanded, persuade themselves that they are also in this state before death, and that when the cold is taken away, they would return to their original nature, if in fact they had been made to change by the cold.

Since, therefore, the matter under consideration is problematic in all kinds of ways, I shall actually state on what evidence I think that both the movements of the arteries are actions. It is clear to all beforehand that when the capacity is still strong, the dilatation becomes large, just as conversely, when it is weak, it is small. **[171K]** It is necessary, if, as some think, the contraction of the arteries is an action, whereas the dilatation is an automatic return to the natural separation of the walls, first for the swelling of the dilatations to always be equal, and then that the size should no more be an action of a strong capacity than a weak capacity. But neither of these is so.

And there is another not insignificant proof pointing to the same conclusion. The largest and highest pulses occur in those who are going to pass through the crisis [of a disease] best. And indeed, under these circumstances, no one, even if mad, would say the capacity is weakened. For if the crisis occurring well is due to deficiency of capaci-

---

273 *De usu partium* VII.20 (III.594K).

274 See the detailed note on textual issues in Furley and Wilkie 1984: 276 – 277n43.

ty, obviously it will occur badly due to strength. If this were the case, we shall also say death itself is an action of strength. What, then, could be more laughable?

But also, the strength of the pulse-beat in the dilatation is more in some but less in others, indicating actions of the pulses to be stretching and loosening (straining and relaxing). If, however, the dilatation were the lesser of the actions, just like expiration, there would not be more and **[172K]** less in this, just as there is not in expiration, nor the whole *tonus* and strength of the pulse-beat. Therefore, the dilatation, due to these things and others similar to them, seems to me to be of the class of actions (functions), and this would again preserve the similarity of the pulses and respiration.

7. What must be considered next is whether, just as expiration is a relaxation (*anesis*) and, as it were, a pause of the action of the chest, while exhalation is an action, and because of this has degrees of magnitude, whereas expiration does not,[275] in this way also the contraction in the pulses is a relaxation (ἔκλυσις) of the action of the arteries, whereas there is something else alongside this analogous to exhalation of breath, which is the action. And this seems to me to be true above all. I form my judgement from these things in common concerning all the physical (vital) powers in us, as I have shown elsewhere,[276] and from the analogy with respiration itself. And in fact, in the conclusions about the powers in each **[173K]** of the organs, it seems they have opposing innate powers. Again, the resemblance to respiration requires the action of the arteries to be analogous to the exhalations. For it would be very strange, or even impossible, for there to be some need in nature of the contraction of the arteries, but no innate capacity were to be given to them, as the demiurge of such a movement.

It is, then, much more reasonable, as I have shown in the writings, *On the Difficulty of Breathing*,[277] that whenever the smoky vapour collected from the burning up of the humors is considerable, then the animal gives itself over to the exhalations, whereas whenever, either because it is *euchymous* or due to the moderation of the heat, no such superfluity is produced, it only breathes out, and in the arteries the analogue to the exhalations is contraction in the excesses of such a superfluity. The other contractions occur when what is emptied out from the arteries is more like vapour than smoke. Clearly, there is much evidence from those things observed in the pulses—an immediate example is those who are sleeping **[174K]** after they have eaten abundantly. For in these the dilatation is weakened, becoming smaller and at the same time also slower, whereas the contraction is increased in both ways, for it appears more rapid than before and to go down inward still more. That these things happen in sleep seems reasonable, and at the same time analogous to the expirations. For the movement of the innate heat is more inward than outward, and because of this

275 The distinction being made here would seem to be between normal, quiet expiration (ἐκπνοή) and a more forcible emission of breath of greater volume than normal expiration (ἐκφύσησις) signaled in the translation by expiration and exhalation respectively.

276 In Furley and Wilkie 1984: 277n46, the reference is taken to be to *De naturalibus facultatibus* III.6 (II.160K).

277 *De difficultate respirationis* I.7 (VII.769K).

much collects around the internal organs and the stomach, working on the humors and the foods, and of necessity the superfluity becomes greater in amount, requiring natural expulsion. Also on this account, in the respirations, the expiration is quicker and larger, and, as is common in those sleeping, is accompanied for the most part by an exhalation, and particularly when they have eaten their full. In the pulses, the contraction is disposed in the same way. And certainly, in this way in children, both claim a greater share, for the working up of the humors is great in that time of life due to growth. Conversely, **[175K]** in old age the contraction seems to be very slow and very small, because this stage of life is faint and weak regarding the concoctions and works up the humors least, as this is not essential.

Analogous to these, they say the seasons, places, and in short, all the changes of the ambient air towards cold and hot change the pulses. In the same way too, each of the psychical affections, both those of the psyche's normal activities and those of its diseases increase the movement of the arteries, either outward or inward. And indeed, it must be the case, if the dilatation were always some relaxation; first, in no way would this become either faster or slower than it was; and secondly, the limit of the movement would always be the same, which the arteries have acquired from the natural constitution. However, that they move further than this is an indication of the action occurring at that time.

But they say it is not by functioning that the artery goes inward quicker or to a greater extent; rather it is because of the softness of the walls. For such walls naturally **[176K]** fall in on themselves more and more quickly than those that are hard, when the dilatation is ceased. But this, we might say, my good fellows, may be consistent with the change of the rhythms related to the stages of life. And indeed, I shall show a little later how this differs. However, it is clearly at variance with very many other conditions in which the change of the rhythms occurs suddenly, when the arterial walls could not, in so short a time, change greatly towards softness and hardness. You are, to no small extent, deceived by what was properly said by doctors of ancient times, having failed to understand this appropriately. For soft arterial walls are more ready for the movements when the capacities (powers) function through them—one might say more easily led by what they might wish (more ductile). However, when they are moved by those capacities, and they allow themselves to return again to their own (natural) condition, those that are hard return quicker than those that are softer, just as the harder plants also **[177K]** yield with greater difficulty than softer plants to those things that are compelling and scarcely follow those that are attracting, but when released return more easily to their erstwhile condition.

But also, this is what reason would point towards prior to the occurrence of what is observed in this way. For there is some specific *logos* of constitution in each of the bodies which is painless, and as Hippocrates said, never rests, and which clearly lies in the middle of the immoderate movements of this and is not displaced from this without some effort. And the labour is greater in both the body being moved and also particularly under the circumstances, in the moving force (capacity), when it is hard and dry. For what is constituted in this way has a stronger state (ἕξις)[278] than what is softer and more moist. And by the amount that each of the things that is going to be affected is stronger, so it becomes more difficult to move by what is disposed to move it. If that which is being moved is affected by what is moving it, and this functions and acts on what is moving it, then it would be reasonable that what is harder is more difficult to affect to the extent that it is also more difficult to move. **[178K]** And because of this, when either any plant whatsoever or one of the organic parts of the animal is moved by some power, what is hard would follow less as it is more difficult to affect and more difficult to move but would be easier if it were softer. However, what is hard returns more rapidly to its normal condition (κατάστασις) than what is softer, as it is constituted with a stronger state (ἕξις). For it is apparent in plants that those which are tender and young, when they are bent, are slower in returning to their natural state, whereas those that are fully grown and hard return with greater *tonus* and greater swiftness of movement.

Accordingly, in addition to this, as I was saying, it overthrows their argument in respect of the suddenly occurring changes of the rhythms, and the hypothesis does not otherwise seem to agree with the observed phenomena, since it was shown that the softer and weaker body more easily follows the moving agent but returns more slowly to its own nature, whereas the harder body, due to συντονία (tension),[279] is scarcely overcome, and on **[179K]** this account, follows more slowly, but when it is released hastens back to its proper nature. In summary, we may say both things happen and affect in turn soft arteries and hard.

8. I have now said everything about the use of pulses—that it is for the purpose of the natural heat in the whole animal, and that this is cooled in the dilatations but purified in the contractions, and that the movements are in every way like those in the respirations, and that they are useful to the psychical *pneuma*, and that in one respect only do respiration and pulse differ. This is, one occurs due to the psychical capacity, and one occurs due to the physical, whereas in all other respects, both for the purpose of use and in the manner of movement, they are similar, as has been stated and demonstrated. For someone who has read this carefully, it is possible to judge very easily what has been well said by those before us **[180K]** and what has not. And it will be easy for such a person to discover whatever is sought of the actions about both individually.

278 Here as elsewhere the Greek term is included in parentheses to make apparent the distinction between ἕξις/*hexis* and σχέσις/*schesis*—a stable and enduring state as opposed to an unstable and temporary state.

279 The meanings given for συντονία in LSJ are: tension, intense application on exertion, intensity and agreement. The term is found, for example, in Aristotle's *Historia animalium* 540a6 (ed. and Engl. transl. Peck 1970: 103). Peck renders it 'rigidity'; Furley and Wilkie 1984 have 'springiness'.

4 *Synopsis of the Pulses* (*Synopsis de pulsibus* IX.431 – 549K)

There are 34 sections in this treatise, summarizing the four long treatises:

1. Introductory remarks. Galen considers the place of synopses in learning and states why he doesn't in general make such synopses. It is better for readers to make them for themselves. In the case of the pulses, he feels compelled to write his first synopsis because of the errors others have made in doing so. In this he covers, in a synoptic fashion, the material in the four long treatises.
2. A brief statement on the names/matters issue and a comment on homonymy.
3. On the 27 possible combinations of the variations in the primary movements of the pulse (dilatation) in the three dimensions of length, breadth and height, with a table. He identifies those which are called strong (ἁδρός/*adros*) and weak (ἰσχνός/*ischnos*) and considers the terms *eutonic* and *atonic*.
4. It is not possible to determine by palpation what is contained in the artery, either in amount or kind. The use of the terms "full" and "empty" receives critical scrutiny.
5. On whether the contraction is perceptible or not. Galen's position is that the contraction is perceptible in some pulses but not in others. Considerable training and experience is necessary to recognize the contraction.
6. On the remaining *differentiae:*
   - rhythm, comparing the times of dilatation and contraction with each other;
   - equality/inequality as destruction of sequential equality in one pulse or more than one;
   - regularity/irregularity, the latter when equality doesn't occur in a cycle.

   He also considers the importance of perceiving the contraction in assessing frequency and infrequency, particularly for prognosis.
7. How to palpate in order to judge the amount of the dilatation, and the *tonus*, and to perceive the contraction. It is not possible to perceive the whole time of either the dilatation or the contraction. On the importance of having assessed the pulses of patients when they are healthy in order to evaluate their pulses properly when they are sick.
8. On large and small pulses understood as relating to the amount of the dilatation. On some differences between Herophilus and Archigenes, and on errors of the latter regarding these parameters.
9. On the causes that create (*synektic*) and change (*proegoumenic* and *prokatarktic*) the pulses. Issues of nomenclature. Dilatation is for cooling, or fanning and cooling the innate heat; contraction is for expelling sooty and smoky superfluities. Consideration is given to specific humors.
10. On the use/need (χρεία/*chreia*) of the pulses. If heat is concentrated in the body of the animal, there is greater need of dilatation; if the sooty and smoky superfluities are increased, there is greater need of contraction. In the latter case, there is rapidity without increase in either magnitude or frequency.
11. Reiteration of the importance of having established the normal pulse when the patient was healthy. Factors other than diseases which affect the pulses (i.e., either in accord with nature or non-natural, as opposed to contrary to nature). These include bodily status, age, nature, seasons and places.
12. A return to rhythms. Galen speaks of Herophilus' comparison to music and *arsis*/*thesis*. A brief account of the views of the Herophileans on pulse rhythms.
13. The four times of the pulse:
   - the major part of the dilatation;
   - the external pause, which may be imperceptible;
   - the contraction, which may be partially or completely imperceptible;
   - the last part of the contraction, the internal pause, and the first part of the dilatation.

   All these times are lengthened with the progression of age. The clinician should palpate the pulses in many healthy people in the various age groups.

14. It is not enough just to compare the time of the dilatation with that of contraction, as with inspiration and expiration in respiration. It is the changes in each of these that are important. A brief comment is made on the terminology applying to the superfluities. On terms: a *pararhythmic* pulse has a slight change of rhythm—i.e., it is close to a normal rhythm; a *heterorhythmic* pulse has a greater change; an *ekrhythmic* pulse has the greatest change.
15. On differences which are in the category of "in accord with nature" (e.g., gender, hotter nature, thinness, age, seasons and places).
16. What to do when called to see a patient not previously seen:
    - establish age, season, place and climatic conditions;
    - place a hand on the patient's chest to evaluate heat.

    A comment on Themison and the diagnosis of fevers.
17. Two classes of pulse inequality:
    - in a single pulse;
    - in a collection of pulses, termed "systematic".

    Four *differentiae* of the pulse in the dilatation: large/small; rapid/slow; strong/weak; hard/soft. Inequality may occur in the first three, and also in frequency/ infrequency. Detailed consideration of the causes of inequality.
18. On inequality of a single pulse. Two forms are recognized:
    - movement divided by a pause;
    - continuous but not uniform.

    On the first, it can't occur in a small or low pulse. There is consideration of the combinations and sequence of strong and weak in the two components.
19. Galen lists the nine combinations: equal/equal, moderate/strong, equal/moderate, moderate/moderate, weak/strong, weak/moderate, strong/weak, moderate/weak, weak/weak. Analogous combinations relate to rapid and slow.
20. It is necessary to look at both *differentiae* (i.e., strong/weak and rapid/slow). An example is the bounding pulse in which the second part of the movement is more *eutonic* and rapid than the first. Mention is made of two prognostic indicators for death or survival—strength of capacity and concoction of the disease.
21. A long section on how many pulses appear in relation to dilatation, from what conditions they arise, and the prognosis from each. Consideration is given to the methods of palpation for assessing dilatation—i.e., varying degrees of compression. On what the findings indicate regarding the nature of the disease and the likely outcome, taking into account the two main determinants mentioned in the previous section.
22. Differences/*differentiae* in pulses due to the hardness or softness of the wall of the artery don't appear in either of the two inequalities—i.e., in a single pulse or a collection (systematic). However, when need is pressing and capacity is strong, compelling dilatation, if the artery is hard, a clonic pulse results. A dicrotic pulse also may arise. Galen considers other relevant factors. He also describes a tonic pulse. He considers the cause of an undulant (wave-like) pulse. When the undulant pulse is weak, it becomes vermicular. Neither vermicular nor formicant pulses are seen as systematic inequalities. He gives details of the formicant pulse.
23. Pulses with inequality in a single beat. First, there is the tapering pulse (mouse-tailed). Galen considers the method of palpation to detect this and gives other variants: recurring tapering and remittent recurrent. The *synektic* cause of frequency is deficiency of dilatation. He considers the causes of deficient dilatation.
24. A brief section on nodding and inclining pulses, giving the presumed causes of these.
25. Galen returns to his Table to consider the 7th and 20th pulses (long/narrow/high and short/broad/moderate) and discusses the genesis of these, taking into account magnitude, rapidity and the wall of the artery.

26. Since to go through all the pulses in the table would be too long a job. Galen offers some general considerations regarding rotation of the limb and placement of the fingers. He considers palpation of the pulses in the different kinds of dropsy and in lethargy.
27. Galen considers the long pulse (e.g., palpation using six fingers). It will occur in a thin and soft person. Look for the features of marasmus. There may be other related conditions. He considers the combinations of long and hard, and high and short.
28. Galen poses the question of why pulses become frequent from moderate harms of the capacity along with heat, although not small and slow. He lists four causes of change.
29. Galen considers changes in a single pulse which occur in fevers, both ephemeral and hectic. Neither a large nor a strong pulse is seen in the latter, so these features of the pulse signify an ephemeral fever.
30. A list of crises and other events which are presaged by particular pulses.
31. On the kinds of pulses associated with affections in different parts of the body. These include the organs of respiration and nutrition, the bladder and uterus, the brain and meninges, the muscles, and the testes.
32. Changes in the pulses due to external causes, including bathing, running, massage and other movements, intake of food, personal customs and habits, and affections of the soul.
33. A review of frequency and infrequency. The former is encouraged by hot superfluities, particularly when abundant, and by heat generally, whereas infrequency is favoured by external quiet, reduced superfluities, and cooling. There is division of quiescences of the pulse into those in which the pause exceeds the time of two pulses, and when it is less. Galen offers some comments on the prognostic significance of intermittent pulses. Extreme frequency is associated with syncope.
34. A summarising statement. As for prognosis, the aim is to know three things:
    - will there be death or survival;
    - over what time will this course be run;
    - what will the mode of death or survival be.

    Galen lists the important determinants of these, both from the pulses and from other aspects of the clinical picture.

1. [431K] The majority of men desire accurate knowledge of those things they learn on each occasion, but they hesitate to go down the road that leads to this. And because of this, since they shun explanatory discussions that are detailed as being long, some read introductions, outlines or sketches, while others read epitomes, synopses or summaries. Then, sometime later, when they encounter men who dispute, they are unable **[432K]** to reconcile the things they say, and because they don't wish to learn the refutations of them, they are kept away from those things that are true.

Accordingly, they must be particularly aware that the detailed teaching has advantages over the others, in that it not only teaches the truth, but also places before them the demonstrations, counterarguments and refutations of these, whenever they depart from mistaken conclusions of such teaching. And they must set themselves apart from proximity to the discourse on those things of which they learned the truth. But they don't do this. And because of this, what happens is that throughout their lives, their opinions are tossed about like a boat driven onward by the waves. For the demonstrations are like anchors against the concepts being forcibly driven on and not being changed. Accordingly, I urge them to learn everything completely, either right from the start, or supplemented by some outline or synopsis being brought in as well. If they have learned completely, then let them then read the epitomes should they wish to be reminded, through brief statements, of those things they previously learned at great length.

Overall, I choose not to make an epitome of my treatises; I think it is better for people, **[433K]** when they have thoroughly read detailed accounts, to abridge these for themselves. In this way, the epitomes and synopses become useful, being written in relation to the specific state in each case. Since I am aware that others make incorrect abridgements of my treatises, by which those who hesitate to read the detailed accounts find themselves harmed, I am, because of this, compelled to come to this work myself, alongside the original opinion, and in fact I undertook a synopsis of the treatises on pulses first of all. The whole detailed description of this was divided into four parts. The first was that on the *differentiae* of the pulses; the second was that on their recognition (diagnosis); the third was that on their causes; and fourth that on prognosis. Thus, in the part on *differentiae* of the pulses, it is taught how one might discover by method the number of all the pulses. In the second part, it is taught how one might recognize each of these, for some of them are hard to distinguish, while some are in fact indistinguishable by palpation. The third part teaches how each pulse arises due to certain causes. The fourth, for the sake of which all these are practiced, is about prognosis by way of the pulses. **[434K]**

2. For each individual part of my four existing books, which are sixteen in all,[280] I shall now, as I said, make a synopsis of these in this work, beginning from the first, which pertains to the *differentiae.* In this, the introduction, placed first, shows it is better to devote one's attention to the matters themselves, giving scant regard to the names. I think it worthwhile to do this, not only in this particular treatise, but also in all the others. For when the discovery by method is from the essence of the matter, knowledge of the *differentiae* in this is secure, no aspect of the matters is left aside, and nothing is said superfluously. In those beginning from the names, there is obscurity, and something is either left out or added superfluously. For often one name signifies a number of matters, or one matter has a number of names. Some matters are not at all named according to the principle, due either to their not being known previously, or due to their being altogether few. There is, then, a danger when things are named homonymously, of some matter being left aside, **[435K]** in that the matter having many names is equal to the multiplicity of names. And because of this, rather than accepting the beginning from the names, it is better to begin from the matter itself and proceed to the investigation of the number of the *differentiae* in this.

3. Certainly the essence of the matter before us is a perceptible dilatation and contraction of a hollow and extended body. What doctors call this hollow body is an artery; what not only doctors but all people generally term the stated movement of this is a pulse. And if someone doesn't wish to call this a pulse, it doesn't matter to us. For we are not here to dispute about names, but to discuss how someone might prognos-

280 There are in fact four sections in each of the four treatises following.

ticate on future events from the perceptible movement of the arteries. Therefore, it makes no difference if someone wishes to call the movement of these a pulse, or a palpitation, or anything else whatsoever, if the prognosis from this is made correctly, which is what we hoped for from the beginning.

Accordingly, since there are certain things existing of necessity due to the movement of the arteries, we need to consider these first. **[436K]** And then next, we need to consider if one of the other things is associated, among those not of necessity. So then, in the movement of the arteries, in relation to the specific *logos* of this, there are two primary things: the intervals during which the movement is suspended (i. e., the pauses) and the times in which there is movement. Thus, the interval of the movement, since it occurs in relation to length, breadth and depth of the artery, will create three simple and primary *differentiae*. One relates to the length, another to the breadth, and a third to the depth or height—the last is named in either way. Therefore, if the pulse that now appears rises up to a height greater than accords with nature, it will be called high, just as also, if it increases to a breadth that is greater than accords with nature, it will be called broad. In this way too, if in length it falls among those that are greater than accords with nature, it is called long by those palpating it. And it is quite clear that the opposites to these are low (shallow), narrow and short. A pulse that is in accord with nature in each of the three conditions will be called moderate in relation to that dimension (distension). Accordingly, there are necessarily **[437K]** twenty-seven *differentiae* of the pulses in all, as tabulated below (Table 2).

Table 2: Twenty-seven *differentiae* of the pulses

	Length	Breadth	Depth	
1			High	Large
2	Long	Broad	Moderate	
3			Low	
4			High	
5	Long	Moderate	Moderate	
6			Low	
7			High	
8	Long	Narrow	Moderate	
9			Low	
10			High	
11	Moderate	Broad	Moderate	
12			Low	
13			High	
14	Moderate	Moderate	Moderate	

Table 2: Twenty-seven *differentiae* of the pulses *(continued)*

	Length	Breadth	Depth	
15			Low	
16			High	
17	Moderate	Narrow	Moderate	
18			Low	
19			High	
20	Short	Broad	Moderate	
21			Low	
22			High	
23	Short	Moderate	Moderate	
24			Low	
25			High	
26	Short	Narrow	Moderate	
27			Low	Small

[439K] So then, in this way, there are twenty-seven pulses in relation to some quantity existing in the three simultaneous dimensions (distensions); two of these have names which are agreed upon by all doctors. The first of these and the last have been written [in the table]; they are called large and small. However, none of the other amounts has a name. For neither those pulses called "weak" (ἰσχνός) nor those called "strong" (ἁδρός) show as one in the table;[281] they are predicated generically of many at the same time. Of these, the dimension of length might prevail over the remaining two; they call all these "weak". The remaining two of these, conversely, they call "strong". As a consequence, also, no. 5 in the table and no. 6, and in addition to these, no. 8 and no. 9, and no. 18 are called "weak". Conversely, of those in the table, no. 10, no. 19 and no. 20, and in addition, no. 22 and no. 23 are called "strong", so that these names are more generic and common than many. [Doctors], however, name these same pulses in another way; "thin" that which is weak and "thick" that which is strong. But this is not in relation to the **[440K]** three dimensions (distensions) of the moderate, which is the only one of these in the twenty-seven in accord with nature. We have a specific name, and we indicate this by a word, either saying it is moderate in the three dimensions (distensions), or intermediate between large and small, or in ac-

281 Greek ἰσχνός/*ischnos:* thin, dry, withered; weak, and ἁδρός/*adros:* thick, stout, bulky; strong, violent, powerful.

cord with nature in the amount of the dilatation, or moderate in the amount of the dilatation, or however otherwise we hope what is said will be most clear.

You will discover such *differentiae* in relation to the time of the dilatation not to be twenty-seven nor altogether many, but the *differentiae* are as many as are conceived in relation to one dimension (distension) generically. For just as in the case of those related to length, there are moderate, long and short; in relation to breadth, there are moderate, broad and narrow and in relation to depth, there are moderate, high and low. So too, in the case of the time of the dilatation, there is the moderate in this class, and the rapid and the slow. There is another class, in addition to those occurring, of *differentiae* of pulses, which is that relating to *tonus*, in which in turn *eutonic* **[441K]** and *atonic* pulses occur, and some third in between both of these, but not like those previously mentioned—in accord with nature and moderate. For *eutonus* is in this class of accord with nature but is increased after food and a drink of wine and departs from an accord with nature in anger. This same pulse we also call strong (σφοδρός), following the custom of doctors, since we know certainly, among the Greeks, "strong" is most customarily said of rapidity and *eutonus*. A fourth class of *differentiae* of pulses occurs in the dilatation of the artery in relation to the actual body of the vessel, when it seems harder or softer than accords with nature. Accordingly, there will also be three *differentiae* of pulses here: moderate, hard and soft.

4. It is not possible for us to conceive of or know by experience any other perceptible affection from the bodies being moved. Consequently, contained within the four classes mentioned are all the *differentiae* of the pulses in relation to the dilatation of the artery. With respect to what is poured into these arteries, it is not **[442K]** possible to recognize by perception what either the amount is or the kind, nor would the issue of the arteries being empty of blood be altogether brought to a conclusion, if in fact it was possible to say that not only do they contain blood along with *pneuma*, but also what kind it is—whether whey-like (serous) and thin, or thick and viscous, or intermediate between these, which is the blood in accord with nature. And certainly, some also claimed there to be a recognizable perception of this, just as in fact there also is of the *pneuma*—what kind in relation to the potency is contained in the arteries.

Some, however, still write of the more unusual nature of these, introducing an additional pulse to those mentioned, which they call "full" (πλήρης). And they make three *differentiae* of this class: the one they term "full", the one they term "empty", and midway between both, one that has not acquired a specific name. If, then, you should wish to refute these people, you should divide the argument against them into two parts. The first pertains to the signification from the fullness, while next there is that pertaining to the recognition (diagnosis) of this. For in each they have been shown to err to no small extent in relation to the signification, **[443K]** not only in this class of the pulses but also in the others, in the second and third books of *On the Differentiae of the Pulses*. Concerning the recognition (diagnosis), they err in the fullness as much as was said in Book IV of *On the Diagnosis of the Pulses*, and all the others before these. For the present, since I teach only what is useful to the art, I shall eschew any refutation directed against those who err in this. We conjecture about this in the epitomes and synopses, as I said. It is time now to move on to the other *differentiae* of the pulses that are useful.

5. We take up the major disagreement among doctors; that is, some think the contraction of the artery is perceptible while others think it is imperceptible. Those who think it is perceptible say the artery pauses for a short time between the end of dilatation and the beginning of contraction, analogous to the pauses of the respiratory organs after inspiration and before expiration. Similarly, there is also a pause after contraction and before dilatation, **[444K]** as also some pause of respiration occurs in the respiratory organs in the expirations before the inspirations. If, then, the times of the pauses are more than accords with nature, they term the pulse infrequent, whereas if they are less, they call the pulse frequent. So there are two infrequencies and frequencies, the one "external" after the dilatation and before the contraction, and the other "internal" after the contraction and before the dilatation. There are those who would say the whole time of contraction is imperceptible to them—for example, when they compare what the time is between the perceptible movements with an accord with nature. If it is less, they term the pulse frequent and if it is more, they term it infrequent. If, however, it is equal to an accord with nature, they term it moderate. I said enough in the first book of *On the Diagnosis of the Pulses* on the discord about this, saying that all the pulses are not perceptible, and showing that some are perceptible and some are not. In addition to these, I showed how someone might train himself in the recognition of the contraction. But I shall speak about these matters a little later. **[445K]**

6. Now I am going to go through the remaining *differentiae* of the pulses. First, there is that of rhythm, which shares common ground with the theory of the musicians. Second, there is that of equality (evenness) and inequality (unevenness). Third, there is that of irregularity (disorder) and regularity (order). The theory of rhythm consists of comparing the times of dilatation and contraction with each other. The theory of inequality pertains to any instance when there is destruction of sequential equality in one class of the pulses or another, or two, or more. Sometimes the sequential equality is destroyed, and because of this the movement seems unequal, as in the case of planets (wandering stars). However, sometimes some equality is found in the cycles, as in those periodic movements themselves, and because of this they have an order, as in the movement of the planets. In this way too, we shall speak of the movement of the equal pulses in relation to cycles, whereas on the other hand, irregularity exists when we don't observe equality occurring in a cycle through an equal number of pulses. For example, if after three large pulses, the fourth pulse seems small, and this in turn **[446K]** follows always, we will say such a pulse is regular but nevertheless unequal. For although the equality is destroyed, some regularity is preserved, whereas in the pulses where this regularity is not preserved, not only are these unequal, but we also term them irregular.

Thus, that we have used the names of the pulses correctly, just as we have preserved in the others the names customary among the Greeks, has been shown in the second and third books of *On the Differentiae of the Pulses*. And we agree that someone might wish to apply a name, preserving the matter of the *differentia* being explained in the discussion, for the sake of providing a detailed explanation and for clarity in teaching. However, the inequalities in relation to one movement were stated in the first book of *On the Differentiae of the Pulses.* Some of the *differentiae* of this remain continuously, while others are interrupted by a pause, which will be spoken about after these matters.

Now, however, I shall come to the discussion of the contraction, being among the most necessary regarding prognoses, since also the teaching regarding the frequent and infrequent pulse goes along with alteration and concurrent change involving the rhythms. Therefore, of the many things said **[447K]** in the first book of *On the Diagnosis of the Pulses*, the most useful will be spoken of here. It is through these that someone practiced in action will learn the force of the teaching, whenever he himself prognosticates on things to come in those who are sick, when each of the other doctors, guided by some prophecy and not by medical theory, is engaged in their prognosis.

7. Begin first from the recognition of the amount of the dilatation. For you, this will consist of making the application of the fingers to the surface and particularly from the parts of the artery. When you palpate from above, you may perhaps sometimes in some way weigh down on the artery and compress it, if you don't focus your attention very strongly so as to place the fingers accurately without any compression. However, the application from the parts below is particularly able to allow conjecture by not compressing the artery at all. To the extent that you do compress it, you will prevent the swelling of the dilatation, and most when the pulse is *atonic*, as in the case of those with lethargy. **[448K].** In those cases, if you compress the artery with the fingers, it seems very weak instead of strong. If, however, as I said, the physical capacity is extremely weak, the opposite occurs in the case of the *eutonic* pulses, which we also term strong, for the artery withstands the compression when the fingers are applied more or moderately, revealing the actual strength of the capacity for movement. Consequently, whenever the skin of the sick person is hard, and the wall of the artery is similarly disposed, the *eutonic* pulses overcome the flesh along with the skin and the applied fingers, just as also when some hard body is applied externally to a soft surface. For in this way certainly, one distinguishes the hard from the soft, because the hard goes down into the soft.

However, the simply *eutonic* pulse, which is without hardness, is overcome, whereas what is condensed and compressed is not. You will understand this a little more clearly in the case of wind falling upon a stone of moderate surface area. For the wind[282] overcomes us, as it often overcomes branches that are by no means small. A stone, however, **[449K]** when placed on the skin, condenses and naturally compresses it, while the whole limb is not able to overcome it. For the overcoming without compression of the area was left out in what has gone before, whereas it is likely that in

282 We presume that ἄνετος/*anetos* in Kühn's Greek text should be ἄνεμος/*anemos*. The Latin translation in Kühn has *venti*/*ventos*.

the condensation by what is harder, the object receives into itself the hardness without leaving the position which it had before.

In this way then, in the case of the artery also, distinguish the hard pulse from the strong pulse. For you will perceive the hardness, as with stone or wood, when what is impacted is overcome by the pressure, unless it is also strong. Conversely, however, the strong seems more *eutonic*, if you apply pressure.[283] Therefore, when you are practiced in judging these things, come to the recognition of the contraction, after first training yourself in the pulses that are large, strong and hard. For it is possible for you, by moderately compressing them, to have the perception of the first part of the contraction. To the extent, then, that there is the likelihood of your overcoming the skin, and as it were hollowing out the impacted artery, it follows that this recedes to its natural seat and condition. As a result, you will perceive, **[450K]** at such a time, the contraction in the amount to which this occurs. However, everything else of this will be indistinguishable to you.

After training yourself beforehand in such pulses, change to those that are large and strong at the same time, even if they are not also hard. And then change to the strong, even if they are not large. However, you must have a soft touch to perceive well naturally, for such a touch is overcome quicker by the artery falling upon it forcefully, and there is greater perception of the affections in this. Still more is said, in the first book of *On the Diagnosis of the Pulses*, on how one might particularly become practiced in the recognition of the contraction. If, however, you never trust yourself to perceive the first parts of contraction accurately, but they seem to you absolutely indistinguishable, or are difficult to perceive due to there being hard skin, you should know that many prognoses will be harmed by this, which is something you learned from the preceding discussion. Nevertheless, I offer you some encouragement regarding the use in the case of these actions. For you will need to store away in your memory how great the whole time in which you don't perceive any movement is in each stage of life. **[451K]**

Since there is no little difference in people, between each other, in the amount of time of the pause, I think it worthwhile for you, in those things that are customary, to expect to have considered them at a time when they are diseased, having often been palpated at time when they are healthy, and in which the body has no external movement from walking around, running, bathing, wrestling, eating, drinking, anguish, or some anger or fear, and to have stored in the memory the amount of time of the pause, and to compare with this what is apparent on each occasion. If, then, the amount of the pause is greater, call the pulse "infrequent"; if less, "frequent"; and if equal to the natural state, "moderate." However, just as it is not possible to perceive the whole time of the contraction, so too it is not of the dilatation. And this was shown in the first book of *On the Diagnosis of the Pulses*. But it is, in fact, possible to recognize the rapid and slow pulses from the manifestation of the dilatation. And

283 There is an erroneous repetition in the Greek text here, not reflected in Kühn's Latin translation.

this was also spoken about at the beginning of the second book of *On the Diagnosis of the Pulses*, and at great length in the third, in which I also showed the teaching of the rhythms to be useless. For it is enough for us, in reference to the **[452K]** prognoses, to recognize only the fast and slow movements of the artery, just as also how great the time of the pauses is, in relation to which the pulse becomes infrequent and frequent. You will learn this through what follows; here my discussion will be about the prognosis from the pulses.

8. Now I shall go over the large and small pulses, and those others that are understood and recognized in relation to the amount of the dilatation, adding as much as is useful, and will pass on to the causes of these. For truly, in the case of these pulses, there is what is taken to be common and moderate in the amount of breadth, and what is specific to each person. In the common, the moderate will be in relation to whatever dimension [is being considered] and will be recognized by the swelling of the artery when it is measured. I call now "swelling" the actual circumferential outline. Similar to this also are the excesses and deficiencies compared to the moderate in each dimension. What is specific in relation to each person comes from prior knowledge and recall of the palpation in that person. Something was also said about such pulses **[453K]** in the second book of, *On the Diagnosis of the Pulses.* Nobody before me wrote about this matter, which I divided off and treated specifically by itself, although a few dealt in some way with subordinate issues on which they disagreed among themselves.

Anyway, Herophilus says, the pulse of the child is sufficient in terms of magnitude, whereas Archigenes says it is small. In this way too, Archigenes says the formicant (ant-like, feeble) pulse is rapid whereas Herophilus says it is not rapid.[284] And the causes of the errors were stated in relation to that matter, so that you must direct your attention all the more to this, if you wish to diagnose and prognosticate on the *differentiae* of pulses—as many as you saw us do. For nothing in the art needs more time to practice, nor has greater force, than the diagnosis of the pulses. For when Archigenes was shown to be in error, not only in the recognition of the rapidity, but also in that of the magnitude, what hope could you have of knowing about these, without first becoming thoroughly versed in theory in the four books on diagnosis, and later, by practice in these, **[454K]** learn what is true? And then, when you do this, I shall tell you the chief points of what was said in those books, just as in the case of the previously mentioned *differentiae*, and in like manner also of the others.

For surely, since the movement of the dilatation is perceptible to all, whenever you touch the surface encompassing the artery without force, and more so whenever you touch the parts from below (i. e., with the limb rotated), the recognition of the magnitude is accurate. It is not possible to know securely and accurately whether [the artery] was moved strongly or weakly, and equally or unequally, in relation to the previously

284 On Herophilus' views on the pulses, see von Staden 1989: 262–288 (on vascular physiology) and 322–361 (on pulse-lore). Archigenes' views are given consideration in this and the other Galenic treatises on the subject.

mentioned application of the touch. But you must compress the artery gently, for in this way you will perceive the dilatation over a longer time than when you apply the fingers to it without force. For the recognition of neither speed nor slowness appears clearly without brief compression of the skin overlying the arteries. By your doing this, sometimes it seems to you to be moving at a uniform rate and sometimes unequally (unevenly), and at times the first parts of the movement, or the middle parts are made slower, and sometimes the converse. In this way too, sometimes the **[455K]** final parts are more rapid, and sometimes those before these, or those first of all. Furthermore, the same argument applies to the stronger and weaker. And accordingly, also, you will perceive the rising up of the artery interrupted by a pause, as it seems to stand still, and the second impact after the stationary period is sometimes stronger or more rapid, but sometimes weaker and slower than the prior movement.

However, there are also many *differentiae* in this class individually; you have them all in the first book of the treatise I have written on the pulses. But here too I shall attempt to state what is useful for you and at the same time those things indicated by these as briefly as possible in the discussions that follow. Similarly, also, concerning the other matters, such as are left aside now but were sought with greater care by doctors, you will hear of these in the prognostic part of the treatise. For the present, it is enough to say as much as would enable someone who reads this book prior to the large treatise, regarding the actions of the art, to need neither the first part of the treatise on the *differentiae* of the pulses, nor the part on the causes, **[456K]** for he will have the force of these included here.

And certainly, the diagnostic and prognostic treatises are most useful in terms of a detailed description of the actions of the art when they are read, and not the epitome. It is good, then, for you not to place your confidence in this teaching alone; it is useful to have it to hand for the recall of things that have been carefully learned in detail from those treatises. Concerning the diagnosis of the rapid pulse, I showed Archigenes erred in this, and that it is not possible to make a firm diagnosis without reading the work, *On the Diagnosis of the Pulses*, (and not just casually). In these treatises it is shown that it is necessary to make a distinction of rapidity from slowness not merely and only from the time of the movement, but also to consider the amount of the rising up. If you direct your attention to the time alone, you will think the so-called formicant pulse is rapid, when it is not rapid but of short duration, inasmuch as the dilatation is very small. For it is not possible to go through a short interval in a long time, even if that which is moving is slow. Anyway, a tortoise covers a *cubit* in less time than a horse covers a *parasang*[285] **[457K]** but there is not equality of the rapid movement of both animals.

285 The comparison is between the following two Greek terms for length/distance: πῆχυς/*pēchus* (the distance from the joint of the elbow to the middle finger) and παρασάγγης/*parasangēs* (parasang) (30 *stades*).

You must, therefore, accustom yourself to recognizing in this way shortness of time also. Concerning this then, as I said, I have spoken at greater length in the third of the books on diagnosis, and in the first part of the second book. I have written about the contraction very completely in the first book, where I have shown this is unrecognizable in the weak pulses, and particularly when the wall of the artery is soft. In the strong pulses, it is recognizable, and in those intermediate in terms of strength and magnitude. It is not necessary to specify as well about softness of the artery in the strong pulses, for this particular pulse is never strong and soft at the same time, but is either hard or moderate, and in accord with nature in this class. In the strong pulses then, when you compress the artery gently, you will sense whether it immediately withdraws towards the contraction, making the external pause short-lived, or pauses for a longer time than accords with nature, and whether its movement when contracting is more rapid than the movement that accords with nature, or slower. **[458K]** You will be greatly helped by such a recognition in regard to prognosticating, which you have in four books.

9. Elsewhere I have gone over the causes that make and change the pulses. I show in these places the primary and, as it were, *synektic* causes of them, the secondary or *proegoumenic*, and the tertiary or *prokatarchon*, which they also term *prokatarktic.*[286] It is quite clear that it makes no difference which of the two—either the feminine form (αἰτία) or the neuter (αἴτιον)—we use. Nevertheless, it is necessary to recall above all how we said sometimes to name a *synektic* cause—that it is not authoritative but misusing the name. For what is called a *synektic* cause authoritatively, no one else named before the Stoics,[287] nor is it agreed upon. And before us, the so-called *synektic* causes were causes of genesis and not of existence. For it has been shown that there is no other prior cause of anything unless it is of genesis. But we do not call every cause one of genesis, as I said, misusing the term *synektic*, **[459K]** but only that which is the effective cause of those things having their being in the coming about, like dancing, boxing, wrestling and running among all the actions. For there are movements that are active, and through these there are passive movements which have existence in the coming about.

There is such a cause of existence in relation to the pulses; three active causes were shown:

- The capacity (power) which moves the arteries;
- The actual body of the vessel;
- The use (need) for which it is moved.

This has been shown in the work, *On the Use of the Pulses*, including everything in a summarizing statement—the use (need) of the pulses is the preservation of the innate

286 For a consideration of the various kinds of cause, with reference to Galenic and other works, see Johnston 2006: 33–35, 81–128. The differentiation here is presumably between the verb προκατάρχω/*prokatarchô* and the adjective προκαταρκτικός/*prokatarktikos*. On *proegoumenic*, see *Appendix on transliterated general terms*.

287 See Hankinson 1998: 238–267 on causes and Stoics.

heat.[288] There are two parts of the pulse—dilatation and contraction. We showed that the dilatation arises for the most part for the purpose of cooling, while rarely sometimes also for fanning and cooling,[289] as also for the sake of some small part of it in relation to the cerebral *pneuma*. On the other hand, the contraction is for the purpose of pouring out, as it were, sooty and smoky superfluities. For some from the colder and moister humors, created by the innate heat or putrefactions **[460K]** that are contrary to nature, flow out to the surroundings through transpiration that is imperceptible to sensation, like smoke, while some from the hot and dry humors, by vapour and thick smoke.

Thus, the cold humor is called specifically "crude", while the moist and cold is called "phlegm". Gathering them together, as the majority do, we call both humors by the one name—phlegmatous or phlegmatic. The dry humor is that of the black bile, while that of pale yellow and yellow bile is not only dry but also hot. The green bile is in between these. Since the black bile is also twofold in relation to genesis and potency, and since the yellow bile also, when overheated, becomes thick and impure blood, it is analogous to the lees in wines. The genesis of the green bile arises in the change of the pale-yellow bile.

10. Therefore, whenever the heat is increased to a greater extent in the body of the animal, the need of the dilatation of the arteries becomes greater. However, when the sooty and smoky superfluities [increase], the need of the contraction becomes greater. When the need has increased, **[461K]** then here too the speed and size of the interval is increased, which goes through the body of the artery, either dilatating or contracting. Consequently, the pulse will be greater while the artery is dilating, since the heat is simply increased, while it will withdraw to the greatest extent in the contractions, when it pours out the smoky and sooty superfluities. The magnitude of the dilatation is, then, very easy to recognize by perception, while that of the contraction is impossible. For it was shown in the works on the diagnosis of pulses that the artery escapes most during contraction. Therefore, the rapidity of the withdrawal alone, along with the brevity of the internal pause, indicates to you the amount of the superfluities in the arteries. For it was shown in the works on the causes in the pulses that Nature hastens towards the function of the contraction when it is choked up with the superfluities. The haste also cuts off the external pause of this (the artery) and increases the rapidity of the contraction. It has been shown also that, when the artery is moderate in the consistency of its body, first the magnitude increases, when the heat is increased, and second the speed. However, when it is heated up sufficiently, **[462K]** it makes the pulses very large and at the same time rapid, and if heated still more, also frequent.

When the abundance of heat is not yet increased, the smoky and sooty superfluities predominate, and there is rapidity, but magnitude is not associated and brought

288 See *De usu pulsuum* 8 (V.179–180K).

289 There is some discrepancy between the Greek terms and their Latin equivalents in Kühn: ἔμψυξις/*empsuxis*, ῥίπισις/*ripisis*, and ἀνάψυξις/*anapsuxis* in Greek, and *refrigerandi, ventilandi, accendendi* in Latin.

into play together, nor the frequency, and the movement of the artery clearly seems more rapid in relation to the coming together in itself. And this is a most reliable sign of an incipient attack of a paroxysm, when nothing existed before in the sick person. This is very clear in those whose pulses we knew before, but more difficult in others, and what is required in many cases is the memory of the common measures of those bodies of which we have palpated the pulses. For inasmuch as the knowledge lies in the amount, this is not the same in all.

11. It is necessary, then, to direct attention to the common measures in those who are healthy, so that, whenever some class of pulse appears that goes beyond these, we know this is contrary to nature, and suppose it to be a sign of some particular condition. Since breadth has a common measure in people, it is best not simply in all cases, **[463K]** but in relation to kinds and *differentiae* to consider the common measures of many people in each class of pulses. There is in general the restitution of the common measures in reference to heat and cold, thinness and fatness,[290] and in turn, of these same things in relation to the stages of life, natures and the restitution of the ambient air in seasons and places. You learned about these in the book I have written for beginners about pulses.[291] And it is better to have read this beforehand. I shall also speak about this now, so I leave out nothing of what is essential in the present treatise. Those who wish to work diligently have the primary and most essential statements in brief in the introductory work, and in detail and a most complete form in the major work; the things now being spoken of are in between both.

12. The matters related to the times in connection with the dilatation and contraction have been written by Herophilus, bringing the discussion back to rhythms as regards the stages of life.[292] **[464K]** For just as musicians establish those orders in relation to certain determinations of times, comparing *arsis* and *thesis*[293] with each other, so too does Herophilus assume the dilatation to be analogous to the *arsis*, and the contraction of the artery to be analogous to the *thesis*. Having made the observation beginning from the newborn child, he supposed the first time to be perceived in the discovery of the dilating artery, while equally he says the time of the contraction is not notably distinguished from each of the pauses. For the contraction of the artery is imperceptible in those, whereas in them the rhythm of the pulse is divided into two times —the one of the perceptible movement when the dilating artery strikes our touch, and the remainder which is compounded from all the following: the external pause and the contraction after this, the pause in turn after that, and the first part of the dilatation, all these being imperceptible. And because of this, they divide the pulse into a beat and

290 Kühn's Greek text has τάχος/*tachos*, while the Latin translation has *crassitudinem* (Greek πάχος). The latter is preferred on the basis of sense.
291 *De pulsibus ad tirones* (VIII.453 – 492K).
292 See von Staden 1989: 273.
293 In this context, ἄρσις/*arsis* is the raising up of the foot in beating time in music. See Aristoxenus, *Rhythmica* II, fragmentum ap. Psell. 12, ll. 17 – 19 (ed. Westphal 1883 – 1893: 2.85). Θέσις/*thesis* is the downward beat.

an interval, placing in the amount of the interval the frequency and infrequency, just as they place rapidity and slowness in the amount of the beat. **[465K]** Therefore, Herophilus says that, to the extent the pulse is of equal rhythm in the newborn, so he seemed to me to recognize the beginning of the dilatation. And in turn, to the extent that the contraction of the artery in those who are aged extends up to ten times that of the first, so it is still not recognized, although the dilatation is recognized by the perceptible movement, which we recognize from the beat against our fingers, while the contraction and all the rest of the movement is placed in the imperceptible. But we must leave aside the art of the followers of Herophilus regarding the pulses.

13. Now, for the purpose of recall, I shall state, from long experience, as much as I take to be useful to the actions of the art. For if I palpate perfectly [the pulse of] newborn children, after a short time I am able to recognize clearly the beginning of the contraction, which I call a retreat and withdrawal of the artery. For under the circumstances, it moves away from our touch and withdraws to the depths of the body. In those in whom I am not able to recognize clearly the withdrawal, during the time of the perceptible movement, which is in **[466K]** our encountering it, the artery makes an addition also to the time of the external pause, which sometimes seemed to me to be equal to the whole remaining time, but sometimes less. In those cases in which the contraction could not be recognized at all, the whole remaining time was very clearly more than the time of the perceptible dilatation, which they call an interval. Next, always in relation to the whole time between the first day on which the infant was born and the end, in which, having gone through all the stages of life, the person dies, all the times are increased, there being four. But these are not the same in terms of substance and regarding our perception.

The times, then, in relation to the actual nature of the matters, are four, comprising two movements and two pauses. Of the movements, there are dilatation and contraction; [of the pauses, there are the external, which is after the dilatation and before the contraction,][294] and the internal, which is after the contraction and before the dilatation. Those in relation to perception are themselves also four. These are:

- The first, consisting of the major part of dilatation, since I showed in the books on diagnoses, that the actual beginning **[467K]** cannot be perceived by touch.
- Next is the external pause, which is wholly perceptible, if in fact some of the withdrawal is perceived.
- The third of these is the perception of the contraction, which was shown after the withdrawal of this to be entirely imperceptible.
- The fourth is after these and is the time comprising the later parts of the contraction, the internal pause, and the first parts of the dilatation.

294 The section in parentheses is missing from Kühn's Greek text; it is supplied from the Latin translation and demanded by sense.

All these times are, as I said, lengthened progressively through the stages of life. It is possible to commit these to the memory through long practice, but impossible to interpret them by reason. For the first time, which I said we discover in the newborn, however big it is, is not increased as much with the progression of age, as is taught by reason. But this is the task of each of those who are diligent—to palpate [the pulses] of many children, many adolescents, many in their prime, many who are past their prime (middle-aged), and also of course many old people. Only in this way will it be possible to commit to memory the common breadth of the measures. After the standard, if there is **[468K]** a sick person [whose pulse] has not previously been palpated, it will be possible to say this is contrary to nature. For the external pause, in respect of aspects of the accession of the disease and the time of the withdrawal, is brought together.

In fact, the dilatation, in relation to the attacks, seems slow ...[295] and whenever you find this, suppose with me that there is an abundance of heat in the heart, and the sooty superfluities, as it were, are many, due to which Nature weighs down the external pause, and hastens the actual contraction towards the excretion of these; the withdrawal is in relation to this. Of the picrocholic fevers (the genuine tertian is of this kind)[296] the dilatation departs greatly from an accord with nature towards largeness and rapidity, while the external pause and the rapidity of the withdrawal are not proportional to the magnitude and rapidity of the dilatation but fall short by a large amount. Consequently, being accustomed to recognizing the beginning of the contraction, you will not only be able to recognize the attacks of paroxysmal fevers, but also the kind of the putrefying humors, due to which on the whole the febrile diseases exist. For the hectic fevers are infrequent, while between both of these are the fevers called ephemeral by Hippocrates.[297] In these, **[469K]** as I said, particularly when you recognize the beginning of the contraction, you will immediately take the sick person to a bath.

14. Therefore, it is superfluous to compare the time of the dilatation with that of the contraction, whether the first parts of the contraction are perceived or not. For we know there to be one use of the dilatation for the animal, just as, in fact, there also is of inspiration, and another of the contraction, just as, in fact, there also is of expiration. It is from the changes in each of these that you will discover the condition of the body. Anyway, in this way too, in the case of those exercising or moving vigorously in any other way whatsoever, the dilatation becomes larger and more rapid. For we learned that the use of the dilatation of the arteries in the pulses is increased, just as that of the chest is in the inspirations, whenever either the consumption of the *pneuma* becomes greater, or abundant heat is collected in the whole animal, or the most impor-

295 A *lacuna* is indicated in the Kühn Greek text but not in the Latin translation.

296 Galen, *De febrium differentiis* II.1 (VII.335K).

297 For Galen's account of ephemeral fevers see his *De methodo medendi* X.2 (X.665 – 671K) and, in relation to Hippocrates, Galen, *In Hippocratis aphorismi et Galeni in eos commentarii* XXXVIII.55 (XVIIB.734K).

tant and largest parts, and particularly obviously in the heart. In this way too, you learned that the use of the contraction in the pulses is extended, just as that of the chest also is in the expirations. For Nature **[470K]** is urged in both these movements to pour out the *pneuma* that has been made turbid, which is sometimes like smoke and sometimes like soot. For the sooty superfluity refers to that from lamps and the smoky to that from green wood. And a sooty [residue] arises from the consumed material that has been burned up in the flame, just as the vaporous [residue] arises from moist substance thinned by use. However, the smoky is a mixture from both of these, being a material mixed from earthy and watery substances, the change not being complete, like a fire that is still getting started. For I think you know there is no difference terming something ἀιθάλη/*aithalē* or λιγνύς/*lignus*, or αἰθαλώδης/*aithalōdēs* or λιγνυώδης/*lignuōdēs*, if you have been brought up in the Greek language.[298]

In comparing the time of the dilatation with the time of the contraction, as Herophilus thought worthwhile, it can be recognized that the sick person has a contrariety to nature, and in addition to this, whether he is contrary to nature to a great extent or slightly. For the greater changes of the rhythms that are in accord with nature to what is contrary to nature signify a major harm, whereas those that are less **[471K]** signify a smaller harm. The pararhythmic pulses, therefore, indicate a slight change, the heterorhythmic a greater change, and the ekrhythmic the greatest change.[299] Recall what was said in the first book of *On the Differentiae of the Pulses*—that pulses are called ekrhythmic that have a rhythm that is not characteristic of any age (stage of life); pulses are called pararhythmic that are near; pulses are called heterorhythmic that have some difference and are not new [to this]. From these, then, it is in the meantime possible to diagnose the magnitude of the harm, but not what the condition is from which the change of pulses has occurred. From the same observation, it is possible to recognize, with some difficulty, the numbering of the times in relation to the rhythm. I now show the path, and it is easily recognized, and teaches what the nature of the condition is. That it not only teaches the magnitude of the harm but is also what the discovery of the remedies arises from, has been shown in the treatise on the therapeutic method.[300]

15. Therefore, of those things I previously defined, let me speak now of those I have written in the introduction about the difference in the pulses in accord with nature.[301] These are as follows. Men **[472K]** in general have a pulse that is a lot larger and sim-

---

298 The Greek terms here are as follows: αἰθάλη/*aithalē* which LSJ has as equivalent to αἴθαλος/*aithalos*, meaning "smoky flame, thick soot", and αἰθαλώδης/*aithalōdēs* as "sooty, black", citing this passage. For λιγνύς/*lignus*, LSJ has "thick smoke mixed with flame, murky fire" (such as is made by burning resinous substances), and λιγνυώδης/*lignuōdēs* as "smoky, sooty, dark" referring to *pneuma* and as opposite to καπνώδης/*kapnōdēs* (smoky, dark, dusky), citing this passage. As Galen says here, the term makes no difference.

299 On these terms, see also Section 1a (*Definitiones medicae*) 222 (XIX.409–410K) = 242sq CMG (92) and the introductory section on terminology.

300 Galen's *De methodo medendi* (X.1–1021K; English translation by Johnston and Horsley 2011).

301 See *De pulsibus ad tirones* 9 (VIII.462–468K).

ilarly a lot stronger than women, and a little slower, and significantly more infrequent. Those who are hotter in nature have a pulse that is larger, more rapid, and more frequent by a lot, but is not a lot stronger. Those who are thinner have a pulse that is larger and significantly more infrequent,[302] slower and a little stronger. In this way, then, pulses differ in nature. They change according to the ages (stages in life) in the following way: the pulse of the newborn infant is very frequent whereas that of the old person is very infrequent. With all the stages in between, the pulse is in proportion to the extent to which they happen to be nearer to the young child or the old person. Similarly, the pulse of the young child is very rapid, while that of the old person is very slow. The pulses of the other stages of life are in between. The difference of the old person in regard to the young child is much greater in relation to infrequency than it is in relation to rapidity.

In the difference relating to size and strength, the pulse of those in their prime is large, as in the stages of life, but very small in the aged, while it is median or a little larger in the case of children. And the pulse of those in their prime is very strong, while that of the aged is very weak; in the middle of these **[473K]** is the pulse of young children. In this way, then, the pulses change in the stages of life.

In relation to the seasons, in the middle of spring the pulses are very large and very strong, as in seasons, but moderate in rapidity and frequency. It is similar also in the middle of autumn. As the spring progresses, there is a reduction of size and strength, but an addition to rapidity and frequency. And at the end [of spring], when the summer takes hold, the pulses become weak, small, rapid and frequent. As autumn progresses, there is a reduction of all [measures]—size, speed, strength and frequency. Consequently, when winter comes on, they are changed to smallness, weakness, slowness and infrequency. There is a similarity of the first parts of spring to the last parts of autumn, of the last parts of spring to the first parts of autumn, of the first parts of summer to the last parts of summer, and the first parts of winter to the last parts of winter. Consequently, the pulses in the middle of summer and the middle of winter are equally distant from each and change similarly. Those of the middle of summer are changed in some way similarly and in some way oppositely **[474K]** to those of the middle of winter, for they are small and weak in each, rapid and frequent in summer but slow and infrequent in winter. They are not similarly small in summer as they are in winter, but less so in summer, and not similarly weak in winter as they are in summer, but less so in winter. These, then, are the changes of the pulses following the seasons.

With places, it is similar to the seasons. In places that are very cold, the pulses are those of the middle of winter. In places that are very hot, they are like those in the middle of summer. In places that are *eukratic*, they are like those in the middle of spring, and analogously also in those places in between. And of the other conditions of the air

302 Kühn's Greek text has a repetition of οἱ δὲ φύσει θερμότεροι μείζονα καὶ ὠκύτερον καὶ ἀραιότερον πόλλῳ here, for which there is no equivalent in the Latin translation and which appears to be an error.

around us, those that are hot are like the hot seasons; those that are cold are like the cold seasons; those that are median are like those in the middle of spring.

16. Therefore, when you are called to attend a sick person, you haven't seen before, consider first whether they are male or female. Second, after this, consider the age; then consider what kind of environment there is; this last can be reduced to season, place and climatic conditions. Next, consider the specific nature of the person. Putting all these together and calculating what kind of **[475K]** pulse this person is likely to have when healthy, you will know the magnitude of the change to a contrariety to nature. If also you add such other things as have been stated in the treatises *On Crises*,[303] assimilating them into the specifics of the pulses, you will come close to a very precise knowledge of the whole condition involving the sick person, for example, immediately in the significations. Therefore, when you first consider someone, determine further the uncertainty in the other signs. One of these is, when you first place your hand on the chest, [you find] it is moderately hot, consider the quality of the heat. If you find this to be mordant, you will say the sick person has a fever, even if you don't yet recognize this securely through the pulse. If, however, the chest seems to be cooled in some way, for sometimes there is cooling of the whole body in certain syndromes, you must not immediately lift your hand and declare the person has an accord with nature; you must allow [your hand] to remain placed on the chest longer and direct your attention in this to whether an acrid heat is rising up from the depths.

Anyway, **[476K]** for Themison,[304] this alone was sufficient evidence of the person having a fever—heat that is carried up from the depths. And a slight cough with some minor roughness, joining along with cooling, a gaping mouth, yawning, pain and some lethargy tending towards sleep will indicate this is not simply an attack of a symptom. These things, then, are external. In the pulses, however, not only are the previously mentioned signs indications of an attack of fever, but they may also have intermittency, or some other inequality or irregularity.

17. Since there are two inequalities in terms of class, one in a single dilatation of a pulse and another in an aggregation, when comparing many sequential pulses in succession to each other, I shall speak first about that in an aggregation, which is called by the majority of doctors, "systematic."[305] For it is more clearly recognized and appears in relation to very many symptoms. There are four *differentiae* of the pulse in the dilatation of the artery: one pertains to largeness and smallness; the second pertains to rapidity and slowness; the third pertains to strength and weakness; and the fourth pertains to **[477K]** hardness and softness. In the last mentioned *differentia* an unequal pulse is never seen, whereas in the other three, most inequalities exist due to humors putrefying in those with fevers. And the *differentia* of pulses in relation to frequency

303 Galen, *De crisibus* (IX.550–768K).

304 Themison of Laodicea (Syria) (ca. 90–40 B.C.).

305 The primary meaning of the substantive τὸ σύστημα/*sustēma*, from which the adjective is derived is, in LSJ: "the whole compounded of several parts or members, system". The English term "systematic" is used in that sense here and in what follows.

and infrequency is itself also very frequently seen in those who are sick. One kind of these changes easily to the intermittent pulse, in each of these two the inequalities, irregularities and regularities are not less.

So let there be some inequality of the class relating to frequency and infrequency, when after four frequent or moderate pulses, the fifth is infrequent. And let this occur sequentially, always preserving the order but with an increasing time of the infrequency. Such a pulse, whether it is irregular or regular, is intermittent after not many cycles, but whether the order is unequal or irregular will not make much difference to whether the person survives or dies. Anyway, we often saw, after four pulses, whether frequent or in between infrequent and frequent, the fifth become infrequent, and subsequent to that same **[478K]** cycle, the tenth, fifteenth and twentieth. Still more, when the fifth pulse generated is infrequent, we see the ninth, or the fourteenth, or some other pulse infrequent, not preserving the equality of the periodicity, but the infrequency intercides irregularly. In such pulses, then, experience has not taught a clear difference in regard to survival or death, or chronic or acute disease in relation to irregular and regular inequalities, and neither has reason. Certainly, in the case of the equal (even) pulses, there is no small difference from the unequal ones, when it comes to safety. One must especially not compare such great differences. For the equal pulse is in accord with nature, whereas the unequal pulse is not only contrary to nature, but is already also a serious problem. Thus, in the progressions (increases) it is more so, and even more in the peaks of the paroxysms, and worst in the abatements. But the intermittency that comes to a pulse that is quiescent for the longest time is an increase in infrequency, then irregularity and then regularity. For I said here that there is no great difference, detected either by experience or by reason. I say by reason now, **[479K]** as a Dogmatic doctor, in relation to which, beginning from the nature of the matters, we make the prognoses. However, I said before that in such discussions, we must not demand very much from compendious treatises. So, if someone should try to understand, along with what is useful to the matter before us, also the causes of what occurs, there is nothing that will detract from the major treatises.[306] In this, there are four books on the causes changing the pulses and four on the prognosis from these.

Thus, infrequency of the pulses is not good, whether it is equal or unequal. However, it is much worse when it is intermittent. For what occurs, as was shown, is that the physical capacity (vital force) is burdened by the abundance of humors, or some arteries are obstructed by thickness and viscidity of humors. And this occurs due to a major inflammation, when the lumina in these arteries are narrowed, or there is some external compression, or they are weighed down, or are confined in any way whatsoever. And because of this, in relation to the symptoms particularly, inequalities always occur in the pulses in many cases when the blood withdraws to the depths.

These **[480K]** pulses, then, indicate the conditions mentioned just now, while the infrequency indicates strong cooling. As a consequence, what occurs is not otherwise

---

306 This is taken to refer to the major treatises included in the following sections.

fatal. For such pulses occur due to the ambient air in winter, and due to nutriment in the abdomen when it is made phlegmatous in all those with bulimia.[307] In the fevers, however, infrequency of the pulse is not without danger, when it is a sign of quenching of the innate heat. Some of those in such a state desire food, and if you provide it, they eat enthusiastically, so it seems to those observing that they are healthy. In fact, they die immediately, some at the time of the symptom having become cataphoric,[308] but some without cataphoria, having been cooled strongly and not yet heated thoroughly.

However, as I said, not only does this class of inequality happen in the attacks of the febrile paroxysms, but also those in relation to large and small, rapid and slow, and strong and weak. You must pay attention closely in all these to whichever of the two is greater. For it is quite clear that many have the more grave danger, like when the fifth **[481K]** follows four that are weak, and is strong or moderately *eutonic*. Better by far is a fifth that is weak, intercident on four that are strong. In fact, when all weak pulses appear in sequence, if you discover an inequality in one or some of the other classes,[309] palpate the artery for a long time, looking to see if somehow one appears strong. However, when the pulse is equally bad in all the classes, recognise that such patients are not far from death. A pulse that is equally bad is one that is very weak and very frequent. For if it does not become very rapid at any time, along with very weak, but becomes very slow at the same time as being weak, I would be surprised, if the sick person in such a state were not already cold in the extremities. In fact, a pulse that is weak and large at the same time is necessarily also soft and is customarily seen in lethargies and the other cataphorias, such as are brought about by much moisture, either of the stomach and particularly its cardiac orifice, or the brain, or some other important part, or also of the whole body.

That in the stomach and oesophagus is easy to cure, whereas that **[482K]** in which the pulse becomes weak and soft at the same time due to the brain is very difficult. Still more is this so with that due to the whole state of the body, or one of the important internal organs, coming to an oedematous condition. Sometimes one frequent pulse, either regular or irregular, follows many infrequent pulses in succession. And such a pulse often occurs which we call specifically intercident, which exists in the opposite mode to the intermittent,[310] for the latter is when the infrequency extends, whereas the former occurs when the frequency reduces. And when in fact such a pulse occurs along with strength, it is a critical sign. Understand that for me "crisis" in all such discussions [signifies] a sudden and quick change in a disease, absolutely not ending well. Such a crisis occurs due to the expulsion or movement of humors from one part of the

307 The Greek term βουλιμία/*boulimia* is understood in its primary sense of ravenous hunger, unaccompanied by the psychological associations of the customary medical term, bulimia nervosa.

308 The English form of καταφορά/*katafora* is given here. The medical meaning of a deep sleep or lethargic attack is to be found in the 1933 edition of the *Oxford English Dictionary.*

309 Reading ἓν for ἐν, following Kühn's Latin translation.

310 The respective verbs are παρεμπίπτω and διαλείπω.

body to another.[311] You will recognise such a crisis from other signs mentioned in the work, *On Crises*—both in those that are good and those that are bad.[312] All these pulses, then, occur in relation to the symptoms, and there are some of them that relate to one dilatation of the artery. Let me speak about these now.

18. Of the unequal dilatations of the artery, some have the movement cut in two by a pause, while some are continuous, although not uniform. Concerning the first, I shall speak of the cutting in two by a pause. For surely the diagnosis of this is clear to anyone whatsoever, if they are not altogether asinine, since it is possible to recognize when, in the dilatation, the artery happens to stand still for a short time, then adds the rest of the dilatation. This cannot exist in small or low pulses. It is clear, then, from this itself, that neither the time of the dilatation is diminished nor the strength of the capacity, in this alone, which we are now speaking of—the change occurring to a contrariety to nature. It is necessary for you to remember this no less than all the other things, for what is taught on each occasion alone is diagnosis and prognosis, since all the other things having an accord with nature have had a very small change. In fact, to begin with, what we now say is the truth, whenever the wall of the artery has an accord with nature. For when it has been made immoderately soft, of the kind **[484K]** that appears in lethargies and in some other cataphorias, it is not true that the very large pulse is brought about by strength and capacity, and that the use (need) is either increased or altogether diminished. For if a large pulse has arisen due to these two things—increased need (use) and strong capacity—nothing would need to be added to the discussion.

For the present, however, since it is necessary to consider the wall of the artery as well, the discussion will not be true when stated simply. For you must call to mind the things said in the first book of *On Causes of the Pulses*,[313] which it is appropriate to read often, since it contains the foundations of the whole theory of pulses, without which it is not possible for other matters to be established and known. For you must pay attention to the genesis of the pulses in respect of the causes which someone might call *synektic*.[314] I showed the genesis lies in these three things: the need (use) through which the arteries are moved; the moving capacity (power); and the wall [of the artery]. For all the other things that may change the pulses effect the change through the mediation of these things. Therefore, whenever the wall of the artery has an accord with nature, but in the dilatation the occurrence of some pause creates the inequality, there is strength of the capacity **[485K]** and need of a large dilatation.

So then, it was shown that need of the genesis of the pulses is increased due to abundant heat or depletion of psychical pneuma. This is depleted by exercises in the service of the movements involving the muscles. Therefore whenever, as I said, the need of the genesis of the pulses is increased due to whatever cause, or the capacity

---

311 This use of the passive of ἀποσκευάζω is listed in LSJ, citing Pseudo-Dioscorides, *Theriaca*, praefatio.
312 *De crisibus* (IX.550 – 768).
313 *De causis pulsuum* I (IX.1 – 54K).
314 On this Stoic term, usually rendered ‘containing’, see *Stoicorum Veterum Fragmenta* 2.121, and for medical relevance particularly, Johnston 2006: 81 – 128.

in the heart which moves the arteries remains strong by reason of itself, but is weighed down by an abundance of humors, or the arteries are greatly obstructed or compressed by thick humors, the inequalities in one pulse arise, the continuity of the dilatation being cut in two by a pause. Here then, I ask you to attend closely to the two parts of the movement, considering in it weakness and strength, rapidity and slowness. For the *differentiae* occur in relation to each of these, as you learned in the first book of *On the Differentiae of the Pulses.*[315] This indicates, in addition to what was said pertaining to the conditions, how the physical (vital) capacity is. For when both movements are weak in the pulses, the danger is extreme, whereas when they are strong, there is hope of salvation. **[486K]** If, however, both are somehow intermediate between strength and weakness, they indicate that the person is situated between death and survival. But if, when the first movement is weak, the movement after the pause seems strong, this is worse than when the two movements are strong but less bad than when both are weak. If, conversely, it seems to be the case that when the first movement is strong, the one after it is *atonic*, this is not good. For it is better for the second movement to seem stronger than the first.

19. Next I shall speak about the order of these nine inequalities, beginning with the most mild of them, then going gradually through the intermediate ones, until coming to the worst.[316]

- Most mild of all is having both movements *eutonic.*
- Second is when the first is intermediate and the second *eutonic.*
- Third is when the first is *eutonic* and the second intermediate.
- Fourth is when both are intermediate.
- Fifth is when the first is weak and the second strong.
- Sixth is when the first is weak **[487K]** and the second intermediate.
- Seventh is when the first is strong and the second weak.
- Eighth is when the first is intermediate and the second weak.[317]
- Ninth, and the worst of all, is when both are weak.

Analogous to the conjunctions relating to the strength and weakness of the two parts of the movement mentioned are those relating to rapidity and slowness, for rapidity is better than slowness in such pulses, but not to the extent that strength is better than weakness. The same qualities of the pulses directly indicate the differences of the capacity in terms of strength and weakness. It is quite clear that the greatest hope of survival lies in the strength of the capacity, while the least is in its weaknesses. But let the comparison be for you in the analogy of the things mentioned in the case of the *differentiae* pertaining to slowness and rapidity. For the first conjunction that occurs is that in which both movements are rapid. The second is that in which the

315 *De pulsuum differentiis* I (VIII.493–565K).

316 Tabulation is not present in the Greek text. It is added for clarity.

317 This follows Kühn's Latin translation. The Greek text duplicates (vi) apart from the substitution of ἄρρωστον/*arrōston* for ἀσθενῆ/*asthenē.*

first is intermediate and the second rapid. The third is that in which the first is rapid and the second intermediate. The fourth is that in which both are intermediate. **[488K]** The fifth is that in which the first is slow and the second rapid. And after this, the sixth is that in which, when the first is slow, the second is intermediate. In the seventh conjunction the first movement is rapid and the second slow. In the eighth the first is intermediate and the second slow. Last of these is the ninth which has both the movements slow.

20. It is then necessary for you to look at both the *differentiae* and consider in your mind the combinations of these. Sometimes, the conjunction involving slowness and rapidity is troublesome; better is that involving strength and weakness. From such a combination, then, the so-called "bounding" (gazelle-like) pulse arises. In this case, the second part of the movement is more *eutonic* and more rapid than the first. And this would be the most moderate of the pulses in which the dilatation of the artery is cut in two by a pause. When some concoction of a disease has come together with it, this is a harbinger of a good crisis. This is just as in the **[489K]** conjunctions involving strength and weakness, and rapidity and slowness. Consider in such pulses what is most essential to prognosis, and in this way too, that relating to size and height. For as was said before, it is not possible for the low pulses to be cut in two by a pause; it is appropriate to consider the extent to which [the pulse] comes to a height. For it is possible for some to have gone away from the low pulses and still be intermediate, or a little higher than intermediate, without, however, being completely high. This, should it be the case, altogether announces that Nature is urged on towards a crisis. It indicates in fact that the crisis will not be entirely good, unless the disease should have the other indications (rhythms)[318] towards a good crisis, which we stated in the work, *On Crises*.[319] In these, the greatest and most safe indication is taken from the concoction of the disease. After this, there is that from the strength of the sick person. You learned there are these two indications in the prognosis concerning survival and death—strength of capacity and concoction of disease. Sick people are most free of danger when there are **[490K]** both of these present, whereas they are most at risk of death when there is neither. They are in between when one or other [is present].

The conjunctions in relation to more or less of these also increase and reduce the hope of survival and the expectation of death. Therefore, what I am accustomed to say is that it is useful for you to look at the weakness and strength, and the rapidity and slowness of the parts of the movement in such pulses, and in this way also look at the largeness and smallness of these, and more, the lowness and height, and whether or not these occur apart from the other distensions (dimensions). Because of this, then, in the excessively cooling attacks of the paroxysms such pulses don't appear, for

318 There is disagreement between the Kühn Greek and Latin texts; the former has ῥυθμοὺς while the latter has *scopos* and is regarded as correct in the light of what follows.
319 *De crisibus* (IX.550 – 768K).

they are reduced in height at the time and made small. Thus, those things that urge on the actions of the art are stated for those who read the detailed accounts.

21. Now, taking up the discussion from the beginning again, we shall state how many pulses there are that appear in relation to the dilatation of the artery, what the condition is from which each of these arises, and what the prognosis is from each of them. I want the **[491K]** fingers you place on the arteries to be gentle, not touching the surface at first nor pressing forcibly. Then, if you feel the beat of the dilating artery strongly, press on the surface more, but if, when doing this, the movement becomes imperceptible, touch the surface altogether. For the pulse which withstands the compression indicates a strong capacity, whereas with a weak capacity, the pulse drops away and is overcome. Anyway, you will find some pulses are very large, when you touch the skin surface, while others are small, but become apparent, if you press. Sometimes they are not still preserved at all but are altogether lost. Consequently, from the amount of pressure, the amount of weakness and strength in the capacity is made apparent to you; of those which are in the pulses very large is good. At all events, if you find at some time a febrile disease in a person who is sick with the signs of concoction, and at the same time the capacity is strong, you should know this person is not in danger because the disease will be resolved quickly. If, on the other hand, the capacity is strong, but there is no sign of concoction in the humors, he will be saved over a long time, since we recall that **[492K]** they know those things they have in mind without my reminding them, as all things relating to the prognosis are stated without error, in respect of those who are sick, or to whom some external harm occurs.

I said a large pulse is strong in the febrile diseases, the strength of the capacity moving the arteries, which we also call vital (life-maintaining), since often, when the danger is due to the sinewy class, the *tonus* of the vital capacity doesn't contribute anything much. But we consider, under the circumstances, how the psychical capacity has strength. It is better, then, for both to be strong, and in addition to these, the third capacity called physical, of which the principle is the liver, just as of the vital (life-maintaining), it is the heart, and of the psychical, the brain.[320] The strength of the specific capacity of the affected parts holds sway in relation to each kind of disease, when it comes to the reliability of the prognosis, Therefore, what the kind of *tonus* of the strong pulse is, when you have diagnosed by compression, indicates at the same time with this, the accurately recognized magnitude of the dilatation. Also however, reliable knowledge of the large pulses arises from touching the surface.

Recognize, however, that a very large pulse is the one which **[493K]** you would touch in many parts of the artery, whether of length, breadth or depth, as also long duration, [is the one] which prevails in respect of length, and broad, [the one] which prevails in respect of breadth, and high similarly, when it prevails in respect

---

320 On Galen's tripartite division, see his *Ars medica* 5 (I.318–319K) and *Ad Glauconem de methodo medendi* 2.4 (XI.97K).

of height, which is also called depth. Just as in the whole first book of *On the Diagnosis of the Pulses*, there was demonstration of the perceptibility and imperceptibility of the contraction, in like manner, in the second book, there was demonstration of the difference pertaining to largeness and smallness. There I show it is necessary to compare the amount of the dilatation of the artery circumferentially; it is by this you discover what you seek.[321] It is not to be sought in the magnitude to which the artery is increased in breadth compared to the narrowness.

For the same reason, then, some say the pulse in children is small, not comparing the dilatation with the lumen of the artery, but with the lumen of those in their prime. So if in fact someone who calculates how big the artery of a child is in circumference takes the pulse as large, he will know that Herophilus spoke correctly when he said the pulse of children is sufficient for their size. Certainly, for you these two classes of pulses in the dilatation **[494K]** of the artery will be subject to the stated manner in which you apply your fingers; the one is in the quantities of length, breadth and depth, while the other is in the *eutonia* of the beat. And it is possible for you to call what falls strongly on the touch by whatever name you might wish. For surely, if you use the terms ἰσχυρός, σφοδρός, εὔτονος, ῥωμαλέος,[322] you will do no harm to the actions of the art. We call this a strong pulse, following in similar fashion those before us; and particularly because it is a composite concept of strength combining strength and *tonus*. Since to prognosticate correctly does not arise from the knowledge of the names but from a precise knowledge of the matters, each must use the names he wishes while zealously pursuing knowledge of the matters.

However, just as the strong pulse always indicates strength of capacity, so too hardness either makes the artery tense or has given rise to hardness. If then, the body (of the artery) also seems tense, some hardness has arisen from this tension. So, in this way, hardness will always indicate a quality in respect of the state in the wall of the artery. Therefore, such a pulse appears following inflammations, **[495K]** or obstructions, or repletions, or tensions of the sinewy parts, or following indurations, not only of the sinewy parts but also of the others, if they are important. For surely liver, stomach, spleen and uterus create such a pulse when they are indurated. It occurs also due to severe cold and dryness, and whenever in fact the pulse seems to you soft in its first entrance, not having the kind of impact it has when in accord with nature; then think of there being one of the affections mentioned involving the person. You judge what this is at the time from the other symptoms and from the antecedent (*proegoumenic*) causes, like a cold drink contrary to custom, or spending an overly long time in cold water, or in cold air, and particularly if there is no prior preparation of the body, but also, when you strike some slave with your hand and bruise the tendon

321 There is uncertainty about Kühn's Greek text here. The Latin translation is followed.

322 The point here is obviously about the interchangeability of terms. The Greek terms transliterated above are, with their primary meaning in LSJ, as follows: ἰσχυρός (strong, mighty); σφοδρός (vehement, violent); εὔτονος (well-strung, vigorous); ῥωμαλέος (strong, mighty). "Strong" is an appropriate rendering in the present context, remembering that Galen makes a distinction between strength and *tonus*.

of the middle finger at the joint, or by chance have a fever from other causes. I find the pulse of this to be a clear indication of tension in the artery, and seeing it, also ask if the hand has been involved, and enquire as to what has involved it, **[496K]** or if the person says nothing exists in relation to this. For when blows afflict someone from this, I apply to the person a salve customary for him. Therefore, when I discovered no other cause of the tension in the pulse, I ordered a change in the treatment of the finger, and to use the treatment used for similar affections of the sinews (nerves). If the person ignores my advice, he suffers from spasms during the following night.

Also, in the *pleuritides* and the *phrenitides*, and those cases in which the diaphragm is inflamed, the hard pulse appears, indicating tension of the artery. Consequently, when there is doubt in these affections and the diagnosis is still pending, you will have no small part of the recognition of what this is from such pulses. When you attend someone already chronically ill, who is now brought in for the first time, if his pulse seems hard to you, consider there already is a scirrhous inflammation of the spleen. Indicative of this also is the colour of the face, just as in the case of the liver being affected in such a way, for the colour is specific for each of these internal organs, clearly indicating to someone practiced which of the two is affected. **[497K]**

But the discussion is not now about this; it is about those aspects of a pulse that are useful for prognosis. It is these that now lie before us to go over. Therefore, if you see in a sick person whom you have not seen before, a disposition to a febrile disease without the vital organ (principal) above having been affected, and when you return the pulse seems hard, then in the interval while you were absent, he either drank cold water or partook of cold fruit. You should know, then, that these things occur. When you discover the pulse is strong, equal and large, and at the same time the wall of the artery is in accord with nature, take this to be a major sign of the impending salvation of the sick person. For certainly, in those having rapidity or slowness [of the pulse], this will now indicate nothing of great significance regarding survival or death. But you will find the rapidity occurring with greater heat and the slowness with cooling. It is not possible for these same things to occur in a febrile disease in relation to the pulse mentioned. Rapidity is, however, very common in the least dangerous fevers, among which are the genuine tertians. For **[498K]** the pulses are large in these, and rapid in relation to the peak of the paroxysms, becoming so due to the abundance of heat. I spoke, in the third and second books of *On the Diagnosis of the Pulses*, of the primary things regarding the diagnosis of rapidity, when I showed that those erred who did not consider the magnitude of the dilatation as well, making a declaration on the slowness or rapidity of the pulses from the time of the movement alone. For it is not simply when the dilatation of the artery occurs in less time that the pulse is more rapid than when the dilatation is effected over a longer time, for in a very brief dilatation, it is possible for the movement of the artery not to be rapid, although the pulse occurs in a short time, just as conversely, in a very large pulse but with a rapid movement, it is possible for it to occur over a long time. Therefore, I said it is better to attend to the rushing motion of the movement, and not the amount of time of the whole dilatation. The rapidity, then, arises from the abundance of heat when the capacity is obviously strong, while the slowness is from cold. And in fact, whenever the pulse is slow, and at the same time small and *atonic*, people are approaching death.

You must bear in mind, in all things said in this way, **[499K]** the differences according to the stage of life (age). For not only is the beating of the pulse of a child bad compared to that in an old person, but also that of an old person is bad compared to that in a child. Anyway, of the ten primary times, as Herophilus measured in respect of the pulses,[323] if sometimes the interval between two beats is generated like in a young child, it is a sign of extreme cold, and because of this also, of mortification. This is as if the pulse of children is like that of an old person, in which the time of dilatation is equal to the time of contraction, this signifying excessive heating in respect of the nature. But I have never seen this, whereas in a young child such a pulse is often seen. Whenever the pulse of an old person is rapid in this way, so that in the time between two beats there is a doubling of the beat, this is also rare, but does not progress still further, for they die beforehand. If, however, it comes to this, and there is progression to the pulse characteristic of the stage of life of the sick person, in succession it will be more infrequent. Here also some guidance arises for those unpracticed in the pulses.

Whenever the pulse comes to an immoderate rapidity for the stage of life due to very great heat, **[500K]** and the heart in such a condition is dried so much that there is also a deadly δυσκρασία/*dyskrasia*, it is necessary for the heat itself to be quenched to achieve an accord with nature. In all likelihood, then, in this time the pulse becomes slower, and someone, as I said, will think it has returned to an accord with nature, but the person is journeying towards mortifying cold, not a geriatric *krasis*. In fact, the distinction between the pulses returning to an accord with nature and those proceeding to death is not difficult. For the pulse is always made weaker than before and particularly more in those who are dying. And in this especially, attend to the sign in all the ambiguous conditions, not considering three or four dilatations of the artery only, but whenever it is weak in such cases, wait for others. For the fifth or sixth after these is sometimes seen to be *eutonic*. Whenever you find this, know the capacity is not weakened by reason of itself, but by abundance, obstruction, compression, or in general one of the spatial constraints or burdens preventing it from moving strongly, **[501K]** similarly to strong men when they are either fettered by bonds or weighed down by a heavy burden.

And whenever it makes that one dilatation *eutonic* and high, know at that time Nature is urged towards the separation of those things that are distressing. If, then, a sufficiently strong fifth or sixth pulse intercides, at the same time as signs of concoction are apparent through the urine, it is possible for the person to be saved. If, however, there is no concoction, Nature will set out towards the separation, and you may judge that the person will die. In the *peripneumonias* particularly such pulses appear.

323 For a detailed account of what is known of Herophilus' views on the pulse, see von Staden 1989: 262–288.

I speak of inequality not only in relation to strength and weakness, but also in relation to largeness and smallness, and rapidity and slowness. And you must give these same people much oxymel, more acidic in terms of *krasis*, along with such other things as cut the thickness of humors[324] without kindling a fever. For when the arteries near the heart are blocked up by such things, they bring about the unequal pulses, and not only the systematic inequality, but also that relating to one **[502K]** dilatation. I have already spoken about this before, but I shall also bring these to your attention now. As our discussion is about the pulses accompanying attacks of febrile paroxysms and increases, I also made mention of the inequalities which frequently occur at such times in relation to large and small, rapid and slow, and strong and weak.

22. The difference of pulses due to the wall of the artery, in which softness and hardness are found, appears in relation to neither of the two inequalities. I say neither in respect of one being in a collection of many pulses in a sequence, and one being in one pulse only. Nevertheless, some inequality occurs due to the wall being hardened in different parts of the artery, whenever the need of the genesis of the pulses urges and the capacity is strong. For if this itself compels dilatation, while the wall of the artery is hard, the clonic pulse is produced. **[503K]** Sometimes the pulse itself becomes dicrotic, when the apparent part of the artery, in so far as it is not weighed down by the bodies lying on it, is raised to the previous height, whereas by those bodies on either side of it, it is slowed in its journey and is drawn back inward again due to the hardness of the wall, and then in turn raised up. On this account also, it appears such a pulse particularly arises in those cases in which the length is contracted. If this itself is long, the dicrotic pulse is never seen, just as the clonic pulse is not when it is short. For certainly clonus seems like the motion of javelins and staffs in relation to the small parts of the artery, when they move in this way. A *clonic* pulse falls either on the four fingers or at any rate on three. However, when one or two fingers only perceive the movement, the inequality of the parts on either side of the artery is concealed. In addition to this also, what happens is that the part not held back beforehand is carried in a contrary direction by those things weighing it down. For all those parts similarly exposed are similarly freed from the weight of the **[504K]** bodies lying on them.

Therefore, inasmuch as this pulse indicates both strength of capacity and abundance of heat, in all likelihood it will sometimes be a harbinger of an impending crisis. And by as much as the clonus is more, so the hope of a crisis is more. Whether the crisis is good or bad—for always remember this—you will learn through the work, *On Crises.*[325] The most important in these is the concoction of the disease-producing humors, at least in the febrile diseases. For surely you must also bear this in mind continually in regard to the diseases of the principal [organ] above (i.e., the brain)—concoction of the humors in the veins doesn't confer any great benefit. How you must have

324 There is a discrepancy between Kühn's Greek text and the Latin translation, the former having τάχος and the latter *crassitudo*. The latter is followed on the grounds of sense.

325 Galen's work, *De crisibus* (IX.550–768K).

prior knowledge of the danger from these, you have in the treatise on prognosis and in the things stated in *On Crises*. The extent of the benefit from the pulses for the diagnosis and prognosis of those diseases, and how much there is to take from hardness and softness, are also matters I have written about before.

What is taken from the clonic pulse in relation to the inequality in different parts is very similar; the opposite relates to the quality of the wall. **[505K]** For the clonic pulse is hard, while the wall is not hard, but is still itself softer than accords with nature. We call this "undulant" (wave-like) because it has the inequality like waves. Sometimes it seems moderately e*utonic* and sometimes between *eutonic* and *asthenic*. It was said that we call the *eutonic* pulse, "strong" following our predecessors, just as we also call the *asthenic*, "weak".[326] This undulant pulse in general proclaims sweats, and more so to the extent that it is softer, but not *atonic*. If the height in it is interrupted (intercided), you will have a very reliable sign of sweating. For this always signifies some separation, along with the undulant pulse, or simply with the large, but not with the hard, of critical sweats. With the undulant pulse, or simply a hard pulse, it signifies greater menses or hemorrhages from the nose, or via haemorrhoids, or a gastric disorder.[327] Whatever of the separations will be more, distinguish from those things that have been written in the treatise, *On Crises*.

Whenever this undulant pulse is weak, it becomes vermicular (worm-like), **[506K]** and has taken this name from the likeness it has to the movement of worms.[328] For those before us have given names corresponding to animals to three of these pulses: bounding (like a gazelle), vermicular (worm-like), and formicant (ant-like), applying the names due to the similarity of the movement in this way. But concerning the bounding pulse, this has been shown at the end of the discussion; I have spoken just now about the vermicular, since it is smaller than the undulant, always weak, and because of this, easily undergoes a change to the formicant, indicating a very complete weakness of capacity, just as in fact also, in the first part of the artery, the first movement seems like a creeping creature.

But in this, nothing specifically notable establishes a name. Thus, this itself occurs also when the capacity is weak, and if it is not strengthened beforehand, it easily undergoes a change to the formicant, which of all pulses that are very weak and small is also the worst, showing a characteristic weakness of capacity not due to an abundance weighing it down, or compression, or obstruction. Let this be for you the greatest sign **[507K]** for distinguishing a capacity which, due to compression, obstruction or excess, is not able to raise the artery in magnitude, which is a difference in relation to equality and inequality. On account of the specific reason for the weakness itself, since the pulse

326 In this passage, εὔτονος/*eutonos* is equated with σφοδρός/*sphodros*, and ἀσθενής/*asthenēs* is equated with ἀμυδρός/*amudros*.

327 The verb ἐκταράσσω in the passive voice is listed in LSJ as, "to have a bowel complaint", citing Hippocrates, *Aphorismi* 4.60 and *Epidemiae I* 15.

328 Here there is a hiatus in Kühn's Greek text which is filled in the Latin text by *gressus* (step, course). Presumably the missing Greek word is κίνησις/kinēsis.

remains equal during all the dilatations in sequence, when it is hindered in movement due to another cause in respect of the natural movement, an unequal movement arises.

In fact, the vermicular, and much more the formicant, are never seen as the systematic inequality, being inequalities due to a specific weakness of capacity. It is just like the impression of a small animal going crawling forth; it is necessary to show what kind of pulse it is, if the *pneuma* sent forth by the heart is carried through the arteries and has become a cause of the dilatation. It was, however, shown that distribution of capacity, which is not a *pneumatic* material, occurs from the heart through the walls of the arteries.[329] And all the things partaking of this, similarly to the heart, dilate themselves, drawing from all sides, that by which they would be able to fill the dilatation of themselves. As Erasistratus said, the pulse arises from the passage **[508K]** of the *pneuma* sent by the heart through the lumina in the arteries. Of the pulses that are seven and twenty (in the table), one of these which is the least, is called formicant. What happens with this, as I said, is that of all the pulses, it is the weakest, and on account of itself also the smallest. And because it is the smallest, it also becomes most frequent due to this. For when the capacity is extremely weak, the dilatation becomes least in the three distentions (dimensions), since length, breadth and depth of the artery are reduced to the least in extent. The cause due to which frequency follows in the small pulses in general was stated; it is due to the deficiency in the capacity and the need of the organs (arteries). However, just as the difference, in terms of more and less, is not small in the small pulse, it is the same also in the others, six and twenty, about which I shall speak a little further on.

23. Now let us add the pulses in relation to one dilatation, which have the inequality in different parts of the artery. Of these, there are also the so-called "tapering" ("mouse-tailed") pulses, in relation to **[509K]** the part nearer the heart, which makes the dilatation rise up to a swelling, while in the parts more distant it is diminished. As a result, when three fingers have been applied to it, by the amount to which the movement seems less to the second finger than to the first, in both breadth and depth, so to the third it seems even less than to the second, since clearly the *tonus* of the capacity moving the artery, which the heart supplies to all these, is dissipated. Sometimes the pulse falls similarly on the applied fingers, but the second movement seems weaker than the first, and after this, the third in turn weaker than the second, and then in turn the fourth and fifth, and all diminish in sequence to the point where they come to complete immobility as regards our perception. Then, when the fingers are removed and after a short interval applied again, there is perception of movement. Sometimes also, apart from the *arsis* of these, the movement returns again and so becomes perceptible.

And such things also happen concerning the pulses due to weakness of the capacity, which does not bear the application of the fingers, but as if **[510K]** overcome by a

---

329 This is the Herophilean position: the capacity (power) which dilates the arteries is distributed to them from the heart via the arterial walls.

burden, drops away. On this account also, it is appropriate to then palpate most moderately, particularly applying the fingers from those parts of the artery below, while rotating the arm of the sick person. If you again palpate the surface precisely, you will not perceive the movement in another manner due to the smallness of the dilatation. It is therefore necessary that, under the circumstances, you make the application of the fingers precisely to the amount of the dilatation, and in this way, sometimes to alternate letting loose and lightening a little, and sometimes compressing, since the correct diagnosis of such pulses lies in precise moderation of the application of the fingers to the surface, due to the touch of the surface not yet recognizing the dilatation of the artery, since, when it occurs slightly, the more forceful touch of this obscures and completely destroys the recognition of the movement. Sometimes in fact, in such pulses, when the reduced dilatation is gradually destroyed and returns again, we call these pulses "tapering (mouse-tailed) recurring".

Therefore, with regard to extreme brevity of movement coming again to increase itself, **[511K]** we now have this same stated name—"tapering (mouse-tailed) recurring". If, however, they come to an end at complete immobility and stay there, and they will no longer make a perceptible movement, we term such pulses "remittent". If, on the other hand, after the immobility, they start to move again, we call them "remittent recurring". It is quite clear that all such pulses occur due to weakness of capacity—sometimes when there is a specific weakness of this alone and sometimes due to some excess, or compression, or obstruction of the arteries near the heart, such as bring about a mixed condition. And if the pulse has not become altogether weak and small in these cases, an inequality would clearly appear in one dilatation. Now this does not appear due to the lowness of the movement which the artery makes in the dilatation. On which account also, such pulses are in general frequent, in that class of frequency which appears in those which have the contraction imperceptible. For we said this is twofold and is called the frequent pulse: either when we perceive the first parts of the contraction, or if this itself **[512K]** is not perceptible at all.

If we do perceive the first parts of the contraction, the time between the end of the dilatation and the beginning of the contraction brings about the infrequent or frequent pulse. Since it is also necessary for the artery to have some time and a large dilatation for perception of the first parts of the dilatation, it is impossible in the case of weak and small pulses for us to perceive these at any time. As a consequence, in these the time of the internal pause will be unrecognizable. And because of this, the whole compound from this and the contraction, the pause after the contraction and the first parts of the perceptible contraction become imperceptible, making the pulse either frequent or infrequent. But the matter of infrequency and frequency due to the amount of the external pause has been clarified in what was written prior to this. It is already now time to speak about what is said on frequency and infrequency, bearing in mind the things we have previously shown in the treatise, *On Causes of the Pulses.*[330]

330 *De causis pulsuum* (IX.1–204K).

As someone might say, we showed the *synektic* (containing) cause of frequency of pulses to be the deficiency of the dilatation. Seeing that, whenever there is **[513K]** need of the function, the artery is quiescent for a moderate time while it is being filled; if the need is not fulfilled due to the deficiency of the dilatation, the second movement is compelled to begin sooner. Accordingly, since, in the very small pulses, the need of the dilatation becomes unfilled to the extent necessary, Nature, being compelled to complete the filling, begins the second movement sooner. However, to the extent that the time between the dilatations becomes shorter, so the pulse seems more frequent. The need (use) of the deficient dilatation arises in the small pulses due to the smallness, and of necessity this follows continuously, although sometimes it is not like this in the case of the large pulses. For there is necessarily an abundance of heat in these, or a consumption of psychical *pneuma*, so the pulse becomes frequent. For this reason, in those in whom the contraction of the artery is perceptible, the time between the end of the perceptible withdrawal and the beginning of the dilatation is short. This time, then, contracts, since Nature is hastening on to the dilatation, where there is need of the artery to draw in more of the external **[514K]** air, and this occurs, as I said, due to an abundance of heat or consumption of *pneuma*. However, the other time, which is that of the external pause, is shortened in those conditions in which Nature hastens on to evacuate the sooty and smoky superfluities collected in the arteries. And because of this, I said in the writing prior to this, that the most reliable sign of an attack of paroxysmal fevers from putrefaction of humors occurring is a short time of the external pause, while the movement of the contraction is more rapid.

24. These things, then, are enough about infrequency and frequency. Let us return to the congeners of the pulses we spoke about a little earlier. There is certainly something which is higher than the median, as considered in terms of length, while those on either side of this are lower. We call this pulse "nodding" and "inclining".[331] It seems this pulse arises from weak capacities, whenever the constitution of the body is such in nature that some skin is small [in amount] and light when pressed on in the middle of the length of the artery, while that **[515K]** on either side of this in the length is weighed down by the parts above and below. Because of this same thing certainly, the pulse of those who are obese seems short in length, even if it is strong, while the pulse of those who are thin seems long, even if it is weaker.

25. And somehow the discussion now comes again to what we cast aside a little earlier—the seventh and twentieth pulses in the table. The genesis of these, as from elements, is from the nine pulses. Three of these are considered in relation to the length of the artery, three in relation to the breadth, and the other three in relation to the depth. It is from the mixture of these with each other that all twenty-seven conjunctions of the pulses arise. One of these—that relating to length—which we began to speak about just now, I showed appears due to the difference of the body in thickness and thinness, and weakness and strength of capacity, while from this, in addition to

331 These are Archigenes' terms.

these, there is the difference of much heat. If you wish to reduce the discussion to two chief points, [they are]: due to the dilatation of the artery becoming less or more in circumference, and **[516K]** due to thinness and thickness of the body.

So far as it relates to the nature of the matter itself, the pulses of all become equal in length, unless at some time the limbs of those already approaching death are cooled. For they do not appear equal due to the inequality of the bodies overlying the arteries. At all events, the artery adjacent to the spine of those who have wasted away in chronic diseases appears to pulsate when a hand is applied to the skin in the hypogastrium (lower abdomen). And in fact this is more pronounced in those who happen to be thin in nature, and much more when they are febrile or drink wine, as in them the dilatation is increased in magnitude. Contrariwise, in those who are obese, even if they are strong in capacity and overly hot, the pulse becomes short in relation to the length of the artery. As a result, the thinness of the body contributes to the length of the pulse most of all the other things, while second after this is strength of capacity, and third, heat. The opposites of these produce an opposite pulse, which I said is called short.

The narrow pulse arises due to a confined space of the regions lying around **[517K]** the artery, weakness of capacity, cooling and hardness of the organ (artery), and rugosity and thickness of skin. The high pulse is due to the opposites: strength of capacity, much heat, softness of the organ (artery), and thinness and tension of the skin. It is quite clear from the things said that the fourteenth pulse in the table, which is moderate in terms of length, breadth and depth, whether it is wholly in accord with nature, or also some mixture from opposites, leads to a moderate state in the dilatation. For it is possible in the sick person for the capacity to be weaker at a certain time to the same extent that the state of the body is thinner, so the pulse comes to brevity due to the weakness of the capacity to the same extent that it is added to from the thinness. It is also possible due to abundance of heat of such a degree as to make up for what is defective. In turn again, in relation to the opposites, when the capacity becomes stronger to the extent that the pulse has increased in length, it is reduced to the same extent by plumpness (fleshiness). It is also possible for the increase in length to become shorter as far as it follows other reasons **[518K]** due to cooling.

Also, we must take into account to some degree the hardness of the arterial wall in investigations such as these. But if this is, in amount, as much as accords with nature, and no change to either being well-fleshed or thin seems to have occurred, nor is there some change to weakness or strength, but in these there is accord with nature, and if length, depth and breadth seem moderate, it will indicate to you that there is neither greater heat in the body nor consumption of psychical *pneuma*. In respect to these same things then, and from this itself, it is very easy to recognize there is no strong function activating the sick person, for the *pneuma* is consumed in the voluntary movements, while the heat is increased in the depths of the body. If all the other things are moderate, it is then possible to prognosticate through the pulse alone. The change of respiration does not clearly occur in such conditions in which a little more of the heat that accords with nature has collected in the heart, other things remaining the same.

But you must attend here to the magnitude and rapidity of the pulse. For first, it **[519K]** increases in magnitude when the wall of the artery is either softer than accords with nature or moderate, then in rapidity, when the heat is increased. Whenever, because of hardness, the wall of the artery cannot be distended to a significant magnitude, at that time the increase of the pulse in rapidity occurs sooner. For undoubtedly when the need of the genesis of the pulse is increased, up to the point where the magnitude of the dilatation is sufficient for it, nothing at all perceptible changes the movement to rapidity, or only a little. However, when the need requires a greater dilatation, but the wall of the artery is not able to extend due to hardness, at that time a large increase in rapidity occurs. And when this is not quite enough, a clonic pulse is brought about, when the capacity clearly remains strong.

26. Therefore, to go through all the pulses in the table, treating each individually, would be a long job; with them, it would be better to follow the lead of the method used with exercises, which is what I did in the first book of *On Prognosis from the Pulses.*[332] However, before coming to this, it is better to have realized something **[520K]** about these—that sometimes an increase is very short, whether in fact it occurs in breadth or height, or the opposites, while it appears much due to our comparing a pulse observed now with one existing before. This, then, is seen to happen in the case of crises, when from an apparently high pulse, if many sweats break out, the change occurs to the opposite kind and the pulse seems low and at the same time broad. On this account also, some doctors have written that the broad pulse is often seen in those who have had a recent crisis. For the same thing appears to them, as was said, to be judged broader, when it has not at this time in truth become broader as much as it seems to have. This is due to the previous height being increased still more than accords with nature in the time after the crisis, and the accord with nature no longer being preserved. For due to the emptying, the spaces lying around the arteries, which were previously constraining them, become loose, and because of this the pulse seems truly broader. For it seems much broader than, in truth, it actually is when the **[521K]** height falls to low due to the dilatation not rising to a significant swelling at all and due to the loosening and wrinkling occurring in the overlying skin.

If, at this time, the limb is rotated, and you place the artery in the parts below it, the pulse does not still appear low in the same way as before, nor in the same way broad. And in dropsy, the whole bodily state (ἕξις/*hexis*) being swollen beyond the proper size, as in dying bodies, differing greatly from the so-called tympanitic and ascitic [dropsies], the pulses are broad, and particularly when we palpate the artery in relation to the parts above and not the parts below, when the limb is stretched out. Such a pulse is sometimes seen in those with lethargy who are in a very bad state. For in them the state of the body is weakened, so due to the considerable moisture and softness, it

332 *De praesagitione ex pulsibus* I (IX.205–270K).

becomes loose and unstable, and the back of the artery is no longer able to sustain itself nor rise up in height so as to be in accord with nature. In these cases, then, as I said, if you palpate from the parts of the limb below, the pulse no longer seems broad to you.

In this, however, you need to attend closely, and distinguish and consider how, if **[522K]** the limb has assumed a certain position, you made the application of your hand. For in many instances, we palpate the artery at the wrist, and sometimes at one of the temples, or in the arms. However, we need to include comparison of those pulses that existed before with those that are present now, and to consider whether or not something more has changed through those, or in truth the changed pulse is an illusion. Bearing these things in mind, I shall make the following discussion for you.

27. Let the pulse in the wrist of the arm of some sick person now being seen by you for the first time appear longer, such that the beating movement is perceptible to six or more fingers.[333] It is then, of necessity, already clear that this is a thin person, for it is impossible, when the body is well-fleshed for the pulse to become so long. Next then, examine the consistency of the wall of the artery as regards the stage of life (age), and look at the state of the body of the sick person. For the body is softer in young children than in those who are fully grown, and softer in women than in men, and in those who are **[523K]** soft and thin in nature compared to those who are the opposite.

Therefore, if the pulse seems to you to have become much harder than accords with nature, examine the swelling of the dilatation. I call for the moment "swelling"[334] the circumferential movement of the whole artery, which we divide conceptually into height and breadth. For just as in cuboid bodies, we do not differentiate the height from the breadth in the case of the artery. If there is such a pulse, it is altogether necessary for there to be one of the conditions that make it hard. You learned the conditions: strong cold such as to congeal the body of the artery; excessive dryness as in the marasmuses; induration (scirrhosity) of some internal organ; a major inflammation; or a spasmodic (convulsive) state in one of the sinewy (neural) parts. But in fact, the face immediately indicates to you whether it is a marasmus or one of the other things. For it is fitting that the marasmus is incurable whenever, due to it, an artery takes on such dryness. Thus, you see hollow (sunken) eyes in those in this kind of state, such that the **[524K]** bones forming their boundaries in a circle exceed their tunics—and you learned the various other things in the book pertaining to marasmus.[335]

But also, it is possible for you to recognize those in such a state from the face alone, before palpating the pulses. If such a thing is not apparent, you will in general

333 This is the first mention in these texts of two hands being applied together in the palpation of the pulse.

334 The Greek term is ὄγκος/*ogkos* which has a primary meaning of bulk, size, or mass in relation to a body. For the Methodics it came to be used as a term for the fundamental particles of matter—ὄγκοι/*ogkoi*.

335 *De marcore* (VII.666 – 704K; Engl. transl. Theocharides 1971: 369 – 390).

discover the hardness of the pulse due to there being an induration (scirrhosity) involving the spleen, and already due to some other important part which you will readily recognize by palpation. And symptoms that follow clearly distinguish these from the others, for if the liver comes to so great an induration (scirrhosity) as to make the pulse like I said, the symptoms of the dropsies involving the face already make this apparent to you. The diarrheas show the indurations (scirrhosities) involving the stomach and bladder, or some of the intestines being affected. So too, if the diaphragm or pleura are affected, the dyspnoeas are characteristic of these. You have learned their characteristics in the treatise on the dyspnoeas.[336]

If none of these is apparent, consider whether or not there is some spasmodic (convulsive) condition. In the case of women, consider whether or not there is this scirrhous affection in the uterus. If no such thing is apparent, [consider] strong cooling that is recent, **[525K]** either from cold water or a drink of wine, for in those cases, the artery seems to have grown stiff in relation to the movement. It seems probable in those so disposed, that unmixed wine dries their bodily state, just as cold water congeals. And fruits, cold in relation to touch and potency, make a pulse hard. But these harms from external causes are quick to subside together when we warm the sick person, and because of this also, the pulses subside together, whereas the conditions remaining in the body are difficult to cure.

Therefore, when the pulse is long and at the same time hard, you will make the enquiries, as I said. However, when it is moderate, in terms of the difference (*differentia*) in hardness and softness, consider first the amount of the dilatation. For should the artery be dilated significantly in each of the distensions (dimensions)—that is, in length and height—and this has not occurred due to recent food, or a drink of wine, or a bath, or some other movement, it is, I presume, quite clear that the person is febrile. If, however, the pulse is rapid along with such strength, and frequency is added, it is clear that the fever is greater. If, on the other hand, **[526K]** the pulse is moderate in relation to height and breadth, while along with the rapidity there is besides significant frequency, it indicates much heat but weakness in the capacity. For if it were strong, there would be largeness in the pulses. But also, due to this itself, significant rapidity arises as well as frequency because the need is not fulfilled due to smallness of the dilatation.[337] Such a pulse seems to you altogether weak, for this sign is characteristically inseparable in pulses when the capacity is weak. As a consequence, it is advantageous for all the other signs of weakness or strength of the capacity to be considered. In fact, what are useful for precise scientific knowledge of the theory pertaining to the pulses and were witnessed and taken as evidence of prior knowledge, are what arise from

---

336 *De difficultate respirationis libri III* (VII.753–960K).

337 Kühn's Greek text has μακρότητα/*makrotēta* while the Latin translation has *parvitatem* (which would be the translation of μικρότητα/*mikrotēta* in the Greek text). As in the previous instances, the Latin is preferred on the grounds of sense.

weak and strong pulses. Moreover, as was said above, following the custom of doctors, I call those that are *eutonic*, "strong", and those that are *atonic*, "weak".[338]

Well then, in someone having a moderate amount of flesh, the pulse appears moderate in magnitude, when the wall of the artery itself is moderate in the **[527K]** *differentia* relating to hardness and softness, the condition of the arterial wall being called by the general name. In the case of such a pulse, you say the capacity is altogether strong before you know in what manner it has strength. For the strong pulse is not recognized immediately in every beat that is palpated, but only when we compress the artery. For touching the surface will accurately diagnose the magnitude, whereas it won't altogether diagnose the strength. However, when you compress, the strength is clearly apparent, just as in fact the weakness also is. The magnitude is also displayed together. For when the artery rising to a height is prevented by the compression, it makes the resistance *eutonic* at the same time, while also at the same time there is some impression of the extension to the breadth. In these, there is an indication of the hindering of the rising up to a height through the compression. But when the wall of the artery is hard, a very large dilatation does not occur up to this point. And because of this, when there is inflammation of an important internal organ, and there is also due to this a hard pulse, if in addition to these things the power (capacity) dilating the arteries is strong, the pulse is made *clonic*. For whenever the **[528K]** need of the dilatation has increased and strength is present in the vital capacity, clonus is produced when the artery is forcibly dilated.

However, in many instances, in such conditions the capacity is not strong, but someone with a body that has become thin due to prolonged illness would be content to have moderate strength of the capacity. Because of this, then, the pulses would not become *clonic* in them. In fact, if they drink wine that is more unmixed or copious in amount, clonic pulses sometimes appear in them, when the capacity is strengthened, and the feverish heat increased. Therefore, these pulses frequently appear, although rarely do they preserve the height along with the breadth, while they are short in length. Very rare among these are the contractions that are long as well as broad. Rarer still among them is the pulse that is high and at the same time narrow and short in those who have a long pulse, and who are neither healthy nor able to become so. In these, brevity exists due to abundance, since the capacity is strong.

On this account also, in general an inequality appears in relation to the two classes of inequality. For just as the strong pulse is a sign of a strong capacity, so too, after that is a high pulse, **[529K[** when it is shortened in length. Thus, it is clear, under these circumstances, that it rises to a height due neither to thinness of the body nor softness of the wall of the artery. For if it were lifted up due to these things, the pulse would not be short. So, for the most part, then, the high pulse is immediately also broad; if not, it is moderate in breadth. There is an inability for this pulse itself to be narrowed and to be

338 There appears to be a deficiency in the Greek text for these two sentences. The translation above is made from a combination of the Greek and Latin texts in Kühn, and is tentative.

higher than moderate, apart from a systematic inequality. And this is less in number in respect to the high pulses compared to the moderate in relation to depth. But also, in relation to one dilatation such a pulse is unequal, apart from being a pulse that is not very frequently seen. And indeed also, the high pulse is curtailed in length, although it is not in fact curtailed to such a degree as occurs in the formicant pulse. Anyway, it is filled in respect of the application of one finger, since the whole filling of the formicant pulse is not complete in the third part. Therefore, the pulse that is high and brief in length, since it will become at the same time harder, clearly avoids being dicrotic.[339] To begin with, I have never seen a pulse that was long and dicrotic. But in fact, when it takes on length **[530K]** following the causes mentioned, it becomes clonic. The causes are an abundance of heat, strength of capacity and hardness of an organ (artery).

Sometimes also, when due to an affection in diseases, the position of the artery is changed or distorted, a non-uniform (unequal, anomalous) δυσκρασία/*dyskrasia*[340] occurs in the walls of the artery, or in the surrounding bodies—for example, if the part was dried or not, and was hardened or not, or has become tense due to the attraction of one of the surrounding bodies, or was forced out of position. But these things are rare, whereas those due to heating or cooling of the whole limb with the surrounding parts, or moistness and abundance, or deficiency, or dryness often happen. It is, however, also necessary to recall those things occurring rarely, so that when you come to a sick person whom you have not seen before, beginning from those things occurring very commonly, you move on to those you have not often seen, and then in this way to the others, right up to the most rare, just as of course you also do in the case of the short pulse in relation to the long, and the high pulse in relation to the low. These are examples of the **[531K]** transition from those things often seen to those more rarely seen.

It is reasonable then, whenever this particular pulse is seen, to look first at how the person has gained weight. For if he is well-fleshed and drinks enough, the pulse appears in proportion; that is, it will be small. On this account, in the immoderate enfleshings, the flesh spreading around prevents the vessel dilating to the extent it naturally would. But also, its length becomes shorter and its breadth narrower being compressed by the abundance of the flesh. Of those who are thin in the bodily state (ἕξις/*hexis*), some are strong in other respects, and the pulse is sufficient in terms of breadth and length, while the increase in height is notable. Still more, in those whose skin is lax, a manifest separation exists between this and the wall of the artery. From this it is clear that, because the lighter bodies more easily rise up, the arteries are dilated, whereas with the heavier bodies, this is more difficult. Those that are very heavy do not move at all. **[532K]**

---

339 There is a defect in Kühn's Greek text. The translation is from Kühn's Latin translation.

340 On this type of *dyskrasia* see particularly Galen, *De inaequali intemperie* (VII.733–752K; Engl. transl. Johnston 2020: 282–325).

28. Now let us say, about the pulses, why it is they become frequent from the moderate harms of the capacity along with heat at the same time, although not yet small or slow, but moderate. At all events, there are four conjunctive *differentiae:*

- The pulses are changed by heating alone or by cooling. And in relation to being weaker and hot at the same time, along with heat, if the capacity is prevailed upon by both the causes still more, or is sufficiently weak, and the heat fiery, the pulses become small, slow and very frequent. However, when the harm is moderate, the pulses are similarly frequent, but moderate in relation to the distension and movement.
- The second conjunction is compounded from weakness and *katalēpsis.*[341] In this, the pulse becomes weak to such an extent that matters pertaining to the capacity are in a bad state, while the pulse is small and slow, just as that in the case of the previous conjunction also is. It does not in fact come to an extreme similar to that, unless the capacity has suffered to an extreme degree. It is clear that in such cases the pulse will be very small, and in many instances intermittent and remittent.
- A third conjunction of the changes in the pulses is **[533K]** when the heat and capacity are increased together. In this, the pulses become very strong and very large, but not yet similarly very rapid, while for the most part they are not clearly very frequent, but sometimes are clearly beyond what is in accord with nature. And this occurs when the heat is increased to the greatest extent.
- The remaining conjunction of the changes in the pulses relating to this class is whenever strength of capacity and lack of heat come to the pulses. These pulses are moderate in magnitude, slower and significantly infrequent, and particularly when the cooling prevails to the greatest extent. However, the amount of the distension of these is not clearly hindered due to the strength of the capacity, for truly the slowness does not prevail to a great extent due to the same cause. When the capacity is strong[342] it preserves the natural dimensions of movement, even if the need does not call for this. In fact, it is necessary for need to be greatly reduced when cold strongly prevails, so that clearly the pulses become smaller. Nevertheless, to a great extent, it preserves the natural amount of distension or to a small extent, what is not at all less. Certainly, it appears to a much greater extent slower than smaller in the coolings.

---

341 The Greek term here is κατάληψις for which LSJ gives a medical use of compression or stoppage. The term also came to mean catalepsy, the disease or symptom defined by Stedman as follows: "A condition characterised by waxy rigidity of the limbs, ... lack of response to stimuli, mutism and inactivity, occurring with some psychoses, especially catatonic schizophrenia." (*Stedman's Medical Dictionary* 2000: 299). Exactly what it meant in ancient times is not clear.

342 This marks the end of the Greek text in Kühn. The translation of the remainder of the *Synopsis* has been made from the reconstituted Greek text of Cortés Gabaudan and Martínez Manzano 2013, with reference also to Kühn's Latin.

29. Let us return to the prognoses in the case of fevers so that we state here the kind of recognition of each of the pulses. Thus, all the ephemeral fevers **[534K]** change the pulses to large, rapid and frequent in an amount proportional to themselves. In the same way too, the hectic fevers change pulses in the same *differentiae* by an amount proportional to themselves, although with them the rapidity is changed more. But those fevers which arise from putrefaction of humors have both movements rapid in the peaks—that of the dilatation and that of the contraction—and they change both frequencies similarly. In the attacks, however, [they do not change the frequencies] similarly, but increase the rapidity of the contraction more. In the increases, however, they also work on the external pause to the greatest extent so that a very frequent pulse appears. And these alone are the characteristic pulses of the fevers, by reason of themselves. Others follow other conditions more. There are two classes of such conditions—the badness of the capacity and that of the organs (arteries). Looking at these three things, one ought to discover the causes which change the pulses. The three are: the organ (artery), the need (use) and the capacity.

One must apply one's mind diligently in this situation because knowledgeable doctors differ among themselves regarding the characteristic pulses of a fever, and they are all wrong in thinking the changes of the pulses due to the capacity itself, or the badness of the organs (arteries) are characteristic of the fevers themselves. However, it is inseparable from, and especially characteristic of, an attack of fever due to putrefaction of humors for the contraction of the artery to become more rapid. And anyone who is practiced in recognizing the contraction will hold this to be an invariable sign of the beginning of the paroxysm, whenever the pulse has some *tonus*. How someone might accrue this practice is stated most fully in the first book of, *On the Diagnosis of Pulses.*[343] At any rate, on many occasions, with patients in whom I had never palpated the pulses previously, neither while they were healthy nor while they were sick, relying on this sign alone, I was never mistaken. Therefore, when the pulse has not been changed in the other factors but has become more rapid in relation to the contraction, consider this an invariable sign of the beginning of a paroxysm. Such a pulse is also immediately more frequent in the external pause, relying on Nature to get rid of the smoky superfluity, on account of which the greater rapidity is brought about.

However, in the ephemeral fevers, **[535K]** this is not present because they are without putrefaction of humors, and not with the hectic fevers, because these never begin unless by chance some other condition should come together with them. If, on the first day, when a person has just undergone a change from health to disease, we palpate the pulses, this sign will immediately indicate to us the class of the fever. Those fevers which arise from inflammations are distinguished from those which arise from putrefaction of humors by the hardness alone, for the hard impact is [characteristic of] those fevers that follow inflammations. If, therefore, you are able to distinguish the hard beat of the artery occurring due to tension from that occurring due to cooling

343 *De dignoscendis pulsibus* I (VIII.766–822K).

and hardness, you will have from these three prognoses of previously occurring causes: a cold drink, overly pure wine and inflammation, [the last of] which we say is not only a cause which existed before the paroxysm, but which also still remains. We say a drink of cold water and an excess of unmixed wine are not themselves previously occurring, but the condition from them is still preserved. And the walls of the arteries are hardened by harmful drinks that are cold, while they are dried out by excessive drinks of wine. One must distinguish these in this way.

Those fevers which, on the first day, never have the described predictive sign must be judged to be either ephemeral or hectic. Very rarely a hectic fever exists right from the start, although it is not impossible that there is such a generation of this—anyway such things were also seen by us. But we consider all the other signs of these in the treatise, *On the Differentiae of Fevers*;[344] those which are taken from the pulses will be spoken about now. Neither a large nor a strong pulse ever come on together at the same time with hectic fevers. If either one of these is present with them, the fevers are not hectic. If all are present at the same time, clearly the fevers are ephemeral.

30. These [pulses] being so defined, the next thing that must be considered is what each pulse among these signifies in relation to the crisis. Thus, if the pulse itself becomes undulant and markedly soft, it behoves us to anticipate sweats. A pulse that is hard customarily presages vomiting more than sweating. A high pulse is a sign of all excretions just as a strong pulse also is. However, if the pulse is large, it is a sign of movement to the exterior more than **[536K]** to the interior. This is twofold—movement externally through haemorrhages and sweats, and internally through the belly (bowels) and the stomach (oesophagus).[345] An undulant pulse presages sweats; a large pulse presages simply a haemorrhage. If the movement inward predominates, when the signs of vomiting are absent, such a fever will be resolved by diarrhoea. When vomiting is present, it will be resolved more through vomiting. If in fact vomiting is present and the belly is drawn off markedly, it is through both. An unequal pulse exists in very many crises, and especially when, to some degree, the person is involved in contention and danger. This is much more so when bilious humors flow to the belly and at the same time other signs of vomiting are present, which was what Hippocrates said.[346] In general, the pulses are unequal when the stomach is stung and burdened.

31. Since the essence of the capacity of each of the parts is located in the *eukrasia* in each of them, then it discharges its specific activity strongly, whenever it is most *eukratic*. And of course, it also carries this out badly, when it becomes *dyskratic*, and badly to the extent that it is *dyskratic*.[347] This is why we propose to say next what kind of pulse follows a condition in each of the other parts of the body. We spoke

344 *De febrium differentiis libri II* (VII.273 – 405K).

345 The Latin terms here are *per ventrem et stomachum*, that is, by defaecation and vomiting, respectively.

346 We have been unable to locate this reference.

347 See *De praesagitatione ex pulsibus* I.4 (IX.244K).

about the heart in the treatise on fevers.[348] Adjacent to the heart are the respiratory parts of the animal; these are the lungs and thorax. Thereafter, there are the organs of nutrition—that is to say, liver and stomach, intestines and spleen, kidneys and both bladders (i.e., gall bladder and urinary bladder).

Anyway, if at any time the lung becomes hotter than it normally is, it immediately also heats the heart, and even if it is colder, this is not put off. Then, in summary, it imparts the other δυσκρασίαι/*dyskrasias* more quickly to the heart than to other organs. The cause of this is because they are nearest of all [to it] and to the vessels which are most similar in kind, and flow together through the largest openings into both chambers. Therefore, if the heart is jointly warmed pathologically with the lungs that have been warmed, this produces pulses that occur due to the need of the pulses being increased. If this happens over a long time, there is a danger of it of taking on the hectic fever. When the lungs are cooled, the heart being jointly affected from the beginning will bring the pulses to that change which we showed previously exists, when the need is dissipated. **[537K]** In the fluxes, the inequality of the pulses is common to all, and blockage, obstruction and compression of the most important arteries readily arises from such fluxes. Not less than these, however, there is also the excess regarding the capacity, as the heart lies near to the lungs in this way, and due to there being the confluence through large openings, following thick, viscous and copious fluxes, a greater inequality arises, for these latter obstruct, compress and weigh down more. Accordingly, the pulses become harder than those which accord with nature in the dry δυσκρασίαι/*dyskrasias* and the inflammatory swellings (tumors) or indurations of the lung, just as they also become softer due to the oedematous conditions.

And in the pulses themselves another *differentia* supervenes alongside those previously mentioned; in this, the flux sometimes falls on the rough arteries (trachea and upper airways) themselves alone, but sometimes on the smooth arteries and the veins, or also takes over all the spaces between these. For due to the rough arteries alone being filled, no change in hardness occurs to the pulses. However, when the smooth arteries in the whole animal are stretched, they create a harder pulse. On the other hand, the oedemas render the pulse softer, creating no tension, as is the case with the inflammatory and scirrhous swellings (tumors).

Similar to such δυσκρασίαι/*dyskrasias* are those of the thorax that make the same changes; tumors (swellings) in relation to the amount and kind of each of the humors creating them. The heart is sympathetically affected more slowly with these; more rapidly when the lung is affected. Those with pleuritis always have a harder pulse than those with peripneumonia, for the tension is greater when the hard and dense parts are raised into a swelling (tumor). The pulse is for the most part equal with the swellings (tumors) in the ribs (chest wall), whereas they are totally unequal when the swellings (tumors) involve the lung. When the liver is diseased due to inflammation, the pulses become less hard than with inflammations in the thorax (thoracic cage) by as

348 *De febrium differentiis libri II* (VII.273–405K).

much as they are more hard with lung inflammations. Moreover, the heart is more readily jointly affected due to the liver, although it is situated more remotely than the thoracic cage, on account of the commonality of the inferior vena cava. When the diaphragm is adversely affected without fever, the pulse becomes hard and small, although not to a great extent **[538K]** frequent, but comes to such a change either altogether slightly, or not at all. Moreover, a major sign of pleuritis that is fatal is if the pulse becomes hard, and because of this, small, and on account of both these things and the heat, very frequent. For no one with pleuritis who had such a pulse was saved. I said in the work for beginners that the pulse in those with pleuritis is rapid and frequent, but not excessively large, and that because it makes the artery somehow more sinewy and harder, as it changes it to strength, it deceives the inexperienced, who are unable to distinguish a hard beat from a strong one.[349] But it is not necessary to prolong the discussion on these matters in this place.

The stomach, when adversely affected by a δυσκρασία/*dyskrasia*, will change pulses analogously to those things previously stated, but if a tumor arises in the cardiac orifice of the stomach, the pulse will be harder due to the abundance of nerves, although it will not in fact be hard in the way it is when the diaphragm is involved. However, just as with the pains in relation to the opening of the stomach, the capacity very quickly suffers. By the same token, it doesn't tolerate excess, not even for a little while, but is weighed down and afflicted, making the pulses unequal. Since the cardiac orifice of the stomach has a capacity which is either weighed down by a great amount of material, or weakened, this with good reason makes the pulses sometimes weak, frequent and small, but sometimes unequal. So also, in the other organs of alimentation, you will find the changes of the pulses, recalling those previously spoken of. For these organs are hard and strained and are provided with many hard nerves (sinews). They readily make the pulses hard, while those that are the opposite make them soft. The bladder and uterus do not differ from one another in position, but the bladder is harder and more sinewy. Consequently, when inflamed, it makes the pulses harder more than the uterus does. On the other hand, the uterus will change them quicker and to a greater extent to largeness and smallness and the other *differentiae.*

Now I shall go over δυσκρασίας/*dyskrasias* of the other organs, and first for the sake of practice, of the brain. It is necessary, then, to attend to the consistency of the artery and whether the wall appears hard, either from dryness or tension. If it seems to be due to dryness, one of two things is considered to exist in relation to the brain—either a hot and dry δυσκρασίαι/*dyskrasias* or an excess of bile. If it is due to **[539K]** tension, it is an inflammatory condition. If, however, it is due to both, we must consider the condition itself to be a compound from bloody and picrocholic humor. Those who suffer such a thing due to seething black bile from putrefaction are affected like this. They have hard pulses on account of dryness and tension, and these pulses are, of course, immediately also small due to the hardness, and more

349 *De pulsibus ad tirones* XII (VIII.477K).

rapid than those in accord with nature because the patients are febrile, and frequent because the need is not fulfilled. The moist δυσκρασίαι/*dyskrasias* render the pulses soft over time and the person lethargic. If, however, feverish heat is joined with these δυσκρασίαι/*dyskrasias* as well, the disease is called lethargy, while the pulses not only become soft but also large. If, however, such a condition occurs in the brain due to putrefying phlegm, the disease would also in this way be lethargy, while the pulses will have a certain appearance of tension of the artery and yet it will be soft. In all such diseases, the pulses are large to the extent that the heat is increased. The rapidity on the other hand is either altogether least or not at all increased in these, and the frequency similarly. If morbid cooling occurs in the brain, when the heart comes to a sympathetic (joint) affection, the pulses will become smaller, slower and more infrequent than accords with nature. And if it is without material, with only the δυσκρασίαι/*dyskrasias* of the brain, it will be without any tension. If, however, it is with material, and if this is black bile (melancholic), it will happen that the arteries are under tension, and because of this, the pulses appear harder. On the other hand, if the material is phlegmatic, and not simply so, but if the phlegm is thick and viscid, they will be made tense. On the other hand, if it is moist and airy, the pulses will be soft without tension, while if there is a mixture, they will be soft and have at the same time a certain tension.

Moreover, the thin meninx (leptomeninges or arachnoid mater) does not seem to me to be able to suffer any of the previously mentioned diseases at any time separately from the brain, just as the brain does not without the meninx. And the hard meninx (dura mater) being most distant from the brain can be affected alone. You must know the significant condition of this that occurs. The heart also draws the arteries into a sympathetic (joint) affection, and the pulse will be changed in proportion to the conditions involving the brain, except that in softness and hardness, if it is not soft or hard to the same extent as the brain, but more hard and less soft. For those bodies that are naturally hard suffer **[540K]** hardening conditions, coming to the greatest hardness, just as those that are soft in nature, when they receive moistening and loosening conditions, become completely dissipated and soft. If, therefore, the affection involves a part that is dense and hard, it makes the pulses hard. If, however, the affected part is fleshy, and is either inflamed or indurated, it makes the pulses hard but less than in those previously mentioned conditions. If the conditions are oedemas, the pulses are soft.

Fevers readily follow the inflammation of muscles in the epigastrium and thorax, and because of this the pulses also become large, rapid and frequent, although the pulses never becomes significantly hard in these cases or spasmodic. However, when the muscles in the forearm or shank are inflamed, and particularly in that part of them from which the tendons take their origin, then they are readily seized by spasm also, while the pulses are very easily strained and hardened. Still more, if tendons that are related to the extremities of the hands or feet are inflamed, the pulses in these readily become hard but readily spasmodic. However, pulses become significantly hard also due to induration of the liver or spleen. But in those hard in this way, there is no fear of them being made spasmodic. And certainly, when there is induration (scirrhosity) of the spleen, we have often seen that in no way does hydrops supervene, whereas it necessarily follows in [induration (scirrhosity) of] the liver, and then the pulses have signs of both conditions, the walls of the arteries being more moist and softer, clearly indicating tension arising from the scirrhus, and due to this, also harder. However, dropsy in and of itself, without inflammation of the internal organs, never creates tension involving the pulses. How those things in relation to the uterus change the pulses was stated when we were comparing it to the bladder. If the tunica vaginalis is in the grip of inflammation, the pulses become hard and smaller. When the testes are affected by inflammation, the pulses are not similarly hard, but are large on account of the fever. And these changes are, for the most part, brought about by internal causes.

32. Oftentimes also, changes of pulses arise from some external cause. One must distinguish these as follows: The change occurring from baths, running, massages, and other movements is of short duration and in this way there is a quick return to the original state, **[541K]** so it is possible for you to palpate the pulses, and then leave a short interval before palpating them again, to recognize the change very clearly, However, the largeness of the pulses from wine and food persists for a longer time, and is distinguished by the associated strength, for the pulses occurring with these are strong no less than they are large, but now are assumed only to have become larger. They are not, therefore, changed from these. And in fact, the change from anger also occurs along with strength, and would not otherwise escape someone of diligence paying attention to the eyes and the whole face. The pulses in someone who wishes to hide his anger become unequal and are more unequal in those contending and ashamed. The increase to largeness from semi-digested humors in the body, when being digested and nourished, increases the nature similarly to those who have taken nourishment, and the strength. In the same way, if someone is warmed in the sun or beside a fire, the skin also will be parched in them and not moist as from baths. They are, then, to some degree oppositely affected to those who have bathed to an extent indicated by softness and moistness of the skin. Midway between both are those who have been massaged and have exercised.

Certainly, then, these are the systematic diagnoses of the causes changing the pulses to largeness in both those who are healthy and those who are diseased. The causes that are external are [discovered] in part from enquiry, as when you subject to questioning those who are among enemies, or those caught in an ambush, or those attempting malicious acts. Sometimes, however, when the pulses have attained a large size, it is either because the patient provides for himself a heating medication or by reason of the accession. In these, there is first a need to examine the time of a paroxysm. Next, there are the customs and habits of the patient. For if at some other time, and not at the time of the paroxysm, the pulses take on in addition a largeness, it is more plausible for the change to have occurred from the heating medication and not by reason of a paroxysm. If, however, it is at the time of the paroxysm, it is more plausible that it has occurred by virtue of the symptoms. You will distinguish this precisely if, after leaving a short interval, you again palpate the pulses. For what happens is that the heat from the medication, due to which the pulses were increased, is always dissipated more with the progression of time, whereas the heat from the paroxysm is increased.

In relation to custom **[542K]**, and habit, one must be aware that some are unaccustomed to and fearful of drinks of medications; some on the other hand are accustomed to them. It is similar with customs. For some set traps and refute everyone, criticizing them as not striving after sound knowledge. Others, however, are simple and honest. So then, in the case of the latter, one must not suspect them of having done anything covertly, whereas with those who set traps, it is appropriate to suspect everything. It has been shown in the first book of the work *On the Causes of Pulses*[350] in those in whom the heat is increased slightly, the pulses obviously increase in largeness, but not obviously in rapidity, whereas when the heat is increased still more, the pulses take on still more magnitude, but also clearly increase in rapidity and gain frequency as well, just as in fact also, if they increase to the greatest extent and the need of the genesis of the pulses exceeds what is the largest dilatation, the pulses do not increase in magnitude, because they have increased to the limit, but they add as much as possible to the rapidity and at the same time to the frequency, and when the rapidity cannot be further increased, they hasten on to frequency.

In any sick person, who has pulses that are smaller than accords with nature, which afterwards become larger without external movement, the heat in this person has also become greater than before. Concerning the smaller pulses, the same reasoning applies. Pulses become smaller than they were, either when the capacity is rendered weaker, or the arteries are made harder, or the need is dissipated (is brought to an end). But the capacity has not now become weaker, or at all events weakness would have come together with the smallness; nor has the body of the artery become harder, for in this way not only would the pulses be smaller; they would also be harder. What remains, then, is that the need alone has been changed. Furthermore, it is also quite clear to those who recall the things that have been demonstrated in the first book of those *On the Causes of Pulses,*[351] that such a change has occurred from pulses that are larger than normal, or in accord with nature. For the change from an accord with nature to smallness due to cooling also immediately makes the pulses more infrequent rather than slower, while the reduction from pulses which are large and increased above the moderate to the moderate is also immediately accompanied by rapidity and frequency, due to the fact that the pulses which are increased beyond an accord with nature change in the three classes. Accordingly, the pulses are reduced only in magnitude, while in other respects they continue unchanged, **[543K]** and are slightly larger than accords with nature. For this is associated with a slight increase of heat only.

---

350 *De causis pulsuum* I (IX.1–54K).
351 *De causis pulsuum* I (IX.1–54K).

However, of the other changes to largeness, none comes about without also taking on rapidity or bringing about so much frequency. That which is changed to such a degree by heat as not to involve magnitude alone, but also to change in rapidity and frequency, will have its return path through these same changes according to class. As a result, it will not only be smaller, but also slower and less frequent than before. However, in the heating, the largeness is clearly added to it first, then the rapidity, and then the frequency. Conversely, when there is cooling, the infrequency is clearly added first, then slowness second, and smallness last. For there are as many causes which bring about cooling, as there are those that heat. For cold air contacting externally, as also cold water, or a bath, or a medication anointing the body cools; also prolonged quietude or hibernation, as it were, make the body colder. And of the things effecting change internally, there are inspired air that is cold, nutriment that is phlegmatous, some medication that is cold in nature, and immoderate use of cold water. Of the affections of the psyche, there is prolonged fear. Of the humors contained in the body, there are those that are cold in terms of *krasis*, which were first quiescent and are now moved, or have somehow become altogether thinner than the hot ones. In the case of the heating causes, we must distinguish these analogously to those mentioned. For the actual essence of the large and small pulses exists in the circumference of the artery, while the length is kept equal for the most part, although it does not seem equal, because some part of it is hidden. I have already spoken about this in the second book of *On the Causes of Pulses*.[352] It is not necessary to speak twice about the same matters.

33. Since it has been stated what large and small pulses signify, we should now move on in the present circumstances to the discussion on frequency and infrequency. That is to say, the pure and least excrementitious heat makes the dilatation larger and at the same time also more rapid, while the contraction[353] is made turbid by the putrefying superfluities. However, when the dilatation is earlier or large, the pulses become more frequent, **[544K]** which will indicate that the large amount of pure heat in the body has been increased, whereas, when it is the contraction that begins earlier, this will indicate an abundance of impure and excrementitious heat. In like manner too, the pulses also become more infrequent in relation to the external pause, if it is with shortness of the dilatation, and will show the heat in the animal to be cooled, whereas, if due to the contraction being slowed, it will show the smoky superfluities to have become less. Conversely, in relation to the internal pause, if due to the contraction being delayed, the pulses are made frequent, it will show an abundance of smoky superfluities building up. If, however, it is due to an earlier beginning of the dilatation, it will show the heat to be increased. Conversely, if the pulses are infrequent in relation to this same pause, due to the contraction being diminished, it will show the superfluities to have been reduced, whereas, if due to the slowing of the dilatation, it will show the heat to have been diminished.

---

352 *De causis pulsuum* II (IX.55 – 104K).

353 *Synopsis:* τὴν διαστολήν; *De praesagitatione ex pulsibus:* τὴν συστολήν.

The pause of the pulses is twofold: Prolonged is what exceeds the time of two pulses, while short is what is within the time of two pulses. The pause involving the times of one pulse is the most moderate of the intermissions and there are many who have been saved from this, both old people and children, although none of those in their prime. But some, due to ignorance, think a long infrequency is an intermission, and believe some in their prime are saved from such a *kakosphyxia* (pulse abnormality). This is, then, also serious, and particularly in those in their prime, but is not altogether fatal; of course, in children and old people it is more mild. Death occurs suddenly with the intermittent pulses analogous to apoplexy. In both, the heat in the heart is quenched, being deprived of respiration.

Infrequency is distinguished from intermittency of pulses by the length of time, while sometimes infrequency also coexists with equality of all the beats, although the intermittent pulses of necessity become unequal. Sometimes they are found intermitting three beats, sometimes four, and sometimes five or more and pausing one time or more. Therefore, if the major cooling involves the actual body of the heart to such a degree, it makes the pulses weak, slow and small to the extent **[545K]** to which it is dominant. The stirring up from one of the other factors draws out the infrequency more than the other classes mentioned which depart from an accord with nature. Therefore, when the actual body of the heart itself takes on the cooling, death is more inescapable, therefore infrequency of the pulses is, in fact, fatal up to a point.

Frequency is the opposite to this, just as intercidence is to intermittency. Of course, the intercident pulses arise from unequal frequency, due often to the heart carrying out its function more deficiently than need requires. This in turn occurs, either when the heart is weighed down by excess, or when the arteries have been obstructed or narrowed, which I think are also causes of the intermittent pulses. Of course, the intercident pulses often brought a crisis, for it is as though Nature were preparing for an unexpected contest—and some people die from such crises. This doesn't mean the condition of the intercident pulses is much more serious than that of the intermittent pulses. Many more people die from the intermittent pulses, and in the same way from those that are very infrequent than from those that are frequent. For surely all the burning fevers have frequent pulses; some of these are without danger. On the other hand, the frequent pulses are dangerous in the collapses (syncopes), due to the weakness which is inseparable from all collapses.

It is good to consider the extremes in all the classes of the pulses. Thus, extreme frequency exists no less in those who are collapsing than in those with burning fevers, who are not altogether fatally disposed, whereas there is no time when infrequency is without danger, just as there is not with weakness or smallness. For these three *differentiae* of the pulses, when they come to an extreme, are continuously dangerous; However, the extreme largeness, whenever it coexists with morbid softness, is dangerous, although when it exists with a healthy consistency of the arterial wall, it is not unnatural. For it seems also that the class of rapid pulses is less dangerous than that of slow pulses, and the same applies certainly to the peak of rapidity compared to the peak of slowness. Thus, the peak of slowness, as also that of infrequency, comes about when the innate heat is quenched, while an increase to rapidity arises due to the need urging things on and altogether **[546K]** by the capacity being strong.

And even Archigenes errs in this, taking the formicant pulses to be rapid. For such pulses on the whole are neither rapid, nor come to a peak of frequency. All the pulses deviating from the moderation of an accord with nature to a great extent are not good, but the weakest, slowest and most infrequent are worst of all. Next in order to these are the smallest, softest and hardest, and then the most frequent, although not the most rapid in fact, nor the largest, but these are moderately dangerous. Of all the extremes, only the strongest is good. And, in fact, the bounding pulses, which are from the class of those that are interrupted in a pause, sometimes supervene in such affections of the heart, just as the dicrotic pulses are also from the class of those that are drawn in the opposite direction inward, and this itself presents together sometimes in the non-uniform δυσκρασίας/*dyskrasias* of the heart, and more if an abundance of sooty superfluities in this is noteworthy in many of its parts. The inequality itself sometimes exists in the position of the parts, comparing up and down, forward and backward, and left and right, as the artery seems to be moved from place to place. In the movement, there is quicker or slower, earlier or later, stronger or weaker, in a greater or lesser time, or always, or not at all, when it is moved. In relation to each part in which there is clearly interruption is the class of those to which the bounding (gazelle-like) pulses belong. Similar also is the recurring, from which there are also the double-beating (dicrotic) pulses. And the bounding pulses are those unequal in a single dilatation, when the second movement after the pause is more rapid and stronger than the previous one. Moreover, we have related the reason for the name in the treatise, *On the Differentiae of the Pulses.*[354] Therefore, the intermittent inequality in one pulse is the most grievous of all inequalities, for it does not allow there to be simple movements in accord with nature. Rather, it stands as a hindrance to, and deprives the arteries of the movement, not through three, four, or any greater number of pulses, but immediately in the very first beat itself and it warns of impending death.

34. Therefore, the whole matter put before us now has an end, for reviewing all the conditions in each part, **[547K]** we have shown what kind of pulse each change creates. Now we shall try to bring together the synopsis of these [matters] under a few headings. For surely the diagnoses of those conditions that presently exist is nothing other than the discovery of the previously occurring things that caused the conditions. If, therefore, we have adequately distinguished all conditions, and all the effecting causes in other treatises, nothing more should remain of those things necessary for prior recognition of either past or present [factors]. [Because also the prior recognition of future events is from present events, as we have often shown through those things stated earlier, it is now necessary for us to run through the whole discussion under headings,

354 *De pulsuum differentiis* I.28 (VIII.556K).

taking the beginning from the recognition of those things we expect to occur in the future.][355]

First and foremost, then, is to what end the disease will come—whether to death or to recovery. Another and second to this is in what time particularly this will be. And third in addition to these, what the manner of death or recovery will be. The prognosis of all these things is from the conditions. From these, as we showed, the kind of the disease and the weakness and strength of the capacity will be recognised. When we compare these with others, we shall prognosticate those things that are going to come about. For the knowledge of the disease is from the affected place and the condition in this, whereas knowledge of the capacity is from the εὐκρασία/*eukrasia* and δυσκρασία/*dyskrasia* of the solid bodies. When we compare these with others, if they are of the kind where the capacity is able to bear the peak of the disease, the person is inevitably saved, as long as no mistakes are made externally. If, however, it is not of this kind, the person will invariably die.

The prognosis of the time in which each of the things mentioned will come to pass is recognized through the following things. Those who are going to be saved will be saved in that time in which Nature completely overcomes the morbid condition. On the other hand, someone will die in that time in which the disease will be to such an extent stronger than the capacity that one of the functions essential to life is destroyed. Furthermore also, the manner of death and that of salvation will be prognosticated from these same things. For when the functions essential to life are destroyed, some bring a suffocation, some a collapse of the physical (vital) capacity, and some what is analogous to either. Those which bring about suffocation are from the respiratory conditions, **[548K]** while some also coexist immediately in the heart itself. Some bring on collapse from the so-called stomachical conditions,[356] from which patients collapse, and besides this, from chronic conditions involving the liver from which patients suffer atrophy, and from chronic or severe pains, and further, from all the hectic fevers.

The weaknesses of the psychical capacity follow the major δυσκρασίαι/*dyskrasias* of the brain, while the suffocations as it were, follow the filling up of the cavities (ventricles) of the brain, and the obstructions of the channels coming to them. Therefore, whoever diagnoses all things with particular care will also have the most complete prior knowledge of future events. Above all, it is necessary to recognize the state of the vital capacity which creates the pulses; this is how the diagnosis should be made, for sometimes it is weak by reason of itself alone and sometimes it is weighed down by an excess. Three classes of inequality exist: in relation to largeness and smallness; in relation to rapidity and slowness; and in relation to frequency and infrequency. Therefore, if the pulses are strong and large more than the opposites, the capacity

355 There is a lacuna in the Latin text. See Cortés Gabaudan and Martínez Manzano 2013: 108n25.
356 This term has a range of meanings. Here these conditions should be considered to include at least those of the oesophagus and stomach.

has suffered less due to the excess, whereas if they are small and weak, it has suffered more. Of the inequalities in relation to a single beat, some were said to occur when the movement is cut through by a pause and some when it remains continuous. All these indicate the capacity is stronger compared to all those that make the movement weaker and slower. Accordingly, in such a state of the pulses, high pulses will show nature to already be superior to the excess. Pulses that are higher and at the same time stronger indicate both a strong capacity and an unhindered movement. It is necessary, under these circumstances, to examine the movement of the contraction[357] and here to distinguish the character of the crisis, for it is better for the crisis to customarily follow, when the rapidity in the contraction is brought to an end to the greatest extent. If, however, the other signs of concoction exist, (I have written all these in the work, *On Crises*)[358] such a condition being present is a prior indicator of the best crisis. If, on the other hand, the rapidity of the contraction is dissipated only a little, or also the disease appears to have not yet been concocted, the crisis will occur [through excretion][359] and will not be altogether good.

What must be known is that the same capacity is not vital, nutritive and physical (natural). For this alone generates the pulses. The physical capacity then draws what is familiar for nutrition, **[549K]** separating out what is not of this kind, while of course retaining and obviously altering what is familiar. That is not to be found yet in plants, or in animals which are cold, like oysters and those things that grow on rocks, or in some other solid bodies. However, those animals that are hot, and particularly those that are sanguineous, have such a capacity that is very obvious and very strong. The vital (physical) capacity brings about the crises, and as far as the arteries are participants in this, they get rid of what is superfluous. However, it is through the other capacity, the pulsatile, of which only they partake, that they eliminate the smoky superfluity in the contractions, just as in the dilatations, they draw in what cools themselves.

The dilatation is always equal in magnitude to the contraction, except when, by some chance, we apply our fingers suddenly, the change should occur. It differs, however, in that sometimes the movement outward prevails and sometimes the movement inward. When the inward movement prevails, it signifies a certain smoky heat to be in the arteries. When the outward movement prevails, there is a fiery and pure heat; on account of this cause, a high pulse is a harbinger of excretion. However, the abundance of the elimination is prognosticated from the plethoric signs occurring beforehand and from the fiery heat present. Due to this same heat, the pulses appear large and high.[360] So, for example, the worst of the eliminations are those in which the pulses are not large and appear rapid in the contraction. When an unconcocted disease exists, if

---

357 The Latin translation has "contraction", while the reconstituted Greek text has "dilatation". Sense suggests the former.

358 *De crisibus* (IX.550 – 768K).

359 This is in the Latin translation, but not in the reconstituted Greek text.

360 The Latin text has *et alti* whereas, in the corresponding passage in *De praesagitione ex pulsibus*, "and high" is omitted.

the pulses are devoid of tension, there is danger of syncope. If the disease is more inflammatory, when the change is brought about in the unequal pulses to greater strength and size, the crisis is due to abscess formation. When the pulses are completely eliminative and there are present in them the so-called haemorrhagic signs, patients will bleed from the nostrils. However, when these signs are absent, look for the elimination from some other place, as was stated in the treatise *On Crises.* These are the matters we proposed to raise in the *Synopsis of the Pulses.*

III *On the Differentiae of the Pulses* (*De pulsuum differentiis, libri IV* VIII.493–765K)

Book I is the key book for understanding Galen's system of pulse diagnosis. It comprises 30 sections as summarised below:

1–2. Introductory, including a statement of intent, some historical observations, and the importance of focussing on matters rather than names.

3. The basic statement of what can be evaluated in the arterial pulse/pulses:

- change in the movement, considered in the three dimensions of length, breadth and depth;
- *tonus* of the artery (strong/weak);
- hardness/softness of the arterial wall;
- a full or empty arterial lumen;
- equality or inequality of any of these in a single pulse;
- the relation of the time of dilatation to the time of contraction.

Two aspects apply to a series of pulses—equality/inequality and regularity/irregularity. These are termed "systematic".

4. A tabulation of the 27 possible combinations pertaining to the changes in the three dimensions (distensions).

5. On *tonus* and the body of the artery, and particularly the distinction between full/empty and hard/soft.

6–7. These two sections deal with the distinction between rapid/slow and frequent/infrequent. The first applies to the rate of the dilatation, the second to the interval between dilatation and contraction.

8. On rhythm—what it is and what terms apply. Six terms are defined: *eurhythmic, enrhythmic, arrhythmic, pararhythmic, heterorhythmic*, and *ekrhythmic.*

9. A detailed and rather complicated discussion of equality/inequality and regularity/irregularity.

10. Some terminological issues relating to inequality—tapering (mouse-tailed, dying away gradually), deficient tapering and recurrent tapering.

11–15. Further considerations of inequality in both a single pulse and a series of pulses (systematic), culminating in a tabulation of inequalities.

16. A detailed discussion of dicrotic pulses.

17. Further consideration of unequal pulses.

18. Combinations of intermittent pulses.

19–21. Inequalities in the amount of dilatation, relating to different parts of the artery (19); physical/vital *tonus* (20); and the body of the artery, (21) respectively.

22. The issue of fullness and emptiness.

23. On judging the heat of a fever by palpation of the pulses.

24. On inequality due to the position of the artery.

25–28. Consideration of several types of pulse: undulant and vermicular (25); formicant (26); clonic and spasmodic (27); bounding (28).

29–30. A rather complicated pair of sections on inequality in relation to frequency/infrequency and rhythm.

Book II contains 10 sections, and is predominantly a general criticism of what Galen sees as erroneous views on the pulses.

1–2. These sections offer general criticism on the misuse of existing names and the creation of neologisms.

3. Predominantly directed at Archigenes and largely in dialogue form. The issues considered are the formulation and relevance of definitions, and the importance of understanding homonymy.

https://doi.org/10.1515/9783110612677-009

4. On the qualities identifiable in the pulses, with particular attention given to Archigenes' use of the term "full pulse".
5. Further discussion of naming and defining, with considerations of variations in dialect.
6. On the qualities of the pulses, primary classes and *differentiae.* Specific criticism is directed at Archigenes, and to a lesser extent Agathinus.
7. Further criticism of Archigenes, and particularly his perceived confusion about both names and matters in describing the qualities of the pulses. Galen focusses on the variations of the magnitude of change in the overall distension of the artery—the three distensions/dimensions, vertical, horizontal and longitudinal.
8. A summary of Galen's own views on the primary qualities observed and their division into classes and *differentiae.*
9. A brief discussion of the establishment and recognition of categories and classes, and the general importance of familiarity with the procedures and criteria involved.
10. Galen considers the term *diaphora* and what he identifies as the 10 *differentiae* in relation to the classes and kinds. He refers particularly to Archigenes' misuse of terms, with a sideswipe at Chrysippus.

Book III, comprising 7 sections, is similar in its critical nature to Book II. Again, the main issue is nomenclature and the main target Archigenes, although Magnus and others are included.
1–3. On the relation between the terms *tonus* and strength (*sphodrotēs*). In section 2, he gives three views:
- his own, that they are different;
- Archigenes' view, that they are the same;
- Asclepiades' view, that they are just names.

4. A long and somewhat tedious account of what he sees as the needless proliferation of names due to Archigenes. It is hard to evaluate this without seeing Archigenes' own book. Galen makes the basic point that there is an orderly sequence to the study of the pulses:
- observe the pulses, being mindful of the determinants of class;
- apply names to the genuine *differentiae*;
- describe how to recognise them, including the method of palpation;
- determine the causes;
- identify the prognostic significance.

5. A brief statement on the terminology applied to the strong pulses, and the method of palpation needed to recognise the strong pulse.
6. On the vexed question of the full/empty pulse, with a long digression on metaphor generally. Does full indicate:
- the quality of the wall of the artery?
- the amount of material contained in the lumen of the artery?
- the nature of the contained material?
- the innate *pneuma* and a certain potency or quality?

7. A brief consideration of the terms hardness and softness, and errors in their use, with reference to Plato and Aristotle.

Book IV comprises 17 sections:
1. A brief summary of Galen's own works, and further criticism of Archigenes on the matter of naming the various pulses.
2. A detailed examination of definitions of the pulse with examples.
3–17. These sections deal with various men who have offered definitions. These are as follows:
   (3) Heracleides the Tarentine, the Empirics and Herophilus;
   (4) the Herophilean, Alexander Philalethes;

(5) Demosthenes, with a digression on the constituents of the pulse—dilatation, contraction and the two pauses—and the similarity of the pulses to respiration;
(6) Baccheus and Erasistratus;
(7) Aristoxenus;
(8) Zeno, with a digression on whether the heart is artery-like;
(9) Chrysermus;
(10) Heracleides the Erythraen;
(11) Agathinus, with consideration of pulsation in the abdominal aorta;
(12) Archigenes and Asclepiades;
(13) Magnus;
(14) Athenaeus;
(15) Asclepiades;
(16) Moschion;
(17) Erasistratus and Apollonius, son of Straton.

Book I

1. [493K] I wish I were able to learn and teach the matters apart from the names for them, so that the business regarding their expression would not exist for us, added into the bargain to the art that is already long by virtue of its own theory. Since, however, it is necessary to make clear through names those matters we wish to make clear, there is here an abundance of sophisms for the Sophists as well as useless waste of time. But for others—we who strive after knowledge of things and not experience in using words—there is unwillingly but inevitably **[494K]** added, through the threats of the Sophists, the business of words. For it would be best to come to any agreement whatsoever about terminology and hasten to the theory, as I did in the book I previously wrote on the *Pulses for Beginners*. But because those who are expert in names alone, and ignorant of the actual practices of the art, are always making a nuisance of themselves, never allowing us to breathe without insult, the business of the terminology is always there already, waiting inevitably for us.

It is possible to find the ancients simply using names with only one thing in mind —always to make the concept clear. The moderns, on the other hand, not only argue and dispute over each syllable, but also never cease to contend about the terms.[361] For example, in order that we may come to a judgement on the whole from one part of the matter before us, someone might say the strongly beating pulse is strong. But someone else who comes along might say this is not so, but that the pulse which is large, full and rapid is a strong pulse. Another might say, however, it is the pulse that is strong and rapid at the same time. And yet another, severely criticising them, might say a pulse that is large and rapid at the same time is a strong pulse. **[495K]** A fifth, however, not tolerating this, and reproving still more solemnly, might

361 The addition of "not" in this sentence follows several manuscripts, e.g., F1, F2 and Va1.

say the full and rapid[362] pulse is the strong one. And, by Zeus, some sixth person, in addition to them [all], might say the large and full pulse is the strong one. But if also a seventh and an eighth were present among them, one saying this is no different from the full pulse, and the other saying it is the same as the hard pulse, how great a battle do you think there will be? Will it not clearly be such as we often see in the Temple of Peace in the case of the patients themselves? And the result is that they resort to fists, when abusive speech fails them.

But the statements of those who are now disturbing us are less severe, even though they are very annoying; it is the statements of those who have taken it upon themselves to write that are insufferable. For refutation of the uselessness of the enquiry is not available to us. But they don't see this, all writing as follows: "such a pulse is strong, such a pulse is full, and such a pulse is frequent," when it is necessary to remove the "is" and substitute "most of all", and if it is possible, "is called"; or failing this, "should be called," and now in this way to teach what kind **[496K]** such a pulse they were indicating is in nature, which they called, as may happen, "strong". For we ourselves certainly do not practise the art concerning the pulses, for the sake of calling them something, or to establish their names, or to try to teach others, but that we may recognise what exists at the present, and prognosticate for our patients what will happen in the future. These are things ascertained from the matters, not from the names invented for them. After all, names are of no benefit to us for the knowledge of matters but only for the teaching, which it is also possible to accomplish by consensus. For if someone were not to assign names to matters at all, he would still be able to recognise them, and would know precisely what they naturally reveal, and he would seem no worse off on this account than those who name them would be. And if, in attempting to teach, he assigns names that seem right to him, he would not be inferior. I know, then, if anyone should wish to call the strong pulse Dion or Theon,[363] he would not err in teaching the several causes producing this, what condition it is a sign of, and what it will result in, nor would he seem to be mistaken. **[497K]** For it is enough for someone who is going to teach anything whatsoever in this way to establish names, but it behoves someone who goes back to the ancients, to teach what was customary for those men, without here again enquiring into whether they named things correctly or incorrectly, or daring to change anything, or finding fault with them, as the Sophists do. For all these things are superfluous and outside the scope of our art. It is not, in fact, the actual knowledge of the correctness of names, but of matters [that is important]. Men don't send those who don't use names properly to doctors; they send those who have need of health. Therefore, this discussion does not propose to teach more than the terms customary and familiar to us and to other doctors. I shall be well pleased if it achieves this moderately well.

---

362 ταχὺν: F1, F2 and Va1; παχὺν: K. This is a recurring variation.

363 Generally used as the names of hypothetical people.

2. Of all those we know, Hippocrates was the first to write the name of the pulse, and not to seem ignorant of the art concerning it, although he did not do much more work on this aspect of the art, nor did he assign a name to every movement of the arteries. And Erasistratus seems to have done something similar to him. But here in the specific **[498K]** discussions about these men,[364] I shall say particularly what name they established, and how they differed from or agreed with one another. The book *On Palpitations*, composed under the name of Aegimius, whether this was Aegimius of Heleios himself or someone else, specifically uses the term,[365] names every movement of arteries "palpitation", and this is very much according to the custom, not only of other doctors, but also of laymen. However, the usage of Praxagoras and Herophilus still prevails, even to this day. These men call every perceptible movement of arteries a pulse. All those after them do so too, even if they differ in their definitions. This will be shown elsewhere.[366]

Consequently, following the prevailing custom, let us now consider how many *differentiae* of the pulses there are in all, and what the name of each of them is. And indeed, people contend about the stated term itself, some thinking it right to apply it in some way to the number of classes of pulses, whereas others say they ought not be called classes but kinds, and still others say it should be applied to qualities, and others again to *differentiae.* And whatever name is seen as appropriate for each is applied, and there is prolonged and ceaseless contention **[499K]** about the correctness of each. And as with classes, so too with kinds, *differentiae* and qualities, they introduce as well an unnecessary enquiry, losing sight of the use of such an issue. Everyone who wishes to investigate the pulses comes, I think, to the same thing—what do they signify by their natures? Although they cannot accomplish their objective without investigation of classes and kinds, they do anything rather than this. But here we avoid this nonsense of theirs, and when we turn to what they might wish to bring forward, let us pursue utility, saying this much in advance—to those who wish for definitions in respect of the pulses, or to learn about the correctness of their names, everything is written in the book following this, i.e. the second book.[367] This particular book has a single use only, containing what is necessary but touching on none of the sophistical investigations, so that if someone should ultimately wish to omit those, he could do so. As for the other matters that concern diagnosis of the pulses, their causes, and further, the

---

364 In this clause manuscripts Firenze 74.18 and 74.28 are followed.

365 The Greek term is παλμός may mean quivering, throbbing, palpitation, pulsation, twitching etc. See Aristotle, *De Respiratione* XX, 479b21 (Engl. transl. Ross in Barnes 1984: 761) where it is used to describe an abnormality of the heart. Aegimius is also referred to in Book IV (*De pulsuum differentiis* IV.11 [VIII.751–752K]).

366 For passages referring to Praxagoras on this, see Steckerl 1958: 61–65, and to Herophilus, see von Staden 1989: 262–287. Galen considers various definitions of the pulse in Book IV of the present treatise. See also Lewis 2017.

367 Following manuscripts F2 and Va1 which have μετὰ τοῦτο τὸ β' βιβλίον instead of μετὰ τοῦτο τὸ πρῶτον βιβλίον (K). This book is presumably the second book of the present 4-book treatise on *differentiae.*

prognosis from them, there is only a single book necessary for each. Those have been written to annoy the Sophists, **[500K]** and to follow closely the writings of others, knowing what signification they place on each of the terms. Therefore, let us say now how many primary classes of the pulses there are, and next their *differentiae*, making a start to this particular discussion.

3. Since the body of the artery is hollow, long and round, and has a continuous two-fold movement composed of opposing components, either contracting on itself from all sides, or dilating in every direction, and since this same twofold movement is termed the pulse, it is necessary that two pauses should occur in each. One is after the dilatation and before the contraction, and the second after the contraction and before the dilatation. The touch of those who are practised recognises these two pauses and reason proves them no less.[368] The artery would not begin the opposite movement before the prior movement has ceased. But to cease is to stand still and be at rest, so there is a pause between the movements. It is, therefore, necessary for there to be a certain time for all these movements, specific to each. **[501K]** However there is another, that of the pauses, and a specific time for each of these in turn. And since an artery has three dimensions, just as every other body does—length and breadth and depth—it is necessary that in each one of the dimensions, there should be a certain amount of dilatation and contraction. Furthermore, it is necessary for the artery itself to have some sort of *tonus*,[369] so that it functions either with difficulty and weakly, or readily and strongly; and also for the actual wall of the artery to be either soft or hard, and its internal space (lumen) should somehow be empty or full, and that in relation to all these, there is sometimes equality (evenness) and sometimes inequality (unevenness), and some proportion of the time of dilatation to the time of the contraction. Apart from these, there can be no other difference in a single pulse. For whether it is necessary to examine carefully the time of the movement or pauses, or the amount of the dimensions in which they move, or the quality of the actions, or of the wall of the arteries, or of the lumen, or to compare those things having an equivalence to one another in some way, while those things that are clearly able to be classed together are all stated, **[502K]** and nothing else remains for the discussion in any way. All the *differentiae* of classes existing in a single pulse have now been stated. Two others are added, called "systematic".[370] On the basis of these, many pulses are compared with one another, and we estimate equality and inequality, and regularity or irregularity.

4. There are many *differentiae* in terms of kind of each of these classes, which I shall now go through in detail, beginning from the class of the movement itself. Of necessity, a movement is either moderate and according to nature, or more rapid some-

368 Subsequently Galen refers to the former as the external pause and the latter as the internal pause.
369 *Tonus* (τόνος) is retained as a technical term referring to an intrinsic property of the arterial wall—see *Oxford English Dictionary* 1933: "the proper elasticity of the organs", and "*tonus* is one of the independent endowments of arteries." Also, see Allbut 1890: VII.109, "whence comes true loss of tonus."
370 LSJ defines the term "systematic" (συστηματικός) as "the pulse beats taken collectively." See also *De praesagitione ex pulsibus* II.4 (IX.279K).

how, or slower; so that in this class there are three *differentiae* of pulses in all: rapid, when the movement of the artery occurs in a short time; slow, when it occurs in a long time; and moderate, when it occurs in a moderate time. The *differentiae* of pulses in terms of the amount of dilatation, when conceived of in relation to a single dimension, are nine in number; that is, three *differentiae* for each one of the three dimensions. In respect of the length of the artery, there is what is moderate in this, what exceeds this and what falls short of it. **[503K]** Those that exceed are called long; those that are deficient, short. In relation to breadth, there is what is moderate in this and there are two deviations from moderation; what is beyond moderate is broad, and the opposite to this is narrow. And in the same way too, in relation to depth in the dimension of the artery, there will be moderate, high (deep), and low (shallow). These nine *differentiae* of the pulses are those considered in one dimension. In the three dimensions of the pulses at the same time there are twenty-seven *differentiae*.

These may be clearly recognised in tabular form as follows (Table 3). One will be long and at the same time broad and high (deep); let it be written first of all. And another will be long and at the same time broad, and moderate in depth; let this be written second. And another will be long, broad and low (shallow); let this be written third. Following these, all the rest will be identified by the same method, the first two dimensions remaining the same and the third and last changing in three ways, as follows: the fourth is long, moderate in breadth and deep. The fifth is long, and moderate in breadth and depth. The sixth is long, moderate in breadth and low (shallow). Then, next in turn are three others, the first dimension, length, is preserved, **[504K]** the second the narrow, while the third changes in three ways. There are nine, in which one dimension (length) remains long, while the other two change in every way. Next there are nine others, with the dimension of length remaining moderate and the others changing. And there are again nine, with the dimension of length remaining short, and the others changing.

Table 3: Combinations of Variables in Dimensions (Distensions) of a Pulse

[Length]	[Breadth]	[Depth]	[Name]
Long (μακρὸς)	Broad (πλατὺς)	High/deep (ὑψηλὸς)	Large (μέγας)
Long (μακρὸς)	Broad (πλατὺς)	Moderate (σύμμετρος)	
Long (μακρὸς)	Broad (πλατὺς)	Low/shallow (ταπεινὸς)	
Long (μακρὸς)	Moderate (σύμμετρος)	High/deep (ὑψηλὸς)	
Long (μακρὸς)	Moderate (σύμμετρος)	Moderate (σύμμετρος)	Thin/weak (ἰσχνὸς)
Long (μακρὸς)	Moderate (σύμμετρος)	Low/shallow (ταπεινὸς)	Thin/weak (ἰσχνὸς)
Long (μακρὸς)	Narrow (στενὸς)	High/deep (ὑψηλὸς)	
Long (μακρὸς)	Narrow (στενὸς)	Moderate (σύμμετρος)	Thin/weak (ἰσχνὸς)
Long (μακρὸς)	Narrow (στενὸς)	Low/shallow (ταπεινὸς)	Thin/weak (ἰσχνὸς)
Moderate (σύμμετρος)	Broad (πλατὺς)	High/deep (ὑψηλὸς)	Thick (ἁδρὸς)
Moderate (σύμμετρος)	Broad (πλατὺς)	Moderate (σύμμετρος)	
Moderate (σύμμετρος)	Broad (πλατὺς)	Low/shallow (ταπεινὸς)	
Moderate (σύμμετρος)	Moderate (σύμμετρος)	High/deep (ὑψηλὸς)	

Table 3: Combinations of Variables in Dimensions (Distensions) of a Pulse *(continued)*

[Length]	[Breadth]	[Depth]	[Name]
Moderate (σύμμετρος)	Moderate (σύμμετρος)	Moderate (σύμμετρος)	Median (μέσος)
Moderate (σύμμετρος)	Moderate (σύμμετρος)	Low/shallow (ταπεινὸς)	
Moderate (σύμμετρος)	Narrow (στενὸς)	High/deep (ὑψηλὸς)	
Moderate (σύμμετρος)	Narrow (στενὸς)	Moderate (σύμμετρος)	
Moderate (σύμμετρος)	Narrow (στενὸς)	Low/shallow (ταπεινὸς)	Thin/weak (ἰσχνὸς)
Short (βραχὺς)	Broad (πλατὺς)	High/deep (ὑψηλὸς)	Thick (ἁδρὸς)
Short (βραχὺς)	Broad (πλατὺς)	Moderate (σύμμετρος)	Thick (ἁδρὸς)
Short (βραχὺς)	Broad (πλατὺς)	Low/shallow (ταπεινὸς)	
Short (βραχὺς)	Moderate (σύμμετρος)	High/deep (ὑψηλὸς)	Thick (ἁδρὸς)
Short (βραχὺς)	Moderate (σύμμετρος)	Moderate (σύμμετρος)	Thick (ἁδρὸς)
Short (βραχὺς)	Moderate (σύμμετρος)	Low/shallow (ταπεινὸς)	
Short (βραχὺς)	Narrow (στενὸς)	High/deep (ὑψηλὸς)	
Short (βραχὺς)	Narrow (στενὸς)	Moderate (σύμμετρος)	
Short (βραχὺς)	Narrow (στενὸς)	Low/shallow (ταπεινὸς)	Small μικρὸς

[506K] There are these twenty-seven pulses which pertain to the amount existing in the three dimensions at the same time. Two of them have names agreed to by all—the first and the last written. The first is called "large" and the last "small". All the rest of the large number have no names. For the pulses called "thin" and "thick" do not indicate one of those from the table; these terms are applied generically to many at the same time. Thus, those in which the dimension relating to length somehow exceeds the other two, they call all "thin", while of those in which the remaining two exceed this are all, conversely, what they call "thick".[371] As a result, in the table the fifth and sixth, the eighth and ninth, and in addition to them, the eighteenth are called "thin"; and opposite to these, the tenth in the diagram, the nineteenth, twentieth, and in addition the **[507K]** twenty-second, and twenty-third are called "thick". So these are the more generic names and common to many. Sometimes people name these same pulses in another way, calling weak "thin" and strong "thick". But we do not have a specific name for the moderate in the three dimensions; this alone among the twenty-seven is in accord with nature. But in discussion, we signify this by saying it is moderate in the three dimensions, or the median of large and small, or in accord with nature in respect to the quantity of dilatation, or moderate in respect to the quantity of dilatation, or in whatever other way we hope what we say will be most clear.

This is why it occurs to me to wonder at those who are particularly clever with names only, always demanding what such and such a pulse must be called, if it is not even because of this that they cease from contention, when they see many pulses not having a specific name, but so no harm comes to the teaching from this, since matters can be made clear by description. At any rate, to call the pulse in some instances

371 The Greek terms are ἁδρός (thick, strong, stout, bulky) and ἰσχνός (thin, lean, dry, weak, light).

short, narrow, and shallow (low) is like saying an animal is a terrestrial and two-footed animal, and the name of the matter of which **[508K]** this is the definition, is man. In this way too, in the pulses the name "large" is a description of a pulse that is long, broad, and deep. And again, another description of a pulse is short, broad, and shallow, but there is no name. But we shall speak again about these matters.

5. Moving on now to a third class of the pulses—that relating to *tonus*—let us mention also the *differentiae* of this. There are three: one is strong in *tonus*, and this is called "strong"; one is weak and this is called "weak";[372] in between these there is one with no specific name—this we indicate by a description. The fourth class, which exists in relation to the body of the artery, is also divided into three *differentiae* itself, and is called among most doctors, and especially those of recent times, by the terms "full" and "empty".[373] Here too, what is in between both is without a name. With us, however, this is not so. One of them is called "hard" and the other "soft";[374] and both signify the consistency of the artery. But whether it is more correct to name them this way or that, whoever is interested in this matter **[509K]** will learn from what follows. Now we must bring to an end what lies before us.

There is another class of pulses which indicates, as they say, and clearly,[375] the contents of the artery; this seems to me to be in need of a long discussion for accurate recognition. Therefore, I shall state here what is useful for the present circumstances and sufficient for beginners: three *differentiae* of this class of pulses are supposed, and are named in one instance "full", and in another "empty", while the one in between both has not acquired a specific name. Thus, a "full" pulse, as Archigenes defines it, is one which reveals the artery as firmer when touched, so giving the impression of being filled with fluid. The "empty" pulse, on the other hand, is one which makes the raising of the artery seem like bubbles, so that, under the pressure of the fingers, it collapses, lacking a firm foundation. These, then, are the five classes of the pulses coexisting in a single movement of the artery, and there is equality and inequality in each of those things which we call the *differentiae*, whenever we should speak also about the other inequality, which exists in many pulses.

6. The other *differentiae* of the pulses, for those who say they do not wholly perceive the contraction, **[510K]** exist in the time of the intervals, for so they name the time between perceptible movements, and besides, in the description of the time of the movement itself and that of the pause; for so they name the interval. And for those who say they also perceive the contraction and obviously the pauses between the two, which they call pauses and rests[376] and absences of movement, other[377] *differ-*

---

372 The Greek terms are σφοδρός and ἀμυδρός, which are generally rendered "strong" and "weak" throughout these translations.

373 The Greek terms are πλήρης and κενός, generally rendered "full" and "empty".

374 The terms here are σκληρός and μαλακός, rendered consistently "hard" and "soft".

375 "And clearly" was added following the manuscripts F1, F2 and Va1, which have φασι καὶ σαφὲς τὸ τῆς instead of φασι, τὸ τῆς (K).

376 "Rests" (ἠρεμίας), present in K, is omitted in the manuscripts F1, F2 and Va1.

entiae exist and are many more. Therefore, it is justifiable to speak more clearly about each of these separately, and first about those who say they perceive the dilatation only, so that, having become practised in things that are more readily recognized, we may understand the more difficult with greater ease.

When an impact of the touch falls upon the artery, we have perception of the other things I spoke of previously and the lapse of time in the movement. But again, when the second movement begins, we note another time[378] in between the end of the first movement and the beginning of the second, so there is recognition of two times in the completion of the pulse. First, there is time of the movement, which they also call both beat and dilatation, and second the time of the rest, which **[511K]** they call both interval and contraction. In relation to the 'quantity' of the first time, the pulse is either rapid or slow, or intermediate between these pulses, while in relation to the 'quantity' of the second time, it is either frequent, or infrequent, or intermediate between these. Consequently, a pulse is rapid when the dilatation of the artery occurs over a short time and frequent when the interval between beats is short; it is slow when the dilatation of the artery occurs over a long time and infrequent when the interval between beats is long. It is intermediate between rapid and slow when dilatation of the artery occurs over a moderate time, and intermediate between frequent and infrequent when the dilatation of the artery is completed over a moderate time. Whether we say dilatation, separation, opening up, raising or moving makes no difference, in my opinion.

So it is now clear what the 'frequent pulse' refers to, and how it differs from a rapid pulse; and what an 'infrequent pulse' refers to, and how it differs from a slow pulse. When we compare the time of the beat with the time of the pause, the rhythm is brought forth. This is the term for the time of the beat in relation to the time of the pause. However, we say it makes no difference in referring to these, whether **[512K]** we say beat, dilatation or movement, and pause, interval or contraction. Furthermore, instead of saying the term for the relation of the time of dilatation to the time of contraction is the rhythm, some have said "order", and others that it is the ratio of the two times compared to one another—that is, the time of the beat to the time of the interval. This is the teaching of our predecessors.

7. Let us pass now to those who say they perceive the contraction, and place slowness and rapidity similarly in the time of dilatation and also in the time of contraction, but not the frequency, infrequency and rhythm similarly to those. Rather, they place frequency and infrequency in each of the two pauses, and rhythm in the time of the dilatation together with the pause after this, comparing this with the time of the contraction together with the pause after that. And sometimes also, they compare the time of dilatation alone with that of contraction alone, and say the rhythm is generated in this way. With regard to the frequency, some of them agree with what was said, where-

377 ἕτεραι (K) is followed; ἑτέρωσε: F1, F2 and Va1.
378 Χρόνος: K; χῶρος: F1, F2 and Va1.

as others, like those before them, **[513K]** say that rapidity and slowness are produced in the time of the dilatation, and they say that frequency and infrequency exist in all the rest, compounded from the two pauses and the contraction. Others do not hesitate to say, like those before them, that rhythm exists in the description of the time of the dilatation in relation to all the rest. But they do not agree with each other [on the issue of] whether they perceive all the dilatation or all the contraction. Rather, there are some who perceive all of both, and some who perceive neither the first parts of the dilatation nor the last parts of the contraction; and then they contend with each other about the names. Therefore, so that we may neither omit any of these matters nor delay any longer over what is useless, but state what is true and at the same time useful, let us entrust the application of names to those who have carried out this task, but are neglectful of the tasks of the art, and are unable to perceive either the first parts of dilatation or the last parts of contraction. But how much of each of these is perceptible and why, we show **[514K]** in the treatise, *On the Diagnosis of the Pulses.*[379]

What comes from perception is, however, clearly useful to a doctor. As a consequence, it is also essential for him to recognize the perceptible times of dilatation and contraction. With regard to the so-called "upper pause",[380] which follows dilatation and precedes contraction, what its duration is can be accurately recognized by touch. This is not the case with the so-called "lower pause", because parts of it are added to both the beginning and end of the contraction, and the end and beginning of the dilatation. One must recall these times—both the magnitude of each that accords with nature and their relationship to each other, and be able to recognize how much they alter in disease, and what each change naturally signifies. This is ascertained by the discovery of the causes and by long experience, and what we say about these elsewhere.

8. For the present, having added the *differentiae* of the two classes we are now speaking of, let us move on to the continuation of our discussion—that pertaining to the pause and the rhythm. **[515K]** Three *differentiae* of pulses exist in the class pertaining to the pause of the artery: infrequent, frequent, and intermediate. We learned the definitions of these a little earlier. In the class pertaining to rhythm, it is not possible to state the associated *differentiae* simply in this way. Rather, first it must be made clear that every pulse has a rhythm, so that the misleading name of the so-called arrhythmic pulses does not deceive someone into falsely[381] assuming that such pulses are devoid of all rhythm. For, like a man without a neck or a lutenist without a voice, so also a pulse is called arrhythmic, when it does not preserve proportion, this designation signifying

379 *De dignoscendis pulsibus* I–IV (VIII.766–961K).

380 This is an important issue for Galen, further addressed in the work referred to in the previous note. The issue hinges on whether the contraction is perceptible or not. If the dilatation only is perceptible, the rapidity/slowness is determined by the speed of this and the frequency by the ratio of the time of dilatation to the time of the pause. If the contraction is deemed perceptible, and there are two pauses, the matter is more complex.

381 ψευδῶς: K; σαφῶς: F1, F2 and Va1.

a defect of normal rhythm, but not a complete absence of rhythm. And the opposite to the *arrhythmic* is not the *enrhythmic*, for every pulse is in a certain rhythm, but the so-called *eurhythmic*. For both of these—I speak of *arrhythmic* and *eurhythmic*—are a common class, which is the *enrhythmic* pulse. Thus, the *eurhythmic* remains one and undivided. There are three *differentiae* of the *arrhythmic* pulse: *pararhythmic*, *heterorhythmic*, and *ekrythmic*.[382] You will learn what each of these is through an example.

For each of the stages of life **[516K]** there is a pulse that accords with nature. Of these, what preserves the rhythm is called *eurhythmic*; when the rhythm is destroyed, it is called *arrhythmic*. And of the destroyed *arrhythmic*, that which takes on the rhythm of the adjacent stage is *pararhythmic*, while that which takes on the rhythm of any other stage whatsoever is the *heterorhythmic*, and that which does not preserve the rhythm of any stage at all is called *ekrhythmic*. In this way too it is necessary to give thought to natures, seasons, places and all other things.[383] For each of these, there is a characteristic rhythm; if this is destroyed, you must give thought to the *arrhythmia* and the stated *differentiae* in this. Of all the rhythms, there are those that continue in equal proportion and those that continue in unequal proportion. In equal proportion is when the time of dilatation equals the time of contraction; in unequal proportion is when one or other of them is excessive. Sometimes this occurs in specified excesses and sometimes in unspecified excesses. In stated excesses, it occurs in two ways: either as in multiple proportion, or as of number to number, which is also called "superparticular".[384] Now a multiple proportion is double, or threefold, or fourfold, and so on; **[517K]** number to number is when dilatation is two times as great as contraction, or five, seven, nine, or eleven times greater. And in the unspecified excesses in general, it is threefold. Either the time of dilatation is unspecified, or that of contraction is, or that of both contraction and dilatation are. And each part of them individually sometimes has the unspecified times increased a little, or there are several [parts], or one first, and sometimes more, and sometimes even more. And it is clear that the first time is not according to nature itself, but must be understood by sensation, which is how it is with musical times. This is enough for now about rhythms. How much of this class of

---

382 On the six terms, see the appendix on pulse terminology. In short, Galen seems to accept the classification made by Herophilus; Herophilus divided the pulses according to their rhythm using the name "eurhythmic" for a pulse normal in rhythm, "pararhythmic" for a pulse of unnatural rhythm, "ekrhythmic" for a pulse almost normal in rhythm and "heterorhythmic" for a pulse which is of a different rhythm than what is expected.

383 Galen refers to the "necessary" causes of health or disease: 1. The ambient air, 2. motion and rest, 3. sleep and waking, 4. things taken in, 5. things excreted and retained, 6. affection of the soul. However, in his treatise *De causis pulsuum* he creates a better and more detailed categorization of causes that may alter the pulses.

384 The term here is ἐπιμόριος, translated as "superparticular", a term that has seen a change in meaning over time. LSJ has: "a number containing a whole number and a fraction with one for its numerator". The *Oxford English Dictionary* (1933) has: "a term applied to a ratio in which the antecedent contains the consequent once with one aliquot part over."

the pulses is practicable and useful, or impracticable and useless, will be stated in my books, *On the Diagnosis of the Pulses* and *On Prognosis from the Pulses.*

9. But let us pass now to the two remaining classes [of pulses]: one in which we consider equality (evenness) and inequality (unevenness), and the other in which we consider regularity (order) and irregularity (disorder). Equality and inequality occur in one pulse and **[518K]** in a number of pulses. However, regularity and irregularity only occur in a number of pulses. Since inequality is problematic and difficult to observe in one pulse, let us speak first[385] about that existing in a number of pulses, recounting together this and the discussion about regularity and irregularity, since it is associated and characteristic, and in this way also going back to the inequality in one pulse, we shall endeavour to speak about all the *differentiae* of this.

Accordingly, equality and inequality observed in a number of pulses, which they also call systematic,[386] are evaluated by comparing all the *differentiae* in the first pulse with those observed in the subsequent pulses. If they remain equal in all aspects, such a pulse is called "absolutely equal"; if they are unequal in all aspects, it is called "absolutely unequal". If there is equality in one or several aspects and inequality in one or several aspects, the pulse will be called equal in those aspects in which it is equal and unequal in those aspects in which it is unequal, there being a great difference in saying "absolutely equal" and "absolutely unequal", or in relation to one class, whatever it is. For the absolutely equal pulse has nothing unequal, neither magnitude, rapidity nor anything else. **[519K]** The equality in relation to any one aspect whatsoever is, along with the name of that, if it is wished to state this—i. e. equality in relation to magnitude, rapidity, strength, or anything else whatsoever. And of course, the inequality also relates to magnitude, rapidity, or one of the other parameters, and sometimes two, or three, or more, or often also all together, whenever it is called absolutely equal or absolutely unequal. Such are the equal and unequal pulses. However, the regular (orderly) and the irregular (disorderly) pulses, when the inequality is divided into the periodically equal and the altogether unequal, have their genesis in each of the divisions. The regular is equal in relation to the equality of the periods, while the irregular pertains to the inequalty in all kinds of ways.

So, taking it up again, let us speak about the whole matter, in order that nothing obscure still escapes our attention. If all the *differentiae* in a single pulse are equal to the *differentiae* in those pulses that follow, in relation to each class, such a pulse will be called "absolutely equal". But if all the *differentiae* happen to be unequal in all respects, such a pulse will be called "absolutely unequal". However, if one or more are equal, while one or more are unequal in one or more respects, the pulse will be equal in one or more respects and unequal in one or more respects. **[520K]**

---

385 Πρότερον: K; ἀρτίως: F1, F2 and Va1.
386 The Greek term is συστηματικός which indicates a sequential collection of pulse beats—see also II.4 (IX.279K).

The pulse which is absolutely equal, or equal in one *differentia*, or even in several, does not admit of division. However, in the case of the absolutely unequal, when we divide this into one or more forms, we shall find in the equality and inequality of the periods, both regularity and irregularity, either absolute, or in several *differentiae*, or in one. That which is equal in terms of period is, I think, already clear by being differentiated from the equal pulses that follow. But so there is nothing lacking in my discourse, let this be added as an example, proposing one class of pulse in terms of magnitude. Thus, let the first, second, and third pulses be equal in magnitude, and the fourth following these be unequal. It is clear that such a pulse is unequal in magnitude, but what is not yet clear is whether it is also irregular, but it is necessary to wait for four other pulses, so that, if the three that are second are equal, both to each other and to those generated from the first period, and after them the fourth equal to the fourth, such a pulse would be called regular in terms of magnitude, if in fact in like manner it maintains the first two periods and all the subsequent ones equal. But if all those from the second period **[521K]** are not equal to those arising from the first period, then the pulse may already be termed "irregular" in magnitude. And the same applies in respect of rapidity, frequency, and the other *differentiae*.

Sometimes, two or three periods of pulses are unequal to each other in everything, but those that follow are no longer so. It is very clear that one must call such a pulse neither "absolutely regular" nor "absolutely irregular". Rather, it is necessary to investigate and differentiate whether or not any one of them is regular in terms of periods. Sometimes, three or four periods are unequal to one another, and those that follow have the equalities in proportion. So if it should happen that the fourth is equal in every respect to the first, the fifth to the second, and the sixth to the third, we are accustomed to call such a pulse regular in terms of periodicity, being intermediate between those called "absolutely regular" and those called "absolutely irregular". Each of these partakes in no way of the opposite—neither the regular of the irregular nor the irregular of the regular. However, the pulse which is regular in terms of periodicity is in some part regular, but in some part also irregular, being regular in all the periods, whereas irregular **[522K]** in each of them. Thus, the absolutely regular pulse has all the periods equal, but the regular pulse in terms of periods does not have them all equal, but only a certain number. Contrariwise, the pulse which is irregular has neither all the successive periods equal, nor a certain number. The pulse which is regular in terms of periodicity does not have each of these, but one or the other. In this way it is in the middle of both—that is, of regular and irregular. The former has something in common with each, but the latter does not. It would also be possible for someone to believe that such pulses were absolutely regular, if he were to count together the periods individually, which, through a certain number, are equal to those that follow in one large period. But how many and what variety and intricacy of the periods existing in different numbers there would be, it is not difficult to see, and particularly for those practised in the periods in fevers interwoven with one another. It will not, therefore, be necessary for us to go further into the *differentiae* of these, but coming to those pertaining to inequality, to speak first about the systematic (collection of pulses), and next about that in one pulse.

10. [523K] Of course, the systematic inequalities exist, as we said, in a collection of a larger number of pulses and have a first *differentia* in which this occurs sometimes equally and sometimes unequally—and one should not be surprised, if we call a certain equality, unequal. For it is not only like this in pulses but also in almost all other things which, changing their initial nature, either in magnitude or quality, make the increases either equal or unequal. For inasmuch as the deviation destroys the equality, one might say they are unequal, but inasmuch as there is always preservation of the equality of the form, one might say they are deviated equally. But those that neither remain as they previously were, nor always take on an equal deviation are called "completely unequal". The same sort of thing, whatever it may be, also happens to these pulses, since their equality is lost, either gradually and equally, or all at once and to a great degree.

The discussion would become clear in one class by way of an example. Thus, suppose the second pulse is a little smaller than the first, while the third is smaller again than the second by the same amount, and also the fourth is equally smaller than the third. **[524K]** And suppose this occurs for several subsequent pulses; people call such pulses "tapering" and "dying away gradually",[387] transferring the name from figures that come to an end in a point. Those of them that diminish continuously and there is not yet cessation of this affection so as to end in complete absence of movement, we call "deficient tapering". Those that do stop have two *differentiae*; some of them end first in diminution and smallness and maintain this continuously, whereas others increase again—people call these "recurring tapering". Of these same pulses, some return to a magnitude equal to that at the beginning, and some to less. And some are diminished in terms of their previous excesses, while others in turn are increased in these; some use lesser or greater additions. Such are some of the equal *differentiae* of the inequalities.

11. Some of the unequal inequalities preserve no equality at all, whereas some preserve it for three, four or five pulses, but lose it afterwards. In each of these, either regularity or irregularity coexists: **[525K]** regularity, if they preserve some equal periods; irregularity if they also lose these. The movements are often deficient in these, just as they are in those moving equally. But they call the former "deficiently tapering" and the latter "deficiently unequal". Certainly, in the inequality related to largeness and smallness, the name "tapering" is very customary for doctors. However, with the inequality related to strength and weakness and for the other classes, these terms are not similarly customary. Of course, the proportion of matters in each class is the same, for they lose the equality either equally or unequally. And in respect to the oth-

387 The two terms—μειουρίζοντας (μυουρίζω) and μειούρους (μύουρος), Latin *mutilos, decurtatos*—are essentially the same, tapering like a mouse-tail or dying away gradually in relation to this passage and elsewhere in these treatises—e.g., *De praesagitione ex pulsibus* II.10 (IX.314K).

ers, there are individual *differentiae*, most of which are unnamed. Apart from "intercident" and "intermittent", there is no specific name for any of them. Of those that are intercident, the only inequality is in the frequency; in the intermittent, it is in the infrequency and smallness, and in the opposite way in each. In the inequality relating to the infrequency, the pause extends as long as possible whereas in the inequality relating to smallness, the dilatation **[526K]** is as short as possible. The deficient ones somehow occur more with the small pulses, whereas the intermittent ones occur more with the infrequent pulses. The so-called "recurring deficient" ones are associated with the intermittent pulses whenever, after being immobile for the time of two, three or even more pulses, the artery seems to begin to move again. These are the kinds of systematic inequality.

12. It now remains to speak of the different inequalities in a single pulse. And here also the discussion is similar in each of the classes and in all of them together. Thus, we shall call one "absolutely equal", another "absolutely unequal", in whatever class it happens to have lost the equality. Therefore, let us state first the *differentiae* in the case of the class pertaining to movement. For if we are practised in this, we shall already be easily aware of the remainder.

It is, then, necessary to establish the first two *differentiae* of the irregularity relating to movement; one has its origin in one part of the artery, the other in several. Each of these has a further natural division into several, which **[527K]** it would be reasonable to say all begin in one part from the coexistence in one part, but obviously not at the same time. For it is impossible for one and the same part not to appear to move similarly at one and the same time, but at different times the movement of one part often does not seem to be the same, when at first it is faster and finally slower, and conversely at first slower and finally faster.[388] Since such an inequality is found very marked and often in pulses, it naturally also signifies many things, as is stated elsewhere. I shall now explain all the *differentiae* of this individually. There are three primary *differentiae:* sometimes a pause cuts off the movement of the part; sometimes the movement seems continuous but not of equal speed; and sometimes it is recurrent. In each of these in turn, there are other *differentiae.*

13. Let us speak first, then, about the first matter mentioned, in which it happens that the movement is interrupted by the pause. Nine *differentiae* of pulses will be found in this particular change. When the first movement is rapid, **[528K]** let the second change in three ways, becoming either fast, slow or moderate. These are the three *differentiae* of pulses. There are, in turn, three others, when the first is[389] slow while the second changes in a threefold way. And there are another three in turn, when the first is moderate while the second changes in a threefold way.

388 The translation here follows manuscripts F1, F2 and Va1 instead of K. These have, after ὅταν: ἀρχομένη μὲν ὠκυτέρα, τελευτῶσα δὲ βραδυτέρα, καὶ ἀνάπαλιν ἀρχομένη μὲν βραδυτέρα, τελευτῶσα δὲ ὠκυτέρα, …

389 Reading μὲν οὔσης (F1, F2 and Va1) instead of μενούσης/*manet* (K) here and in the next line.

14. Of those pulses that are unequal in movement and not intermittent in dilatation, but always in succession have one inequality not interrupted by immobility, some begin by moving rapidly[390] and end by slowing, while others conversely begin slowly but always become more and more rapid, this occurring either equally or unequally in each of these: I said before how some inequality becomes an equality. Therefore, of those which diminish or increase in their movement equally, some pass from extreme slowness and come to moderate movement, while some go from moderate movement to extreme slowness. Some pass from the first rapidity up to moderate movement,[391] while others pass from this to extreme rapidity. Others **[529K]** in some cases depart from moderate movement in each of the two possible directions, sometimes becoming faster and sometimes becoming slower. As a result, there are in all six *differentiae* of these, and indeed more can be thought of. But let us state those that are apparent.

For of those having their inequality equal, we do not find any pulse which progresses either from extreme slowness to extreme rapidity or conversely from the most rapid movement to the slowest. But in those which make the change unequally, this occurs rarely. So great a change does not naturally occur due to a chance condition, so that an artery which begins to move very slowly finally comes to extreme rapidity, or an artery that is moving very rapidly at the beginning changes to extreme slowness. And when this does occur, and the change under these circumstances is of necessity unequal and all at once in large changes, and does not occur gradually, but as if the artery leaps when it undergoes a change from slowness to rapidity. And again, as if hampered, constricted and restrained by something, **[530K]** when from moving rapidly, it sometimes changes suddenly to slowness, it passes over the moderate in between. What, then, the conditions are, what the causes of them are, and to what end such pulses come will be stated when we consider all the other things—either the causes or what they naturally signify. At the present time, we are only going through the *differentiae.*

15. Having stated adequately that, when the unequal pulses in one part make the inequality equal, there are in all six *differentiae*, and that we do not find any such pulses passing either from the slowest movement to the most rapid, or from the most rapid to end in the slowest, let us now speak about those making the change unequally, in which sometimes two and sometimes three *differentiae* of the movement clearly appear. However, to us a fourth is sometimes also indistinctly revealed. Perhaps someone spending even more time, attending closely to the matter, and practising his touch, would also be enabled to recognize[392] four clearly. **[531K]** But now we shall speak about those that often clearly appeared both to us and to those who have best practised frequently the art pertaining to the pulses, whenever the movement of the artery cre-

390 Reading κινεῖσθαι (F1, F2 andVa1) instead of κινήσεως/*motum* (K).
391 Reading ...ἀπὸ τοῦ πρώτου τάχους ἄχρι τῆς συμμέτρου κινήσεως. τινὲς δὲ... (F1, F2 and Va1) instead of ...ἀπὸ τοῦ πρώτου τάχους τῆς συμμέτρου κινήσεως ἐπὶ τὴν ἐσχάτην βραδύτητα, τινὲς δὲ... (K).
392 Va1 has διορίσαι instead of γνωρίσαι.

ates two or three perceptible *differentiae*. When there are two, there will be six forms of combinations; when there are three, there will be twenty-four. Why, when there are three pulses in each of the different movements, rapid, slow and moderate, we don't say there are nine of them from their combination, but that there are six *differentiae*, is something you would understand, if you were to know that we assume the same pulse occurs twice, there being no pause between them, creating one wholly equal pulse, and because of this, of the nine combinations, three are withdrawn; in these, there is no longer an unequal pulse; it becomes equal, either rapid, slow or moderate, as is clear from the following table [Table 4]. For in it the second, fourth and ninth are equal; the second rapid, the fourth slow, and the ninth moderate. But if there is some difference of these in respect to one another, between rapid and rapid, or slow **[532K]** and slow, let this be put aside now, for it will provide more obscurity. But that the same are also equal to one another, let the present discussion establish.[393]

Table 4: Combinations of speeds in two sequential pulses

	Pulse 1	Pulse 2
1	Rapid (ταχὺς)	Slow (βραδὺς)
2	Rapid (ταχὺς)	Rapid (ταχὺς)
3	Rapid (ταχὺς)	Moderate (σύμμετρος)
4	Slow (βραδὺς)	Slow (βραδὺς)
5	Slow (βραδὺς)	Rapid (ταχὺς)
6	Slow (βραδὺς)	Moderate (σύμμετρος)
7	Moderate (σύμμετρος)	Slow (βραδὺς)
8	Moderate (σύμμετρος)	Rapid (ταχὺς)
9	Moderate (σύμμετρος)	Moderate (σύμμετρος)

In the same way too, if we postulate three movements are assumed, differing from one another, and in relation to each of these,[394] you combine the three pulses when they change, **[533K]** the total number of forms will be twenty-seven, but three of these are necessarily changed into equality. This is also clear from the following table [Table 5]:

393 From here, up to page 534, all 3 manuscripts have nothing. The text continues from p. 535: "…περαινέσθω. ὁμαλῶν…".

394 Berrey 2015, suggests from the Latin translations that τοὺς is erroneous and that the correct word is τῶ: i. e., "displaying the altered speeds of pulses in three general motions, not three particular pulses." Unfortunately, the three major manuscripts lack the charts and the text between them carries on from VIII.535K. In all cases where the charts are included, along with the text in between, that is in Paris gr. 2161 and 2167, *Vulc.* 43, Modena α. O. 4. 12, *Vat.gr.* 1064 and Moscow 464, τοὺς exists. In any case, we cannot verify or falsify Berrey, because the aforementioned manuscripts are all of minor importance to the correctness of the text.

Table 5: Combinations of Speeds in three Sequential Pulses

	Pulse 1	Pulse 2	Pulse 3
1	Rapid (ταχὺς)	Rapid (ταχὺς)	Slow (βραδὺς)
2	Rapid (ταχὺς)	Rapid (ταχὺς)	Rapid (ταχὺς)
3	Rapid (ταχὺς)	Rapid (ταχὺς)	Moderate (σύμμετρος)
4	Rapid (ταχὺς)	Slow (βραδὺς)	Slow (βραδὺς)
5	Rapid (ταχὺς)	Slow (βραδὺς)	Rapid (ταχὺς)
6	Rapid (ταχὺς)	Slow (βραδὺς)	Moderate (σύμμετρος)
7	Rapid (ταχὺς)	Moderate (σύμμετρος)	Slow (βραδὺς)
8	Rapid (ταχὺς)	Moderate (σύμμετρος)	Rapid (ταχὺς)
9	Rapid (ταχὺς)	Moderate (σύμμετρος)	Moderate (σύμμετρος)
10	Slow (βραδὺς)	Rapid (ταχὺς)	Slow (βραδὺς)
11	Slow (βραδὺς)	Rapid (ταχὺς)	Rapid (ταχὺς)
12	Slow (βραδὺς)	Rapid (ταχὺς)	Moderate (σύμμετρος)
13	Slow (βραδὺς)	Slow (βραδὺς)	Slow (βραδὺς)
14	Slow (βραδὺς)	Slow (βραδὺς)	Rapid (ταχὺς)
15	Slow (βραδὺς)	Slow (βραδὺς)	Moderate (σύμμετρος)
16	Slow (βραδὺς)	Moderate (σύμμετρος)	Slow (βραδὺς)
17	Slow (βραδὺς)	Moderate (σύμμετρος)	Rapid (ταχὺς)
18	Slow (βραδὺς)	Moderate (σύμμετρος)	Moderate (σύμμετρος)
19	Moderate (σύμμετρος)	Rapid (ταχὺς)	Slow (βραδὺς)
20	Moderate (σύμμετρος)	Rapid (ταχὺς)	Rapid (ταχὺς)
21	Moderate (σύμμετρος)	Rapid (ταχὺς)	Moderate (σύμμετρος)
22	Moderate (σύμμετρος)	Slow (βραδὺς)	Slow (βραδὺς)
23	Moderate (σύμμετρος)	Slow (βραδὺς)	Rapid (ταχὺς)
24	Moderate (σύμμετρος)	Slow (βραδὺς)	Moderate (σύμμετρος)
25	Moderate (σύμμετρος)	Moderate (σύμμετρος)	Slow (βραδὺς)
26	Moderate (σύμμετρος)	Moderate (σύμμετρος)	Rapid (ταχὺς)
27	Moderate (σύμμετρος)	Moderate (σύμμετρος)	Moderate (σύμμετρος)

[535K][395] Therefore, since there are three equal pulses in the table (Table 5 above) —the second, thirteenth and twenty-seventh—the second will be rapid, the thirteenth slow and the last (twenty-seventh) moderate. Already there are also twelve other pulses that change relative to the first table. For from the first in the first table, two would be produced which are the first and the fourth of those from the second table, one having more of rapidity and the other of slowness. The same applies to the third of the first table, the two being the third and the ninth in the second table, the former having more of rapidity and the latter more of moderation. Corresponding to the fifth of the first table are the eleventh and fourteenth of the second table, the former having

395 The translation of this paragraph is somewhat free in the interests of clarity. In essence, the point Galen is making is that when a third pulse is added to the nine possibilities listed in the first table, of the twenty-seven new possibilities, three remain equal (2, 13 and 27), twelve have two pulses the same in the parameter of rapidity (those detailed in this paragraph—e.g., rapid, slow, slow in possibility 4) and twelve have all three pulses differing in rapidity (e.g., rapid, slow, moderate in possibility 6).

more of rapidity and the latter more of slowness. Corresponding to the sixth of the first table are the fifteenth and eighteenth in the second table, the former having acquired more of slowness, and the latter more of moderation. And corresponding to the seventh in the first table are the twenty-second and twenty-fifth in the second table, the former having a greater degree of slowness, and the latter of moderation. There are two remaining from the second table (Table 3 above) which relate to eight in the first table (Table 2 above), i. e., twenty and twenty-six,[396] the former having a predominance of rapidity and the latter of moderation.

And here there would be a choice,[397] if there were only two different movements in the dilatation, **[536K]** of putting all such pulses into the first table, while others, taking into account magnitude as well, and in this way, when they are equal, of putting them into the first table, but when one or other is double, into the second table. I would have attempted to adjudicate in the conflict if I expected there would be any great gain. However, since this much is enough to differentiate that sometimes one of the movements is carried to a greater dimension (distension), and sometimes the other, or even both equally, someone bearing this in mind, so that he may again learn what each naturally signifies, may put the aforesaid pulses into whichever of the tables he wishes.

However, it is completely clear that the remaining twelve from the second table, which are wholly unequal, have nothing in common with those in the first table, in the case of which this was written in addition to that of the inequality. It is not that the other twelve of those common to the first table, are not unequal—they are—but that these are clearly unequal in respect to the three *differentiae* of the movements, whereas the others are in respect to the two *differentiae*, and are, to someone reckoning from the first in order, **[537K]** fifth, sixth, seventh, eighth, tenth, twelfth, sixteenth, seventeenth,[398] nineteenth, twenty-first, twenty-third and twenty-fourth in relation to order.

16. Such things seem to me to be enough about these matters. We now return again to the still remaining and third *differentia* of pulses unequal in one part, which Archigenes describes as one and calls "dicrotic".[399] There are, however, a number of these. We shall say a little later how many there are and of what kind. First, let us consider whether these pulses exist in a single dilatation of the artery, or whether there are two pulses unequal to each other in succession. However, it is not possible to judge this

396 The translation follows the manuscripts F1, F2 and Va1 which have, λοιποὶ δὲ τῷ ἐκ τοῦ προτέρου διαγράμματος ἢ δύο τῶν ἐκ τοῦ δευτέρου διαγράμματος, οἱ αὐτοί εἰσιν. εἰκοστός τε, καῖ κστ΄. ὁ μὲν δὲ, τῷ τάχει....

397 Reading ἀν καὶ αἵρεσις ἐνταῦθα... (F1, F2, and Va1) instead of ἀν διαίρεσις ἐνταῦθα (K).

398 Seventeenth added following F1, F2 and Va1.

399 This term is still in use to refer to a twice-beating pulse, in which "there is a palpable second wave after aortic valve closure ..." It is found mainly in "...patients sick with some fever such as typhoid; the peripheral resistance is low, the blood pressure low, the arteries lax, and the cardiac output probably normal." See Wood 1968, and the appendix on pulse terms in the present work.

properly, unless we first learn accurately what kind of pulse this is which has been called "dicrotic" by Archigenes. For what is perfectly clear is that there are two palpable beats. But this occurs also where there is an interval in a single beat, and if someone, anticipating only one beat, thinks the two pulses are one striking twice, he is clearly wrong. For one would not say those with an intermission in one dilatation are two pulses, even though they strike twice. For clearly, the movement in them, **[538K]** has lost its continuity, being cut in two by an intervening immobility.

The class of the dicrotic pulses is not like this; rather, when it has completed the whole dilatation, it immediately goes away for a little while and again beats for a second time, so that a pause of the intermittent pulses is midway between the beats and is the contraction of the dicrotic pulse, or constriction, or subsiding, or whatever you may wish to call it. Because of this, some say the dicrotic is two pulses, not one. But those who say this are particularly those who do not follow closely the contraction of the artery, but say the pulse is a combination of a beat and an interval. These people will say that after the first palpable impact of the artery, there is a brief immobility, and after that, another palpable beat, and they will say such a pulse is unequal in terms of frequency. For they think the pulses in which the interval between the beats is unequal are unequal in frequency, just as we too shall say a little later. But now, in fact, we should try to demonstrate first that in respect to these too, the dicrotic pulses will differ from those intermittent in a single dilatation, agreeing with their hypothesis in which they say the contraction is not perceptible. **[539K]** For if this is also granted, no one would deny that the artery does not seem to be beating at equal intervals. This is clearly apparent in the intermittent pulses. From below, the point from which the artery begins its movement and then pauses, the artery again makes the beginning of a second movement where the first terminated, but the dicrotic pulses do not beat like this; instead, the first movement clearly seems to have been carried from a greater depth and over a greater distance, and the second perhaps from the same place, or nearby. Therefore, we shall have these people admitting this much—that it is not the same thing for a pulse to be intermittent in a single dilatation, or to beat twice, like the dicrotic. One is unable to compel them against their will to call the dicrotic pulses unequal in respect to one beat. They will contest this, saying that two beats are clearly perceived, which are not equal in magnitude, time, or even *tonus*. And they will say that the second occurs a short time after the first. They will admit that such a pulse is unequal, but not in a single beat. Rather, **[540K]** they will fix upon the unequal difference in it in frequency, smallness and *tonus*.

But to those who also clearly perceive the contraction, the dicrotic pulse falls outside the so-called systematic inequality. For what occurs in it is like double blows of the hammer on the anvil, with the first being brought down from a considerable distance and striking forcibly, while the second, as if the hammer doesn't rebound very far from the anvil, falls on it more feebly than before and from a short distance. The dicrotic pulses also happen to be of such a kind, as if the artery were pushed back at the time of the impact, and again returned. And what particularly would most clearly demonstrate the nature of the dicrotic pulses, and would distinguish them from the systematic inequality, is the form of the withdrawal. For the artery does not contract at that time but is somehow carried down as if in agitation, and no clear pause separates its descent from the end of the first dilatation, just as appears at the time of contraction. Rather, as soon as the upward movement ceases, the withdrawal occurs, when it agitates a little and then comes forth again; it is like what **[541K]** happens when branches are drawn back forcibly and then let go. Those too go back to their former position, but do not immediately remain in it; instead they go back again towards the place they came from, although they do not traverse a distance equal to before, but a little less, and again are carried back from this end point to their proper place. And they retreat again a second time and return again, and a third and fourth time, and a number of times, until gradually diminishing the excursion they will ultimately come to rest completely.

This sort of thing also befalls movement of the artery in the dicrotic pulses. As a result, I for my part, have sometimes seen them coming on not once only but also a second time. And to speak briefly, the movement of the dicrotic pulse needs at all events to be clonic, so that, also in relation to the first beat itself, the artery, as if rising up and changing its whole position, is carried up more than it is dilated. Such are the clonic movements. And when, in fact, this affection befalls them severely, the parts of the artery are moved differently, so for instance the first, second and third, and often the fourth and fifth parts of this are carried up **[542K]** and carried down again. As a consequence, sometimes, at one and the same time, some parts will seem to you to be carried up and some to be carried down.

Such then, is the form of the dicrotic pulses. The differences will occur in relation to all the previously described classes: those of fullness, hardness, largeness, strength and rapidity, and their opposites, both in the first movement itself and also in the return after the recoil. For this can be large or not large, strong or not strong, rapid or not rapid, and each of the other forms, and likewise either equal or unequal, so as to create many kinds of such changes of the dicrotic pulses. These are the *differentiae* of the systematic inequality that exists in one part in relation to one[400] dilatation of the dicrotic pulses.

17. Passing to the pulse inequality (unevenness) in different parts, let us now speak of all the *differentiae* of that, and first, of the two in which the movement is either continuous or intermittent to some degree, and the others associated individually from the division of these. The continuous is either equal or **[543K]** unequal. For we learned how even and uneven arise With the even, it is necessary for the rapidity to either diminish or increase in six ways, as was said a little earlier, whereas the *differentiae* of the uneven (unequal) pulse are more and relate to it being long, short, or moderate in terms of length. If the movement impinges on two fingers, there will be six *differentiae*, but if on three fingers, twenty-four; and if on four fingers, seventy-eight. How, then, if the movement does impinge on two fingers, the *differentiae* must necessarily be six,

400 "One" (μίαν) is added following manuscripts F1, F2 and Va1.

and if on three, twenty-four, the two previous tables teach. And how, if it impinges on four fingers, there are seventy-eight, you would learn, if you previously knew that of any four things whatsoever, each having three *differentiae*, of necessity the total number of combinations will be eighty-one. This is not difficult to discover from knowing that of those things having three *differentiae*, [there are twenty-seven combinations. If the fourth is added, having three *differentiae*],[401] it is reasonable that there will be eighty-one **[544K]** combinations altogether. If we deduct from these the three constituting the equal pulses, the remainder will clearly become seventy-eight. In the same way, it is clear from the two previous tables (Tables 3 and 4) that some of these twenty-four will be the same as the first six, just as some of the seventy-eight, will be the same as the twenty-four. And in the same way too, in the case of those, there will be two schools of thought; those who think the tables concur with one another, and those who do not.

18. What remains now is to enumerate the differences of the movement that is intermittent. Four will be the primary movements, with several individual *differentiae* of each of these. The four are as follows: either the first and third fingers feel the movement, but not the second and fourth; or the first, third, and fourth fingers feel the movement, but not the second; or the first, second, and fourth fingers feel the movement, but not the third; or the first and fourth fingers feel the movement, but not the second and third. When the first and third fingers only feel the movement, there are nine *differentiae* of pulses in each class, but when the first, third, **[545K]** and fourth fingers feel the movement, there are twenty-seven. The number will be equal to those also when the first, second and fourth fingers feel the movement. However, when the first and fourth feel it, there are nine. The number in each case is clear from the previously stated methods. This also is the division of the inequality (unevenness) occurring in relation to the different parts. And if the various kinds of both the primary *differentiae* are mixed variously with each other the result is that the inequality is not only in different parts but also in any one part whatsoever. If now someone should wish to mix those admissible of the *differentiae* of one class with those from another, and were to do this without tiring before making all the combinations, the exercise would not be without use. I think it would not be difficult for someone to do this, if he didn't rest until he had run through the things written by me on this point. In this way, someone would learn all the inequalities pertaining to the movement occurring in a single dilatation.

19. The inequality in the amount of the dilatation involving the different parts of the artery obviously occurs frequently and is very easy to observe, whether it occurs intermittently or continuously, equally or unequally; **[546K]** falling into this class are the so-called tapering pulses in relation to single dilatation, and those called by some, "nodding" and "inclining" pulses.[402] An inequality in one pulse would not

401 The words in parentheses are included in K but omitted in F1, F2, and Va1.

402 These are terms coined by Archigenes. See the appendix on pulse terms in the present work.

exist in one part of the artery in relation to this class (part),[403] for prior to the movement reaching its end, the magnitude is not yet clear, but only that this class of pulses needs to come to an end for recognition. For it is possible to recognise strength, rapidity [and hardness],[404] and the opposites of these, and the intermediates while the artery is moving, but it is not possible to recognise the difference of largeness and smallness before the movement ceases. Thus, at the same time as it stops moving, the largeness of pulse is recognized, while before this, it remains unclear how much it will move. How then is it possible for a pulse that is unequal in one part to exist in this class? For it to be possible to recognise what a pulse is, there needs to have been a termination of the movement, so how could it be recognised when movement is still going on?

20. The inequality in relation to vital *tonus* allows of the same division as that in relation to the movement, for it **[547K]** involves either the same part or exists in different parts. Involving the same part it occurs in three ways: the movement is either intermittent, continuous, or becomes dicrotic. Involving different parts, it occurs in two ways: the movement either remains continuous or is interrupted. And in respect of each of these classes, the *differentiae* individually are equal to those stated in the case of the movement. Furthermore, it is inevitable that the inequalities existing in one part are mixed similarly with those arising in different parts.

21. The inequality in the body of the artery, in relation to which the pulse becomes soft or hard, cannot occur in one part, whereas in relation to different parts it clearly exists in two ways: that in relation to movement and that in relation to *tonus*, when the movement either remains continuous or is cut in two. Consequently, the *differentiae* in relation to this class are the same as those previously stated for the others. But it is not possible for one part of an artery to seem soft at the beginning of the dilatation and hard at the end; or conversely, hard at the beginning and soft at the end. **[548K]** As a consequence, there is in fact a difference in this class of pulses, pertaining to hardness and softness, from the previous two pertaining to the movement and the *tonus*.

22. There is a class of pulses relating to the substance poured into the lumen of the artery; in this class there is fullness and emptiness. This admits of the same investigations as previously mentioned in relation to the amount of dilatation. For it will seem here also that an inequality can exist in a single pulse involving different parts, but not the same part. But this whole class of pulses needs a prolonged discussion, and now is not the time for this. It will be covered in my books *On the Diagnosis of the Pulses.*[405] However, there is still need of a longer discussion on the third signification of the fullness, as stated both by certain others and by the followers of Archigenes.[406] So let this too be deferred to another occasion and, having added this much to what has been said, let us come to the next matter.

---

403 Γένος: F1, F2 and Va1; Μόριον: K.

404 καὶ σκληρότητα (K) is omitted in F1, F2 and Va1.

405 *De dignoscendis pulsibus* I–IV (VIII.766 – 961K).

406 See *De pulsibus ad tirones* 10 (VIII.467 – 470K).

23. Because there is a clearer recognition of the kind of heat of a fever from the place of the artery **[549K]** than from palpating any other part, it seemed plausible to some to count this class of pulses with the other things through which we recognise the quality of the heat. Clearly, in this class, equality and inequality will exist in a single dilatation. And *differentiae* of the inequality will arise primarily in different parts or involving any one part whatsoever. When these are in turn divided according to the previously described methods, there are many others individually.

24. There is another class of inequality in a single pulse which does not fall under any of the pulses mentioned; this pertains to the position of the artery. For some part of it sometimes appears to deviate to the right and sometimes to the left and sometimes to be higher and sometimes lower, either in two or more places, when such an affection exists. This is comparable in terms of appearance to the inequality in the amount of dilatation. How this must be differentiated is stated in the books, *On the Diagnosis of the Pulses.*[407]

25. Another class of inequality in a single pulse is the so-called undulant (wave-like) **[550K]**—the very name of this is sufficient indication of the kind. However, it is not a bad idea for me to also say something for the sake of clarity. In such an inequality, the whole artery does not seem to dilate at the same time, but first one part of it, then a second, then a third, then a fourth. And this is not random, but with it the wave-like elevation occurs, the second wave always succeeding the first. And how such a kind of inequality differs from the interrupted movement is not difficult to detect. For the dilatation of the first part of the artery is succeeded by that of the second part, with no motionless part falling between; but the end of the waves is nearly motionless. In fact, you would not detect a pause of any part, for the joining together of the movements to a height is like a wave rising, and is not altogether motionless, but maintains some slight movement. This pulse differs from the vermicular in magnitude. Anyway, if you imagine the largest of the waves always diminishing until it becomes so small that you are in doubt whether it is undulant or already vermicular, for they seem to differ from one another not in quality, but in quantity. **[551K]**

There is, however, something common to them both and to all their forms, and something specific to each. What is common is the derivation of the name for each; in the undulant, it (the pulse) rises like waves in the artery, one after another, whereas in the vermicular it resembles a worm in the pattern of movement since the movement of the animal itself is wave-like, as Democritus says somewhere[408] when discussing those things that wander about continually in an undulating fashion. And it is clear that the beginning of the flow is always from above and that it ends at the termination of the artery. This, then, is in common.

---

407 See *De dignoscendis pulsibus* I–IV (VIII.766–961K).

408 Presumably Democritus of Abdera (440–380 B.C.). None of his works are extant; we have not been able to locate this reference in his fragments. For a brief comment on the movement of worms, see Aristotle, *De motu animalium* 705b25 (Engl. transl. Farquharson in Barnes 1984: 1099).

Their differences which exist in terms of kind are as follows. Some have the wave carrying straight on, while others have it inclining to the side; some attain a significant height over a short distance in respect of length, while others in turn attain a low height over a longer distance; some are broad, while others are narrow; some are equally rapid, while others are not. In the same way too, of *tonus* there is a greater or lesser share among the waves. Sometimes also inequality exists in all the stated *differentiae* at the same time. But they are not so named on account of one or other of the aforementioned inequalities, **[552K]** nor others of these, nor the undulant, nor the vermicular. The name for such pulses is through something else apart from the inequality in the amount of the dilatation. The inequality itself alone, added to the movement in the first and last of the parts, constitutes such pulses. Consequently, the genesis of these is a composite from the inequalities in the amount of dilatation in different parts and from the fact that not all the parts of the artery are moving at the same time. In fact, such an inequality has not been found up to now in relation to itself, so that the first part of the artery obviously moves first; but when it does occur someone might completely mix with it the others previously mentioned. As Erasistratus says:[409] the experience of the senses doesn't always teach what obviously moves first. Actually it is completely the opposite, and he speaks falsely about appearance, saying this for the sake of a dogma, just as the majority of those supporting him in his doctrines do, inventing many other things he wrote that have never appeared, even as a dream. **[553K]**

26. Just as the vermicular pulse succeeds the undulant pulse when it becomes smaller, so the formicant pulse succeeds the vermicular pulse whenever the many movements are lost in becoming one, and this ends up being extremely small. Because of this, it does not seem [altogether][410] unequal. And indeed, although it is believed to be from the class of those pulses that are unequal, the inequality escapes detection on account of the smallness. And this is called the formicant from its resemblance to the animal, the ant, as some say in respect of the smallness, and others on account of the manner of the movement, so this too has been named similarly to the vermicular and the bounding.[411] Those were named for their similarity of the movement to that of the animals whose names they have; the formicating is itself similar in this way. However, some say it is named in this way for both reasons—on account of its smallness and the kind of movement. But we must not contend with them on these matters; we must state what kind of pulse the formicant is. It is extremely small (nothing else is smaller), and likewise, it is altogether very weak and very frequent, but not in fact rapid, as it seems to some. This will be shown in the books of **[554K]** *On the Diagnosis of the Pulses.*[412]

---

409 The apparent reference has not been located.

410 παντελῶς (K) omitted in manuscripts F1, F2, and Va1.

411 The Greek verbs are respectively, σκωληκίζω (*vermiculans*, vermicular, to wriggle like a worm); μυρμηκίζω (*formicans*, formicant, to feel as though ants are running under the fingers); δορκαδίζω (*caprizans*, to bound like a gazelle/antelope).

412 *De dignoscendis pulsibus* I–IV (VIII.766–961K).

What, in fact, now still remains, after adding the *differentiae* involving the pulses unequal in one dilatation, is to pass on to the others.

27. There are still the clonic and spasmodic [pulses], similar to each other in not preserving one position but being carried upward and downward, leaving the natural position. However, they differ also in this same respect, for the spasmodic pulses have little of such an affection, whereas in the clonic pulses, it exists to a marked and very clear extent. And as much as the spasmodic pulses seem pulled, stretched and drawn to either side from their limits, nothing of this sort exists in the clonic pulses. And besides, the dilatation is often large in the clonic pulses, but always small in the spasmodic. Also, it is clearly apparent when some parts of the artery are carried up and some are carried down at one and the same time in the clonic pulses, while such an occurrence is altogether very infrequent and indistinct in the spasmodic pulses. To speak in general terms, the perception in the case of the spasmodic pulse is like a stretched harp-string, and in this **[555K]** it is like none of the others. In the case of the clonic pulses, nothing of this kind is apparent. In particular, someone might liken the movement to javelins which, when hurled forcibly, travel swiftly, vibrating. For so they also call in this way the frequency of these[413] an unequal (uneven) quivering, which does not exist in the spasmodics.

All these pulses, outside those of the first five classes, are unequal in relation to one dilatation, with the exception of these two, in which the kind of inequality derives from the position of the artery and that in which the parts begin their movement too early or too late. In the case of the first mentioned in these, the inequality is due to the position of the parts, and this alone, whereas in the others, there is this and certain other *differentiae* of the inequalities of the first five classes that are present, for the concepts and origins of these are a combination of the undulant, the vermicular, the clonic and the spasmodic. As we said, the formicant is unequal, but doesn't in fact seem so. **[556K]**

28. The bounding pulse, named by Herophilus, is one of those unequal in a single dilatation; but it is also compound, being possessed of neither of the two preceding inequalities in the first five classes. But when the artery is interrupted in respect of its movement in any one part whatsoever, then particularly it occurs, but not simply. For the bounding pulse is not wholly of this class, but [only] when the second movement after the pause is more rapid and stronger than the previous one. For this name is applied only to such a pulse because of its resemblance to the animals, gazelles (antelopes). For those seem to bound in this way, as if making a kind of double movement. They briefly leap upward, then seem somehow to pause in their movement, then make a second unexpected sudden movement, much faster than the previous one, so it is possible for them to leap. All these are inequalities in a single dilatation.

29. Now we shall speak about the inequality in respect of frequency and infrequency, and rhythm. **[557K]** Doctors think that only such inequalities are systematic. They

---

413 Reading αὐτὸν (K) as αὐτῶν (F1, F2, and Va1).

call in this way the inequalities existing in many pulses. Let us speak first about those occurring collectively; then we shall pass to those involving one, choosing first as an example one selected condition of those occurring in frequency, which in fact is also the clearest, considering that this class of pulses exists in the intervals of the beats. Thus, when all the intervals in succession are equal in all the pulses, they say this is an equal pulse. However, if some intervals are greater and some less, it is clearly unequal in frequency and infrequency. Similarly, in respect of the rhythm which exists in the ratio of the time of the beat to the time of the interval, when the next remains the same in succession, then they say the pulse is equal, but if it changes the original ratio it becomes unequal, just as if also the magnitudes of the intervals, which they also call pauses, change, the inequality becomes one that relates to frequency and infrequency. Thus, they state these things correctly, and it is in passing from the stated choice to the others that we recognise equality and inequality in both the classes **[558K]**—that existing, as we said, in many pulses which they also call "systematic".

30. On the other hand, concerning the inequality of these in a single pulse, which all have neglected, there is not yet similar clarity. But so that nothing is lacking in the present treatise I shall attempt to speak about these as much and as clearly as possible, making the beginning of my discussion as follows. Either all the parts of the artery begin their movement at the same time or some of them do so earlier and some later. If all begin to move at the same time, and then, not being moved in equal times of the second dilatation, they begin again at the same time, an inequality of the two classes will exist in a single pulse, and because it occurred, it is known that when the second dilatation begins it does not have its beginning in the second, just as neither the rhythm itself, nor the class in relation to frequency and infrequency do. For truly these also occur in a single pulse, but the second is altogether necessary for their recognition. Before this begins it is not possible to recognise the end of the first. And while the end still remains unknown it is not possible **[559K]** to recognize the frequency, infrequency or rhythm. As a consequence, also, concerning the genesis of the inequality, which I have spoken about, it is no wonder we await the beginning of the second dilatation. But it is not necessary on this account to deem it systematic.

Of course, before the second dilatation is completed, it is recognised immediately at its beginning, so that we need neither the second pulse, nor the third, nor those that follow. In this way it occurs in the first alone. Because of this, then, we said that when all the parts [of the artery] begin to move at the same time in the first and second dilatations, but the time of movement in them is not equal, an inequality of the two classes exists; one in frequency and infrequency in the inequality of the intervals, and the other in rhythm in the ratio of the time of the beat to the time of the interval. For the time when the movement of the very first part of all ceases will be least, and of the last, greatest. Those in between will be in proportion to the extent from which they are separated from the first or the last. And again, in the times of the movements, those of the pauses will be greater concerning those that cease first, compared with those **[560K]** that cease last. As a result the time of the movement in the first part to cease is least compared to the greatest time of the pause in the same part, whereas that in relation to the last is the largest of all the times of the movement, when compared to the least time of the pause in the part itself, and in this way, the rhythm changes in each [dilatation], to the extent that the ratio of the time of the movement to that of the pause does not remain the same in all. However, should the movement of the first dilatation make a start of all the parts at the same time, but there is not equality of time in all of them, the second dilatation will not make the start of the movement of all the parts at the same time, but of the first when the first ceases, and by as much before as the first ceased, and second by as much before as the second ceased, and in the same way in succession, the inequality pertaining to the rhythm would be like this, but that in frequency and infrequency would not still be. When the occurrence of the pause is equal in all parts it will produce no difference in frequency and infrequency. The ratio itself will not be the time of the movement to that of the pause, if in fact that of the movement was not equal in all parts, while that **[561K]** of the pause was equal, so that the rhythm is not the same in all parts.

If, however, when all the parts of the artery begin to be moved at the same time, but conversely neither the second dilatation begins at the same time, nor the first ceases, but there is not equality of the pause in all parts, of necessity a twofold difference exists in such a situation, since an inequality exists in one or other of the *differentiae* pertaining to frequency and infrequency, but not necessarily in that pertaining to the rhythm. For it is possible for the times of movement of all the parts to be in the same ratio to the times of the pauses. And if this happens, and the rhythm is equal, whereas the actual ratio is not preserved in all the parts, for this too can occur, the rhythm would in this way also be destroyed. If, in turn again, when all the parts begin to be moved at the same time, and are moved in an equal time to the second dilatation, but not all the parts will make the same beginning at the same time, some beginning earlier and some later, the rhythm would no longer be equal, nor would there be a difference in frequency and infrequency, for the inequality in respect of this creates an inequality of the pauses, whereas **[562K]** the difference of the ratios in respect of the rhythm is when all the parts are moved in an equal time, but the pauses are not.

Such are the *differentiae* of inequality of both classes in a single pulse when all the parts of the artery begin to dilate at the same time. However, when one of the parts begins to dilate first and another part second, and all pause simultaneously, and in turn they begin the second dilatation together, inevitably the inequality is only in the other class—that of the rhythm. For the intervals between the movements will be equal in all the parts, so that no inequality will exist in respect to this class. But if an unequal time is compared to the equal time of the pause in all the parts, the time of the movement in each of them in different ratios will destroy the rhythm of each part. And if the artery should begin its second dilatation in the same way as the first, so that in each of the dilatations the same part begins equally earlier than another, and this is maintained in all the parts, and they pause at the same time as the first dilatation, **[563K]** it is inevitable that the inequality occurs in each of the classes. For the whole time, composed of the movement and the interval, will be equal in all the parts of the artery. However, the ratios of the movements to the pauses will be unequal, in that the whole time is not divided into equal parts. The time of the movement of the parts beginning to be moved first will be greater than that of the parts beginning to be moved later, but the time of the interval will be less. Contrariwise, the time of the movement of those beginning the movement later is less, while the time of the pause is greater. As a consequence, there will necessarily be an inequality of rhythm in each of the parts and an inequality in frequency and infrequency will exist in them of necessity to the extent that the times of the intervals differ.

If, however, when the second dilatation makes the beginning of the movement of the parts similar to that which the first had made—that is to say, where in relation to each movement, the same part begins first in one case and second in another, the end of the first dilatation would not be equal in all parts, and there will be two *differentiae.* **[564K]** For the part that begins its movement first can stop before but is also able to stop later. If it should stop first by the same amount of time as it began first, in respect of both the classes, it would become in this way an equal pulse. But, if it should stop much sooner than it began, in this way it will generate an inequality of the two classes. However, if that which began earlier stops later, it will in all cases produce an inequality of the two classes, whether it happens to stop much later or a little later. But if the beginning of neither the first nor the second dilatation occurs at the same time in all the parts (of the artery), it is possible at some time for the rhythm to be equal, and it is possible also for the class of frequency and infrequency to be so, if the times of the movements are apportioned in this way in relation to the times of the pauses. As a result, there is the same ratio in all when the rhythm is equal. If, however, the times of the pauses alone come to a state of equality, it is an equality in a different class. These, then, are rare, whereas those of the inequalities are very numerous. For surely all the parts can cease their movement at the same time, and if they don't cease at the same time, some may cease earlier and some later, and some even much earlier or much later.

What kinds and **[565K]** how many the inequalities might be, you will find in those previously described, so there is no need to delay further. The only thing to remind you of is that what was proposed has been shown, which is that some inequalities occur in a single pulse in the class relating to frequency and infrequency, and some in that relating to rhythm. Here it is timely for the first book to come to an end. What I have said on one course of action, that of dividing the pulse into a beat and an interval, I have stated as an example. It is not difficult to also carry this over to the other inequalities, since we are similarly able to demonstrate this in respect of all in these two inequalities, which I have now stopped talking about—that is, the inequalities existing in one pulse.

Book II

1. [566K] This book does not profess to have the same use as the previous one. Rather, what will now be said in this one, has the same relation to what has previously[414] been said as hellebore or scammony have to bread or meat. The latter can nourish healthy souls, whereas the former can clear away bad opinions, just like some diseases, It is reasonable to expect assistance in the establishment of a firm belief in things correctly stated from the refutation of erroneous views. Consequently, if someone should attach this use to the present book, he will find some benefit follows.

2. [567K] My first argument is directed against the Sophists and concerns the use of names. I follow the custom of the Greeks, for I was nurtured in this, and always attempt to explain a concept through the clearest terms. However, I have no objection to those who take this lightly. Even if someone should wish[415] to utter barbarisms in each statement, it doesn't matter to me, as long as he doesn't harm the clarity of the explanation. For the sake of this, we use the dialect we use with each other to understand the things being set out, and think the man who uses speech best is the one who is able to make known most clearly what he wishes to make known. Such is our purpose—a moderate and benevolent one, I think. But the Sophists won't allow me to do this. Rather, just as they contend with each other over names, neglecting the matters themselves, so too they also insult me, always arrogantly and derisively, if I don't use their terms. If, then, we were all agreed on a single dialect, new and sustained by usage, perhaps we could attempt to forget **[568K]** that of the Greeks and learn the one ordained by them. But not even if they established one of the barbarian dialects to use as a single dialect, could we shrink from harming this in order to please them in every respect. However, since they do not devise any pure Greek dialect or simply a barbarian one, but one mixed from all, like a sauce or something more strange and varied than a sauce, as each of them wishes, with elements of Phoenician, Syrian, Egyptian and Thracian, gathered together from one place or another, then mixed with Greek names—and these not genuine but corrupted in various ways, by the gods, I don't know where this would leave me, as I am neither able to recognise nor learn such dialects, nor persuade them to learn the one of the Greeks. Since I am not able to do either of these things, perhaps I could gain a third and remaining concession from them, which is to allow me to use the dialect in the language in which I was brought up. This, then, is the magnitude of the problem concerning the use of names.

3. [569K] Something else about the definitions is much more troublesome than this. Some of them would seem to me not to buy vegetables without a definition, and of course are not in a dream bound by a dialect in these matters. And men such as these, having been brought up in such ways, put forward one of these fine

---

414 The addition of πρόσθεν follows in manuscripts F1, F2 and Va1.

415 ἐθέλοι τις K; ἐθελοίμεν Va1; ἐθελοίεν F1, F2.

names from middle Syria or Cilicia,[416] and bid us define this, which perhaps no native Greek has ever heard in the first place. If he doesn't know this because it is foreign and barbaric, the Sophist now laughs, scoffs, mocks and calls him a child and young lad. However, often the term put forward is Greek, but is not applied in relation to the matter usual among the Greeks, and in this way, the wretched Sophists accept it and then demand a definition. And perhaps some Greek might define it according to the custom of the Greeks, but they would not only not be persuaded—for that would be tolerable—but they would laugh and mock. This is the second form of the problems concerning definitions.

The third is no longer in relation to the name and signification, but is now **[570K]** to be mistaken in respect of the underlying subject itself; this arises when some dialectitians attempt to prove their definitions which in some cases are deficient and in others superfluous, vague, obscure, prolix or false, or are not definitions at all. Suddenly, as if possessed, they appear as shrill accusations of dialectics are produced, and then it seems to them fruitless and bad to pursue this in every aspect of life and in the arts. As a consequence, I have often, by the gods, been astonished if, when we called upon them, they were not persuaded to attempt anything through definitions, neither teaching nor learning. If, however, they consent to make a definition, and then are shown how greatly they are in error, they say definitions are useless, and they do this, not after one or two days have intervened, but in a moment, as heads become tails one might say,[417] at one time putting forward a name and demanding a definition, and at another time, saying the definitions are useless to them.[418] The fourth fault among them in such matters is their failure to realise there are not definitions for everything, and then, as they are ignorant of this, they attempt to define everything, and differ from one another, clamouring in vain **[571K]** over matters which cannot be defined. The fifth and most amazing of their errors in definitions is that before going over the homonymy, they make one definition of various meanings, as if they were not definitions of things (matters) but of names.

At any rate, one of these current Sophists came close to strangling me. To this same man, putting forward the following question, "what is a full pulse?" I said in reply that the term is used in many ways, and because of this doesn't have one definition, unless you determine the homonymy. When I said this, he replied: "I don't know! What nonsense! It is enough for me to learn from you what a full pulse is. Do away with the foolishness of dialectics which so afflicts medicine."

And I said: "I have taken pains to ensure that nothing you ever heard from me is determined by dialectics."

---

416 Syria was a geographic area in northern Greece. Cilicia was a region, now known as Çukurova, often a political unit, on the south-eastern coast of Asia Minor, north of Cyprus.

417 See Plato, *Phaedrus* 242 (Engl. transl. Hackforth in Hamilton and Cairns 1989:489).

418 There is a minor variation here, as follows: εἶναι τοὺς ὅρους φασί K; εἶναι τοῖς ὅρους αὐτοῖς φασί F1, Va1; εἶναι τοὺς ὅρους αὐτοῖς φασί F2. The last reading is followed here.

"This much is certain," he said, "and is the most reprehensible—that you don't dare to offer a definition but don't hesitate to inundate us with dialectical nonsense. Anyway, say something—speak, talk, answer—what is a full pulse?"

"I cannot," I said, "before defining precisely the **[572K]** homonymy."

He again cried out and said that homonomy and prolix nonsense have emerged and these things are afflicting medicine and destroying us.

"Not destroying us," I said, "but destroying you, and those who wish to define everything."

"See here," he said, "you are trying to rebuke us."

"What?" I said, "more than you are trying to rebuke me?"

At this he jumped up and shook his fist, but those present restrained him and he went on his way, laughing derisively. But of those still remaining, some old man with a great beard coming down to his chest, who was otherwise very stern (for this is now regarded as a sign of gravity), said: "Stop! Don't make a disturbance, boys. Allow our friend to make clear what he is saying."

And I, looking at him, said: "What are you asking?"

"This very thing," he said. "What you were arguing about—the full pulse."

"Are you willing, then," I said, "not to be in too much of a hurry, so I may follow you, for I am slow by nature, and don't easily follow those who hurry?"

"Argue," he said, "I won't be in a hurry."

"Do you agree, then," I said, "briefly and by way of example, that I ask you all in common? **[573K]** For then it will be for anyone who wishes to come forth to answer."

He agreed. Whereupon, assenting, I enquired, "What is a tongue?"

After a little time, someone answered: "An animal's organ of taste." And another said: "The organ of articulate speech." And a third said, "It is the organ of taste and of articulate speech at the same time."

I, looking at them, said: "And is the tongue of a sandal and the tongue of a flute an organ of taste of an animal and articulate speech?"

When they were silenced by this, and the very angry-looking old man had become somewhat more angry-looking, I asked him if I might again make one similar enquiry. He, with great solemnity, promised he would just about allow it. I, fearing that he might not tolerate slowness, put forward this question in haste: "What is a dog?"

Someone from among those present answered: "A four-legged barking animal."

"And is the sea-dog a four-legged barking animal?" I said, "and the star in the heavens, and the affection involving the face?"[419]

After these things, the old man already shook his ears like an ass. But to the others present, I seemed to be saying something worthwhile, and they all thought it a good idea, for their sakes, to finish these matters I had devised, since right from the start I promised, **[574K]** as an example, to make the enquiry into those things following.

---

419 Among the several meanings of κύων listed in LSJ is that of a "unilateral facial palsy" (presumably Bell's palsy).

As I had silenced them and they were desirous of listening, I said: "I think you already know how great an error it is to attempt to establish definitions before determining the homonymy. For the definitions are not of the names but of the matters (things). Therefore, whenever many things are signified by one name, there must be as many definitions as there are matters (things). For example, in the case of the tongue, there is one name by which three things are signified, and regarding each of them there is a specific definition. On the other hand, as may happen, there are two names, and often in fact three, in regard to one matter (thing), as in the case of *xiphos, aor* and *phasganon.*[420] There is one definition, not three, as with the names. Always the number of definitions is equal to the number of matters (things). Consequently, you will not, when one matter (thing) has acquired many names, think it worthwhile to state many definitions. Nor when there is one name signifying many matters (things), will you think it worthwhile for one definition to be stated by you. For you will always make the number of definitions commensurate with the number of matters (things).

If certainly the term "full" **[575K]** is stated in relation to one underlying matter (thing), it is possible, when this is put forward, to now state the definition. If, however, the followers of Archigenes most of all do not accept the term "fullness" as signifying one thing, ordaining variously the application, bringing in one upon another the number of significations, we must not be charged with not daring to offer a definition by those who are the originators of such confusion, if anyone is. For I find younger doctors, whenever they should wish to make known what consistency the wall of the artery has, use the terms "full" and "empty", while others do this when [they wish to make known] the substance contained in the lumen [of an artery]. And some think to indicate by the terms the quantity or quality or both of these. These things alone are, however, not enough for Archigenes, it would seem. He introduces for us in addition the signification of the potency (capacity) of the *pneuma.* It is not, therefore, possible to make one definition of a "full pulse"; we must make other definitions indicating the quantity, quality or both of the contained substance, others of the potency (capacity) of the *pneuma*, and others of the quality of the arterial wall.

When I said these things, I **[576K]** was silent, having completed what was put before me to do. They demanded that I go over the argument for them (again),[421] since I began all at once. Therefore, I gratified their needs, but it is not appropriate to say all those things for the present, since I am going to say more about them and the most useful among them in the treatise, *On the Diagnosis of the Pulses.*[422]

4. I shall say as much as is useful regarding the matter before us. I do not think it worthwhile either to define everything, or to make one definition of the many things previously said, which in the case of the "full pulse", Archigenes failed to achieve. Because of this, his definition itself is not clear, nor is there a correlation of what is said

420 The Greek terms are ξίφος, ἄορ and φάσγανον. The primary meaning of ξίφος is "a sword". In LSJ there is: "used by Homer as equivalent to ἄορ and φάσγανον—see *Odyssey* 10.294, 321; 11.48; 24.82."
421 There is a minor textual variation: αὖθις K; αὐτοῖς F1, F2 and Va1.
422 *De dignoscendis pulsibus* I–IV (VIII.766–961K).

either about the body of the artery or about the substance contained in the lumen. For there is the danger of creating a definition of a name and not of a matter (thing). At all events, if he had viewed the matters, he would have made many definitions, but because he considers the name only, it is possible to learn from what he says right at the start of this work on the pulses that eight qualities are constantly associated with the pulses: size, strength, rapidity, **[577K]** frequency, fullness, order/disorder, equality/inequality and rhythm. If the term "fullness" is said of many qualities, I don't now how they would still remain eight.

Well then, if someone were to add to the other seven the quality pertaining to the body of the artery, that pertaining to the so-called content, and that pertaining to the *pneuma* itself, he would say there are ten qualities of the pulses in all. What shall we say on this? That, by the gods, if someone were to divide the contained substance into two, the one would be from the the amount of this, and the other would be by assigning the quality to the kind. In this way, Archigenes seems to have fixed on the names and not to have distinguished the significations. And now it is right to state the commonplace of the maxim sufficient for the present, when it says: "If water chokes, what shall we swallow?"[423] If Archigenes fails in such matters, to whom shall we turn? Who will correct the errors in these for us? And it is not these things alone, but if we must speak the truth, he is obviously mistaken in many others. In the first place, at any rate, in relation to the eight primary qualities, he makes no mention of hardness and softness, although subsequently **[578K]** he does discuss them.

First, he discusses magnitude; second, strength; third, fullness; and fourth, hardness. If, then, someone made this a combination of simple qualities, he would perhaps have spoken in his defence, saying that he necessarily omitted them at the beginning, when he made mention only of the simple[424] qualities, reasonably later having written about them in the combinations. But now he made the discussion about the fourth itself, placing it particularly as one in the middle of all the simple qualities. However, he does explain what it is on occasion, speaking of it in relation to formation or dissolution of the artery reciprocally;[425] from these things he obviously wished it to be among the simple qualities. These are mistakes made by Archigenes right at the start concerning the enumeration of the primary qualities. He didn't think it worthwhile to demonstrate them, thinking it sufficient to say such is the number, but simply threw them into the discussion in this way.

---

423 Aristotle, *Ethica Nicomachea* 1146a35 (Engl. transl. Ross and Urmson in Barnes 1984: 1810).

424 K: ἁπλῶν; F1, F2, Va1: ἁπάσων.

425 The three Greek terms are σύστασις, διάλυσις and ἀντίληψις. The last, which means laying hold of, receiving in turn and reciprocation, among other things, is omitted in Kühn's Latin text and has other attested uses in relation to the pulse.

Eight qualities are said to be constant attributes of the pulses—those that are commonly spoken of[426] alongside those that are correct. I, however, am unable to conclude precisely what "commonly spoken of" is, nor have I discovered the term among any of the Greeks. As a result, I don't know to what matter it has been assigned by Archigenes, and **[579K]** this is because he didn't write his book in his own dialect, as Chrysippus did about those things he proposed regarding the dialectics of names.[427] Only in this way could we understand him. Of course, from the whole discussion, someone might suspect that he applied "commonly spoken of" to what is said and believed by all. However, it is impossible to assume this to be simply the case. For he says not among all but among the pure—and again, I don't know what these are. And yet, it behoves me to understand, so that I might consider whether he must be believed in these matters, apart from demonstration, or not.

From Aristotle I have heard notable assumptions that seemed right to all men, or to many, or to the wise. Now, however, I don't know whether we must understand the pure in preference to the wise. It would be much better to assume something, even if there were not a strong demonstration, as long as there were sufficient reassurance from the argument about the eight qualities, so that someone would not, right at the start, like coming upon the way of Moses and Christ, hear unproven laws, and those in which there is least necessity. For since he was going to speak about all the pulses, as someone **[580K]** would infer from those things he wrote previously—how many there are—he did not consider it worthwhile to put before us any logical method, but created an empirical teaching, saying there were eight qualities to be separated in those that were pure. But why, if they are pure, would he so indifferently leave some of them out for the sake of this—that there are many, and not for the nature itself, or say it was enough to state the notable ones in those that are pure as the truth? And not knowing with respect to these, that some are the pure at some time, whether they have learned the writings, or the first teachings, or are sagacious, or prudent, or wise? But also, to term the primary and generic qualities of the pulses' *differentiae* is of little consequence. Anyway, the difference pertaining to largeness and smallness is not a quality unless you also say the *tripēchus* (3 cubits length) is a quality. For I am accustomed to think quantity. But I also know that no little discussion has arisen about the large and the small, and whether they signify a quantity, and in regard to what, or what is also true, in regard to what "one" is, whether it is in the category of quantity, and not at all in the class of quality, just as rhythm is not, nor disorder and order, nor equality and inequality. **[581K]**

Clearly these are also among those terms that refer to something. But how is it not really absurd to think that magnitude is the common class of the large and small pulse, strength of the strong and weak, rapidity of the rapid and slow, frequency of the fre-

426 The Greek term is διηχημένας, contrasted with καθάρειος as a grammatical term signifying pure or correct. Here LSJ defines the former term as "commonly spoken of" in relation to qualities, citing this passage; it is transliterated in Kühn's Latin text.

427 None of Chrysippus's written works have survived; neither has Galen's commentary on Chrysippus.

quent and infrequent, and again in those following, fullness of full and empty, and hardness of hard and soft? If, at all events, he wished to categorise the class as a whole by one name, why didn't he persist in doing this in the case of regular (orderly) and irregular (disorderly), and equal (even) and unequal (uneven)? If he was not accustomed to need a common category in these cases, why was such a use needed in relation to the other *differentiae?* Or if, at a stroke, he was bold enough to transgress the custom of the Greeks, why did he establish frequency as the category for a frequent and infrequent pulse, and not dare to establish the category of equality for an equal and unequal pulse? I would like to learn how these things are not in agreement with each other. For he writes some things as if respecting the custom of the Greeks and others as if despising and confounding that custom. Having provided himself with a point of entry in the case of the one, first class, and indicating **[582K]** he would have used the same method in all the others, he did not retain his daring; he needed from the beginning either not to deviate from the custom at all, or to proceed boldly throughout.

For surely, having said that the largeness (magnitude) of the pulse is stated generically, and that small is a specific instance of largeness, and then that strength generically stated somehow does not relate to the strength alone, but is also to be brought as a charge against the weak, he immediately backs down regarding the fullness, writing thus: "There is some common class for the full pulse and the empty. Sometimes this is called fullness." Thus, he added "sometimes" in this particular class, but did not add it to those before, as if he were already censuring the confusion of the old custom of diction.[428] I think someone who was brought up among the Greeks asked him: "My good man, what fullness exists in emptiness? Why do you transgress the common expressions of the Greeks?" So what is called "sometimes" is not something to be added. And clearly, we shall ask: Why? How? And by whom? For this is what he wishes the "sometimes" to do. But he will have no book of the Greeks to show in which he might confirm the use. **[583K]** Perhaps this alone will be enough for him to say among sailors, merchants, pedlars and bath-attendants. People always seek refuge in such witnesses when the enquiry is about a name.

5. Perhaps someone will ask me: "Do you then honour the custom of the Greeks so much that you won't allow anyone to depart from it to the slightest degree?"

"Not at all," I shall say, "by Hercules and all the gods, but if you wish to call bread *bekos*[429] like the Phrygians, it makes no difference to me. Only first teach me your dialect that I may follow, and don't say, 'it is called'. Anyway, I shall ask you, 'among whom?' For it is certainly not among the Greeks. But how do they say, 'let it be called.' And if having said this, you find me still grumbling, I deem myself worthy of death."

428 There is some variation in this clause, with F2 and Va1 having ὑποτιμώμενος ... ἔθνους instead of ἐπιτιμώμενος ... ἔθους.
429 On βέκος LSJ has: "... bread, Phrygian according to Herodotus 2.2."

But now you say, "it is called by some", making a false allegation against the custom of the Greeks. However, it is not simply "by some" but also along with this, "sometimes", as in the case of the fullness. For what do you mean by "it is sometimes called fullness?" But on fullness, **[584K]** he already wrote this reluctantly, whereas about the hardness, this was not so. Now such speech is that of a fallen and thoroughly defeated man. Perhaps there is some common class of soft and hard pulse that, if acceptable, is called "hardness". And he did not dare here to still add this, "sometimes, it is called hardness." And yet "fullness" was not more honoured than "hardness", such that he should be bold in the case of that but fearful here. But I said he is like those falling gradually. It would be more high-minded, I think, to fall all at once and not gradually, and not make work either for us or for himself. But perhaps, should he be wasting time, he might say, "it is called hardness." By whom, perhaps, would it be accepted? Perhaps by your heirs. For just as we know that in each city there is a token which is a symbol of purchase and sale, and that lawmakers punish those who falsify this, so we also know there are many characters of dialects, and we consider it worthwhile to preserve each of those chosen, in whatever way they are from these. Therefore, we have brought together the common dialect, as it is called, no matter whether it is one among the Attics, **[585K]** for the dialect of the Athenians has taken on many changes, or another one altogether. I shall present elsewhere my opinion on this.[430]

And we attempt to preserve this dialect, and not to misuse it or introduce a false custom of speech, and not to debase it. If, however, you desire to converse with me in it, you must first attempt to learn it, whereas if you use some other dialect, you must disclose it. For if it is one of the Greek dialects, I do at all events understand this; and I have also read the writings of the Ionians, Aeolians and Dorians.[431] If it is none of these but is one of the barbarian dialects, tell me this too, only try to preserve this undefiled, whatever it might be. But don't present me with three terms from Cilicia, four from Syria, five from Galatia and six from Athens. For I cannot learn so many dialects in such a way that I follow polyglot men to such an extent. The ancients called such a person bilingual, and it was remarkable that such a man was accurate in two dialects. But you should not think it worthwhile for me to learn many dialects; you need to learn one, so specific, so **[586K]** common to all, so eloquent and so cultured.

---

430 Galen wrote a number of works on linguistic and rhetorical matters—see the final section of his *De libris propriis.* Included among these *De vocalibus quae apud Atticos scriptores XLVIII*, no longer extant.
431 The Ionians were one of the four main ancient Greek tribes. The Athenians, in the peninsula of Attica, were the only Ionians on the Greek mainland. The Aeolians were another of the four main ancient Greek tribes. They were considered to be the descendants of the legendary Aeolus son of Hellen, the mythological patriarch of the Hellenes. Originating in Thessaly, they moved their location when the Dorians were attacking the Achaeans and then abandoned most of the Mycenaean territories. The Dorians were another of the principal ancient Greek tribes. Traditional accounts place their origins in the north-eastern regions of Greece, ancient Macedonia and Epirus, whence obscure circumstances drove them south into the Peloponnese, to certain Aegean islands, and the coast of Asia Minor.

If haply the man directs his attention to the sounds of the barbarian dialects, he will clearly utter sounds like those of pigs, frogs, jackdaws and ravens, making the form of the tongue, lips and whole mouth unseemly and contrary to what it should be. For the most part they either make sounds issuing from the pharynx, like those who snore, or distort their lips and whistle, or they increase the whole sound, or reduce it to nothing at all, or yawn hugely and poke out their tongues, or cannot open their mouths at all, or have a tongue that is slow and difficult to move, as if it has been bound. Would you then pass over the sweetest and most human dialect in which so much beauty is perceived and grace blooms, and collect names from many strange and awful dialects? It would be much easier to learn one very beautiful dialect than a thousand bad ones. But these same men not only give little thought to learning [one beautiful dialect] but also compel us, who have been brought up and educated in the language, to put this aside and learn their dialect.

Do you not wish, my good man, to learn the dialect of the Greeks? **[587K]** Be a barbarian if you wish. Only, just as I allow you to babble in whatever way you choose to bring forth, so forgive me for having learned my own dialect. My father, who was perfect in the dialect of the Greeks, was also a Greek teacher and pedagogue. It was in this speech I was nurtured; I don't know yours. Do not lay on me the use of the words of traders, tavern owners and tax collectors—I am not acquainted with such men. I was brought up among the books of ancient men, and I say this, which I have never said to anyone else: it is not that you act like a barbarian, or that you commit solecisms, or that you use names badly and improperly, but that I allow everyone to speak as they wish, even if as the helmsman said, "draw in the sail"—it makes no difference to me.

I don't blame Phaborinus (Phavorinus) or Dion[432] for these things. I only want to understand what they say. However, when I am at any time confused in relation to this itself, then I inevitably ask him whom he wishes the name to indicate, and whom he wishes to inform, and am silent, neither gainsaying nor censuring him, if he spoke apart from the Greek custom. For me, there is only one concern—to know **[588K]** the meaning of what is said. But for those men, it would seem such a thing is not enough. Whenever we use the terms of the Greeks, it is they who first find fault; some as dialecticians, some as natural philosophers, some as orators, and some as grammarians. They speak disparagingly in many ways. Then, when someone is compelled to come to a longer discussion, he confutes them as knowing none of those things which, by knowing, they are emboldened; and they immediately forget themselves and speak against the very same things which they refuted at the start as being useful. For who of these insolent fellows does not jump on us when we say the *differentiae* of the classes of pulses are so many, claiming we have stated the *differentiae* badly and the

432 Phavorinus is presumably Favorinus of Arelate (A.D. 110–150), a Roman intellectual who, like Galen, wrote in Greek, on history, philosophy and science. Only fragments remain. Little is known of Dion (Med.), as evidenced by his dates, given as 120 B.C.–120 A.D.; he is referred to by Soranus.

classes much worse. Avoiding their vain contumely, we say, "but if you wish, call these 'qualities' or whatever else might be pleasing to you. Our argument is not about names but about matters. It is useful to know how many the primary classes of the pulses are, lest somehow, immediately in the first differentiation, we overlook a whole class."

But they don't allow us **[589K]** to say these very things; immediately they silence us, saying our ignorance is of the whole matter and not of a name. Then they compel us to say to them things we would not at any time willingly say. And then, when refuted first on the grounds that the enquiry is about a name and not about a matter, and that it is they who have used the names badly and not us, they become terribly annoyed, like those waking from sleep, and say the logic-chopping about names is useless, as if we had from the start been saying something else, or wishing to, or it was not they themselves who had initiated such nonsense.

You say, my good man, that eight qualities are attributed to the pulse. I agree that you have used the term 'quality' properly. Saying nothing about this, I shall show how there are not eight. The actual enquiry is about a matter and on this I call forth the most distinguished Sophists. They, however, so that they provide resolutions from this, were I to speak in the midst of the discussion, were accustomed to seize upon some term. Then, when they were refuted on both counts—on matters and on names—they became annoyed. What is needed, I think, is to set aside the bad opinion about the matters, **[590K]** whereas with the names, just as we agreed to their using them, so too should they allow us to do so. But they don't allow us to. Because of this, I am compelled to speak briefly about their discourse.

6. And now I shall still need to put aside the contention about the names, and they need to assent to discourse with us, according to the custom of the Greeks, and to learn they are wrong when they say there are eight qualities. For, as we showed, not all the primary classes of the pulses are qualities. ["But how must we speak?" they say. And they ask this, not so they may learn but so they might have a beginning of a dispute. Accordingly, you will reply, "It is necessary to speak of the *differentiae* of the primary classes in the pulses," or also, by the gods, in this way, "the primary classes of the *differentiae* in the pulses." Of necessity, they will say, "What kinds are primary classes, and which, and what kinds are *differentiae*?"][433]

You will answer them on these matters as follows: first, my friend, ask to hear what classes are, and then you will hear what are primary; and after this, what are the *differentiae*. What, then, are classes you will learn here. What a large pulse is was described, **[591K]** and opposite to this is a small pulse. And midway between both of these is the moderate pulse. And by the gods, what a long pulse is called, and a short, a broad, a narrow, a high and a low, and the intermediates of all these, which are the moderates. And there seems to you to be the same common class of

433 There is a major textual variation in the final four sentences of this paragraph. What is translated here is Kühn's text. In essence, manuscripts F1, F2 and Va1 do not have the text in parentheses and have a slightly different version of the final sentence.

the pulses mentioned, of the kind they state quite clearly, in an overly bold voice. You, however, are not disturbed by this, taking and reading to them the book of Archigenes, which has as the title of its first chapter, "Concerning a Large Pulse." Then, at the very beginning of the discussion, he has written the following statement: "The largeness of the pulse is stated generically, for both the small and large pulses have largeness." Then, having turned over the pages of the book a little, read again this statement from the discussion about strength: "Therefore, there is a class of strength, and the kinds of this are the strong pulse, the weak pulse, and the intermediate pulse." And then, having again turned a few more pages of the book, read the beginning of the discussion on fullness, which is like this: "There is a common class for the full pulse and the empty pulse; sometimes this is called fullness."

And **[592K]** stopping the discussion for a little while, and reading the book, say to them that we are saying nothing new, but only what was said by Archigenes. That for each of the pulses, there is a common class of many *differentiae*, just as in the same present three stated by Archigenes: largeness, strength and fullness. Then next say he writes of a fourth class, hardness, and read the statement which runs like this: "There is a common class of a soft and hard pulse, which perhaps, if you spend time on it, is called 'hardness'." It is even clearer in the discussion on regularity and irregularity, and equality and inequality, referring to the unnamed classes of these *differentiae*, beginning in the following manner: "Herophilus set out according to class the other *differentiae* of the pulses in this way: largeness, rapidity, strength and rhythm. Without exact correspondence according to kind, regularity and irregularity, equality and inequality were mentioned. Accordingly, he is accused by those who complain of trifles of opposing kinds to classes. These things are at the beginning of the discussion. Further down, no longer **[593K]** expounding Herophilus' opinion, but writing his own opinion in the discussion regarding regularity and irregularity, what does he say? "More, then, must be said because the class is unnamed." What in fact does he mean by this—is it that he named the common class of largeness and smallness, "largeness"; of strength and weakness, "strength"; of fullness and emptiness, "fullness"; of hardness and softness, "hardness"; of rapidity and slowness, "rapidity"; and of frequency and infrequency, "frequency." However, although he conceives of the class of regularity and irregularity, he has no name for speaking of both such things, like "magnitude" in relation to largeness and smallness. For just as in that he said there is magnitude in the small pulse so here he has the temerity to say there is regularity contingently present in the irregular pulse, but not equality in the unequal pulse in the same way, from which the common class of these is not named equality. And because of this he says the classes of these *differentiae* are unnamed. Certainly not only we, but also Archigenes and Agathinus, say there are classes relating to the *differentiae* of the pulses **[594K]** as do almost all those who write treatises about the pulses.

In what way, then, do we differ from them? In that we use more legitimate terms. For when Archigenes says, "in relation to largeness" it is a certain quality; and again, in relation to strength it is another; in relation to fullness it is a third; and in relation to hardness it is a fourth; and there are others in addition to these in relation to rapidity and frequency. We shall use the term largeness in relation to the amount of the dilatation. For largeness is not the common class of large and small, any more than whiteness is of whiteness and blackness. But if the colour is the class of these, we must ask at some point, what the class of large and small is. For it is either quantity or what is relative to this, or a compound of what is relative to this and the quantity at the same time. For we know that in respect of many classes something exists relative to it. For surely both whiter and blacker are in the class of colour. The colour is, however, a quality. And both higher and lower are again somehow in the class of where. And earlier and later are in the class of when, while in relation to the many other classes, categories of relativity exist. Why then is it surprising, if it was also introduced in relation to the category of quantity? **[595K]**

What I would particularly wish for, if it were possible, is to designate the quantity of dilatation with a defined measure, just as, in the case of many things, we speak of a foot, or a cubit, or two cubits and such measures. Since we discover no such thing from within the present matters, we approach as near as possible when defining them. For since we know, I think, the absolutely moderate pulse in relation to each of us, we clearly attempt to name those that exceed and those that fall short, sometimes speaking in this way, or even if we are not speaking to one another, we at any rate know in ourselves that this pulse is a little larger than the moderate, and again in turn, this pulse is considerably larger than the moderate, or again in this way, this pulse is not by a little larger than the moderate. However, we also say the same thing in this way: significantly larger than the moderate, or sufficiently larger, or again, absolutely this pulse is large. Or also in this way: this pulse is sufficiently large. Or by the gods, in this way: this pulse is large to an extreme degree. If we increase the concept still more, we say it is very strongly large, or by the gods, the largest. All these statements indicate it is larger than moderate (normal), **[596K]** making clear the more and less, so that it indicates to someone the most satisfactory moderation regarding the use, but only if he knows what the moderate is. For it is enough, in what is greater than the mean, to make five or, by the gods, six differences, that one above all being the actual one we must know, just as to term the largest the mean, and the intermediate between the largest and the mean simply large, and between this and that, place two others in relation to each, the one a little larger than the mean and the other a little less than the largest, and the one beyond the large, nearest to the largest.

I said these things are, regarding indication of the quantity, the apparently large and small, and they particularly seem to have the category of being relative to something. Because of this, then, as we are particularly concerned about clarity for those who are learning, we say to name all such things in relation to the amount (quantity) of the dilatation, at one and the same time teaching them the use of the terms, and not to be confused when we say there is largeness in the small. For even if someone were a layman, or unpractised in the **[597K]** primary disciplines, he would come near anything rather than this, whereas if he has learned anything before, he has found, I think, largeness and smallness a thousand times in things the Greeks have written. And he is accustomed to think of largeness as being in the large and smallness as being in the small. And I know someone who, on one occasion, when interpreting the book of Archigenes on the pulses for me, took there to be no largeness in the small. For certainly, he says, if there is smallness in something, there would also be both in it (in the same thing). When I remained silent, he said he was able to conceive of quantity, or size, or some other such thing being in the small, but never largeness. And when I would neither discredit him as saying what was not right, nor wished to inappropriately upset those learning for the first time, in teaching them, that since the homonyms said of largeness in two ways, I thought I needed to say what is more correct, which is that, in relation to the amount of the dilatation, many *differentiae* of the pulses exist. For in this way I know smallness exists through custom among the Greeks as something contingent and inseparable from those things that are small. As a result, Plato also somehow **[598K]** made the Ideas (Forms) of largeness and smallness, and by the alternation of these, generated the large and the small by reason.[434]

And if someone were to disregard all these, he would agree there is some largeness also in small pulses, according to the other and rarer of the significations pertaining to the homonymy. But how would he say there is largeness in the long and short, broad and narrow, low and high, as being from the same class as the large and small? For the long is what is increased still more in length than what accords with nature, the broad in relation to the breadth, and high in relation to the depth, none of these being conceived of in relation to the three dimensions of the artery, which do not exist without largeness. Thus, the amount in relation to one dimension is not largeness but length, and in respect to this, we say some are long and some are short, not some are large and some are small. For we don't ever speak of a line in this way, but either as long or short. It is only those bodies that differ in three ways that are referred to as large and small. Anyway, Archigenes himself says the largeness of a pulse is the mass of the rising up of the arteries. **[599K]**

A mass, however, does not exist in a line, but only in those things that vary in three ways (dimensions). Clearly, at any rate, all those who say the pulses relate to one dimension diverge from that man in respect of largeness as a class. Accordingly, let him in turn make a specific class, like the largeness for largeness and smallness, or if he doesn't make this, let him seek a common class, not only of large and small, but also of long and short, broad and narrow, high and low, and of the other commensurables in all these. But he will have nothing in common to predicate of all these apart from the amount. And in this way, the teaching in respect of all these would become clear and be brought together. For who is so dull as not to understand those who speak in this way in respect of the amount of dilatation of the arteries, whenever, in relation to any one dimension (distension) whatsoever which we wish to make clear, we use the names long, for that which exceeds the normal length; short, for that which falls short of the normal length; and similarly, broad, for that which exceeds the normal breadth; and narrow, for the opposite to this; and then high and low anal-

434 See Plato, *Theaetetus* 173b (Engl. transl. Cornford in Hamilton and Cairns 1989: 878).

ogously to these in the **[600K]** dimension in relation to depth. And we term the quantity in relation to the three concurrent distensions in this way. That which is increased still more than what is natural in every respect, we call "large", while that which is diminished in every respect, we call "small"; that which preserves a measure in accord with nature in every way, we call "moderate".

When the three dimensions (distensions) don't maintain their proportion, but one or some of them are deficient or excessive, there are twenty-four *differentiae* of pulses, which we set out in a table along with the previously mentioned three, so we might show there are twenty-seven in all, existing in relation to the three different dimensions (distensions) occurring simultaneously. This is, I think, useful for beginners in the theory of pulses to learn—to know the number of all of them in each class, which our predecessors overlooked. But a lenient judgement is warranted for the others. But what of someone wrestling with those who have completely fallen down? How would one not reasonably accuse the followers of Archigenes? For them, there is need to use dialectics for the discovery of such things. In turn, they must not display themselves **[601K]** badly in respect of this since they only confuse those learning, while teaching nothing. It would be better, having the analytic and synthetic method, to discover with this the number of all the *differentiae* in the pulses, although not of course to display it to late learners, but not to show the same thing to beginners, as it is a method in relation to which the diagram is introduced. To those who have not learned their childhood lessons, you will speak in vain; those who have learned them, discover the method from the actions before you reveal it.

Why, then, does not one of them ever seem to have used the analytic method for the enunciation of the number in each class, nor the synthetic method for the discovery of those that are primary?[435] And yet it is possible to use these dialectical methods alone, at one time going from the quantity of the indivisible and infinite to the one primary class of all through the generic and specific differences in between, and at another time going from that back again to the infinite through those same things in between. But **[602K]** not one of them dared to credit any method with the number, either of the primary classes or of the *differentiae*, arising from their division. We, however, have in fact done this, and as we travelled the road, discovered the twenty-seven pulses stated in relation to the three distensions (dimensions), and the nine of these in relation to the one common class—the quantity in the dilatation of the artery. If it were also possible to recognise by touch the contraction, I think there would be in this the same number of *differentiae*. But we shall enquire into this in the books on diagnosis.

7. Archigenes, for it is right to mention him alone[436] after Herophilus, when he quotes Homer as follows:

---

435 See Aristotle, *Ethica Nicomachea* III.3, 1112a18–34 (Engl. transl. Ross and Urmson in Barnes 1984: 1765).

436 Μόνου: F1, F2, Va1; Πρώτου: K.

> For dead though he is,
> Persephone has left to him, and him alone,
> a mind to reason with.
> The rest are mere shadows flitting to and from.[437]

seems to be confused not only in the names, but also much more importantly in the matters themselves. And because of this, he does not, for a moment perceive that he is contradicting himself. You might learn this from the statement itself, which follows:

"There are some other **[603K]** pulses that are neither intermediate, nor large, nor small when considered in terms of magnitude, or of magnitude in respect of the dimensions." Then next, he says: "There is a pulse when the artery is distended to a sufficient degree in length but narrowed in breadth and raised to a lower extent. Let this be called long. There is another, however, which is broad in the dilatation but is low and subject to shortness in length. Let this be called broad. And there is a third, raised up considerably in height but narrow and short in length and is rare. Let this be called high."

In these, he proposed to speak in relation to a single dimension (distension), not including in the concept the remaining two. And he was led astray regarding those things he thought to write concerning the three dimensions. Thus, he has made the first of them not only long but also narrow and low—and not in concept alone but also in the actual names. What does he say? In relation to length of the artery, when the distension of the artery is considerable, but it is narrow in terms of breadth, and raised up to a lower extent, the narrowness in terms of breadth indicates the narrow pulse, while the lesser degree of raising up **[604K]** indicates the low pulse. Next, he made the second not only broad but also low and short—and not in concept alone but also in his description of it. And the third, in respect of the same things, he made not only high but also narrow and short. These three are among the twenty-seven described in respect of the three dimensions (distensions) which I set out in the table.

The first of these has been written ninth in the table in this manner: long, narrow, low.

The second of these has been written twenty-first in the table in this manner: short, broad, low.

The third of these is twenty-fifth in the table and described in this way: short, narrow, high.

Accordingly, let him change the beginning of the discussion about these, in which he says they are being considered in terms of largeness, or let him write simply the dimensions of largeness in terms of magnitude, or of the associated largeness in relation to the dimensions, and no longer think of them as being from the *differentiae* pertaining to largeness. For what does it matter to him whether they are considered among the three simultaneous dimensions, or in relation to one of these alone? Either way **[605K]** there is a difficulty; if we understand them in relation to the three dimen-

437 See *Odysseia* X.491–495 (ed. and Engl. transl. Murray and Dimock 1995: 394).

sions, since all those that have been passed over in relation to a single dimension are nine in number, while there are eighteen others in relation to the three dimensions at the same time. For in this way, it seems only nine are mentioned; the first three spoken of, and the large, small and intermediate of these, whereas of the three now stated and the three others which are the opposites of these, which he mentions next, and in addition to these, he has still unnecessarily added the statement, "in relation to the dimensions of the largeness." It would be enough to say simply, "in relation to largeness."

Contrariwise, concerning the concepts relating to a single dimension, let us suppose he writes these things. First, it is strange to have also mentioned the remaining two in relation to each; then [second] it seems not only the eighteenth but also the twenty-fourth of those in the table are left out; and third, because it would seem bad to say they were considered in relation to largeness, when he ought not to say "in relation to largeness" but "in relation to the dimension of largeness," if in fact, as he himself has said, largeness is a kind of swelling. Indeed, not only length in itself but also depth **[606K]** and breadth are swellings, but just as a dimension is also termed.

Clearly, a swelling and a dimension of a swelling are not one and the same thing. Archigenes' argument is certainly altogether false in both ways, because he spoke in a confused and disjointed way, and as if open to two interpretations in the concept. But what more he wished to say, if it must be declared by surmise, I shall venture to say. It seems to me, he attempted to write the concepts relating to one dimension, and at the beginning he indicated this very thing; or he attempted at least to state the dimensions of largeness, but erred in the interpretation, so that, in the whole book, he was confused and mixed up in his understanding. He didn't know there are twenty-seven concepts in relation to the three simultaneous distensions, the greatest evidence of this being from the "thickness" in the argument and the inconsistency in the statements about them. That he held to the plan but involuntarily missed the mark is clear from those things he offered. For what is the very next thing he says? "Since two distensions are often concurrent, they can be, at the same time, short and low, or short and narrow, and **[607K]** in the other conjunction, long and broad or long and high." It is clear from these things that he wished to conceive of the previously mentioned pulses in relation to one distension (dimension) alone.

If he writes of this in relation to two distensions alone it is as if he has overlooked the third. For what does he say? Since two of the distensions are often concurrent, they can be, at the same time, short and low. There is a need in this statement to know beforehand what is simply short and what is low; otherwise, what is short and low at the same time will not be clear. But neither would long and broad at the same time ever become clear, if each were not considered beforehand individually and by itself. It is necessary to know first the long individually, and after this to know the broad individually, and then to combine the concepts of these into one, and in this way to say long and broad at the same time. From this it is obvious he contradicts himself in an instant. For it would not have been enough to say long since when the artery is distended it becomes still greater in length. And indeed, this alone would have been enough, but he added superfluously, "narrow in relation to breadth and low **[608K]** when stirred up." And again, it was not enough for him to say the breadth, when the distension was broad, but he added "low and short in relation to length."

But Archigenes, if it is necessary for me to preserve these concepts, how shall I recognise the pulse that is at the same time long and broad, as you write next? For you yourself take away the breadth from the concept of the long, and as someone might say, length from the concept of breadth. These clearly differ from one another, since, in the prior discussion, you remove breadth from the concept of length while in the second, you preserve it. And it is similar for the broad and high—in the first statement you reduce[438] the quantity from each of the remaining two distensions but preserve it in the second. He suffered all these mistakes in his explanation of the names due to being mistaken much earlier in the matters themselves.

It is apparent he has used neither the analytic nor the synthetic method in any of his teachings. I am in danger of saying that he has been affected like the many who are also unpractised in reasoning. As a result, he often attempts to resolve what is compound in nature into the primary and simple, **[609K]** but on the other hand, often, since none of these is able to exist individually and of itself, he stops and runs back again to the compound. And again, in turn, since in respect of what is compound, indivisible and inseparable, he has the teaching that is altogether unprofessional and artless, that it is not divided into primary [elements], and not being able to stand aloof from the division, then here again he returns to the compound, as he would be unable to define precisely each of the simple [elements] in the concepts. And turning and revolving this resolution in the middle of the extremes, he is ambiguous throughout in the concepts, and is inevitably oppressed by the contradictory things said.

It is necessary, I think, for someone who lays claim to being a dialectician to have sufficient skill so that everything undertaken, when it is compound, is resolved into primary and simple [elements] according to concept, not as being at some time individual and alone, for this would be impracticable, but as being able to be conceived of as alone, since we don't consider together in the *differentia* any concept of the things joined together necessarily. For we recognise a man as snub-nosed or hook-nosed without needing to know if such a man is large or small, or pale or dark, or fat or **[610K]** thin. It is in this way, we recognise a pulse as high, without needing to know how it is in terms of length or breadth.

Those who say these things, then, have no concept of a pulse that is simply long, for it must somehow of necessity have breadth and depth. They are uncultivated and ignorant of the primary lessons, which they ought to have learned and become practised in before coming to us regarding them in this way. From the fact that length cannot exist by itself, they say it cannot be conceived of alone. Accordingly, let them not think of the snub-nosed man, nor the hook-nosed man specifically on his own, unless they learn all the other things about him—whether he is large or small, or fat or thin, or dark or pale, or Greek or barbarian, or worthless or good, or healthy or diseased, or young

438 Reading ἐμείωσε (manuscript Firenze 74.28) instead of ὤσε (K).

or old. How is it not absurd to agree that you conceive of something that is long, high and broad at the same time, and call it by the one name, "large", while on the other hand say you are unable to conceive of length by itself and alone, just as if it did not have the same correspondence as the large has to the whole pulse, or which the long has to the large? For as length by itself and alone **[611K]** is nothing without having in some way depth and breadth, so too a large pulse is nothing without rapidity or slowness, or strength or weakness, or has one of the other *differentiae.*

But if you don't conceive of the artery moving with all the *differentiae*, it becomes able on occasion to move by itself. But who saw an artery still pulsating outside the body? It is necessary, therefore, for it to be in the intact body, lying on some sinew, and for this to have quality and quantity, and lying under a vein, and this also to be of some quality and quantity, and to have many bodies surrounding it. And starting here from all these things, it is possible to state what is *sine qua non* for the pulse to exist. But I think a person who is paying attention, if he were to listen to the one saying these things, will ask, regarding some of the present matters, whether he is joking or mad to make such foolish statements! For how is it not truly great and long-winded nonsense, for those things not to be able to exist without others, nor to grant they can be conceived apart from those? No one denies that there cannot be a long pulse without depth and breadth. But this is not **[612K]** what is being sought. Rather it is if it cannot be conceived apart from those things.

If someone were to say this, he has overlooked the fact he has overturned all the arts and every use of names in life. For who among us ever learned the therapeutic art for Deon's pleurisy? No one. For it is not simply of a pleuritic man—the arts consist of specific and generic concepts, first in relation to the affection itself, specific definitions having been accepted. We think either of the pain involving the clavicle or ending at the hyperchondrium and there is no coincident thinking of Deon or Theon here. Then there are the preceding and accompanying factors—is the pleuritic person plethoric or on the contrary deficient in humors, or *euchymous* or *kakochymous*, and each of these things either now or formerly? And if formerly, was this due to regimen or nature? Again, nowhere here is there Deon or Theon. But not even when our discussion is about the *krasis* of the diseased person is there a necessity to call to mind Deon or Theon, or to think about them as a whole, or something less.

How would geometry be undone and what would need to be said, if someone **[613K]** were to say he could not conceive of a line or surface, because apart from a solid body it is impossible for these things to exist? What of all the things pertaining to life which cannot exist apart from one another? How would someone not be absurd, if he were to concede he could not think of them individually? But Archigenes would never tolerate such boorishness. Anyway, he would obviously always do the opposite, breaking each thing down into the primary and simple. But, it seems to me, he errs regarding the objective, and is often unwittingly mistaken due to lack of training. At all events, it is possible for me to speak clearly on the matter now before us based on evidence.

"For," he says, "two dimensions (distensions) often concur, so at the same time there is short and low." But we think of them as a pair of dimensions in which each is one, so the conceptions in relation to the three dimensions are twenty-seven, which was not taught before we were accustomed to think of each one of the dimensions individually.[439] Sometimes there is increase, sometimes decrease and sometimes moderation in which an accord with nature remains, and in addition due to the many things we are able to recognise from one dimension. For as the large pulse **[614K]** signifies some specific nature, so too does the high pulse in and of itself, not considering the other dimensions along with it. The pulses in relation to two dimensions have no use in comparison to those conceived of in three dimensions at the same time and have dropped out of the method of division. Because of this, they are technical[440] in teaching, uselessly wasting the beginner's time, as if someone had taught a pulse to be like this, saying that one was large and at the same time rapid, and another large and at the same time strong, and another rapid and at the same time strong, going forward linking pairs in this way. This is not what Archigenes himself did, as it would be unskilled, very unmethodical and useless. For we wished, and the art sought to know, the combination of all the *differentiae*, as a means of knowing the whole nature of the matter before us and leading to the simple *differentiae*. Without this it is not possible to learn the compound thoroughly and exactly. For if you don't comprehend the whole nature of the matter as one, but **[615K]** remain fixed on a single, simple *differentia*, there will be no art, no method and no use.

8. Because of this, I first analysed the whole nature of the pulses into the *differentiae* constituting this, naming one in relation to the quantity of the dilatation, one in relation to the quality of the pulse beat, another in relation to the quality of the arterial wall, and others in relation to the things subsequentialy mentioned. In these themselves in turn, I sought simple concepts so that, whatever relationship the *differentia* pertaining to the quantity of the dilatation has to the whole nature of the pulse, put together from all the elements, it is revealed to us what similarity the other has to this itself. For as this class is what completes the whole nature of the pulse, so too this in turn completes each of the distensions. However, to conceive of distensions together as a pair is not necessary as regards the completion of the whole pulse, nor as regards the single completion pertaining to the quantity of the dilatation. Because of this, the discussion about these is superfluous, and rightly has been deliberately overlooked by us, whereas Archigenes attacked it, as it were, badly, and speaks of six conjunctions only, **[616K]** omitting what are the other twenty-one. It is necessary for all the *differentiae* of each of the distensions to be combined with all the *differentiae* of the others, so that nine *differentiae* or twenty-seven in all are created in relation to each

439 There is some textual variation in this statement, F1, F2, Va1 differing somewhat from Kühn. However, K and Kühn's Latin translation are followed in the translation.

440 K has ἀτέχνου; F1, F2, and Va1 have ἐντέχνου, perhaps technical in the sense of artificial or overly technical. Either way, they may be considered unnecessary for the beginner.

conjunction of the two distensions. I subjoin one; from this it is possible to learn about the others that remain.

Table 6: Nine Possible Combinations of Length and Breadth in Arterial Dilatation[441]

	Length	Breadth
1	Long	Broad
2	Long	Moderate
3	Long	Narrow
4	Moderate	Broad
5	Moderate	Moderate
6	Moderate	Narrow
7	Short	Broad
8	Short	Moderate
9	Short	Narrow

[617K] In this table, by combining the three *differentiae* relating to length with the three relating to breadth, I made nine conjunctions. There is another table again, combining in the same manner the three *differentiae* of depth with the three of length, and another third table in which the three *differentiae* of breadth are combined with the three of depth. It is clear that in relation to each of these, there will be nine combinations. As a result, there are twenty-seven in all.

Therefore, it has now become clear that Archigenes by-passed many of these by conceiving of them as pairs, just as he also did of the others. I, however, by-passed them for the reasons I stated, and wrote those conceived in relation to one distension, which are nine in number, and those in relation to those distensions simultaneously, which are those in the table and number twenty-seven. I did not make largeness the common class of all these, but the amount of the dilatation. Some of the *differentiae* pertaining to the pulses, I make known by names and some by descriptions, for all do not have names and I think it superfluous to furnish new ones, **[618K]** since the descriptons can do no worse. For the descriptions seem to me [not only] not very bad, but also much better than the names for teaching. They divide the other things and most of the *differentiae* into many others, just as I have said in the case of the large pulse. This itself in turn is conceived of as having six *differentiae* and is to be expounded by description, conceiving and calling the largest in these the one farthest from the moderate, and the absolutely large what is midway between the largest and the moderate. And it is possible to put two other *differentiae* between each of these, one being a little larger than the moderate and one lacking a little from the largest.

441 This table is altered slightly in form to make the correlation with the text more readily apparent. The table is included, in the same form as in Kühn, in manuscripts F2 and Modena α. O. 4. 12. It is altogether absent in the following manuscripts: F1, Va1, *Marcianus* 287, *Parisinus* 2161, *Harleianus* 5625 and *Vaticanus graecus* 1064.

That dividing them like this is most useful for prognosis is clear to everyone who has even the slightest care for the actions of the art. For in the case of hot, cold, dry and moist, and absolutely everything pertaining to the art, to be able to divide into the smallest parts and to attempt to follow with very little in the way of excesses is the work of a man who is wise and, in addition, practised in the actions and about **[619K]** everything done, and is skillful in encompassing the art within as narrow a framework as possible. Truly in fact, in the beginnings of the teaching, it must not be divided in this way. Of those learning, to the unintelligent it will produce no little confusion; to the intelligent will come understanding of what is said. Because of this, we didn't mention these things in the first of the books, and nor should we speak of them now, unless we are required to demonstrate the whole range of the names and concepts of the pulses pertaining to amount (quantity).[442]

From the beginning, it is possible to divide the excesses and deficiencies of the concepts relating to one distension, as we showed in the case of the large. In each of the classes, the moderate alone is one and indivisible. The others, however, are able to receive many divisions, for the nature of these is limitless, in respect to more and less in every class that has the excess and deficiency, it is necessary for there to be countless *differentiae*—although not in this way as regards perception. If one is practised at comparing them in terms of better and worse it will also be possible, in one way, **[620K]** to recognise the *differentiae* pertaining to smallness, and in another way only those that are notable.

For surely, the musician not only recognises the excesses of the voice up to tones and semi-tones, but also adequately up to quarter-tones, whereas the layman is not able to hear such things as an excess in two tones. Anyway, we are now also relating our own perception, for we follow closely the six *differentiae* of the pulses in the excess and deficiency compared to the moderate in each class. But it is enough for beginners to perceive the primary aspects of one or the other and to be able to separate these from the moderate. It would be time, I think, for those who use them constantly in relation to the matter, to first perhaps make a twofold division of them, and then to go to a third and fourth. All these things were stated particularly to show the many pulses common to one class, and are useful to know in and of themselves—and we say such things to those already advanced in the teachings and only of those that have been heard of. That there is a class of pulses existing in relation to the amount (quantity) of the dilatation has been sufficiently shown. **[621K]**

9. It is also of the primary classes that you would particularly learn, for I know I promised to show this too, if you should attempt to carry out the same exercise in relation to each of the other classes, which I have now carried out in the case of this one. You will especially do this if, in dealing with each of them, you consider what common class it is possible to establish uppermost and to put forward with this in relation to the

442 There are minor textual differences between K and two of the manuscripts (F1 and Va1). The meanings are however not significantly different.

amount of the dilatation. Well then, since a certain pulse is rapid, slow or moderate in the movement of the artery, and this common class is the quality of the movement, as was shown in the treatise, *On the Diagnosis of the Pulses*,[443] we must consider what common class we will at any time be able to find of quality and quantity. If this cannot be done, we shall assign the quantity of the dilatation and the quality of the movement as primary classes. But also, since the quality of the body of the artery was another class of pulses, in which were softness, hardness and moderateness in the consistency of the wall of the artery, we must in turn consider which of these is a common class; **[622K]** I speak of this now stated and of that pertaining to the quantity of the dilatation.

If, however, nothing should be apparent, we must also place this among the primary classes. That in the consideration of such things, it is necessary to have become practised in recognising the categories is, I think, quite clear, so that when the things being investigated both come under the one category, but there is not in fact some common class for them, we shall not be confused by the commonality of the category, like the majority of those who come late to learning and are unpractised men, feigning skill in dialectics, are. When there is a common category, you think you will discover here a common class, as you would be unable to distinguish the homonymy. Anyway, this exists, even in the quality relating to movement, and to the body of the artery. There is, in fact, a common category of quality in relation to both the classes, but there is not in fact also a class which is common to both. Both are called qualities homonymously but not synonymously. The quality is not one and the same in both, as in white and black, or in cold and hot. Rather, in movement, the quality is specific and of another class, while [the quality] in the body of the artery is also specific and of another class. **[623K]** Movement and the consistency of the body of the artery are not in this class. But something more will, perhaps, be said about these matters in what follows.

Concerning the class of the pulses pertaining to quantity, it is already clear how this must be considered—whether it is among the primary classes or the secondary, there is obviously a difference between the class pertaining to the movement and that pertaining to the body of the artery, if the quantity is different from the quality. And because of this also, there is a difference from that relating to *tonus*, in which some pulses are strong and some are weak, for they are associated with the quality of either the pulse beat or the vital (physical) *tonus*. Furthermore, there is a difference of rhythm too, relating to regularity and irregularity, and equality and inequality. All these things come under the category of relative. What remains is to compare this to the class relating to frequency and infrequency. It is obvious that this is associated with the quantity of the time of the pause. But even here again, it will be brought under the same category, whereas the class will not be the same for both. For **[624K]** there is the difference of a whole class between the quantity of a body in distension and the quantity in terms of time, and it is clear there is no higher class in pulses than that

443 *De dignoscendis pulsibus* I–IV (VIII.766–961K).

in relation to the quantity of the dilatation, just as there is not for any of the other things spoken of.

Whoever wishes to follow these things accurately must have trained himself previously in the categories. The statement by Arcesilaus is good: "No one carries wool to a fuller's shop." For as order is like the weaving of wool, so too is the teaching of sciences. And no one is taught to read before learning all the syllables, while no one is taught the syllables themselves before learning all the elements of speech (the letters), nor the vigorous wrestling methods before those that are simple and easy; nor those before being massaged properly, nor this before being anointed for gymnastics. In each of these, there is a first, a second and a third; this applies in all the arts. So certainly, in relation to dialectics, it is impossible for someone, prior to being practised in the primary, and as it were elemental aspects, to follow the subsequent demonstrations of these. This is enough on these matters for the present.

10. What remains is to speak about the term *diaphora.*[444] Beforehand, I shall show that I am not introducing anything new about the term. Then, I shall explain what kind of thing it signifies among the Greeks. Archigenes is the first witness that we are not introducing anything new, writing as follows at the beginning of the classes pertaining to regularity and irregularity (order and disorder), and equality and inequality (evenness and unevenness). Herophilus set out all the other *differentiae* of the pulses according to classes[445] in this way: largeness, rapidity, strength and rhythm, and mentioned without exactly matching by kind, regularity and irregularity, and equality and inequality.

In this statement, it is clear Archigenes not only mentioned that there are *differentiae* involving the pulses, but also that some of these relate to class while some relate to kind. Not only did he do this at the beginning; after this, he forgot about it in the whole work. Countless times he mentions the term, *differentia* of classes and kinds, most correctly thinking and recognising that some of the *differentiae* are generic and some are specific, and of these same terms, some are more generic and some more specific. **[626K]**

Therefore, I shall quote a few of his statements to demonstrate what I said before, and first this one from the actual discussion about regularity and irregularity (order and disorder): "Some have understood equality in certain *differentiae* only, whereas I understand it in all those to do with order." Then to follow: "Inequality (*anōmalia*) is inequality (*anisotēs*) of pulses in relation to some of the *differentiae* following these closely, whereas irregularity (*ataxia*) is a loss of temporal cohesion in relation to some of the *differentiae* of the pulse." In the same statement, he twice wrote the term *differentia.* Then, a little later, [he writes]: "They have not said about these, whether it is necessary, in relation to every *differentia*, to say equal (*ison*) or equal (*homalos*)."

---

444 The term διαφορά has eight separate meaning groups in LSJ. That listed second is "difference" which includes the use in Logic as *differentiae.*

445 Here γένος (K) is emended to γένη following the manuscripts of Firenze 74.18 and *Vaticanus graecus* 281.

Again in turn: "And in what has been compared, this according to us will be equal in magnitude, strength, rapidity, rhythm, and in any other such *differentia* that, when being in this state (*schesis*),[446] is observed in relation to all; and in addition, to what is observed in relation to some one *differentia* of these in state (*schesis*), or in relation to two or more."

In the same statement, he again mentions the term *differentia* twice. Connected to this in turn is the statement, as he pushes on to speak thus: "What is absolutely separate of all the *differentiae* **[627K]** is equality (*isotēs*)." And at one time, thus: "An unequal (*anōmalos*) pulse is said absolutely and unequal (*anisos*) is said in relation to one *differentia*." At another time, thus: "Sometimes, as I said, the others correspond in three. The strong is like the quotidian [fever], being even and equal; next comes some other of the remaining *differentiae*; or another." Connected to these statements, he next says: "When pulses are in proportion at one time, it is in all the *differentiae*, or in one, or some." Then next: "Sometimes all the others are regular, and by the gods, equal, whereas one *differentia*, such as largeness, or two, or more are completely irregular." Then, at the end of the discussion, "We need to consider," he says, "those things said about rapidity and slowness in the case of every other *differentia* of the pulse." Let such things be quoted from this discussion. What more might someone mention? Let me move on again to something else, so we don't seem to be taking evidence from one place. Let us, then, quote from those that followed, and in turn from those that came before. Next, then, is such a title: What is *anōmalia* (inequality/unevenness) in one pulse?" **[628K]**

Immediately after the start of the book, he writes in this way: "There are clearly a number of *differentiae* of this inequality in relation to one pulse." And then, next: "For the most part, they coincide with these *differentiae* and with the inequalities relating to largeness." And again, as follows: "However, there is another *differentia* relating to magnitude of inequality of one pulse." These are things to take from this book. Let us quote with these a few things said, still right at the start of the book. One statement in the discussion about a strong pulse is this: "In relation to the strength, there are other such *differentiae* to be encountered." Another is, "someone might establish this other *differentia* of a weak pulse." And then there is another: "If it is not slackened but, as it were, impeded and having the weight inclined inward, it is a hobbled and bound *differentia* in relation to the strength of a pulse."

Subsequently, in the discussion about fullness, on one occasion he writes in this way: "Of so many *differentia*." And on one occasion in this way: "The *differentiae* should only be considered simply in relation to what is nodding/inclining."[447] Again, in the discussion on frequency and rapidity, he writes somewhat as follows: "A few **[629K]** seem to shrink from *differentiae*." And at the end of the whole discussion [there is]: "These

---

446 As elsewhere, *schesis* indicates a transient state, whereas *hexis* indicates a relatively permanent state.

447 We are not clear what Galen means here.

are the *differentiae* of frequency and infrequency."[448] And again, on the question of the equal and unequal pulses, [he writes]: "This is not a specific *differentia* of a pulse but lies either in the kind of rhythm or the kind of order."

But enough now on *differentiae*, lest we somehow unwittingly lapse into a digression. For if someone wants to read the whole book by Archigenes on pulses, he will himself be able to fill another book on the term *differentia*. For we now have selected these things from not even a fifth part of the book. And we have quoted many statements, not in this way being zealous about the term *differentia*, but that we might demonstrate the variety that is in its signification. For he has often called the classes, *differentiae*, both the primary and secondary, and the kinds of the classes no less. And there are those that constitute neither classes nor kinds, but are *differentiae* in their own right alone, and he has called these *differentiae*, in which there is also the misunderstanding of the pseudo-Dialecticians. For among all the other Greeks, and among the **[630K]** ancient philosophers, when said in relation to these three things I spoke of just now, it is more common in relation to the classes and kinds, whereas in those things discovered in the divisions, being more specifically neither kinds nor classes, they called to mind this alone as the signification, overlooking the common one altogether, but being shamed by Nemesis, they do not stand by their own decrees.

Because of this, I do not now remind you of the significations of the names among other Greeks; it would be easier for me to choose from all, rather than from those of the pseudo-Dialecticians themselves. I am accustomed to name in this way those who declare they are going to explain the concepts behind the names and establish this as the very foundation of dialectical theory. However, they don't explain; they only make laws for themselves. And I know Archigenes would sooner choose anything whatsoever rather than agree that the name of the *differentia* is also stated in relation to kinds. In other respects, also, they would sooner forsake a city than a dogma they say is necessary. But if you are silent, while they are laying down laws, and utter no word of contradiction, and then trust yourself to discuss something, **[631K]** they will immediately speak against you with their own laws. And this is much more so with the great-grandfather of their sect, Chrysippus. It was he, more than Solon, who set up for the Athenians names and standards for their coinage. However, he was the first to bring disorder to these. And if you were to ask the successors to his decrees, why they do not now stand by his proclamations, they say they are misused.[449]

Is it possible then, my fine fellows, for those doing this to misuse them and not to be wrong? "It is possible," they say. What else is there for them to say, when they are trapped in a well, as the proverb goes? So, what then do they not permit for other men?

---

448 Kühn's Greek text has: πυκνότητος καὶ τάχους. Kühn's Latin translation has *de crebritate et raritate.* The latter is followed on grounds of sense.

449 Of the people mentioned in this and the following paragraph, Chrysippus was the renowned Stoic philosopher, Solon was a famous Athenian legislator, descendant of Codrus, who in turn was the son of Melanthus and the last king of Athens; he died in 1068 B.C. Erechtheus or Erechthonius was the Homeric king of Athens.

Is it possible for Chrysippus alone and his followers to do this? Why, by the gods? Clearly because he was born of the Attic families of Codrus and Erechtheus on both sides. But if he had really been of such people, he would not have counterfeited, like some coin, the custom of the ancient tongue. Now, however, and most terrible, being neither born nor brought up among Athenians, but just yesterday and for the first time coming from Cilicia, before learning anything of the Greek tongue, he attempts to make laws about names for Athenians, **[632K]** a jay imitating a Siren, so that we say jay and not jackdaw or raven, nor anything else that it would be more proper for someone so bold to say.

Those things, then, that Chrysippus insolently inflicted on the Athenian dialect, might perhaps at some time become possible for me to go over. However, these very errors of Archigenes, which now lie before us, you wondrously see as agreed to, since he is afraid to say primary classes, and primary *differentiae*, and generic *differentiae*, and classes of *differentiae*; and indeed, all those things that are customary among the Greeks, said countless times by Herophilus and almost all the Herophileans, since he finds the common name underlying all these is that of "quality". And further, it is possible to say the term *differentia* in relation to all these, and kinds, and amounts, and those things that are relative, and in a word, everything. For the term *differentiae* has arisen and is conceived from the difference of one thing from another. However, things differ from each other straightaway in the first classes of the categories, such as man, white, two cubits and right side; one of these is essence (substance), one kind (quality), one amount (quantity) and one relative. Some, however, are in the classes under these (subsidiary classes), just as **[633K]** the white and the triangular, for colour is the class of the former and form is the class of the latter. But both these have in turn fallen under the category of kind (quality), so that white will be said in relation to the first category, similar to triangular, for each quality will be in a different class—the one of colours and the other of forms.

There are, however, other differences in turn in relation to kind (species), without differences in class, as with white and yellow. For these are the same in class, being colours, but have a certain specific (as opposed to generic) difference. For as black, red and dark blue are kinds (species) of colour, so too are white and yellow. But there is in turn a difference in relation to indivisibility, and in the number one, just as there is the white that can be shown compared to whiteness that cannot be shown, these differing in respect of a certain specific difference, although they still fall under the same kind (species), which is "white". Also, however, the class of these is the same, being colour, and there is a common category, which is quality. In this way, everything differs from everything, and in short, one thing from another altogether in some respect. Since, therefore, the *differentia* is named and conceived from the differing, and the differing will be discovered even in the indivisible and in the kinds, and in the classes reciprocally, and concerning **[634K]** everything that simply exists in any way whatsoever, it is necessary also that the *differentia* is in all the essences and not in fact the quality. Because of this, Archigenes wrongly termed the primary classes of the *differentiae* in the pulses, "qualities", whereas we correctly called these and all the others, *differentiae*, from which we also made the title for the whole book, *On the Differentiae of the Pulses.*

What now still remains of those things that were deferred? Nothing other than to show what was customary concerning the term *diaphora* (difference, *differentia*) among doctors of old and the other Greeks. For the earlier book by Archigenes himself created the beliefs, as of course did all his books, and not solely the one on the pulses. So, if someone were to go to the other writings of Archigenes, he will also find the man speaking of differences of fevers, and affections, and regimens, using the term "the difference" countless times in relation to those things we term matters. For to write in addition about the uses among other doctors and Greeks generally would seem to me to be a very long business, such that it would be possible to fill very large books. **[635K]**

Therefore, it is best to select from each of these. For one will also immediately become accustomed to the other terms of the Greeks, and it will provide fewer issues for us in the discussion to follow. And someone who wishes has my two treatises—the one on Attic terms and the one on medical terms.[450] Perhaps, at some time, I might also produce a work on Ionian terms, so that from those, he might learn those things he seeks about speech and names in general. However, for the present, let me attempt the continuation of the discussion, after I put an end to this here, and add in the subsequent books those things that remain.

Book III

1. [636K] That we differ about matters will perhaps seem to one man to be inevitable, to another to be reasonable, and to someone else, to be worthy of forgiveness. As the tragic poet said:

> If this was called good and wise by everyone,
> there would be no contentious strife among men.[451]

For the present, since the same thing does not seem to be good, wise and true to all, it is necessary, I think, to differ. According to the philosopher's statement: "Even if most important things **[637K]** have been fully worked out," he said, "I nevertheless do not know that the opinion has been established in all. Due to this opinion, it is reasonable to differ."[452] According to those who think knowledge is certain and strong, but difficult

450 The work on Attic terms, referred to earlier, is no longer extant. *De nominibus medicis* exists in Arabic and has been translated into German by Meyerhof and Schacht 1931.

451 Euripides, *Phoenissae* 499–500 (ed. and Engl. transl. Way 1912: 384).

452 The Greek term δόκος is read here as δόκησις (opinion, idea, appearance; Kühn's Latin translation has *opinio*). Also, we are unsure whether the verb τελέω is used in the specific sense of "to assign to a class" (LSJ, τελέω 3), nor what the source of this apparent quote is.

of discovery for the majority, one who does not attain it is to be forgiven. However, to differ about names, and to do this immoderately, should it so happen, and to fill very great tomes with hair-splitting pedantry about these things is neither necessary, nor reasonable, nor worthy of forgiveness. This seems to me to be so anyway, for there is so much obscurity in the matters that, if we were to live a life three times as long as the one we now live, probably not even then would all things become precisely known to us. This is why Hippocrates said: "Life is short but the art is long."[453] So how is someone not absurd who wastes time on matters that are not relevant? It is as if he has so much time that he learns this in his spare time and seeks the essentials at his leisure. Isn't this what those who write such things do—people such as Archigenes in his discussion on the strong pulse?[454] The exposition is prolix and everything written there **[638K]** is strange. Nevertheless, let someone who wishes to, read from the treatise on the pulses. There is a fourth heading from the beginning which starts in the manner: "Magnus says strength of the pulse is not among the simple qualities."

So therefore, to you reading this whole discussion about the strength, I shall now attempt to state those things I know. For I am surprised that he contends about names, for this is a mistake customary to nearly all now, and not only to doctors but also to philosophers, but that he is unaware he is doing this very thing, that he contends about names, and because of this, I suppose, thinking he is discoursing about a matter, extends discussions to great lengths. This, in addition to the other absurdities, not to mention also the things proposed obscurely, requires another book by no means small to explain them, so that we always exhaust ourselves, quite clearly disputing about an ox's shadow, not being helped to such an extent by the dialectic, that we are able to recognise what enquiry is about a name, and what is about the matter itself.

For when some pulse is forceful in its impact, so as to thrust against and stir up the touch, **[639K]** to enquire how someone might diagnose it accurately, and besides, diagnose the causes from which it arises, and what such a pulse naturally signifies, is to investigate the matter, but to enquire into what we must call it, is about a name. So that we might clearly recognise this is about names and investigates nothing, but to look into the matters themselves will increase the arts, let us call the previously mentioned pulse "Dion".[455] And then let us say how it is accurately diagnosed—that is to say, by compressing the artery, not touching it superficially—and then that it arises due to the strength of the physical (vital) capacity, as we showed in the books, *On the Causes of the Pulses*,[456] and then, that it is a good sign in diseases and naturally signifies among other things the kinds of natural crises.

Since the whole theory about the pulses has been distributed into four parts—that about the *differentiae* of them, that about the diagnosis, that about the causes, and a fourth about the prognosis from them, none of these things is damaged incidentally,

453 Hippocrates, *Aphorismi* I.1 (ed. and Engl. transl. Jones 1931:99).
454 Presumably a section in Archigenes' book on the pulses, now lost, as is Galen's commentary on it.
455 An arbitrary name.
456 *De causis pulsuum* (IX.1–204K).

if we have called the strongly-beating pulse, "Dion". For we have learned in which way it differs from the others; whether it is called Theon or Dion or something else, and how **[640K]** it will be diagnosed, and due to what causes it arises, and what it signifies.

So, for the rest, let Archigenes dispute about this with Magnus, who himself also perhaps transgressed the custom of the Greeks regarding names. Nevertheless, he is in fact linked to such a degree that with him the major part of the dispute directed against the ancients is about a name. For what does Magnus say in this passage? "It is also necessary, moreover, that there is a notable largeness and fullness in the pulses and that they strike the fingers with speed, if someone is going to properly apply the term, 'strong pulse'." Clearly, then, Magnus knew he was disputing with the other doctors, not about a matter, but about whether the naming was legitimate or not. It is very clear he is doing this from what follows. For he expects Demetrius, to whom he dedicated these books about the discoveries after the time of Themison,[457] to consider together with this also, as being a philosopher and knowing more accurately what is a legitimate name and what is not such a name. He speaks in this way, for it is no bad thing even here to quote Magnus's actual statement: "How then, they established this properly in the simple *differentiae*, **[641K]** you must decide for me, for it falls to you particularly to judge the proper usages and from the names to form a judgement on the substance of the significations."

Then, after this, he makes it even more clear, speaking thus: "Up to this point, I am not changing my own opinion. I say the term 'strong' does not signify a simple *differentia* of pulses, but is a measure for size, speed and fullness." Even here, therefore, he agrees he is disputing about what the term "strong" signifies. And overall, if someone wishes to read the whole treatise (he has written in the third book of the discoveries after the time of Themison), he would find himself becoming aware that the investigation is about a name.

Perhaps then, it will occur to someone to wonder at the innovations of Magnus. For what need was there to change the significations of the names, whether the doctors previously used them correctly or incorrectly? To discover some different theory pertaining to the actions of the art is without offence, not only for Magnus but for everyone capable [of doing this], whether such a person is of the present day or existed countless years ago. **[642K]** To give so much consideration to the contention about names in these is not right.[458] Instead, if someone wonders at and finds fault with these, perhaps he acts properly if he seems ignorant of the choice of men. For all the so-called Pneumatics are satisfied with the doctrines from the Stoa. Consequently, since Chrysippus was wont to dispute with them about the terms relating to philosophy, they themselves did not hesitate to dispute about the terms relating to medicine. And Zeno of Citium even earlier dared to innovate and transgress the custom of the

457 Themison is regarded as the founder of the Methodic sect. See the introductory section on individuals and sects. It is not clear who Demetrius was. To judge by the reference, he should have been a contemporary of Magnus.

458 Following the manuscripts F1, F2, and Va1 in reading περίσκεψις rather than πέρι ἡ σκέψις as in K.

Greeks in the names. Therefore, there is no longer anything remarkable in the younger doctors and philosophers falling into conflict among themselves to such a degree. But would it not be worthy of wonderment in the case of these men, if, when they come together, they did not contend about a name?

Therefore, as Archigenes, who was certainly wise in other matters, did not think it worthwhile, having indicated this very thing, and because Magnus having made changes regarding the term "strong", passes on in this way to more useful matters, he nevertheless goes over and over the same ground, and among his other mistakes, **[643K]** and because he doesn't compare on each occasion the actual statement of Magnus, what he himself writes is not very much clearer.[459] However, almost all those who have written about the pulses after Herophilus have made another and greater error—that of over-elaboration on names.[460] For it is necessary, as we made the distinctions a little earlier, distinguishing them in this way, to teach specifically about the *differentiae* of the pulses, and specifically about the diagnosis of these, and again specifically about the causes of them, and then next specifically about the prognosis through them. I do not know how they confound and confuse these all together, neglecting some things altogether, or paying little attention to mentioning them, and dragging others out uselessly and superfluously, gathering them together irrationally, and treating them as one, although they differ a great deal.

Therefore, in fact, in relation to this same statement about strength, Archigenes, when he writes that the strength arises from the separation in relation to one manifest [beat] being, as I said, the *tonus* of the movement of the arteries, thinks he has shown the cause of the strength calling it actual **[644K]** strength itself. For if strength truly is this, Archigenes, answer us on its cause! But you will not be able to. Or alternatively, you will be compelled to say it is this *tonus*. And so to you the *tonus* of the movement of the arteries will be, at one and the same time, the cause of the strength and the strength itself. Accordingly, much better are those writing about pulses before Archigenes, some calling strength (*sphodrōtēs*) the firmness of the beat of the arteries, some the forcefulness, some the strength (*ischuros*) and some the upsetting. And Archigenes himself knows this. What does he say anyway?

"It seems to some to lie in the impact to the touch. In relation to this, some also say it is in fact the actual impact from the artery." Then, I know not how, he thinks to overthrow their opinion, writing as follows: "The firmness of the impulse is apparent in relation to the whole dilatation, and the beat, if we apply the fingers, falls more solidly, not during the end of the dilatation, but occurring at that time from lower." This is the statement. Considering this at the present time, the refutation accomplishes nothing. For what is the point, if what **[645K]** makes the beat quite clearly occurs before the

---

459 This is a very complex sentence. What Galen seems (to us) to be saying is that Archigenes, in going over the same ground as Magnus without including the latter's own statements, doesn't in fact make matters much clearer regarding the strong pulse.

460 The Greek term is μικρολογία, which LSJ defines as hair-splitting or logic-chopping. See Plato, *Hippias Major* 304b (Engl. transl. Jowett 1989:1558). Jowett speaks of "pettifogging arguments".

end of the dilatation. This cannot but be the beat from an artery which is strong. For if those before Archigenes said this, i. e. that strength is the opposing beat occurring in the end of the dilatation, he would properly refute this, calling to mind that the force of the impact already appeared before the end, since they simply said the firm beat is the strength, so it is useless to point out that such a beat already appears before the end of the dilatation.

2. But let this be what Archigenes understood—strength is a certain *tonus*—whereas the strength (force)[461] of the beat is not strength. But let us ask him what the cause of strength of pulses is, for surely, strength and cause of strength are not the same. The former is apparent and is what every doctor seeks in those who are diseased. In fact, someone who is altogether unpractised immediately recognises the pulse that is strong or weak, although the cause is not apparent. And because of this, there is no agreement about the cause. Thus, Herophilus says, strength of the physical (vital) capacity in respect of the artery is the cause of a strong pulse. **[646K]** Athenaeus says it is the strength of the vital *tonus.* Asclepiades derides them both, saying *tonuses* and capacities, and all such things are new-fangled names. He attributes the cause of the strength to the abundance and thinness of *pneuma*, just as Erasistratus does, I think. For he does not attribute part of the tonic capacity to the actual walls of the arteries, but says that when the heart strongly compresses the *pneuma*, the firmness in the beat is generated by the passage of this through the arteries. But why is it necessary to mention these things? For certainly Magnus, who himself lays claim to being of the Pneumatic sect, says the strength of the beat arises differently from what Archigenes says. But now is not an appropriate time for such discord. For the present, let us call to mind just this much: all are agreed that the strength of the beat lies in the thrust, resistance and upset in respect of the touch, whereas there is discord about the cause from which this arises. It is, however, ludicrous, in the discord about what a certain pulse is, not to teach this, but to write that it arises due to something. What it is ought to be apparent, **[647K]** whereas what it arises from is discovered by reason.

It is clear, then, that Archigenes does not know distinctly, as the strong pulse was not differentiated from the cause from which it arises. And because of this, he places the strength in the *tonus* of the movement of the arteries, whereas what he should say is that this is the cause of strength, but a certain force and resistance exists in relation to the quality of the beat. However, that he is confused about the matters and says nothing distinctly is clear from these things he brings forward, when he chooses to define the pulses. What does he say? "A strong pulse is one having a greater *tonus* of the movement and is rushing. A weak pulse is one having the *tonus* dissipated and a beat that is slack."[462] He did not simply say: "A strong pulse is one having a greater *tonus* of

---

461 In this section Galen uses three different words for "strength" (σφοδρότης, ἰσχυρός, and ῥώμη). Whether he intends to signify something different in each case, or is simply a matter of *variatio* is not clear. We are inclined to favour the latter view.

462 The word ἀσύστροφος has three meanings listed in LSJ: (i) not forming a solid mass; (ii) not condensed; (iii) slack, careless (referring to ἕξις). In addition, there is a fourth entry as follows: "in a dubi-

the movement," but added also, "it is 'rushing'."[463] What he means by this, I do not, by the gods, exactly understand. For the term, "rushing" (ῥοῖζος) is poetic and not befitting an ordinary citizen; it is not something frequently said, nor does it show clearly what, at any time, it signifies. But if one must **[648K]** believe the grammarians when they explain what is said by the poet about the Cyclops: "With a great rush, he turns the fattened sheep towards the mountain,"[464] the "rush" signifies swiftness of movement. Archigenes himself did not in fact mean this when he gainsaid Magnus, unless he is acting like the orators, who often speak against those they don't understand. But I, for my part, don't think so. What is it, then, that he wishes "rushing" to signify, if not the rapidity? Does it rather mean the strength of the beat? But in this way, he will again have made two definitions at the same time—of the effective cause and of what arises from this—and the perplexity will be greater.

Whichever one of these he wishes the strong pulse to be—whether the magnitude of the *tonus* or the force of the beat, or both together—he would not be right because it is not possible, as was said, for the cause to be the same as what arises from it. Furthermore, to say the strong pulse has greater *tonus* of the movement is unclear. I don't understand what he is saying here—whether **[649K]** the strong pulse arises from greater *tonus*, or the greater *tonus* is the strong pulse. For according to the first of the significations, it would have been shown, not what the strong pulse is, but what it arises from, whereas according to the second, the teaching would be of what is unclear and contentious in relation to the sects. If, however, it must be something else, and the strong pulse is one of the perceptible matters, but the *tonus* is not perceptible, it would be altogether agreed to by everyone. Some say they do not entirely understand what is signified by the term *tonus*, whereas they certainly understand and perceive a strong pulse. Some say they agree to give thought to what is being said, but do not agree it is so.[465] In this way *tonus* of movement and strong pulse are not the same. In fact, while one of these is apparent to perception, the other is not only unclear to the senses but is ambiguous and altogether obscure.

3. Why, then, is the teaching of the strong pulse obscure and false in relation to all the things signified, while that of its opposite, the weak pulse, is clearer and truer? Perhaps from the antithesis, **[650K]** we shall be able to understand some of the things in the previous discussion. What, really, does he say? "A weak pulse has a dissipated *tonus* and a stirred-up beat." Here, certainly, he has clearly set what is dissipated against the large *tonus*, this not being necessary, but to set in opposition the small [pulse]. Or he said "the large" wrongly in the strong pulse, and set the slackness of the beat against the rushing. If I knew what "slack" (ἀσύστροφος) is, perhaps I would understand the

ous sense said of the pulse" citing this passage. In the context, might it mean slackness in respect of the state (*hexis*) of the arterial wall?

463 The term ῥοιζώδης has, in LSJ, the meaning of "like a rushing noise" citing this passage.

464 Homer, *Odysseia* IX.315 (ed. and Engl. transl. Murray 1919: 324).

465 The translation of this sentence follows the manuscripts F1, F2, and Va1 in including νοεῖν after τινὲς δὲ.

"rushing" (ῥοιζώδης). But now this very thing is in turn much less clear than the rushing. As a consequence, neither of these is explanatory of the other. For whether we think he says slack meaning slow, or weak, or soft, or easily upset or lax, I am unable to interpret.

I think only one thing more occurs to me in this matter; it is when he arrogates the quality of the beat to the concept of strength, and certainly at the beginning, when he contradicts Magnus, who says *tonus* is only the movement of the arteries. But now he also says the beat is slack. And a little earlier, he spoke against those who think it lies in the impact to the touch. Now **[651K]** he certainly seems to me to be confused about these things in every way and to have said nothing accurately. However, if these things don't seem to you to be tedious nonsense, I know that those who follow will seem so, if you are not in fact a Sophist yourself, but fond of the actions of the medical art. The actual statement of Archigenes has, then, the following form. In strength it is possible to encounter such *differentiae* and others in which the weakened beat falls and is without weight, which some already called a faint pulse. Another beat which has weight but is light, you would place in the *differentia* of a weak pulse, although it is not light, but has weight as if it were impeded and inclined inward, compressed and sinking in. The *differentia* in relation to strength would be of a pulse and in relation to strength. The beat would be resistant, thrusting against the touch more moistly, of the kind that occurs particularly from fresh nourishment. The 'hard to break'[466] is more like a struggling movement, that in some affections appears following the surpluses of flesh.

Are these, by the gods, wonders he speaks of or just nonsense? Anyway, life is short in relation to the magnitude of the art, which is not short,[467] **[652K]** but he, in fact, does nothing with respect to the shortness, introducing to the art such a mass of false theory. At all events, from where do these pulses come to us again—the heavy (weighty) and the weightless? And why did he not make known the difference in the beginning, when he said there are eight qualities in the pulses? For as there were largeness, rapidity, strength and the others, what was it that prevented him writing heaviness as well? From where did this fresh tragedy come—what is a heavy pulse? Explain to us, Archigenes, how many *mna*,[468] if in fact you have not just written a name and have discovered no matter. There is some strong pulse, as those not exaggerating wrote, resistant in relation to the beat; recognition of this is by compressing the artery and not by touching it superficially. There is one cause, according to someone else, and another according to you Pneumatics; this is the strength of the *tonus*. Tell me also about the heavy similarly—the nature of this, its recognition and cause.

You would be most amusing, if you were to explain and define what a strong pulse is, and a rapid, and a large, which we should know without definitions, if in fact **[653K]** we are not entirely unaccustomed to Greek speech. But what is this heavy pulse, which

466 The term is δύσθραυστος (Kühn's Latin translation has *aegrius frangitur*).

467 A paraphrase of the famous Hippocratic aphorism quoted earlier (*Aphorismi* I.1 [ed. and Engl. transl. Jones 1931:99]).

468 One *mna* (μνᾶ) is equivalent to 100 *drachmae*.

you bring in for us new, not explaining but assuming, if you write the name only, something more is said? This is not speaking but chattering. If you wish, then, according to the comic poet, to be "excellent in chattering but absolutely unable to say anything,"[469] this would be another argument.

If, however, it is not chattering but speaking, and you want to teach something, make clear to us with which of the senses and how we recognise this heavy pulse. For I am not able to recognise the weight of a stone or a piece of wood, unless I pick it up with my hands, or put it on my shoulders, or hang it around my neck. Should we, then, lift up the artery of the sick person with our hands and raise it up—will it be necessary to determine the weight of the pulses in this way, or what do you direct, Archigenes? Concerning these matters, someone of his school, a revered old man and a nonagenarian, said: "O my good friend, my boy, a pulse would be weak, sluggish, slow and delayed." In return, we said: "O father, do you speak of something weakened? What about **[654K]** the slow pulse or the infrequent? For we don't understand." Then, after a short silence, he replied: "The slow." And I said: "I would not ask you about the sluggish, for it is clear you would also say this is slow. But also, the delayed," I said, "is I think slow."

The old man, being apprised of these things—that from all the four terms, one which signified the matter was indicated, which was the slowness of a pulse—became confused and contentious. First, he said that the pulse called by him "delayed" lay not in relation to the slowness of the movement but in relation to the length of the pause. For this "to be delayed" is to hesitate to begin. After this, he also said the weak and the sluggish are like the feeble. For to be weakened, to lack strength, and to be feeble signify the same thing.

Accordingly, replying to him, I said: "No matter! And if you wish to set out a second and third about these, I shall agree with you, for we are seeking to discover the truth and not to dispute continuously. Let there be then," I said, "the delayed in relation to the length of the pause and consider the weakened and the sluggish in relation to the feebleness. Moreover, remember that **[655K]** you said the heavy pulse is weakened, sluggish, slow and delayed." And when I demanded to learn more clearly about the significations, in the beginning it seemed to indicate nothing more than all being slow, whereas at the present time, we said there is weakened, sluggish, and as it were, enfeebled, while the delayed is thought of in relation to the length of the pause, whereas the slow now still remains slow. Again, answer me—do you not say the infrequent is weak?"

"But I do say so," he says.

"What of that which pauses for a long time? Will you not also say this is infrequent?"

He agreed.

---

469 The quote is from Eupolis (446–411 B.C.), son of Sosipolis, and an Athenian poet of the old comedy. An older contemporary of Aristophanes, he is said to have surpassed even the latter in wit.

"It would then come about," I said, "from what we agree on, that the heavy pulse is at the same time also weak, slow and infrequent."

After these exchanges, the revered old man was utterly silent.

"But if this also obtains," I said, "would there not be some other quality compared to all those they say of weight, but with a name based on a coming together of three qualities? For we shall say the heavy is the weak, and at the same time slow and infrequent."

"I don't know" he said, "what you are saying, and you are contending in vain." And having said this, he was done with, regarding me somewhat censoriously and **[656K]** damningly.

But a few says later, at the home of a patient, when that same old man was present, as were many other well-respected doctors, when I and all the others were confused by the symptoms existing in the patient, I predicted he would be greatly troubled in the belly.[470] Then, being asked by them from what source I had predicted this, I said, "No one takes raw wool to a fuller's shop," showing them they were greatly deficient in learning the major things, and did not know the most minor.

A few days later, when I predicted haemorrhage from the left nostril, and then again, that there would be parotid swellings, and then another thing, and then another, prognosticating in some instances and treating in others, since those men were unable to do so, it was necessary for them to ask me to say from which of such theories I prognosticate and act, I remained silent throughout, saying nothing to them in reply. However, I say this from the comic poet to you yourselves:

> That neither crooked wood is made straight,
> nor an old person transformed into a young calf.[471] **[657K]**

For someone would sooner convert the followers of Moses and Christ than the doctors and philosophers who hold fast to their sects. So, I finally realised that nothing at all would be gained by engaging in prolix and idle talk with them. And this is what is done and will be done in the future. I know clearly that these very writings will be useless to all but a few—that is, apart perhaps from some one or two who, by virtue of a remarkable nature, will make use of teaching and practice, and before all these things, will stand aloof from the madness involving the sects. Nevertheless, I write for a pastime and not unrefined amusement, and at the same time towards the forgetfulness of old age, as Plato says,[472] preparing treatises for myself, and at the same time for

---

470 For κοιλίαν ἐκταραχθήσεσθαι as "troubled in the belly" see Hippocrates, *Aphorismi* 4.60 (ed. and Engl. transl. Jones 1931: 151).

471 In this quote, γεράνδρυον is taken as an old person, rather than the more usual "old tree stump"—in fact, the whole line could be taken more literally in the botanical sense: "Crooked wood is never made straight, nor does an old stump produce green shoots." We haven't been able to locate the source.

472 Perhaps *Respublica* 1.328d (Engl. transl. Shorey 1989: 577–578).

some of my friends who might need them. But concerning superficial ignorance along with love of contention, these things suffice.

4. Let us return again to those things we proposed to discuss, demonstrating to those who are able to understand that the heavy pulse is only a name and not one entity (matter). For if, as is customarily understood, the wisest of those Archigenians among us compelled him **[658K]** to explain, it will be weak as well as slow and infrequent. If, however, they say such a pulse indicates the nature weighed down by excess, and because of this ought to be called a heavy pulse, it would be time for them to call it not only heavy, but also "unconcocted" and cold, which could indicate nothing less than an excess of crude humors and cooling. In this way, I think, we shall not hesitate to call some pulse bilious or phlegmatous, if we should once and for all become accustomed to applying the names of the effecting causes to the pulses themselves. This then, as I always said, leads into such absurdities as the following: not to distinguish among them separately in the treatise on *differentiae* of pulses; or separately in the treatise on recognition; or separately in the treatise concerning the causes in them; or separately in the treatise on prognosis.

Therefore in fact, Archigenes, writing about the first class of all the pulses, which is that relating to largeness, failed to mention either the effecting causes of these, or the prognosis from them, but did teach the *differentiae* of the existing pulses in this class, and the name of each of these, interposing briefly in some way the diagnosis for them, although **[659K]** he did not know this, as will be demonstrated in the subsequent books. When he mentions the second class, that relating to strength, he is accustomed to mix together the three teachings; not only those concerning *differentiae* and that concerning diagnosis, but also that concerning their effecting causes. And in that concerning fullness, he is still more confused. Next, in those concerning rhythms, he attached a little of the diagnostic theory, while the majority remains in the first. Only in that about regularity and irregularity, and that about equality and inequality was he able to preserve the first part of the theory about the pulses pure, not speaking there about diagnosis of them and prognosis from them and saying nothing about the causes. In the discussion about the heavy and weightless (to use his own term),[473] among other things, he remarkably wrote in this way: "which some already called an indistinct (obscure) pulse." You must say to whom you are referring, my good man. From what kind of Scythia do they come?[474] In this case, there must be a knowledge that the term "indistinct" ("obscure") signifies nothing more than "weak", for the Greeks say both in relation to this, but "weak" (*amudros*) is more customary for them.[475] **[660K]**

---

473 The term is ἀβαρής which is found in Aristotle, *De caelo* 277b19 (Engl. transl. Stocks in Barnes 1984: 460) and Chrysippus, *Stoicorum Veterum Fragmenta* 2.143.

474 Scythia is a term used by Greeks and Romans to characterise the lands to the north and east, roughly from the Danube to the Don, Caucasus and Volga.

475 The two terms here are ἀμαυρός (dim, faint, obscure, gloomy) and ἀμυδρός (dim, faint, weak, applied to the pulse).

So then, these two pulses, heavy and non-heavy (weightless), Archigenian gifts, have reached the point of being talked about but have no diagnosis. The pulses subsequent to these are compounded still more strangely. For what the *differentia* of pulses having the weight turned inward is, I do not know. And because of this, I was compelled in turn to ask the followers of Archigenes what in the world the enigma is. When they replied more enigmatically and strangely than he himself wrote, I knew for sure they were men astonished at what they didn't understand. And it seems to me that Archigenes, knowing this same thing well, deliberately wrote obscurely in many places. I, therefore, am unable to mention "the weight turning inward" without laughing. For it is clear that weight turning outward is something different. And the latter is, as I think, heaviness (weight) on the skin of the person, while the former is weight on the bones. For how otherwise might he speak of inclining outward and inclining inward? But naming a certain *differentia* of a pulse "impeded", "squeezed" and "sinking into",[476] do they not also need those who will explain them? What is the "impeded"? **[661K]** For the slower must be impeded in movement, as must the more infrequent than is appropriate, the intermittent, and the remittent,[477] and every unequal, irregular, and arrhythmic [pulse], and others—the smaller than it should be, and the weaker; these are altogether, I presume, in some way among those that are impeded.

But being impeded is more indicative of a cause than a *differentia* of a pulse. And again, with the squeezed, is this either what is weak or what is small? And what does he wish to differentiate in the case of the "sinking into"? For the "sinking into" seems to be something small. But perhaps we are all dullards, whereas Archigenes is all but Hermes in intelligence. It was, then, necessary for him, as a god trying to teach men not only to write names, but also diagnoses, as he did in the case of those that were clearer. But if he attempted to distinguish them in general by perceptible signs, he would not have spoken in this way in vain and not been aware, just as we too would not have been able to understand many others without reason.

And so, I speak first about the "swooping" pulse,[478] for **[662K]** is it not possible for him to speak to us of this "sinking into" or "swooping"? And, by Zeus, another is "nodding". Now I shall also say viscid, brittle, black, bright, pure, turbid, terrible and joyous. And somehow also, I shall speak of a pulse "like a harp string" and "four-stringed", if it is necessary to state names only. But, they say, "like a harp string" and "four-stringed" are altogether without signification (unintelligible). Clearly, they are terms of Archigenes. Why do you deliberately speak nonsense, my good fellow, for *blituri* signifies some note while *skindapsos*[479] is not only a house-slave but also the name of some in-

---

476 The three terms are παραπεποδισμένην/παραποδίζω (to hinder, impede); πεπιεσμένην/πιέζω (to press, squeeze); and δεδυκυῖαν/δύω (to plunge or sink into).

477 The two terms here are διαλείπουσα and ἐκλείπουσα (see Galen, *De causis pulsuum* II.3 [IX.66K]).

478 The term for this pulse is ἀναπεπτάμενος, the perfect participle passive of ἀναπετάννυμι (to unfold, spread out, swoop)

479 These are two terms (the second used earlier) which are used as "nonsense" words, although they do also have specific meanings, as Galen indicates.

strument. What in fact do I think someone will say regarding the pulses. And what are the names you state regarding them—those mentioned a little earlier.[480] *Apokekrēmnismenos* (mixed), *truzōn* (murmuring), *hygrophanēs* (moist appearing*), *karōdēs* (drowsy), *bombōn* (rumbling), *ektethambēmenos* (alarming), *analutheis* (dissipated), *atenēs* (strained), *adranēs* (feeble), *apopepēgōs* (congealed), *diapephusēmenos* (inflated), *diēgkōnismenos* (disrupted), *egkaluptomenos* (concealed*), *exōstēs* (expulsive*), *embrithēs* (weighty), *egkrekōn* (falling away*), *notulos* (moist*), *katanenagmenos* (stamping*), *dianenagmenos* (caulking*), *lagarizōn* (nudging*), *hypokleptomenos* (concealed), *hrakōdes* (ragged), *sesobēmenos* (agitated), *skuthrōpōs apoxōrōn* (sadly withdrawing) and that which befalls a drawn-out artery, **[663K]** and a very great number of other such names which signify something specifically and of themselves, but there is nothing common between them and the pulses, just as saying some pulse does not thrust more moistly.

For here, when he mentions moisture, Archigenes again wanders away from the teaching of the pulses into the prognoses of a useful matter, but not from the theory of pulses. This (i.e., moisture) he mixed with the actual "captious" beat,[481] as named by him (which is no different from the strong pulse), professing to teach that this is another *differentia*, although it is not, but just a strong pulse with moisture.[482] Then he deceives someone unpractised in the theory, as if describing something major—a kind arising particularly from recent nourishment. For the "captious" beat, he says, is itself associated with moisture arising from nutriment that is particularly recent. And what reason says, Archigenes, is that what you say is true, but in a disjointed and confused fashion, and because of this, obscurely. What he ought to say is that from recent nutriments a strong pulse arises with moisture. **[664K]** But he should not say this is some other pulse beside the strong pulse, and something else, which is altogether not a pulse, in respect to which it is also inopportune to attach it to the common theory and the part about the causes in the pulses, and about the prognosis from them. For in those it would be better to say what kind of pulse arises after food. And, Archigenes, we do not speak lightly in this way, as you do now, but along with the specific distinctions. For the strong pulse with moisture does not simply arise after all food, but only after that which is moderate in amount and moist in quality. For more copious food does not produce a strong pulse, but an unequal pulse from those that are weak, strong and intermediate; and dry food does not produce a moist pulse.

As a consequence, the skin at some time becomes more moist due to the moisture of the food, for it is not always due to this, but there is need of some distinction, whereas in turn, with food that is moderate in amount, a strong pulse altogether follows, it is better not to mention moisture at all, which does not at any time change the pulses by reason of itself, while those who mention moderate nutriment, say **[665K]** the strong

---

480 Galen then lists 24 terms, for which a possible English translation is given, and which presumably Archigenes applied to pulses. An asterisk indicates those for which the meaning is given in LSJ.
481 The term used here is ἐξεριστικός, meaning "captious" or "disputatious".
482 There are some textual issues with this sentence. The translation follows Kühn's text.

pulse arises from this. But this, even if it were true, is first of the teaching of the causes in pulses and next of the prognostic. I don't propose to speak to you about these. If, however, I am permitted to write vaguely and not at the appropriate time, you should know such a pulse occurs not only after recent food, but also sometimes even in baths, exercises and rages. For it does occur sometimes even in these. And if we were satisfied to say only how and in what manner it occurred, we would have taught nothing more, whereas, if we provide a definition, we shall move away from the subject before us.

Accordingly, let us not keep on chattering, but if you have some *differentia* of pulses to teach us, other than those we know, then teach this, not only speaking about names but also the signification of these names and the recognition of the whole subject. Since, certainly, on the contrary, someone might leave these aside, I attach to them next [the name], "elbow-leaning" pulse, but leave aside the following: *diezōsmenon* (recovering), *apopeplegmenon* (separating), *probebēkota* (advancing), *apobebēkota* (reaching), *laktizonta* (kicking), *gonatizonta* (knee-bending) and *gastrizonta* (gorging).[483] **[666K]** For what law (of naming) calls something "elbow-leaning" but leaves out the others?

"But," he says, "we say 'elbow-leaning' like we say the *tethumōmenon* (being angered) pulse."

"Therefore," [I say] "you are in turn wrong to leave out those like the *lupoumenon* (grieving), the *orgizomenon* (provoking) and the *mēniōnta* (being wrathful)."

"Don't joke," he says, "for we call the strong pulse like this."

"I don't joke, my dear friend. However, you direct us to say 'elbow-leaning' when it is in fact a strong pulse. For if you had said strong, I would have understood you. What at any time is this strong pulse, and how shall we recognise it? We shall recognise the strong pulse by its resisting us when we compress the artery, while the weak pulse is the opposite to this, disappearing when we compress the artery. If, however, you say what the strong pulse is, and how we shall recognise this, you will teach us something more. But if not, and even here you add only a name and not a matter, [you will not]."

"When it forcibly upsets," he says, "and thrusts against the touch. But the strong pulse also upsets, although less strongly than this."

"And therefore, do you say the strong pulse and the 'elbow-leaning' pulse are juxtaposed on the basis of strength?"

"Yes," he says, "I do say this."

Accordingly, it is time for us to also call the pulse increased in **[667K]** rapidity, not the most rapid but something else; and the pulse increased in magnitude not the largest but some other thing; and the others likewise. However, it is enough for those to say most rapid, largest, slowest, most infrequent and most frequent, and enough even here to say strongest, and not to think to mention something else closely comparable

483 Here and immediately below, Galen lists another eleven terms presumably used by Archigenes for pulses.

in strength as another *differentia*. For the increases in intensity in each of the classes are not outside the class. Thus, the very white is not anything other than white, and the very black is not anything other than black. Therefore, the very strong is not anything other than strong. Thus, all such things are just protracted nonsense, which follows inevitably from the failure to divide these things properly—that is, neither the concept of each of the pulses, nor the essence, nor the diagnosis, nor the causes from which they arise. For it is necessary, as I have already said often, for one who devotes himself to truth, to distinguish how many and of what sorts these are, beginning from the matters themselves, and then next to apply names to them, and after this to speak about their recognition, then after this, the cause, and then the prognosis, **[668K]** which is what we have done. In this present work, we have gone through all the *differentiae* of the pulses; in the work on diagnosis, we have gone through how we must apply the fingers for accurate perception in relation to each of them; after these things, I have put together the treatise on causes and then the one on prognosis. But now in fact our discussion is still about the first, which is itself useful for the actions of the art, as has been shown in the first book. So now let me add to the discussion a section on strength.

5. There is a certain pulse which overcomes the touch; that is to say, it beats strongly and resistantly. This is either strong, or vigorous,[484] or whatever someone might wish to call it. One must know how to recognise this by placing the fingers in this way, pressing them still more, and not touching superficially. A cause of this is the strength of the capacity producing the pulses. If, then, we were the first to be in a position to totally control the establishment of the names in relation to medicine, we would have called such a pulse vigorous (*eutonic*) and not strong. Since, however, it was **[669K]** named strong beforehand by the majority, we did not change the name, even if we certainly knew that among the Greeks the name of strong is always said of actions that are at the same time strong and swift. For wrestling bouts are termed strong when they occur with intense exertion (*suntonia*) and speed; and men, in the same manner, when they are acting vigorously and swiftly at the same time, are called strong. Furthermore, everyone calls the class of lions strong, because they see action with intense exertion (*suntonia*) and speed at the same time, so that, if it were up to me to establish the name, I would call the simple quality of the pulses that resist and upset the touch, power (ἰσχύς), strength (ῥώμη) or vigour (εὐτονία), or some other such thing, whereas I would not call this compound with speed, strength.

Since at the present time, the term “strong” has arisen through custom among the majority of doctors and is said in the case of one of the primary *differentiae*, I know we must not avoid it. And because of this, I called the pulse beating strongly, “strong”, and the pulse beating weakly, “weak”; it makes no difference **[670K]** to say stoutly, strongly forcefully or *eutonically*. I say[485] a cause of this is strength of the capacity moving the arteries, recognised when the fingers press deeply. For in such a manner of palpation,

484 The two terms are σφοδρός and εὔτονος.
485 Reading φημί instead of φησί following Kühn’s Latin translation (*dico*) and the context.

the strong pulse seems stronger, opposing more forcefully the compressing fingers, whereas the weak pulse sometimes drops out and is vanquished by them, so the person seems to be pulseless. But we shall say more about these things in the subsequent treatises. Let my present discussion now come to an end.

6. I wish to come next to the much-discussed fullness, which has given all the more recent doctors many issues and will in fact somehow also give us not a few, and not only in the present treatise, which would be bearable, but much more in that on diagnosis. For we agree on what the concept of this, which is clear, is, but we say there is no perceptible diagnosis. But I shall speak about the diagnosis again. Let us speak now about the concept, since some even confuse this, **[671K]** as we already said.

Thus, to those beginning from the matters, there will either be a certain quality of fullness of the body of the artery or of the contained substance. Someone might begin here from the names particularly. We say the jar is full of wine and the sack is full of barley. In this way too, we say the theatre, or the stadium, or the council-chamber, is full of people, just as we also say [these places] are empty. In this way too we sometimes say the stomach and the mouth are full, and sometimes empty.

> Let thy fair mouth be full of honey, Thyrsis,
> full too of the honeycomb,
> and may you eat of the figs from Aegilus.[486]

All in all, we say every vessel containing an open space within itself is either full or empty. When its open space is taken over by some other body, either one or many, it is full; when it contains only air, it is empty. All people apply the terms properly and primarily in this way. However some, here already transferring to their own arts, no longer obviously properly nor primarily, **[672K]** but secondarily and figuratively, term wool full, or wine, or something else, judging each matter by the use. For what is enduring and stable, they call full, just as those who taste wine, when they have a sense of the quality lasting a long time, whereas if the quality ceases and is destroyed, or the potency, or the use, they call that empty. In this way too, they sometimes term false arguments "empty", but not in fact "empty" when they say something of sense, using transference under these circumstances, just as they do not, when they say thin people have the body empty of flesh, but fat people, full.

But in these instances, they preserve the concept of a vessel and what it contains. In these cases, as was said, they are accustomed to say full and empty properly and primarily. Others again say bodies are porous, not empty, like a sponge or a pumice stone.[487] For they do not think the air is a body, but some kind of empty space suitable for receiving bodies. But those who have given consideration to the matter of the air by

---

486 Theocritus I (*Thyrsis*) 146 (ed. and Engl. transl. Edmonds 1912: 22). The third line here is not included in manuscripts F1, F2, and Va1.

487 The terms are διάκενος and κενός. The former, as well as meaning empty, hollow, void or thin, is listed in LSJ as having the meaning of "porous", citing this passage.

reason, doctors and philosophers, do not **[673K]** assume this to be altogether empty, but some of them, whose opinion we share, consider there to be one body continuous with itself everywhere and nowhere to have an admixture of emptiness.[488] Some others, however, consider many bodies are in an emptiness (void), these bodies being small, and primarily coming together, colliding and rebounding, but not in fact intermixing and combining.[489]

Accordingly, also, it is impossible for us to name the empty pulse in this way, as many do who say it is truly empty, for there is no such thing in the cosmos, but not in relation to prevailing. For, if there is nothing wholly empty in the cosmos, how would something be empty in relation to prevailing? So then, what remains is for us to call a pulse empty, like a bag, or wineskin, or basket, or sack, or all in all any vessel whatsoever. But if we speak in this way, such a thing will be thought of as occurring in two ways: one way is like a wine jar, when it has only air in it, and the other is like a wineskin or bag or some other such thing in which it is possible for the walls to fall in on themselves, just as the inflated wineskin remains always empty of wine but full of air. And if you evacuate this, so it completely collapses everywhere, then it becomes [entirely][490] empty. **[674K]** Accordingly also, we shall speak in this way of the artery as being empty but containing air. As a result, even if it were inflated and distended to the maximum extent, it is called empty, or not so, but as we said in the case of the wineskins, when someone empties the air from them. So that, in relation to the first of the significations, the empty pulse will be such as this, like Erasistratus always understands it to be in accord with nature. In relation to the second signification, it will be no different from the small pulse. In relation to the third signification, neither Archigenes, nor Agathinos, nor Magnus nor Athenaeus, nor any other of the Pneumatic doctors will agree on the truth.

I say, when the artery contains few bodies in itself, the lumen is mostly empty. For they think there is nothing such as this in the cosmos, but that the whole substance is united with itself. What is left then, as I said, we shall call either the small pulse or the airy empty pulse. In this way too, we shall call the opposites to these, "full", both the large and the seemingly moist. But it is these, in relation to the substance that has been poured in, which someone might most legitimately term a full or empty vessel. Otherwise, in relation to **[675K]** the actual body of the artery—that is to say, its walls—it would be better, I think, to term them hard and soft and not full or empty. For no *homoiomerous* body is called empty or full other than metaphorically, which is not necessary to attach in relation to the scientific teachings. But if we were to have proper names, it would be appropriate to use them. If not, however, it would be better to interpret each of the matters more by reason, and not to name from transference (metaphorically), when someone wishes to teach and not babble on, since for someone who

---

488 The term παραπλοκή has two meanings listed in LSJ: (i) weaving in, intertwining; (ii) intermingling, admixture. Erasistratus *apud* Galen, *De respiratione usu* II (IV.475K).

489 The Atomists generally, following the original concepts of Democritus and Leucippus.

490 "entirely" (ἀκριβῶς), present in K, is absent from the manuscripts F1, F2 and Va1.

has learned the matter, it is possible for the sake of succinct explanation to demonstrate what is being said, both through names from transference and through those from analogical explanation. Indeed, the primary teaching of all technical matters, for the sake of being clear and distinct, needs properly applied names. Accordingly, the body of the artery—that is to say the wall of the artery[491]—is either harder or softer than it should be, or clearly moderate and in accord with nature, so that we shall not say again in regard to this itself, that it contains some airy or moist substance. For it is one and continuous with itself in every direction, unless it has somehow **[676K]** acquired certain very small perforations.

But it is necessary also for these [arterial walls] to contain some air-like, or moist, or altogether vaporous substance, or to have collapsed completely in on themselves. But those things that have collapsed in together do not altogether come to an equal state to these. However, containing moisture, vapour or air will make the wall[492] of the artery softer, so that we come to this in turn periodically, which we said from the start, that the wall of the artery is either hard or soft, for the discussion showed there was no other third nature in between, apart from those things they state, [which were] the two *differentiae* of the generating cause. But the matter proposed is not at all useful for our present purposes, as I for my part am prepared to say about all the origins of the hard and the soft. If I am here going to have the benefit of something useful for the present matters, I shall attempt to accomplish this in as brief a fashion as possible.

Of all bodies put together from earth, water, air and fire, in those in which there is predominantly the earthy element, or the watery made solid by cold, there is necessarily such hardness. On the other hand, those in which the fiery **[677K]** or airy element, or the watery element that is not solidified predominates, are all soft. In this way too, if they should have some perforations and pores, and if either the earthy substance or the solidified watery substance is predominant in them, of necessity this contributes to the hardness of the whole body. If, however, it is to some extent fine-particled or there is non-solid moisture, it contributes to softness. But regarding those that are like this, it is already fitting to ask those men by what underlying matter in particular they apply the term "fullness". For all in all, they either call the hardness simply fullness or some of the stated *differentiae* of it. However, there are *differentiae* when either dry earthiness or congealed moisture predominates. Therefore, what is underlying, either has some pores in itself or not. In relation to these things, they call the emptiness either "the softness" or one of the *differentiae* of this. There were, I think, also the *differentiae* of this in relation to the predominance of either a fine-particled substance or a non-solid watery substance. But they don't dare to respond to any of these. For they would not in this way be clearly assuming either another fullness apart from hardness, or another emptiness apart some softness. But they resile from the discussion in every

491 Reading αὐτῆς (F1, F2, Va1) instead of αὐτοῦ (K).
492 Reading χιτῶνα (F1, F2, Va1) instead of σῶμα (K).

[678K] way, some of them quite shamelessly, but others with some shame. The escapee, or at least the honourable one, flees for refuge to the potency of the *pneuma.* As for wine, they say the quality is said to be fullness, and in this way too, there is some fullness of the *pneuma.* But here again, when we ask them to make clear what they call a certain fullness of wine, they say it is ineffable. How then, by Zeus, do you attempt to name the unnameable or expect to teach us? For if you are neither able to teach the matter nor interpret it by reason, how would we learn any of your writings or form a judgement? "But look here," they say "I show you the full pulse in the case of those who are diseased."

You will do well, my man, if it is inexpressible but demonstrable, to write nothing about it but only show us the matter. And yet, you do write as though you are able to interpret but unable to demonstrate. For no teaching can exist in relation to demonstration apart from explanation. Thus, if you direct me to apply my fingers to the artery, as I think, then you say, "Behold the full pulse is such a pulse as you now perceive." However, if this is something simple, **[679K]** and one quality in the pulses, it would be truly stated. But now, since these [qualities] are many according to you, while they all occur at once, it is not possible to thoroughly learn any of them without explanation.

See, at any rate, that I touch the artery, as you direct, but do not know at any time which of the things concurrently befalling it, you call fullness, nor still more do I understand the matter from your demonstration, or any more than before, when you were not demonstrating. Let it be that, whenever I apply my fingers to the pulse, I divide it as far as possible into a length and a circumference. If I also squeeze and compress, it falls more forcefully on the touch, and appears hard and rapid. Then, if I think to myself the forcible impact is called fullness, how does something strange not occur in the demonstration? For what we learned in those things prior to call a strong pulse, we now learn anew to call full. Accordingly, state some other symptom of the dilating artery to which you direct me to apply the term "fullness", or if besides these things, there is nothing else, you introduce a name to the pulses, but teach no matter. **[680K]**

If, however, it suffices for you to say the matter is ineffable, you will presumably permit me to say there are other apparent qualities in the pulses in addition to those stated by you[493] during the dilatation of the artery. These are three in number, incorrectly neglected by you. If you ask me what these are, I shall say they are inexpressible by proper names. I shall submit to teaching by metaphor, if you wish. For just as we say, in respect of sound, one is the white but another the black, in the same way too, of the pulses, I shall say this one is white and that one black, and immediately this primary quality has been left untold by you, while another and second, in relation to which something of these is perceived as sweet and something as bitter, we must also under-

493 Kühn's Greek text has ἡμῶν; his Latin translation has *vos* here and in the final clause of Galen's sentence.

stand these names from transference, for the matter is ineffable, although it can be demonstrated. In fact, look and I show you this pulse is bitter.

But the third quality is also ineffable; however, it can be demonstrated. I shall call this, for the sake of an example, sourness. For just as in taste, we recognise a sour juice, so too even in touch we recognise **[681K]** the kind of sour beat of the artery. If someone were to use this method, he would say that not only three, but thirty *differentiae* of pulses have been left aside. And above all, pay attention to me on that—some sour juice is properly named but no wine is properly full. But if someone were to name it in this way, he is using a metaphor in the manner I expounded a little earlier. So, it is possible for all men to transfer from those things properly named, but not to make another transfer in turn from these names that have already been transferred.

For this is in fact something that must not be permitted to poets, nor to those who profess to teach matters that are scientific or technical. As a consequence, the transference of the term sour from juices to touch is allowable, since it begins properly from the juices. However, in asking to transfer from the fullness relating to taste to the fullness relating to touch is not to make the transition according to rule, but also goes way beyond the authority of the poets, for if this were to be permitted once, making another and second transfer from the first, he would be so bold, I think, as to make a third transfer from the second, and a fourth **[682K]** from the third, and a fifth from the fourth, so truly to wear out through doing this those afflicted by these. And indeed, it is not permissible for the poets, should they happen to transfer from the proper [names], so even if someone were a Pindar, saying the leaves of the ocean are the fountains, he would not be praised. And much more so, should he say he forges a true tongue on an anvil, or some other such thing. But, as it seems, nothing is as productive of loquacity as the poetic man who dares venture into those things he doesn't know. Anyway, to discover *differentiae* of each of the matters—how many there are in number and of what kind—are tasks for a man made perfect in theoretical logic. And next, to place names on these, is a task for some rhetorician, or to speak more boldly, for a dialectician. Therefore, when men neither perfected in logical theory nor adequate in interpretation, dare to tackle matters much too big for them, they themselves are prolix, and compel those who refute them, even if they are particularly practised in brevity, to fall into the same prolixity as them, which, I think, is my problem now. For certainly, because of the number of things in which they err, I am divorced from **[683K]** the brevity which is customary for me. But even here, I shall attempt to cut things short as far as possible.

But this would be when I am refuting the general and common errors of all, but leaving aside the individual errors of each. The common and general errors are these: whether the full pulse signifies a certain quality of the wall of the artery, or the amount of the substance that has poured in, or the nature, or of the innate *pneuma*. Also, as some dared to say, whether this is a certain potency or quality. For the qualities of the actual body of the artery, as I demonstrated a little earlier, are not fullness or emptiness but are named more properly hardness and softness. However, someone would be right in calling the amount of the actual substance poured in fullness and emptiness. These differ not at all from largeness and smallness. If not the quantity but the quality of the substance contained in these, they would say is indicated by the names, this would not be naming properly. Nevertheless, they will say there is some existing difference not in fact recognised by perception, as **[684K]** I shall demonstrate in the books on diagnosis.[494]

However, the fullness and emptiness of the *pneuma* is even more imperceptible in and of itself, if someone were to take this as a cause for indication and would place the strong pulse as a sign of the strength of this, and the weak pulse of the weakness. One who assumed these things would be from the Pneumatic sect. Consequently, it is even now clear that it is not a common concept, nor will someone say it is an affection of sensation coming to recognition but will give first place to his own doctrine. It is not necessary for dogma to be given an account of in any of these [issues] of the concept of the matter, or encompassed in the diagnosis, for it is common to all, but only in speaking of the causes, as I shall do in the treatise, *On the Causes of the Pulses.*[495] For in those, when I go through all the causes changing the pulses, I shall be compelled to touch on doctrines. However, in what is presently to hand, the exact opposite applies —it is not necessary, in respect of the discussion, to touch on doctrine to the slightest extent.

I think these things are enough about fullness and emptiness, at least for present purposes. Still more will be said about these in the treatise on diagnosis. Furthermore, those things which **[685K]** Archigenes babbles on about at great length will be stated in my judgement of his book about the pulses. For I had decided from the start to be satisfied with these four works, which I have already mentioned often: the first is the present one, the second is on diagnostics, the third looks into causation, and the fourth is on prognostics. But many others have forced and compelled me to write about Archigenes' book on the pulses, and to show what is said by him correctly and what is said incorrectly, and these make my present treatment more concise.[496] For in this I stay away from all the things individually and go through in the subsequent discussion those things in general, not as in the second book before this. And in this book, itself, in the discussion about strength, I quote the statements of Achigenes for the purpose of showing the things he says badly.

7. Having already passed over to the discussion about hardness and softness, I shall attempt to show here too the fault of men concerning the names, due to which **[686K]** they were necessarily mistaken about diagnosis and prognosis. For if one significication is proper for each of the names, I don't know how they make metaphors from metaphors. Anyway, who doesn't know that we say iron, stone and wood are hard bodies, while oil, honey, milk and water are soft? For I think all men have touch as a criterion

494 *De dignoscendis pulsibus* I–IV (VIII.766–961K).

495 *De causis pulsuum* (IX.1–204K).

496 The reference here is to Galen's commentary, in eight books, on Archigenes' treatise on the pulse; this is no longer extant. See Galen, *De libris propriis* 5 (XIX.33K).

of a hard and soft body. Therefore, when we apply this from without[497] to underlying things, those we find receiving and as it were going down with it, and somehow withdrawing, we call soft, while conversely, those forcing and thrusting against our actual skin, which receives the forcing things and withdraws into itself, we call hard. For in simple terms, if something draws back and turns in on itself in response to what forces it, everyone terms what is turned back "weaker", while what does the turning back, they term "stronger".

For in this way too, a strong wind naturally not only turns us back and throws us down, but also does this to very large trees and boats. But we ourselves also turn aside bronze, iron, stone and wood and any of those things that are even harder, when they are moderate in mass. And the **[687K]** strong differs from the hard in the same way. For the former, as I said just now, displaces from the original position and turns away what is approaching, whereas the hard, if it is not also strong, does not displace but drops down into what is softer, hollowing out, excavating and compressing it. Certainly, these are things Plato also said—with those things that are hard, our flesh draws away, whereas those that are soft, draw away from the flesh. However, Aristotle, much more accurately, did not term what yields "soft", but added to this sometimes "into itself" and sometimes "not changing position"[498].

Whether we should say in this way, "soft" is to withdraw into itself, or in this way, "soft" is to withdraw without changing position, we shall be defining properly. And indeed also, if we add both together at the same time, as Aristotle did in the second book of his *On Generation and Destruction*,[499] and we say "soft" is withdrawing into itself and not changing position, we shall be defining still more clearly, indicating the "soft" through two characteristic symptoms. For "into itself" is to withdraw and not change position. Each is what happens to it alone, as to withdraw (yield) is a common symptom of softness **[688K]** and weakness, while the "into itself" is of softness alone, just as "not into itself" is of weakness alone. Furthermore also, as changing position into what is surrounding is characteristic of weakness, so is not changing position of softness alone. Therefore, the withdrawing into itself and not changing position, all Greeks term softness, whereas what is customarily disposed in this way, they call hard. It is not as if, with certain other terms, there are more significations for the one naming both hard and soft properly and not figuratively in this way.

For Aristotle, the most skillful of all in the division to signification, said these things, just as of course, subsequent to these, he discoursed on moist and dry in the second book of *On Generation and Destruction*, and taught that each of these is said

---

497 Kühn's Greek text has ἐντός; Kühn's Latin translation has *extrinsecus*, while the manuscripts F1, F2 and Va1 have ἐκτός.

498 Aristotle, *De Generatione et Corruptione* II (Engl. transl. Joachim in Barnes 1984: 538–554).

499 The reference here and following is to Plato's *Timaeus* 62b (Engl. transl. Jowett in Hamilton and Cairns 1989:1187) and Aristotle, *De Generatione et Corruptione* 2 (Engl. transl. Joachim in Barnes 1984: 538–540).

in various ways.[500] But he doesn't say in that book that hard and soft are terms used variously, as he does not in another book, just as Plato doesn't. But both these men discovered one signification of each of both the terms, at least whenever, as I said, someone names properly and not figuratively, since in relation to **[689K]** metaphors at least, there will be a countless number of significations, not only of these terms but also of all the other significations.

Certainly, at least to begin with, Plato himself says in "The Sophist" that you say men are hard and resistant, not using each of the terms in reference to the body in some way, which is not hard or resistant, but to the matter of the soul, and of this, particularly of the *ethos* and manner. In this way too, he said someone terming wine hard metaphorically, and sound hard, and practice and life, and again custom and law, and wind and water, or harshness, or the hard to mix, or the hard to deal with, or the hard to please, or the hard to work, or the hard to overcome, or any other such of the names from transference. However, the body called properly and primarily hard is resistant and doesn't yield at all to what is adjacent. In the same way, I think they say a man is hard, who is solitary and self-willed, and above all resistant to everything. And a law is hard, and a judge who partakes of no forgiveness. Indeed, wine is hard, when it is strong in quality and **[690K]** forcibly strikes either the head or the gustatory senses. Water in turn is hard, when it is hard to digest, slow to pass and stays a long time in the hypochondria. And each of the other things, we call hard not properly and primarily, but when we name them in this way contingently and metaphorically from some similarity.

However, since many things happen contingently to the hard body, the metaphors arise in relation to each of these. But it is not appropriate on account of this to forget the primary and the proper of the significations. For the hard is certainly tangible, and hardness is a tangible quality. Therefore, whenever the one speaking transfers the name to tasting, smelling, seeing or hearing, it is necessary under these circumstances for the one listening to be mindful of the likeness to the reality. For truly, the actual interpretation of the metaphors that are stated is either right or wrong. We commend those that are clear and obvious in respect of the similarity, whereas all those that are obscure and unclear, we leave aside. But for the younger doctors, **[691K]** they know nothing else good of interpretation, on which account they are extremely unclear, and not that they must use the metaphors at an appropriate time.

If, then, only their books are unclear, perhaps someone might blame them less. Now, however, since they corrupt with inappropriate metaphors the actual truth of the matters which they attempt to teach, it is justifiable for them to be blamed more. When, in addition to this, someone comes to such a degree of stupidity that when he applies the actual primary and most proper of the names to specific matters, he takes the significations from transference, how would someone endure it? For while someone might form a concept of a white voice from a metaphor, it is not possible to

500 Aristotle, *De Generatione et Corruptione* 329b–330a (Engl. transl. Joachim in Barnes 1984: 540).

similarly hear the colour white from a metaphor. For if this white is not said properly, no other of [the whites] is said properly. In this way too, we may call a man sour metaphorically, while wine is not still sour metaphorically but clearly primarily and properly. Therefore, just as if someone named Theraean wine sour, then he would be refuted as holding a false opinion, calling this a sweet wine which partakes of no small measure of sourness, so he would seem unbelievable, if **[692K]** he were to say in regard to what is signified, that there are those who speak of "sour wine," just as we also speak of "sour men", in this case being equivalent to "disagreeable".

In the same way too, if anyone says "hard" of any tangible body whatsoever, directing us to understand the term in the same way as wine is called hard, he would not, I think, seem to be in his right mind. Since all tangible qualities have names, it is great foolishness to introduce other names for them from transference. In the relation to odours, there are not actually names for all the qualities, and someone would assent to the use of strange names through transference. Nevertheless, in relation to touch, everything is named. This is just as Aristotle also taught somewhere, setting out all the names of these in order, which I shall also now state: heat, cold, dryness, moistness, heaviness, lightness, hardness, softness, viscidity, friability, roughness, smoothness, thickness, and thinness.[501] In this way then, all the tangible qualities are stated, as in the same way also are the bodies partaking of these. For truly, of these there are seven conjunctions in terms of opposition, and fourteen terms. Thus, the first opposition **[693K]** is that of hot and cold; the second, of dry and wet; the third, of heavy and light; the fourth, of hard and soft; the fifth, of viscous and friable (brittle); the sixth, of rough and smooth; and the seventh, of thick and thin.

Accordingly, my friend, do not be capriciously foolish. Don't apply significations to all of your own terms that are Greek and common, but say first and foremost that the body of the artery is either hard or soft. On the other hand, it will be possible for you to call the pulse, indicating how this is not properly or even primarily, but incidentally hard or soft. For if the pulse is the actual movement of the arteries, someone naming properly would not speak of a hard movement, but either of the beat from this in relation to our touch, or the perception of the beat, and this would not be said properly or primarily hard, but clearly the hard pulse is named derivatively (paronomastically) of the hard body. We only name that hard, not derivatively from anything else, or incidentally or metaphorically, or catachrestically, but properly and primarily. Therefore, whenever someone says a hard pulse, **[694K]** such a pulse is apparent either in relation to the quality of the *pneuma*, or in relation to the particular nature of the heat, or in relation to any other of the powers which can be spoken about but not taught, we clearly shall not tolerate someone talking foolishly and wittering on about such things, just

501 Presumably the reference is to Aristotle, *Categoriae*, particularly 8b25–11b6 (Engl. transl. Ackrill in Barnes 1984: 14–18).

as we don't tolerate Archigenes in his work on the pulses and in the second book on the visible signs of fever.[502]

Book IV

1. [695K] I also said, right at the start of the second book, how much of use to the actions of the art there was in the first completed book. For it was shown how great the whole number of the pulses is, and we went over names to the extent that there are names. I thought it worthwhile for someone who wished to go into the usefulness of the theory, to read through the treatise on the diagnosis of the pulses, which is subsequent to this one and has been written in four books; then, after that, the one on the causes **[696K]** in four other books. In this way, we came to our work, arising from these, on prognosis, itself also written in four other books.[503]

Those things said after the first [book] in this work on the *differentiae* of the pulses, just as also those things I shall write now, were said on account of the superfluous nonsense of more recent doctors, and were requested by some colleagues, although I have resisted them as far as possible. For I thought it to be in no way better to write useless and superfluous things, talking nonsense, either willingly or unwillingly, against those who talk nonsense. But really the audacity of some is unbearable—people who are ignorant as to how many problems there are in matters pertaining to definitions, who readily define not only the pulses but also everything else. And indeed, there is no need of definitions, when the matters are clearly understood prior to them, as in the case of the name of the pulse, which not only doctors understand, but also all men.

Anyway, whenever people stretch out their hands, presenting their wrists to the doctor, directing him to palpate the pulse, is it to be thought that in this, they must say the name of the pulse, as if they **[697K]** said "so and so",[504] signifying some matter, having uttered this sound? It seems to me better to say that those who understand some matter should lay down the name for that pulse. Should they then apply a name like horse-centaur, or Siren, or Scylla, or a name in relation to some matter that actually exists? And here it seems to me better for the name of the pulse given by them to be said in relation to some existing matter, and this matter should be the movement of the arteries, which they see moving throughout life in all people, in every part of the body, wherever they should happen to be. And further also, in in-

502 This is a lost work by Archigenes, *De febrium signis*, which Galen criticises, for example, in *De febrium differentiis* II.8 (VII.365K) and *De crisibus* II.8 (IX.669, 672K).

503 The first book of the 4-book treatise on the *differentiae* of the pulses contains a clear and detailed account of Galen's own views on the pulse. Books two and three are discussions of other views, discursive and largely critical, particularly of Archigenes. The fourth book is an historical account of attempts to define the pulse. The three further 4-book treatises are on diagnosis, aetiology and prognosis respectively.

504 The Greek term, also used elsewhere in this sense, is σκινδαψός.

flamed parts they say they perceive a certain pulse, and a pulsating at the temples in burning fevers, by drinking too much unmixed wine, and with certain headaches, particularly those from sunstroke.

And when they explain these things to each other, they don't need long accounts of the kind more recent doctors have filled books with in the remarkable refutations they make towards each other regarding the definition of the pulse itself, and of all the *differentiae* of this individually, like the strong, **[698K]** the large and the full, and of equality and regularity of the others, about which much more is said by them and in the treatise in which I explain and at the same time evaluate Archigenes' book on the pulses.[505] This man, along with other doctors of recent times, was consumed by the disease of "love of defining",[506] for I have assigned this new name to a new matter. And yet, in fact, I think Archigenes, a man who has not been negligent concerning the actions of the art, did not deserve to suffer the evil. But just as some people "enjoy the benefits" of itchings[507] and ophthalmias unwillingly from those with whom they associate, so too was this man dragged down into the foolishness about definitions, and yet he was contending more than the others about the actions of the art.

Accordingly, I show them briefly that they lay claim to those things they are most ignorant of, and ask them to propose a name for those things they wish to, and then, when they do this, I direct them in turn to recite a number of verses in as much of a metre as they wish, about the definition of what was proposed. For they wound themselves with this kind of display, and indeed come to a great degree of lack of awareness, for as it is possible to make a whole **[699K]** book about each of the names, the verses will be numbered in the thousands. With respect to these things, I am going over all of them similarly, but now I seem to be persuaded of a terrible thing, compelled by myself to propose something in relation to the pulses to make a display of their vain foolishness.

2. So then, let the pulse itself be the first and clearest thing. For in the case of this, if I shall show how much idle talk they made, the discussion will no longer remain unclear about other things. Anyway, some define the pulse as movement of arteries. Some, speaking in opposition to them, said they erred in the primary definition, since "and of the heart" is omitted, for it is necessary to define it in this way:

– A pulse is a movement of arteries and heart.
  Some, however, say there is no need of "heart", but, "of the arterial part of the heart" should be added, so such a definition becomes:
– A pulse is a movement of arteries and the arterial chamber (cavity) in the heart.
  Or, by Zeus, in this way: a pulse is a movement of arteries and of the arterial chamber of the heart.

---

505 All Archigenes' works are lost. Galen refers to his own eight books of commentary on Archigenes' work on the pulses in *De libris propriis* 5 (XIX.33K).

506 The Greek term is *philoristia* (φιλοριστία).

507 The Greek term ψώρα *(psora)* refers, among other things, to a cutaneous disorder in humans, itch, mange, scurvy and perhaps scabies.

Others, in turn, censure them, claiming they erroneously say the pulse is simply movement. For there are also other movements of arteries and heart—for example, the concoctive (peptic) movement of the **[700K]** blood contained in them, and attractive of this, and nutritive, and separative of the superfluities.

It is, then, necessary, they say, to add to the definition the kind of movement of the pulses, so that it is specific to these and not common to many movements. And accordingly, they created the following definition:

– A pulse is dilatation and contraction of arteries and heart.

And clearly also as many corrections of this same definition as of the previous one have arisen in turn, since some add to it also, "of the heart"; others not simply, "of the heart", but "of the arterial class in relation to it", and others, "of the arterial chamber" [or] "of the left chamber".[508] There are those who also added this, so such definitions as follow arose:

– A pulse is a dilatation and contraction of arteries and heart.
– A pulse is a dilatation and contraction of arteries and of the arterial part of the heart.
– A pulse is a dilatation and contraction of arteries and of the arterial chamber of the heart.

However, there is another chorus of men in conflict with all these, saying not only these [structures] but also the [lepto]meninges pulsate; some also say the brain pulsates; **[701K]** others say both do. Some say the brain pulsates primarily, whereas the [lepto]meninges pulsate contingently. Others in turn think neither the [lepto] meninges nor the brain pulsates primarily, but this begins in the arteries and contingently involves them. For they say these are raised together by the arteries in them, which are very numerous. If, then, they are only raised together and have no capacity for the previously[509] mentioned movement of this sort, nothing will still need to be added to the already stated definitions. If, however, they have some power of their own, similar to the arteries, it will be necessary to mention brain and [lepto]meninges in the definition, so the whole definition becomes as follows:

– A pulse is a dilatation and contraction of heart, arteries, brain and [lepto]meninges.

If, however, only one of these has such a power, we shall mention only that one, so making the definition in this way:

– A pulse is a dilatation and contraction of heart, arteries and [lepto]meninges.

Or in this way:

– A pulse is a dilatation and contraction of heart, arteries and brain.

---

508 The translation follows the manuscripts F1, F2 and Va1, all of which have an additional (to K) κοιλίας following ἀριστερὰς.

509 The manuscripts F1 and Va1 omit προειρημένης.

And in relation to each class of the **[702K]** definitions stated previously, a similar definition will be made. For some, as I said, were satisfied to say: A pulse is a dilatation and contraction of heart and arteries. Some, however, did not simply say this, but added either, "of the arterial chamber in the heart"; or "the left chamber"; or "of the part of this that is arterial". To each of these definitions, there are those who would add brain, or [lepto]meninges, or both. And there will be by no means a small number of definitions.

There is no less dispute about the chambers of the heart than about these. For some think the right chamber (ventricle) does not pulsate, while others think it does pulsate, but not primarily or through itself, rather moving together with the so-called *pneumatic* [motion]. As a result, those who refer to the left chamber of the heart, or the arterial, or those who name it in any other way whatsoever will differ from those who think the right chamber pulsates. Another still greater difference about the arteries arose among doctors from ancient times. There were are those who thought they pulsated by themselves, having an innate capacity to do so, like the sort of capacity that exists innately in the heart, Praxagoras among them. Some, however, thought **[703K]** the actual walls of the arteries pulsate (i.e., dilate and contract just like the heart), although they don't have an innate [capacity] by which they do this, but take it from the heart; Herophilus holds this opinion as well. Neither, however, suffices for Erasistratus, for he would have it that the arteries pulsate contrariwise to the heart. The heart is dilated when it is filled by what follows as a consequence of the emptying, whereas the arteries dilate when they are filled; he says these are filled by the *pneuma* sent from the heart. It is necessary, then, that both occur at the same time—the dilatation and the filling of these. But by reason of cause, one must precede the other—the dilatation in the heart and the filling in the arteries—as is also seen in external things. For the bellows of smiths are filled because they dilate, whereas sacks, bags and wineskins dilate because they are filled.

If, then, the definition is going to show accurately the whole substance of the matter, it is fitting that it leaves out nothing of what is most important. However, since the substance is in dispute, it will be necessary **[704K]** for there to be a specific definition in relation to each sect. As a consequence, it must be defined by the Erasistrateans in such a way as to show the specific opinion of Erasistratus concerning the reality (essence) of the pulses, and by the Herophileans to show what is specific to Herophilus, and similarly for all the others; for each, the definition must be made specific to their own opinion. Straightaway there is this which almost none of these men know. And more than this, it is necessary to know there are two primary classes of definitions. The one expounds clearly the concept of the matter which those who name it have, while the other teaches the essence, as was said. Of each of the two definitions, it is best if specifically expounding the concept is agreed upon by all who speak the same language and doesn't touch on the essence of the matter, while the one that teaches the essence agrees with the one relating to the concept but is different from it.

Therefore, you should take it as the first sign of them rashly talking nonsense, whenever you hear them stating one definition of the proposed matter. For in some cases, it is better to make not only two, but three or four definitions. The first **[705K]** is that agreed upon by all speaking the same language, which declares nothing about the essence of the matter, staying within the bare concept. Another [and second] in addition to this, departs a little from the concept but lays hold of the essence to a slight degree. And a third, after this, already expounds more of the concept on the one hand, and more of the essence on the other. And a fourth, after these, teaches the whole essence. This, then, ought to show it has reference to the third, this latter in turn to the second, and that again to the first, while the first is to be relied upon in its own right, since it manifestly contains a clear explanation of matters. On this account also, Aristotle has called this definition "nominal",[510] as if saying the definition also explained the name. Some are accustomed to call the second definition essential, saying a definition is what is clarifying. He says what it is, for each of those defining it, explaining what this is, going through together the notional in respect of things specifically contingent to this. However, he also assumed another definition to be that which teaches the essence of the matter.

But why do I need to speak at greater length about such great matters, which **[706K]** I showed in two books—the third and fourth of the treatise I prepared on the second of Aristotle's two *Analytics.*[511] In the case of the pulse itself, I shall attempt to show you each of these with the greatest brevity possible. Anyone who previously attended to those treatises I spoke of will follow more easily what will be said. In some parts of the skin, we perceive a certain movement, not only when we apply the touch to it, but also sometimes with the eyes. And this particular movement appears in all those who are healthy in many parts of the body, one of which is in the wrist of the arm. Clearly, then, we perceive something sent up from the depths to the skin and striking us, while after the beat, it sometimes clearly goes away and then pauses. Sometimes, the appearance of the pause occurs immediately from the start, and after this, it again rises up and strikes, then again goes away and pauses, and does this continually from the day someone is born until they die. All people call such a movement "a pulse". **[707K]**

Anyone who is able to interpret this quickly is the sort of person who will best make the nominal definition, clearly neither omitting nor adding anything superfluous beyond what I went through just now. However, if in fact he does add something, even if it is small, if what is added is useless, he will be talking idly in vain. On the other hand, if it is useful, he will be departing in that from the concept and going to the essence of the matter, as if he were saying such a movement of the arteries is a pulse. For such a definition not only explains the movement but also what it consists of. And further, although laymen clearly perceive there is such a movement in the body, they are ignorant of the moving body itself. However, if someone not only saw the artery by itself after stripping the skin, but also closely observed not only its external outline, but

510 Aristotle, *Analytica Posteriora* 93b29 (Engl. transl. by Barnes in Barnes 1984: 154).
511 See Galen, *De libris propriis* 15 (XIX.47K).

also the walls, how many there are and of what sort, he will advance beyond the concept, and much more than that person previously who only saw this from the outside. So then, if he should discover the capacity by virtue of which the arteries pulsate, and whether they have the beginning of the movement from themselves, or from the heart, and if from the latter, how in fact **[708K]** that is—whether by sending *pneumatic* material into the internal cavity (lumen) of the vessels, or by some power through the walls themselves—and in addition to these things, what at any time the use of their movements is, this person will know precisely the whole essence of the genesis of the pulses.

It is not in fact possible to encompass all these things in a definition, for those who delight in definitions do not want them to become too long. Consequently, not every definition of pulses is accurately substantial or it will be very long. "Accurately" is added to the statement because some of the more recent doctors adding one or other of those things relating to the essence, think they have stated a substantial definition of the pulse. In that, they have made the greatest mistake. The only definition stated by them is the essential definition; they have not previously stated that pertaining to the concept, which in our reasonings about these matters has been shown to be a criterion of that pertaining to the essence. But since they don't know this very thing, in all likelihood they declare their own illogical opinion.

Let us then begin again from the notional definitions, which we said explain nothing more than those things all men **[709K]** know. It seems, however, that no such definition has been stated—not by the Empirics, for whom it would be most of all fitting to use such definitions, which those skilful regarding terms do not consider worthy to be deemed definitions; they call them outlines or sketches. Such outlines would have been something of this sort: An involuntary movement, interrupted by a pause, existing in some parts of sanguineous animals. And furthermore, if some were to say a natural movement, or involuntary, or non-appetitive, it would make no difference. However, if he were also to include that it occurs in the small parts of the skin, or that it is of short duration, in the statement of the outline, he will deliver the notion more completely. If also he were not to maintain the false claims of the Sophists, he will add "the perceptible" to the name of the movement. And yet, in fact, such additions are more a sign of irrational cowardice than rational caution. For all men are accustomed to say the term "movement" in relation to what is clearly apparent to perception, since certainly there is some credible argument in respect of those contemplated by reason, believed to have been shown to always exist **[710K]** in all bodies, even if they don't appear to be moving.

And this term itself, "the perceptible", signifies two things: one, the matter now already falling on perception at the time one might say this, and the other, in respect to that which can become perceptible, even if now it is not yet so. And in relation, in this way too, we say a stone in deep water is perceptible as much by itself because it clearly is of the nature of perceptible things. Accordingly, the perceptible movement in certain parts of the body, which is apparent in us whilever we are alive, is termed a pulse. However, it is interrupted by a pause, as I said. For some doctors think this is safer to say than that the whole movement is compounded from dilatation and contraction; and indeed, the appearance is clearly like this, when the movement is extended in height, breadth and length in those who are thin and have large pulses. Others, less bold than these, say there is for them perception of specific affections of touch, fearing to declare about this that it is among external things, being in fact led to this by the sect of those called sceptical and doubting.[512] We must then pass over these. Perhaps they will not dare to proclaim that any movement is apparent to them, if **[711K]** they are persuaded by the doubters on all matters. Anyway, some of those men say they do not certainly know their own affections—these are men people reasonably call "crude Pyrrhonists".[513]

Passing on to those who allow themselves to be persuaded that they see clearly a movement under the skin of some body extending in length, breadth and depth, since they are more moderate than the others, we shall demand of them, when the skin is removed, whether by their own choosing or at some other time by chance, to consider the appearance of what is being moved under it. For it is clear they will see a vessel very like the vein, except that it has a double wall which is much stronger, and which we call an artery. And they will see clearly what was seen by them indistinctly before—that this vessel is distended in a circle (circumferentially). In the same way, the heart, when exposed, appears to be moving, and the brain, so that they will venture to say at that time that the movement seen in the generation of the pulses arises in the previously mentioned structures. But whether the right chamber of the heart seems to pulsate, **[712K]** primarily and of itself, or to move jointly with the left in the same rhythm to that, they will decline to seek what, in relation to the first change from the layman's concept, goes forward to the essence. Similarly, also, in the case of the apparent pulse in the brain, they will decline to look into whether it has this function primarily and of itself, or rises together with the arteries in it, which are many and large. However, in the case of this, someone particularly emboldened to declare about the essence of the pulses will also add this.

Perhaps someone would create the perfect definition of the pulses by adding what is said to be the use of their generation, which we said in the treatise *On the Use of Pulses* is preservation of the innate heat,[514] if the fashioner of such a definition knows clearly in what manner the preservation of the balance of the innate heat occurs—whether by being fanned in the movement or by being brought together by the impact of the cold air so as not to be dispersed easily, or by purifying the smoky and sooty superfluities. The use of the *pneumatic* substance in the arteries was shown also in regard to the genesis of the cold *pneuma*. **[713K]** It is of the essence

512 On the Sceptics as a philosophical school, see Hankinson 1995.

513 Pyrrho was the founder of the school; those named Pyrrhonists may be considered "hard-line" Sceptics.

514 *De usu pulsuum* (V.149–180K), particularly II (V.156–160K). This work is included in the present volume.

of the pulses also, whether the actual materials happen to be drawn in when the arteries dilate, or when they contract.

The same argument again applies concerning the expulsion of something from the arteries. For we have demonstrated there is drawing in during dilatation and expelling during contraction. However, those from the so-called Pneumatic sect think there is drawing in during contraction and expulsion during dilatation. Someone else will add to these, defining the pulse and its effective cause. What was sought concerning this itself was whether it is some physical (vital) or psychical capacity, or some other capacity altogether, as Asclepiades maintains. It is clear, therefore, that the so-called essential definition, which is the didactic statement of the essence of the matter, will state the class of the pulse and the *differentiae*, the use of the genesis, and the effective cause, and further, in addition to these, will state the material or instrumental cause, or however someone might wish to term it. For example, the class is either movement, or function, or affection of the arteries, while in respect of the *differentiae*, that they relate to dilatation and contraction, and that they are involuntary, and with respect to the use, that it is for the sake of the innate heat, or **[714K]** genesis of physical (vital) *pneuma*, or by Zeus, psychical [*pneuma*], while in respect of the effective cause, that it is the force (capacity) dilating the heart and arteries, and is either the physical or psychical [capacity], or both. The material and instrumenal causes relate to the structure of the arteries, which I said showed suitability for function.

Obviously there will be a specific definition in relation to each sect that is rational. Erasistratus will say the pulse is a movement of arteries in relation to dilatation and contraction occurring due to a physical (vital) and a psychical capacity for the purpose of filling arteries which have in themselves physical (vital) *pneuma.* Asclepiades will say the pulse is a movement of arteries involving dilatation and contraction, being filled with *pneuma* by the passage towards the fine-particled,[515] and emptied by the collapse of their walls. We, however, shall say there is a specific function in respect of the pulse, primarily of the heart and secondarily of the arteries, involving dilatation and contraction, being moved by the vital (physical) capacity for the purpose of preserving the balance of the innate heat, and for generation of psychical *pneuma* in the brain.

Analogous to these definitions, those from **[715K]** other sects also define the pulse, each obviously in relation to their particular hypotheses. If, then, they define perfectly, they will agree with each other; if, however, they define deficiently, or superfluously, or unsoundly in respect of something, they will be differentiated. The greatest assurance of this for you comes from the followers of Herophilus. Those who first attempted to define the pulse differed from one another, and the followers of Erasistratus attacked them, as both these teachings were at their height after the death of Herophilus. The Erasistrateans in turn created a different definition of the pulse, disagreeing with each other and with the Herophileans. For the ancients themselves made their teach-

515 The term is λεπτομερής. See Vallance (1991) for a discussion of Asclepiades' theory in general and this term in particular.

ings without definitions, demonstrating the significations from the names[516] which they said in relation to the explanation, are in the Form (Idea) itself. Obviously, in relation to this also, the grammarians make clear the statements from Homer and all the other ancients. What is not yet clear in this particular interpretation becomes quite clear from another, as in the case of *pisures.*[517]

For example, in this: "*pisures* horses with arched necks" **[716K]** the signification is not yet clear. However, in this: "Thereupon they took four great shields, but eight spears and four helmets"[518] it has become very clear. For since there were four being fully armed, just as he says they take four great shields, so too do they take four head coverings (helmets). Similarly, then, when we read in some book of the doctors of ancient times, that only the inflamed part pulsates, or they call only the movement of the arteries in inflammation a pulse, never using the term in the case of a body that is in accord with nature, we infer that not every movement of arteries, but either the large and strong, or the one perceptible to the sick person himself are called a pulse. And Erasistratus seems to me to have used the term pulse in this way too. But when in turn we read Aegimius' book, *On Palpitations,*[519] we discover that what we now call a pulse is called by him a palpitation.[520] Contrary to him, you would find Herophilus, right at the start of his work on pulses, distinguishing a pulse from a palpitation. **[717K]** For this man obviously terms every movement of arteries, which we see existing in us from beginning to end, "a pulse". It is from this we make the diagnosis of present things and the prognosis of things that will come about, without needing a pulse in the heart, brain or [lepto]meninges.

Therefore, the most authoritative concept of the pulse for all doctors now is the one stated. For surely, they need this, as I said just now. However, everything in addition to this is already sophistical. Wherefore, it is not appropriate to define such aspects of the matters, nor for oneself to put in arguments closely connected with the Dialecticians themselves, which need a great training in order that someone might accurately correct them. In addition to this, it is not necessary to make the teachings through essential definitions. However, the so-called notional definitions are superfluous for those who already know the matter, while they are not necessary for those who don't know, since in fact the teaching can become clearer through longer discussion. For if someone is able, with a brief and at the same time clear discussion, to encompass accurately the concept of the indicated matter by the **[718K]** name, he would be an ex-

516 The manuscripts F1, F2 and Va1 all have παρὰ before τῶν ὀνομάτων.

517 *Pisures* (πίσυρες) is an alternative form of *tessares* (τέσσαρες). In Homer it means simply "four". Four other later specific uses are listed in LSJ.

518 See *Odysseia* XXII.110–112 (ed. and Engl. transl. Murray and Dimock 1995: 352).

519 Aegimius (fl. 325–300 B.C.) was perhaps the first to write a book on pulsation/palpitation, a work no longer extant (see Galen, *De pulsuum differentiis* II.1 (VIII.498K).

520 The distinction is between σφυγμός and παλμός.

cellent teacher, although one who is deficient in such a practice is not useless, just as someone who does not explain it properly is not.

It is best to interpret clearly in this way, as the rhetoricians following Lysias and Demosthenes did,[521] and those who now speak in the law courts who are not useless. I do not, therefore, stand in the way of those who are able to explain to those who don't know the signified matters, in respect of each of the names clearly, and at the same time quickly, as I myself am obviously eager to do in my work. But because it is superfluous to act in this way in which everyone already knows what is going to be explained, it is altogether unnecessary to state the essential definitions for those beginning their learning in any way whatsoever. But whenever, beginning from the concept, we introduce them to the knowledge of the essence, it would now be useful to encapsulate the whole matter in a short statement for the sake of easy memorising. Accordingly, those doctors and philosophers making the beginning of the teaching to those learning anything at all from the essential definitions, even if they are particularly capable in speaking, and even if they explain clearly, nevertheless become obscure because **[719K]** the listeners are not able to understand matters stated in brief definitions; they need long explanations for understanding.

It is necessary to be demonstrated beforehand through many discussions, the essence of the matter being investigated, so that being known and believed, it may be understood at some later time in a short statement. Therefore, I naturally always censure those who attempt to teach all these things through definitions—for example, I come to this particular argument, often demonstrating in practice, that whatever name someone might put forward, he makes a whole book demonstrating how the matters being indicated must be defined by him. Therefore, taking it up again, let me speak about the definition relating to the pulse, mentioning by name those who seem to have defined it properly.

The Herophileans, as I said, began such over-elaboration. Some of the Erasistrateans succeeded them. Then there were the so-called Pneumatics and some of the Methodics. If I mention them all, there will need to be two books, not one. For me, who wishes to put an end to the unnecessary prolixity in one book, it will suffice to go over as many definitions, together with the men who formulated them, as **[720K]** one book would be able to accommodate. From these it will be possible to judge all the others.

3. Heracleides the Tarentine,[522] in making an outline suitable for an Empiric, says the pulse is a movement of arteries and heart. We know to begin with that those of the Empiric sect are not at all in a hurry to make a definition; they use sketches and outlines. They call in this way the statements with which they briefly expound the concept of the matter, the name of which we utter. He (Heracleides) believes people have this concept about the term pulse, when sometimes they extend the arm and ask the doctor

---

521 Lysias (ca.459 – 380 B.C.) was a noted Attic orator renowned for his mastery of ordinary language. A substantial body of his work remains extant. Demosthenes (384 – 322 B.C.) is regarded as the greatest of the Athenian orators and, like Lysias, has a considerable amount of extant work.

522 The manuscripts F1 and F2 have ὁ μέν οὐν Ταραντῖνος omitting Ἡρακλείδης.

to palpate the pulses. Whenever we say someone is pulseless, we think to signify from this term, in respect of the person so disposed, that no movement of any artery in any part of the body happens. If, however, there is some other movement in the arteries, as they increase, nourish, fill and empty, he doesn't care about this, because **[721K]** this class of the movements doesn't impinge on any of our senses, but is apprehended by reason.

According to reason also, the natural scientists (physiologists) say all plants are always moving, and in fact more than this, also all other things up to the hardest stones, for truly something of these is continually flowing away. Some also say there is assimilation into them from the ambient air. Therefore, according to those investigating the natural causes of these phenomena, such movement necessarily exists, not only in a person who is still alive, but also in one who is already dead, so that we would not be correct should we say that the arteries are immobile in those who are dead, nor that they appear pulseless in some of those still living. If all were to speak in this way, it would be quite clear that we conceive of and name a movement not in respect of apprehension by reason but by someone coming to a clear knowledge of a condition.

From such conjectures then, Heracleides the Tarentine and some others of the Empirics, having convinced themselves, wrote that the pulse is a movement of arteries and heart, as they were also accustomed to see this (i.e., the heart) perceptibly moving like the arteries. Some of the Dogmatics however gainsaid them, not deeming it right to judge the movement by perception only, but sometimes **[722K]** also be reason. Anyway, it is agreed by all people that the sun, moon and all the stars move, for they would not come on such a great journey from sunrise to sunset without movement, although they don't appear to be moving. It was necessary, then, they say, even if nothing else, that this at least should be added to the argument of the Empirics, and for them to say the pulse is a movement of the arteries and heart that is perceptible.

However, some who speak against them say, with respect to perceptible movements in relation to the heart and arteries, that not only the pulse-like movements are apparent, but also the palpitation-like, and it is therefore necessary for these to be included in the outline of the pulses, whether they were simply to say the pulse is movement of the arteries and heart, or were to add "the perceptible". Some, however, also mention spasms and tremors, and that some such movements sometimes occur in arteries and heart and are clearly contrary to nature. And because of this, they think it worthwhile to add "the natural" to the definition, making the whole definition as follows: **[723K]** "A pulse is a natural movement of arteries and heart."

Speaking against them, those who join in advocating the definition of the Tarentine say such movements never occur in arteries or heart, being only of sinewy bodies. But that spasm and tremor only occur in sinews (nerves), I too agree. However, palpitation is clearly seen to involve the skin also. And obviously also in the fleshy parts, which are parts of muscles, and in relation to fear and anguish, the heart clearly appears to palpitate. No little opposition regarding these affections arose in Herophilus towards his teacher, Praxagoras,[523] who wrongly declared palpitation, tremor and spasm to be affections of arteries, not differing in class from the pulse-like (pulsatile) movement in these, except in magnitude. Being in accord with nature, the pulses occur without any particular circumstance, although when the movement of these is increased to become contrary to nature, first spasm is produced, second after this tremor, and third palpitation. All these affections differ from each other in magnitude.

Because of this, then, **[724K]** Herophilus, immediately at the start of the work on pulses, attempted to refute the opinion of his teacher, but as was customary for Herophilus, through an obscure explanation, which his followers, changing his explanation for the purpose of clarity, wrote in the works they produced on the school of Herophilus. Consequently, if I mention now the actual statement of Herophilus, which he wrote at the beginning of the first book about pulses, or those things he said to his followers, it will come to be as big as one book, if I go carefully over these things alone, even without mentioning those things in my whole treatise on tremor, palpitation, spasm and rigor.[524]

It is sufficient for me here, having demonstrated the majority of the arguments that will be introduced besides for the proposed enquiry, to leave this aside, saying only this much—when we are in accord with nature, the movements in relation to the heart and the arteries occur when there is dilatation of these in a circle (circumferential dilatation) and in turn contraction into themselves. When we are contrary to nature, however, and the movement of the arteries appears, palpitating, or clonic, or spasmodic, or tremulous, there is nothing strange in the pulse sometimes being called **[725K]** spasmodic, sometimes clonic, sometimes palpitating, and sometimes tremulous. For truly it is also customary for us to call a certain pulse clonic which is not infrequently clearly apparent. I interpreted the nature of this in the first book of this particular treatise, so that also for this reason, someone would not be right, if he were to add [the term] "natural" to the definition of the pulse. However, concerning "perceptible", since some also think it worthwhile to make this addition, it is worth saying what was often said by Aristotle. For that man, sometimes after going over [a matter] sufficiently, says what needs to be added to this, due to the sophistical annoyances, is something else from without, which in each case seems to benefit the definition. And because of this, I also in general write additionally that the pulse is a movement of arteries and heart that is perceptible, and particularly that we call "pulseless" those in whom the movement is not perceptible, although the movement is often preserved as regards the specific essence. For if the movement of the heart and arteries were altogether destroyed, the person would not be alive.

---

523 For Praxagoras' position on this matter, see fragments 26–28 in Steckerl's 1958 collection which include this passage and also passages from two other Galenic works: *De tremore, palpitatione, convulsione et rigore* (VII.584–642K; English translation by Sider and McVaugh 1979) and *De placitis Hippocratis et Platonis* (V.181–805K; English translation de Lacy 1978).

524 See previous note.

4. In fact, the Herophilean, Alexander, **[726K]** called by the surname Philalethes, made a twofold definition of the pulse due to such calumnies. One was from the proper nature of the actual matter,[525] which he named as foundational (assumed *ex hypothesi*), while the other was as it appeared on observation. Thus, the pulse, as assumed *ex hypothesi*, is, he says, an involuntary contraction and distension of heart and arteries which is apparent. The pulse as it appears on observation is an impact on the touch from the continuous involuntary movement of the arteries and the interval occurring after the beat. And he states certain reasonings for these, which he thinks are credible, in the fifth book of his, *On Opinions*, just as also the Tarantine, Heracleides does in those works in which he speaks against Herophilus' *On Pulses*. No less did counter-argument arise in regard to the definitions of Philalethes from those who came after him. Some [of the arguments] were very unconvincing and conspicuously contentious, while others were less so and more credible, to which some of his followers responded. And, in fact, a long discussion was carried out between both sides. Perhaps I shall speak generally about these a little later under the chief points. First, however, I shall add the **[727K]** definitions of Demosthenes, surnamed Philalethes like his teacher.[526]

5. He also has three books about pulses, well thought of among the majority. He divides the pulse similarly to Alexander, and says it is fundamentally a dilatation of heart and arteries or a natural contraction which can be apprehended. However, the one from which we make the observations is a natural beat of heart or arteries against the touch and the interval occurring after the beat. In the definitions, he uses the disjunctive conjunction "or", thinking that "and" being conjunctive, defines only the joining together from both, while each [part] is not individually defined. However, he understands the movement of the arteries to be a pulse in and of itself alone, while clearly also the movement of the heart is of itself alone, and both together are the pulse. Similarly, also, he terms the dilatation alone "a pulse", and the contraction, and both together. He is persuaded that the pulse is named in this way since **[728K]** it is called large in relation to the dilatation, and small, and strong and weak, and rapid and slow, while it is called infrequent and frequent in relation to the contraction. Therefore, when he speaks of the parts of the pulse in relation to what is theoretical (assumed), he uses the terms dilatation and contraction, whereas in observation, he uses the terms beat and interval, thinking the contraction to be indistinguishable by sensation.

Here again a large difference of view has arisen among doctors, no longer about the name only,[527] as with some of those mentioned, and not meddlesome and superfluous like many of these, but the greatest and most important consideration of the theory. And I choose to judge impartially, just as also with respect to the other things about

525 K: πράγματος ἐξ οἰκείας; F1, F2, Va1, πράγματος τῆς οἰκείας.

526 Of the three men mentioned in this section, two (Alexander, 20 B.C.–25 A.D. and Demosthenes, 50 B.C.–25 A.D.) were Herophileans while the third (Heracleides, 95–55 B.C.) was an Empiric. None of their writings are extant.

527 F2, Va1, μόνον omiserunt.

which doctors have disagreed, and in this way brought to an end many years of difficulties. Quite a long time later, I discovered the cause of their disagreement and wrote one book about this alone—the first book of *On the Diagnosis of the Pulses.*[528] In this, it is distinguished that some pulses have a contraction that is perceptible and some don't. It seems, then, in that book a whole discussion is brought in besides for those who think the pulse **[729K]** is compounded from a beat and an interval. For some also, apart from making the distinction between the definition which is foundational and that which is observational, have simply declared the pulse to be compounded from the beat and the interval. And yet as it is also said to be compounded from dilatation and contraction, some question arises about the pauses which the heart and arteries make—are they after the dilatation and before the contraction, or after the contraction and before the dilatation. And even if we sometimes overlook the word "heart", and only mention the arteries, it is necessary to understand that the same things are said about that, just as when we speak about the heart but overlook the arteries, we must think the same thing has been said about these.

And the double pause of the respiratory organs has the same uncertainty in relation to respiration. For the chest and lungs rest in the time between the inward and outward passage of the breath (*pneuma*), and in relation to the time of these pauses, they call the respiration infrequent or frequent. The recognition of these pauses is not the same as those in the pulses. For clearly, in the case of respiration **[730K]** they always appear, whereas of the pauses in the pulses, by some there seems to be no recognition at all, while by others only those that are external. So what we say is perceptible is not in all the pulses. Therefore, as in respiration, frequency and infrequency are among the convenient signs, so too a pulse is frequent and infrequent, as you will learn in the prognosis through these. But there is dispute as to whether the pauses are part of the whole respiration and of the pulses. For truly, those who think they are not primary in this way but incidental (contingent) say the pauses are parts in these functions. For in the inspirations and expirations, the organs of respiration clearly function, while in relation to the pulses, when the heart is dilating and contracting, the pauses between these functions (actions) are like some matter in those who are mortally afraid, just as those who say the pulse is a function of arteries and heart, or a movement of arteries and heart, or dilatation and contraction of arteries and heart, think to leave aside the pauses, which are no small part **[731K]** in regard to prognosis.

Anyway, the frequent and infrequent pulses are in this class, along with the equality (evenness) and inequality (unevenness), the irregularity and regularity in these, which show not a little, just as also the intermittent and the intercident which exist in this same class; these are by no means small signs of things that will come about. Therefore, if someone, when defining the pulse, should wish to bring into this the discussion about the pauses, whether these must either be established as parts of the puls-

528 *De dignoscendis pulsibus* I (VIII.766 – 822K).

es or not, and if they are parts, whether primary and by specific reason of themselves, or incidental (contingent), this will consume not a little time. But let us, as in the case of other things, having come to the point of showing the prolixity in relation to each, move on to something else, and now we will do just that, returning to the definitions of Alexander and Demosthenes, which they made of the pulse. For it is apparent, even if not in the statement, at least in the force, Demosthenes means the same things as his teacher. For Alexander said the action of the pulses is involuntary (spontaneous), while Demosthenes said it is natural. And the former said, "of the kind to be apparent" while the latter said, "able to be apparent". In this way, then, in the definition of the pulse related to what is fundamental, **[732K]** Demosthenes seems to agree with his teacher, and likewise in that relating to the observations. For it is quite clear that it makes no difference to say beat or impact,[529] just as it does not with involuntary and natural. It was said previously that he differs from his teacher only in admitting the conjunction "or" into the definition, and on the basis of what reasoning he did this. Therefore, let us next consider the definitions by the other Herophileans.

6. Baccheus, in the other lectures, says the pulse is contraction and dilatation occurring at the same time in all the arteries. Regarding this being a definition, the Erasistrateans spoke against it derisively, making at the start of the accusation their own error. For it seems to them that the heart dilates first and there is contraction of all the arteries when the *pneuma* is sent from the heart to the arteries, and dilatation of these occurs in relation to the filling—first of those near the heart itself, second of those subsequent to these, then next those after these, and in this way, on the basis of the continuity, **[733K]** the movement reaches to the ends of all the arteries, since the arteries themselves have no power, either of their own or sent to them from the heart, which they use to dilate and contract themselves, as the heart does.

However, it is acceptable to the followers of Herophilus that the arteries, being continuous with the heart through their walls, have the power flowing down along these, which they use like the heart itself to dilate and draw from all sides, from which they would be able to squeeze out by contracting what filled them during dilatation. And because of this, it is clearly apparent that all these dilate and contract at one and the same time, maintaining the same fixed time of both movements with the heart. However, it follows, according to Erasistratus, that the heart and arteries appear not to dilate and contract at the same time. But the appearance is not like this. Consequently, the Erasistrateans were wrong to accuse Baccheus for stating what is apparent. Those contradict better who say only the pulses in accord with nature come into being in respect of the definition of this. For in the case when the animal is contrary to nature, sometimes it appears that all the arteries do not dilate at the same time. **[734K]**

7. Aristoxenus the Herophilean not only criticised this, but also that the class of the pulses was left out in the definition, from the division of which, something has been

529 The Greek terms are πληγή (*ictus*, blow, stroke) and πλῆξις (*percussus*, stroke, percussion).

said about dilatation and contraction. For he himself thought it right to say the pulse is a function (action) of arteries; this is the first class. Then, since there are also certain other functions of the arteries—those of growth and nourishment—he adds "the specific", so the whole definition is as follows: "A pulse is a function (action) specific to heart and arteries." Therefore, Aristoxenus correctly said that the class of the matter, which we happen to be defining, must be stated first of all in the definition. Since, however, the class of this is what is proximate, for they term it thus, being through no other intermediate, or through one or two intermediates, or also the highest of all, no small enquiry has arisen as to what sort of class must be stated first of all in the definition—whether in fact the highest, next to which there is nothing more generic, or the proximate, or also one of those in between, when of both things said, there is one that is clearer.

To some, then, it is likely to suffice **[735K]** to say the proximate class of the pulses is movement. However, since there are many movements, for carrying along and changing, growth and decay, and in addition to these, generation and destruction are believed to be movements, it was consequent to say which movement they meant the pulse to be. Obviously, they would say a carrying along. Then, since the carrying along has many differences, it was necessary to say next what carrying along of the arteries producing the pulses they meant there to be. Someone, beginning above from the primary class, if he should wish to follow the dialecticians, would define the pulse as a function of heart and arteries in the movement relating to carrying along occurring in the dilatation and contraction. Perhaps someone will think it right to add to these also the pauses, just as some do think it right. However, some dispense with them, since they occur *per accidens* (contingently) and not primarily.

But for Aristoxenus these are also sufficient, as he wishes to define dialectically, without himself preserving the dialectical laws. For those who are clever deem it right regarding the definitions, that we begin from the primary class, then proceed through those in between, to make the complete definition. Perhaps I should have written the **[736K]** example of such a sort of the definitions, if there were not two books by Plato in which he teaches and trains those using the *diairetic* method to make the definitions. He titles the book in one case "Sophist" and in the other, "Statesman" (Politicus).[530]

8. Let us come to the definition stated by Zeno, a man second to none among the Herophileans. He says the pulse is a mixed action of the arterial parts comprising contraction and distension in which the same order exists and is followed in all the parts, whether that is completed in equal or unequal times. And in this way then, when he said the action is mixed, not only did he give an account of the class of the pulses, but also, as he thought, the proximate *differentia*, passing over that the action might occur

530 The Greek term is διαίρεσις (διαιρετικός) with a basic meaning of "a dividing" or "division". For the use in logic concerning division of genus into species see particularly, Plato, *Sophist* 267d (Engl. transl. Cornford in Hamilton and Cairns 1989: 1015) and Aristotle, *Analytica Priora* 46a31 (Engl. transl. Jenkinson in Barnes 1984: 74)

in relation to movement, and of the movement in relation to being carried along, making mention of the kinds or *differentiae* of the carrying along—contraction and dilatation. For it is quite clear that "the distension" is said instead of "the dilatation".[531] He said the movement is not badly mixed nor simple, **[737K]** nor of one kind, for it is compounded from opposite parts, like that which pertains to respiration.

It is quite clear also that he does not wish to call only one part of this compound movement a pulse, but part of a pulse. For to consider, because of this also, that the whole dilatation is to be called a pulse because a large pulse is named in relation to that, is unthinkable, as sometimes the whole takes the name from those things that are contingent (*per accidens*) in the parts. Anyway, in this, a person is snub-nosed or hook-nosed, grey-eyed or bright-eyed, or bald-headed, and in these and many others, a name arises for the whole from the condition pertaining to the part. Although almost all the other Herophileans term a pulse the dilatation and contraction of arteries, or heart, or both, he did not say of heart and arteries, but of arterial parts, due to the inquiry into the heart, which I spoke of previously.[532] For the largest of the arteries in the animal grows out of the left chamber of the heart. However, the actual body of the chamber is not the same as that of the arteries—this is acknowledged also by those who call this "artery-like". **[738K]** For as we say some body is "nerve-like" being different from a nerve, in the same way too, they called this chamber, "artery-like", although it is obviously different in substance from the arteries. But it is not accepted by those who speak against them that this is correctly termed "artery-like", for it has the same nature as the right chamber, which is flesh-like in substance. On this account, some have even reckoned the heart to be among the muscles. But how these are mistaken has been shown elsewhere.[533]

But they ventured to speak in this way because of the similarity of the bodily substance of these. For the body of the heart is very similar to that of the hard muscles. And indeed, there also clearly appears to be pulsation in both the chambers, during the time it still lies within the animal and for some time after its removal. But also, because Zeno defined only the pulse in accord with nature, he has reasonably been accused in relation to this. That the same order of the movement of the arteries exists in all the parts is true in the case of those who are in accord with nature, whereas, in the case of those who are contrary to nature, it is sometimes false. Aristoxenus reasonably accuses Zeno, as he also does many others, for the superfluous addition to the statement of "if **[739K]** that movement is accomplished in equal times and if it is accomplished in unequal times."

For in relation to the common and generic concept of the pulses, neither this, nor anything else at all of those things pertaining to the division of the matter before us, is

531 The distinction is between the Greek terms διάστασις (rendered in the translations variously as distension, expansion, separation and dimension) and διαστολή (rendered in the translations consistently as dilatation).

532 This is the issue of the chambers other than the left ventricle.

533 Galen, *De usu partium* 6.8 (Engl. transl. May 1968: 1.293–295).

included in the definition. Anyway, in saying man is an animal, then adding to this the *differentia* terrestrial, then next two-legged, there is no longer a need to add the *differentiae* in the men themselves, so the definition becomes this: A man is an animal that is terrestrial and two-legged, while among men, there are the skilled and the unskilled, and the wise and the foolish. For such *differentiae* are really only of particular men and are superfluous regarding the proper definition. It is sufficient to say man is either a terrestrial, two-legged animal, or a rational, mortal animal. It is not necessary for the matter before us to have been divided for the definition, as I said, but the class of this should have been divided up to such a point that we reach the matter before us. However, when the concept of this has been filled up completely, and the discussion pertaining to the essence sufficiently clarified, every addition after these is useless, and particularly that relating to the division of the actual matter that **[740K]** is going to be defined.

For it is appropriate that what is common to all pulses is shown in the definition, which is that by which pulses are conceived and described, and not in relation to that by which some particular pulses are conceived and described. So, if in fact the argument relating to the definition should touch on some difference which is subordinate to the proposed matter, it will provide an occasion for those who wish to lay a charge that the others have been left out incorrectly. For, as Zeno added, "if [the movement] is completed in equal and unequal times," indicating the difference relating to time in the pulses, in the same way, another will add the difference relating to magnitude, and another, that relating to strength, and another, that relating to frequency, and another, that relating to number, while in respect to all the remaining differences, another will add one, and another, and another [and so on]. Therefore, they have reasonably found fault in Zeno's definition, which goes to excess superfluously. And they would have grounds to accuse, if they wished, also in respect of those things said by the Dialecticians concerning the excessive definitions in relation to the statement. For those men show the increase of the statement creates a reduction of the matter relating to the definition, just as if someone were to say man is a terrestrial biped who practises geometry, for then those who don't practise geometry will not be **[741K]** men. And now that their refutation against this definition has become clear, to the extent that it can be prolonged, let us move on from Zeno to Chrysermus, who defines the pulse in this way.

9. "A pulse is a distension and contraction of arteries, the wall rising up on every side and again gathering together into itself due to a psychical and physical capacity, being a constant and continuous attribute in health and disease, apprehended by percepion." This man also seems to have left out the heart, terming the rising up of the wall of the arteries circumferentially and a collapsing a "pulse," which the majority called contraction and dilatation. What some of the other Herophileans left out, he added. He said the stated movement of the arteries occurred due to a psychical and physical capacity (power), adding superfluously, if someone should give an account of the so-called notional definition, but not superfluously to those who think it right to state the essential definition. For of those matters involved in genesis, the effective **[742K]** cause seems to be the best definition. And the pulse is one of those things that has its essence in the coming-to-be, just as every movement does. Whether he spoke truly or untruly, when he said the pulse occurs due to a psychical and physical capacity, is no small issue that awaits a discussion which one who wishes to say all things in all cases will bring in as well to the present speculation, although in fact it is not necessary.

The definitions explaining the essence of the matter arise in accordance with the doctrines of those who are defining; it is towards these doctrines that the refutation must be made specifically through the definition itself, so those that have arisen are not subject to pettifogging criticism. It is in the speculation regarding the truth of the things said that the greatest prolixity of all is included. If someone should wish to make an accusation because both definitions are brought together into one statement—the so-called notional definition—he proceeds in relation to the common concept, and the so-called essential definition, which expounds the essence of the matter. In doing this, he will rightly lay a charge against many of the definitions stated and will identify those things responsible for such a charge. It is appropriate for the beginning of every teaching to arise from one of the definitions relating to the concept, while from that, **[743K]** it is appropriate to pass little by little to the essential definition or straightaway to the accurate explanation of the substance of the proposed matter. I said right at the beginning of this work what kind of precise notional definition is appropriate, and little by little what kinds of things depart from this. On this account, it is not necessary to spend more time on this discussion. For just as this has been said before, so too does it apply to everything else Chrysermus said relating to the definition.

10. The Erythraean, Heracleides, who became the most famous of his pupils, doesn't seem to accept very much of his teacher's definition. For he said the pulse is contraction and dilatation of arteries and heart brought about by the dominant physical and psychical capacity. To say dilatation instead of distension is to disagree on a word only, not on a matter, whereas to take away from this "to follow along continuously in health and disease" is to disagree on a matter. He omits this as superfluous, just as he also does "the stated movement of the arteries and heart apprehended by perception," even though **[744K]** that the pulse appears continuous in health pertains particularly to the concept of this, and of course also to most of the diseases, although certainly not all.

But I stated before, concerning the movement of these being apprehended by perception, what kind of distinction is needed in relation to that statement, when I considered the stated definitions by Alexander and Demosthenes. Heracleides also took away the rising up of the wall in all directions and its gathering together into itself again, as has been shown in the dilatation and contraction itself by him. However, he added "the dominant" to "the pulse occurs due to a physical and psychical capacity," as had been said by Chrysermus, since Herophilus himself, and all those called Herophileans from him, say that certain other factors contribute jointly to the genesis of the pulses. In fact, Aristoxenus, and with him also Apollonius Mus, is faulted for adding to this, "the dominant." For it was not necessary, he says, to recount what is general to all natural functions as specific to the pulses. He finds fault also with the wall of the artery

[745K] being distended in a circumferential fashion in every direction having been said in the definition, not only, like Heracleides, as being superfluous, but because also some of the pulses contrary to nature do not make the dilatation of the artery equally, but only a part of themselves.

However, he is clearly speaking disparagingly in this, for if Chrysermus himself had written "equally" in the definition, he would reasonably have censured him, bearing in mind that the pulses have unequal movement. But since he simply said, "the wall rises up in all directions and gathers together into itself again," his accusation is no longer with good reason. It would be better, if he was generally seeking fame by censuring Chrysermus, to come in the discussion to those pulses in which the wall of the artery seems like a stretched lyre-string, the artery as a whole being carried up and down and not in fact dilating. For this is advantageously said by some towards an explanation of such a pulse, but not in fact by way of precise knowledge. Anyway, it is impossible for the arteries not to dilate and contract fully while the animal is still alive. It is, however, possible for them not to be clearly affected in this way due to the brevity of the dilatation **[746K]** and the associated clonus in such arteries.

At all events, we said that under these circumstances, the arteries appear to be rising and falling like stretched lyre-strings, but not clearly dilating and contracting. So that someone would not reasonably accuse Chrysermus in the discussion as being deceived about the dilatation and contraction of the arteries in a circumferential fashion, nor as making the definition about their perceptible movement, but about the essential (true) reality of the pulse. But if someone should wish to know about these also, he should himself read what has been said by these men, turning to the notes in the seventh book by the Erythraean about the sect of Herophilus, the twenty-ninth of Apollonius, and the thirty-third of Aristoxenus. It is enough for me, in relation to each refutation, to indicate the chief points of the discussions, and to pass on to something else. For truly, by acting in this way, the whole book will be filled.

But perhaps it would be better not to leave this out, as Aristoxenus himself has, having been a pupil of Alexander Philalethes, refuting all the definitions of those older than himself, as he thought, regarding it as best **[747K]** to have said this: "A pulse is a specific action of heart and arteries." All the same, he says next: "Nothing prevents such a definition being made as: 'a pulse is a specific action of heart and arteries that is mixed and involuntary'." He praises this definition as clearer than the one before, taking as evidence for that its brevity and practicality. For he thinks involuntary movements often also still have something that is voluntary, which is quite clear, however many counter-arguments and questions there are. But such things are disputed by those from other sects, both with him and with the other Herophileans, whereas he would seem to differ from Herophilus himself in thinking the pulse is an action of heart and arteries, placing as parts in it the dilatation and contraction, and at times also the pauses. For if he precisely follows the doctrines of Herophilus, the contraction is an action of the arteries, while the dilatation is a return to the proper and natural state of the body of these. He means that just as in those who have died, the wall of the artery is seen to be distended, so too in those who are living **[748K]** it is distended as much in itself, although Asclepiades thinks the opposite.

This man thinks both the heart and the arteries dilate when filled with *pneuma* which flows into them being fine-particled,[534] and which they have within themselves. However, when they are filled, what previously flowed no longer flows and the wall falls back again to the previous state existing in them naturally. And this investigation can come back in a subordinate way to the Sophists who wish for endless prolixity. But just as here someone can provide himself with a hypothesis of many words, in the same manner, if someone takes in hand the things that have been written by Philonides the Sicilian, in the eighteenth book of his work on medicine, which Agathinus also mentioned briefly, censuring him first as declaring wrongly that the pulse occurs in the arteries alone, and second that also Baccheus, being incorrectly drawn along with him, obviously he will, in such a manner, increase the prolixity by the discussion. For first he will need to state the dialectical proofs[535] by which Philonides was persuaded to think the pulse is of arteries alone; second to refute these; and third **[749K]** to show that Baccheus has spoken in this way in his lectures—"the pulse is a contraction and dilatation occurring in all the arteries." In the epitome on the pulses, he writes in this way word for word: "A pulse and every affection of a pulse occurs only in arteries and heart." And a little later again: "For the pulse, as was said, is a distension of the arteries, or of the arterial part of the heart. And remember that above all we, almost alone of the Herophileans, have delivered up the definitions, although many have been made by the Erasistrateans and the Asclepiadeans." And the same in fact applies to the followers of Athenaeus the Attaleian whom they call a Pneumatic; and if it is not of such definitions, it is nevertheless not of such statements.

Anyway, Agathinus, who was mentioned some time earlier, although censuring those who attempted to teach everything through definitions nevertheless himself wrote one whole book, the first about pulses, explaining each of the names in the pulses along with the specific definitions of the matters. But about the others, I am silent. For what is proposed in this particular book **[750K]** is to make mention only of the definition pertaining to the pulse. I have pointed out often, nothing less than this: that it is possible for a book to be filled in relation to each *differentia* of pulses similarly, if someone defines the large pulse, along with those things associated with it, in relation to how great the dilatation is, and if he does the same with the small, along with the things associated with this, and in this way also all the others. But now, as I said, what is proposed is to go over the pulse itself alone under the chief points

11. Accordingly, Agathinus, censuring those who wish to define the pulse completely, nevertheless himself unnecessarily fell into prolixity along with also being clearly mistaken, thinking "pulse" is said homonymously. For in one way, he says, "we call a

534 On Asclepiades' view of the pulse, see Vallance 1990: 80–82.

535 The term here is ἐπιχείρημα which is appropriated by Aristotle as a logical term for a dialectical proof as opposed to a demonstrative syllogism (LSJ). See *Topica* 162a16 (Engl. transl. Pickard-Cambridge in Barnes 1984: 273).

pulse the movement of the heart and arteries" and in another, "the joint perception occurring in the sufferer in the inflammations." "For truly this is also called a pulse homonymously," he says. "But also the dilatation alone of the artery is called a pulse, when we say it is either large or strong, since these [features] do not occur in the contraction." And yet, **[751K]** he says Athenaeus thinks the contraction too is named a pulse; a signification he exhorts Herodotus to direct his attention to. For he addressed his book to this man, being a pupil of his and a very prominent practitioner of medicine in the city of the Romans.

But he says the combination of many dilatations is called a pulse, and besides these, not only that from the dilatations but also that from the contractions. "But", he says, "is this not something specific in relation to the wrist also called a pulse? It seems to me that also the movement of the stretched-out arteries that are to hand is also called a pulse." In addition to all these, he also calls to mind the palpitation occurring in the hypochondria when sought, asking if it is better to call this a pulse also, for it occurs in some of those who are delirious and is of the thick artery in relation to the spine (abdominal aorta). And he says, if we call this same movement not a pulse but a palpitation, it will be some movement of arteries—not a pulse but a so-called palpitation. In relation to this signification, Aegimios of Elis also wrote a book *On Palpitations* in which he discoursed on the pulses.

And even here again, another discussion is put in as well for those **[752K]** who wish to go on at great length, considering whether it really is a genuine book by Aegimius, or whether Aegimius wrote a book about pulses first of all, or someone else did before him.[536] This is truly superfluous, although perhaps it will not seem to be so. And yet it is superfluous to consider whether some one thing is signified by the term pulse, or many. For Archigenes also seems to have followed his teacher in this, thinking to use the term "pulse" in many ways. However, in order for someone to judge such arguments, it is necessary for him to have been practised in the division of classes into kinds and *differentiae*. Only in this way is it possible to discover the specific and general significations of each of the terms. In many places, then, those having a great reputation in dialectics seem to me to have been mistaken, thinking there are as many significations of the generic terms as there are specific *differentiae* in each of them.

I say, however, if someone thought to use the name "animal", it doesn't signify one thing, but is said variously of immortal animal, mortal animal, rational animal or irrational animal, and furthermore, wild and **[753K]** tame, and gregarious and solitary, and other such things, so that, if he should enter into this particular discussion and train his pupil in many examples to readily see the common in the many and the specific in each, it will take quite a long time. And it is not possible, apart from such teaching, to show that the term, "the pulse", signifies one thing, just as the term, "the animal" does,

536 Aegimius (Aigimios) of Lucania was a doctor, said to have antedated Hippocrates, and the first to have written a work on the pulse.

and also that, particularly in the division of each of the terms, many *differentiae* are discovered.

It is already quite clear that even here, the length of the discussions imposes the need of a whole book for someone wishing to go through this, starting from what these men think who speak of the pulse in many ways, and what it is that has led them astray. Therefore, I spoke in brief about these things also in the first of the commentaries I have prepared on Archigenes' work, *On Pulses*, which are eight in number. And now it will suffice, as in the case of previous matters, to have shown once the length of future discussions, if someone were to pursue all such matters. It seems clear, then, that Agathinus also particularly disapproves of those who attempt to define the pulse. Judging by those things he said, **[754K]** he thinks it is sometimes an action of arteries and heart, sometimes a movement of arteries and heart, and sometimes dilatation and contraction of arteries and heart, the stated definitions not being in conflict with one another but, as someone might say, some being more generic and some more specific.

12. Anyway, Archigenes, thinking he had completed the discussion about the pulses in one large book, wrote the following definition at the beginning of it: "A pulse is a natural dilatation and a natural contraction of heart and arteries", adding obviously the "natural" to the statement of Agathinus, for the reason I stated previously. All the Pneumatics think the pulse is compounded from dilatation and contraction as parts, believing both movements to be actions; this is something which the followers of Herophilus and Asclepiades do not seem to similarly accept. Thus [the followers] of Herophilus have an investigation about dilatation and contraction of arteries which is specific and longer. For sometimes it will seem to you they think both the dilatation and the contraction are actions, although for the most part they think only the contraction is. **[755K]**

Asclepiades, however, since he assumes there is no capacity[537] by which the heart and the arteries are moved, is varied in his statements, just as he is on all the other actions, physical and psychical, so too is he on this. Sometimes he will seem to you not to regard them as actions, and yet in fact this is at times a consequence of the hypotheses, while sometimes he regarded them as actions,[538] as in applying them or acceding to the custom. However, all the followers of Athenaeus, as I said, consider both movements actions—that occurring in the dilatation and that occurring in the contraction, and in this way, we make this assumption, not only in the case of the arteries and heart, but also in the case of the chest. For truly, we also showed this to function in the dilatation and contraction. As in the case of the chest, I distinguish in this way also in the case of the arteries, the motions from the extremes to the median state, sometimes occurring apart from action and sometimes with it. If, then, someone were also to introduce into the discussion the demonstrations about these, he will carry this same discussion to a greater length. If, however, there were also to be the change of the names,

537 The manuscripts F1, F2 and Va1 all have εἶναι following ὑπολαμβάνων.

538 K: ἐνεργείας αὐτὰς τίθεσθαι, ὡς καταχρώμενος; F1, F2 and Va1: ἐνεργείας ὡς καταχρώμενος.

[756K] which he made in the definitions, recounting in detail those reasonable causes they added in turn, then next how they criticised some of these, and then how they defended what they said, another superfluous length will be added to the discussion from this.

13. Magnus says the pulse is not dilatation and contraction but also perceptible swelling and collapse of the heart,[539] and of those structures affected sympathetically with it.

14. Athenaeus himself, the leader of their sect, defines the pulse as natural movement in relation to dilatation and involuntary movement of the heat in arteries and heart, this being moved from and into itself, and moving heart and arteries with it, making the definition not only essential, but also with the reasoning of the cause, in the way the Dialecticians direct us to define those things which have attained existence through becoming. And more concisely, he says the pulse is a manifest inflation of heart and arteries enough in itself to be apprehended by perception. This is also the definition of Athenaeus: **[757K]** A pulse is dilatation and contraction in relation to inflation of heart and arteries. It is, however, possible for someone who so wishes, to also extend the discussion regarding these definitions, by refuting and confirming to the greatest possible extent. And certain others of the Pneumatic sect made definitions of the pulse, deviating slightly from those previously stated,[540] as I said, so that it is not necessary to make mention of these for those who are in haste to run through the discussion as much as possible.

15. Similarly also, there was disagreement of those among the followers of Asclepiades, one with another, some adding to those things said by Asclepiades, some subtracting and some changing them. For Asclepiades himself says the pulse is contraction and dilatation of heart and arteries, while no small number of those who are his followers add, "not once only but occurring often in relation to one inspiration," so the whole definition is as follows: "A pulse is a dilatation and contraction of heart and arteries not once only but occurring often in relation to one inspiration." They were thoroughly deceived by **[758K]** his having written the statement in this way in his book about respiration and the pulses, forgetting the discussion in which he wrote this. For his purpose was to show that certain things occurred similarly in the respirations and the pulses while certain things occurred otherwise and differently. Accordingly, in this, he points out the differences, and the previously set out statement has been written to add the distinction existing between respiration and pulses—"not once only but also often occurring in relation to one inspiration."

16. Some, however, also added certain things to these same statements, like Moschion, the praised editor, when he corrected some of the things that have been said by Asclepiades, not being altogether persuaded by the man, just as Alexander was

---

539 The two terms substituted for διαστολή and συστολή are διόγκωσις, also used for διαστολή by Marcellinus, and συνίζησις, meaning subsidence or collapse. This has relevance for the debate on whether the movements of the arteries are active or passive.

540 Τῶν προειρημένων (K); omiserunt F1, F2, Va1.

not by Herophilus. Therefore, having stated what the definition of Moschion was—"A pulse is a specific movement of heart, arteries, veins, brain, and the membranes (leptomeninges) surrounding this, in relation to one inspiration, occurring not once only but often"—let us now deliver ourselves from the Asclepiadeans. And for you also, **[759K]** I don't think there is need of the explanation of each of the things said in the definition, if you remember what has been said up to this point.

17. Therefore, it is perhaps worthwhile to also make mention of the followers of Erasistratus, who differ among themselves no less than the others do. For surely what is remarkable is that not only from different sects, but also among those trained by one man, the majority have differed; some have also written two or three definitions. An example is Apollonius, a follower of Straton. One of his definitions is this: "A pulse is a distension occurring in the artery relating to the refilling of the *pneuma* being sent out from the heart." A second definition is this: "A pulse is the distension from the artery and a contraction from the capacity present in these and is in proportion." Another and third is: "A pulse is a distension in which the artery strikes against the touch." Therefore, that this third definition has the same force, in considering the definition of the pulse, as that previously stated by the followers of Alexander, is clear to everyone. However, the first written by him is essential and **[760K]** is in agreement with the opinion of Erasistratus, since, as I previously said, he also thinks the arteries dilate in response to the refilling of the *pneuma* sent from the heart. For they expect us to say "refilling" and not "filling", so someone does not suspect that all the *pneuma* is evacuated from the arteries in the contractions.

The second definition also touches on the discussion of the capacities. It is better put together by other Erasistrateans—those who said the pulse is dilatation and contraction of arteries and heart brought about by a physical (vital) and psychical capacity. For Erasistratus himself, in his writings on fevers,[541] clearly declared that in the heart there is not only a physical (vital) capacity, but also a psychical. For some of them it was sufficient to say the pulse is a common action of heart and arteries. Others, however, said more elaborately that the pulse is a counter-dilatation[542] of the arteries towards the heart, thinking that in the time the heart is being dilated, the arteries are being contracted, and when the former is contracted, the arteries are dilated. The first of the definitions articulated by Apollonius signifies a pulse to be the dilatation only, which he himself called "distension",[543] **[761K]** writing as follows: "A pulse is a distension involving the artery occurring during the refilling of the *pneuma* sent from the heart."

Erasistratus, in fact, seems not to call a pulse the natural movement in arteries, but only the movement in inflammation. It is clear from his statements in the first book on fevers, that in these he calls the movement of arteries—what we call the

---

541 Erasistratus wrote a work on fevers, now lost.

542 The term here is ἀντιδιαστολή for which LSJ has a medical meaning as above, citing this passage; the primary meanings listed are opposition and distinction.

543 The different terms are διαστολή (dilatation) and διάστασις (distension, expansion, separation).

pulse according to nature (the normal pulse)—which he himself says is the change from inflammation, and appears to call a pulse, thinking it is not necessary to apply the term "pulse" to every movement of arteries, but only to that in which there is some departure from an accord with nature, due to the condition in the inflammation. Therefore, it is now quite clear what the number of the discussions will be, if someone discusses all these things and mentions those things said by the Erasistrateans against each other, and those said against others, and what in turn those others said in refutation against them.

Therefore, since I have now filled a whole book stating these same chief points alone, not even going through all the stated definitions, would it not be clear beforehand that, for someone wishing to discuss everything, three or four **[762K]** books will be filled? Accordingly, suppose such books have been filled, and that some young man has fallen into silly sophisry, zealous to say everything, and leave aside none of the things said by doctors over a long time, and diligently to learn these things. Suppose he then happens to meet, in the case of some patient, another young man from a different teacher, who has learned how he ought to recognise each of the pulses and what they naturally indicate. Does it seem to you, he will discover anything useful concerning the art of prognosis in the case of one who is sick? Or that he will be highly thought of by the patient himself and by those present? For one young man will go through how the pulse must be defined in a long-winded way, while he will be altogether ignorant of whether the patient is in danger or not, and whether [the disease] will progress in him over a long time or will come to a crisis quickly. The other young man will prognosticate on these things accurately and predict to those present, while not knowing how the pulse must be defined. In this way, by comparison it will be clearly recognised what kind of teaching is useless and what kind **[763K]** is useful.

But as the proverb says, "war is sweet to someone who has not experienced it."[544] This is what those who prefer to define everything, not knowing at all the enquiries that have been produced by the Dialecticians about how it is appropriate to define, seem to me to do. Therefore, the explication by definition of what is signified by the stated name, whenever in fact all do not have a very clear understanding of the proposed matter, is not only useful but necessary. It is also necessary, even when someone either purposely through mischievousness or unwittingly through foolishness, changes the significations of the names, applying these to matters other than all people do. However, when neither of these is the case, it is not necessary for the matter to be indicated by a definition, since the specific name sufficiently signifies it. The wish to define every matter completely is an action of the sophistical teaching. And it is possible for me to say these things, having particularly given myself sufficient experience in day-to-day problems.

For always about every matter for definition, I have shown myself able to speak all day about something. It is not, then, due to lack of suitable words for definitions that I

544 Pindar, fragm. 110 (ed. and Engl. transl. Sandys 1915: 574).

accuse those who have much to say, but that I have shown myself of all men **[764K]** to be very well equipped to deal with such prating. I thought it to be humane, for those learning the art for the purpose of being beneficial, to demonstrate the truth, and not, like the majority of those who have knowledge of anything whatsoever do, attempt to exalt this—those who think themselves to appear more noble—but always of the things known by myself to distinguish the useful from the useless.

For apart from a few, all those who teach the arts go to great lengths on useless matters, one of which is this one which I now undertake in a whole completed discussion. Love of definitions is a disease that not only spreads to doctors and philosophers, but also to orators, musicians and grammarians. To all then, if they love wisdom, I think I have made, through one example, a useful aid. However, if again at some time, there should seem to be plenty of time to go over and decide on everything disputed regarding definitions, I shall do this without hesitation. For some say the same things are useless that others say are useful, differing on how they must be defined. And now there are already two books **[765K]** about these matters, which are the third and fourth and the commentaries on Aristotle's work on demonstration.[545] The same book is given the title, *Posterior Analytics* by the majority, just as that on syllogism is given the title, *Prior Analytics.*

545 Galen wrote several commentaries on the *Organon* of Aristotle. See *De libris propriis* 15 (XIX.47K). None of these is extant.

IV *On the Diagnosis of the Pulses* (*De dignoscendis pulsibus, libri IV* VIII.766–961K)

Book I comprises 10 sections, as follows:

1–2. These are introductory sections dealing briefly with the following topics:

- the relative interest in and the main advantages of the four treatises—this one is specifically for doctors (as opposed to philosophers, including Sophists);
- the need for doctors to train their sense of touch and learn the techniques of palpation, with mention of his own long and diligent training;
- the four components of a single pulse—dilatation, external pause, contraction, internal pause—and what can be perceived;
- general comments on what is perceived—e.g., Empirics say it is just the impact of the beat while Dogmatics say it is the dilatation of the artery, involving change in length, breadth and depth.

3–6. On the difficult issue of contraction. Can it be perceived or not? Galen uses the analogies of bladders, sacs and bellows. He says a palpable contraction is favoured by a strong, large, hard and slow pulse. Vital *tonus*, quality of the arterial wall and movement are important. He gives details of the kinds of pulses in which it is possible to perceive contraction by palpation.

7. On methods of palpation. There are two divisions:

- above and below (the latter with the limb rotated);
- with compression, with superficial touching, and intermediate between these.

8–9. Consideration of how much of dilatation and contraction is palpable. In general, the beginning of dilatation and the end of contraction will escape detection by touch. Galen also considers visual examination of the pulses.

10. On the pauses: the external after dilatation and before contraction, and the internal, after contraction and before dilatation. He says neither are of great importance.

Book II: This book consists of four relatively long sections. Much of the material might be considered theoretical. The aim is to discuss the components of the pulse set out in the previous book.

1. This considers three aspects of the dilatation:

- rapidity/slowness;
- equality/inequality;
- variability in duration (frequency/infrequency).

Galen makes the distinction between the first and the third, as he does elsewhere. He introduces the terms *isochronous* and *isodromous* to develop his argument and stresses the difference between relative and absolute in the relevant terms, like slow and quick compared to slower and quicker.

2. This focusses on large and small as compared to larger and smaller. It is necessary to identify the moderate or median to which these terms are referred, and to accept the need to take into account attending circumstances in determining this.

3. This section develops the argument. In establishing the "normal" (moderate, median), one must start with the best constitution of the body and add freedom from external and internal influencing factors. Also, in relation to such factors like locations and seasons, there are individual variations. Rhythm is considered, as well as magnitude.

4. In this short section, he offers a definition of the mean: there is an adequate dilatation of an artery when no part of it is unduly stretched/strained or relaxed. In summary, in the movement of the pulses, there is recognition of rapidity and slowness; in relation to the pauses, there is recognition of frequency and infrequency; in relation to the *tonus*, there is recognition of strength and weakness; in relation to the body of the artery, there is recognition of hardness and softness.

https://doi.org/10.1515/9783110612677-010

Book III comprises three long sections, as follows:

1. In recognising moderate and immoderate pulses, starting with the movement, attention must be given to the manner of palpation (the three kinds, which are with compression, superficial touching without any compression, and something intermediate between the two, have been considered elsewhere).
2. Consideration of the quality of the movement, which occurs in the three dimensions (length, breadth and depth). Influencing factors include the structures surrounding the artery. The three key factors are:
   - the nature of the person palpating (i.e. how sensitive);
   - the degree of practice/experience;
   - the manner of application of the fingers. He covers some of the difficulties and the need for perception to be supplemented by reason.
3. Rhythms are considered, determined by the comparison of the time of dilatation with the time of contraction. He says the rhythm of the pulse can only be accurately determined in the very strong pulses, although with sufficient practice, a general idea can be obtained in other cases. He considers three opinions on rhythm depending on what is deemed palpable; particular consideration is given to Herophilus.

Book IV is also divided into three long sections, as follows:

1. Galen begins by saying he is moving on from things in pulses "by specific reason of the essence" to things which "cannot be recognised apart from movement of the artery but are not specific to this", starting with the forcefulness of the beat. This is related to hardness and *tonus*. He questions what *tonus* is, and examines the term "strong" with reference to Plato and Aristotle.
2. Here Galen comments on the explanation of hardness and softness in atomist terms, contrasting this with the explanation in terms of the vital *tonus*. He moves on to the term "fullness" with reference to the pulse, referring in detail to Archigenes and Agathinus, and also considers the antithetical term "emptiness". Basically, he is very critical of both men and of the application of the terms "full" and "empty".
3. The consideration of the category of full and empty is continued. Galen sees three possible classes of application:
   - the composition of the actual body of the artery;
   - the quality of the substance contained in the arterial lumen;
   - the capacity of the innate *pneuma*;

Archigenes and Agathinus apply the terms to all three.

Galen says the name for the first is really hard and soft, while there is no way of assessing the other two. He quotes Herophilus, who says pulses differ from one another in terms of quantity, magnitude, rapidity, strength/*tonus* and rhythm, and examines what the first term may signify. He also questions the appropriateness of the term "fullness". This, he says, concludes his survey of names.

Book I

1. [766K] The speculation about the pulses has been distributed into four parts; the first enumerates the *differentiae* of these in terms of kinds and classes; the second, which is the actual one now lying before us, is about the recognition (diagnosis)[546]

546 Throughout the work the term διάγνωσις is predominantly rendered "recognising", "discerning" or "distinguishing". In certain instances, when the context is favourable, the English term "diagnosis", understood in the medical sense, is used.

of these; the third teaches the causes of the genesis of each; the fourth and last is of such a nature as to show doctors what is most esteemed of all, so that they practise the others because of this, while it is not altogether without esteem to philosophers, or at least those who investigate science of nature. **[767K]** From this, in all likelihood the first part of all is equally common to doctors and philosophers, whereas the second and fourth are more specific for doctors, just as the third is for philosophers. For this reason, then, the third is pursued very zealously by both, to the one group being itself the most esteemed and to the other being very close to the most esteemed. The fourth, however, is less esteemed than this, since only doctors have written about this, and they esteem it more than the others. The first is rather immoderately pursued by the Sophists, while for doctors it is pursued sufficiently by those whose concern is more for the actions of the art than for foolishness in disputes. The second and remaining part was not something written for philosophers, for it is not in fact useful for the third part but is for the fourth, nor is it for doctors except to an altogether slight degree for the followers of Archigenes.

Recognition (diagnosis) has the least need of a long discussion but does require an altogether longer time than all the others. Regarding the need to practise the touch, so that it perceives accurately small differences, the training of this is not in our hands. The use of implements and instruments, and of dissections, and **[768K]** experience in regard to bandages and ligatures, and those other practices that are necessary for doctors, someone might perhaps learn thoroughly apart from the actions. And in fact we do many things in this way, making use of the imitations of the real things, through which, although it is not possible to attain perfection, it is for the most part possible to advance and come close to perfection. However, in the practice regarding the pulses, what might someone discover as an imitation of the bounding, or the vermicular [pulse], or one of those that occurs rarely, but is of great significance? Nothing comes close, but by reason one person may be more able to explain the nature of these than another. Consequently, if you were to train yourself to recognise the things encountered very easily and very accurately, apart from the practice in the actions themselves, you would discover no such instrument devised by doctors of the kind used for other practices. Because of this, then, it requires a long time, and even more so due to the precision of the nature of the matters. For if sculptors and painters with respect to sight, wine makers and cooks with respect to taste, those who are involved in the preparation of perfumes with respect to smell, and musicians with respect to hearing, and indeed those having the material of the art in abundance, **[769K]** practise towards precision not for days or months, but for years on end, how much time do you think the doctor must need for the training of the touch? At any one time he must recognise very many affections of the artery. He is not like sculptors who must recognise forms only or painters colours; he must first recognise the amount of the distension, then the quality and time of the movement, and then in this way, the *tonus* of the capacity moving these, and then how the wall of the artery is itself. For it is necessary to recognise all these at the same time, when the artery dilates.

Furthermore also, the impact of the artery due to the movement makes known the quality of the heat more clearly, and if all the things mentioned occur equally or unequally in relation to some one part and to different parts. It is necessary to recognise these things in the dilatation, and things similar to them in the contraction, and the time in between both in which there is a twofold pause; one is after the dilatation and before the contraction, while the other is after the contraction and before the dilatation.[547] And it is necessary to recognise the proportion of these to one another—that is, of the time of one movement to the time of the other and of the time of one pause to **[770K]** the time of the other, and of both together to both together. It is necessary to have recognised all these when the first pulse is completed before the second beat of the artery. For certainly, when the second pulse begins, it is necessary to observe these things we have indicated in the case of the first and to compare the things mentioned in that one with those mentioned in this one, so that you recognise all the differences in equality and inequality of these, and then look at three, four, five and many pulses in succession, so you recognise the regularity and irregularity.

How much practice, then, do you think all these things require? For to me, if it behoves me to speak without deception, it seems to require the whole lifetime of the person for the most complete knowledge, which in us is already so great that [it required] not only six months, in which the most sapient Methodics profess to teach the whole medical art,[548] but very many years. In fact, from when I first began to frequent doctors while still a young lad, I had a remarkable enthusiasm for the art concerning the pulses. Then, **[771K]** although I have paid close attention to it from that time to this, I do not consider myself to have learned the whole art thoroughly. For many years I did not know clearly whether it is possible by palpation to recognize the contraction of the artery, but was in doubt as to whether the followers of Archigenes and Herophilus or those of Agathinus and almost all the Empirics are telling the truth, as I fell so far short of knowing clearly whether it was possible to perceive the whole of the contraction, or some part, and the amount of it. But also regarding the aspects of the dilatation, I was similarly no less at a loss. And yet these are in fact the primary, and as it were elemental, aspects of the art in the pulses. If one is ignorant of these, it is not possible to know any of the others with certainty. This much alone I knew to recognize by perception: that the artery is dilated, and does not move up and down as the so-called Empirics say is possible, and that this is true. For it is not dilatation but only a beat that is perceived by the touch; whether there is some rising up or dilatation is not yet apparent. This was almost the only thing I learned in many years, although there was one thing subsequently, which was the **[772K]** recognition of the strong pulse. However none of the others did I yet know; rather, in regard to those, I wallowed about in deep uncertainty for many years, while not in fact abandoning the investiga-

547 Galen terms these respectively the external and internal pauses.

548 See Galen's detailed and virulent attack on the Methodics, and Thessalus in particular, in *De methodo medendi* I (X.1–82K; Engl. transl. Johnston and Horsley 2011). For a more balanced account, see Galen, *De sectis ad eos qui introducendos* (I.64–105K; Engl. transl. Frede 1984).

tion, but always pursuing these same things more indefatigably, until at some point the contraction became clearly apparent to the touch. Henceforth, I practised still more until the impression was no longer faint; indeed, the recognition of the contraction became no less clear than that of the dilatation. After this, I have nothing to say about the speed of recognition of the other things, for everything flowed forth, and the remainder appeared clear, as when the bright light of a lamp suddenly shines forth in the darkness.

If someone were to accept this discussion as a true account and not a myth, he would benefit greatly by not despairing beforehand of the discovery of those things being sought, nor abandoning the search prematurely, especially if, even after seeking for a very long time, he was still ignorant. For in the seeking itself some hidden state is nurtured and progress is made gradually. What happens is that it hardly seems to have been started before it is completed, while when it is completed, it appears to distribute the fruit of all the prior labours suddenly and **[773K]** in haste, and if I may speak more crudely, with actual progeny. If, however, I pass on to those men who follow me, no small part of my zeal for these matters, up to the point of attempting to explain by reason all the experiences of the senses, in less time and with less labour the art concerning the pulses might be practised. For I shall attempt to demonstrate the other things and the paths by which someone might particularly attain the objective, if only I could release men from the insatiable desire for wealth and reputation which now holds them in thrall. But nothing more will come of this particular matter whilever there is a lack of those eager for the truth—and there will be a lack, so long as pleasure prevails and rules, while the good is discarded and dishonoured. For this particular discussion requires those who are sufficiently resolute and zealous and not those devoted to pleasures. Such men are, however, very few at the present time. Rather, as Heraclitus [said], there is for me one in ten thousand for whom it would be a pleasure to create my discussions, rather than for the ten thousand who, compared to the one, are worth nothing. For the truth is not judged by number but by wisdom. **[774K]**

Therefore, let me now say to this one person desirous of the truth, whoever he might now be or in turn will become, first that it is not impossible to explain by reason many of the experiences of the senses, and then, that even if it were impossible, it is not necessary on this account to pass judgment against our zeal. For even if someone were to come through reason to the knowledge of these things that was not accurate in every respect, at least he would come nearer than not at all. Anyway, who is able to teach by description the outward appearance of each person precisely? Obviously, no one. Nevertheless, every day people claim to identify runaway slaves according to the signs of their particular features. And to some, this alone is enough to hear:

> He was round shouldered, swarthy and curly-haired.[549]

549 See *Odysseia* XIX.246–249 (ed. and Engl. transl. Murray and Dimock 1995: 252).

Although I never tasted the sweet Phalernian wine before coming to Rome, I was not ignorant when I first encountered it, recognising it through the signs I discovered. And a friend of mine readily recognised the sweet wine produced in Lydia, which the locals call Caryinum, **[775K]** only having heard this much from me—that it is like both the sweet Cretan wine and the so called Scybelline wine in terms of colour, sweetness and thickness, while possessing, besides those qualities, a certain remarkable and pleasant astringency.[550]

Such things, then, we also claim to state as signs of the recognition of the pulses. And besides, we state how someone might apply the hand in each case, for in some instances it is necessary to press forcibly, while in others it is necessary to palpate softly, as in the strong and the weak [pulses], for you would not gain knowledge of the former, if you were not to compress the artery, whereas you would not discern the latter pressing in this manner, in that it will seem to you to be altogether immobile when you do this. Furthermore also, with those pulses that are large and at the same time weak, as in the lethargies, you would not discern their size, unless you were to palpate them very softly. Because the discussion will teach the distinction of the high pulses from the swollen, large, hard, and strong, if I were to say to you that this pulse is large, which is increased in proportion in the three dimensions—that is, length, breadth and depth—while that which is swollen is increased in two dimensions only, **[776K]**—that is, breadth and depth—and that pulse is high which is increased in the one dimension only—that is, upward, and that pulse is hard which makes a strong impact due to the quality of the wall of the artery, while the strong pulse is due to the strength of the vital (physical) *tonus*, you would not say it is not enough for someone who is intelligent and not altogether indifferent to acquiring secure knowledge of the recognition of these same pulses beginning from these things, if only he has trained himself sufficiently regarding the primary and, as it were, elemental aspects of the art. Since I have spoken about these first in this way, let me pass on to the rest in the appropriate order.

2. There are these four things. The first of all, which we said the Empirics say, contending rather than telling the truth, is that they perceive only the beat, but don't recognise dilatation of the artery. And because of this, some of them define the pulse as recollection of the experiences of touch from movement of arteries. They did not say from dilatation of arteries, but from movement, for no other reason than they do not admit that they perceive the manifestations of the dilatation. But, they say, someone might perhaps believe in this dilatation by reason, but it does not **[777K]** in fact become obvious to the touch. What they ought to have said is entirely the opposite, if they were wholly bent on contention—that is, touch perceives dilatation of the artery, but if it is also dilatation in truth, this is not yet known with certainty. For to make false claims about things that do not appear is part of perception. If someone should ask them what kind of thing the spasmodic (convulsive) pulse seems to be,

550 Galen gives a detailed account of wines in his *Hygiene* V.5 (Engl. transl. Johnston 2017: 32–39 with additional references).

what else would they have to say but that, if the artery, when stretched out in both directions, is like a harp string. What impacts the touch occurs due to it not preserving precisely the original position, but somehow being gently carried upwards and downwards. If, however, we also ask about the clonic pulse, they will concede this: if the experiences of the touch were observed very closely, what appears to happen is that the artery leaves its proper position. For clearly, we perceive it being carried upward and downward in such pulses. Furthermore also, the inequality in one pulse is when parts of the artery seem to lie comparatively higher, while other parts seem lower. Unless they agree to recognize this, they will be convicted of not learning from the experience which they always hold in the highest regard. For such a thing is not one of those shown by reason, **[778K]** but appears clearly to the touch. However, the doctors called Dogmatics, when seeking the cause of this, differ amongst themselves, and would differ also about the appearance, unless it were very clear. But also in magnitude, a pulse is sometimes unequal in the one dilatation, the distensions of it being greater in some of the parts and less in others, which would not occur unless the dilating artery was a true appearance—but perhaps I should realize that I am prolonging the discussion with them too cleverly rather than discussing more clearly.

For it is enough to say that pulses, whether long or short, high or low, broad or narrow, large or small, and full or empty, and clearly those in between these, would not have occurred unless the arteries appeared to dilate in a whole part in which they think movement alone occurs;[551] it would only have been either high or low or the intermediate of these, but none of the others. Therefore, what kind of pulse the rarely occurring spasmodic pulse is, is a question that all such people who don't acknowledge the dilatation as perceptible always seem to ask. For truly, in the case of the spasmodic pulse, the perception of the dilatation either does not occur at all, or only briefly and weakly, **[779K]** when the artery is carried upward and downward again, like a harp string.

What is said by Archigenes—that also the arteries in the fleshless parts of thin people seem to the sight to be dilating and contracting—is correctly said among modest people, whereas for others, when it seems to them, not having considered the appearances, it is not correctly said. For they will say that either this does not appear, which is also to deny this of those same things, or if they do agree, they will say the artery does appear to dilate to sight, but not still to touch, which is what the enquiry is about. For this reason, we do not direct the discussion to sight, but by making mention of those same appearances which relate to touch, we shall attempt to put them to shame. If, however, the truth must be told, it is not so much for the sake of those men, but to make a correction to the theory for beginners in respect to the recognition that we made mention of the perceptible affections, both of those pulses having a clear dilatation of the artery and of those where it is not clear, like the spasmodic pulses. Further,

551 Following τῆς ἀρτηρίας, K has καθ' ὃ δοξάζουσι μόνην κίνησιν γίνεσθαι which is not present in the manuscripts F1, F2, or Va1.

they should not somehow make the consideration of the existing pulses confused in relation to the amount of the dilatation, **[780K]** but should always bear in mind that there are three distensions (dimensions), comparing the pulse now occurring with that which existed previously as to whether it is the same in the three distensions (dimensions), or in some of these, or in one. For we set out diagrammatically, in the first book of *On the Differentiae of the Pulses*, the twenty-seven pulses which exist in relation to the amount of dilatation. In this way, someone might best recognise them, when he calls to mind the amount of each of the dimensions (distensions) individually. Because that book must precede these actual things said, and all the others that are going to be said subsequently. I have said this right at the start of this particular book, so that whoever comes to these matters before being thoroughly versed in them is going to be altogether mistaken. So then, let this statement be in our minds throughout the discussion.

Now, however, let us bring to an end the matter before us, which is that what clearly impacts on the touch is the dilatation of the artery, and not only the beat, as the Empirics, argumentative about all these things, say. So if they are persuaded by me, and say the opposite to what they previously said, apart perhaps from certain Sophists,[552] and the beliefs they seem to be clinging to, [and accept] that the dilatation of the artery is in fact apparent to **[781K]** the touch, and not in fact if it is recognised in relation to nature in this way. For would these things be more acceptable to this sect of men, or why else, being disposed to accept the appearances in all the other arguments, are they at a loss when it comes to nature and do not still do this here? And if they do this, it will not be possible to say: "perhaps the artery dilates but it doesn't seem to," but conversely perhaps, "the artery doesn't dilate but nevertheless seems to." For them to follow only what is apparent, is not, of itself, consistent with their concept, if in fact the appearance is in every way strong and not as might be wished according to Timon.[553] Why then is the concept they now form not strong? And it should in fact be, if it is strong everywhere.

Therefore, taking a common starting point, what appears is that the artery dilates whenever it is not moved spasmodically. After this, we might perhaps differ from one another, if we believe that and they don't. Perhaps even here again, reminding them of their own arguments, we shall exhort them to a common agreement. For what in truth do they also wish? Nothing other than that each **[782K]** of the appearances should be revealed, but particularly to distrust whether it is like it appears to be, and then, if it also is wholly so to start with. Let them, therefore, also take this along with us as agreed, for I wish to be gracious to them in every way, and if to be sure it is neither sun, moon nor earth, nor if it is sea, nor if we are awake, but not thinking, or alive but not aware of all the things that we are in nature. I do not know if they desire

552 "Sophist" is mainly used as a general term by Galen, and in a pejorative sense, and does not seem to refer to any particular sect or group, as it did in earlier times.
553 The reference is taken to be to Timon of Phlius (ca. 320–230 B.C.). He may be linked to the Empiricist medical sect.

still more doubt than this. Let them have this from us willingly and let them not labour to show it by reason. Let them abundantly enjoy the ignorance about nature; let them only reply to us briefly, whether they ask us also to be in doubt about the use of the matters pertaining to life, and not when the sun clearly appears to rise, for us to do the things we do during the day, but to lie down on the couch to rest, uncertain whether it is night still or day. And when we happen to be in a boat at sea, not to act accordingly, but to be in doubt on whether to venture to jump into the sea, as perhaps it is the sea and perhaps it is not. **[783K]** And when we have already come down the coast and the ship is in the harbour, carried down near to the land, and we see everyone disembarking onto the shore, should we not trust our senses up to the point of use, but sitting within the ship doubtful and at a loss, say perhaps it seems to us the land, but is not the land. And all such things in terms of themselves are worthy of belief up to the point of use, although there is doubt concerning their nature. And we ourselves, I presume, wish to use them as being worthy of belief. However, what is beyond use, we say is superfluous. Therefore, if they grant us what we wish, I don't know that we still differ about anything. "About what pertains to nature," they say. "Stop there, my man". We shall never say this, even if you bid us to.

But I am ready to call upon you to agree, and for the rest, let it go forward as for other people, just as for us, if you wish, since in the matters of practices and arguments I see we ourselves putting a very high price on not refuting those things agreed with all other people relating to life. So then, as **[784K]** in respect of the actions accomplished by them, let these be accomplished by us in this way too. They should not enquire closely into a phenomenon that is obvious, but believe and follow it without undue reflection. An example is this: suppose someone announces that it has rained, the river has risen and destroyed the bridge. If one of you present were to oppose this simple statement that there has been rain and say rather that rain appeared to have occurred, and not speak of a river, but of what seemed and is believed to be a river, and not say increased but is believed to be increased, and not that the bridge was destroyed, but that it seemed to have been destroyed, how will he not seem to be crazy? Why, then, do those discoursing differently to all other men provide problems for us alone, although we use the same senses in matters relating to medicine and even to the rest of life? Why, then, you say the sack, should it so happen, and the bladder and the bag and the bellows of blacksmiths appear to dilate and contract, leaving aside the artery alone, which has not at any time caused you any trouble, and provide problems and disbelief in the appearances relating to its movement. Why don't you simply say it distends and is dilated and it contracts and is contracted, but that it seems **[785K]** to be dilating, if it is truly being dilated, I do not know.

Stop now, by Zeus, and do not at any time overload the art with inappropriate nonsense; it is long enough as it is. If you keep quiet and your purpose is not to disrupt life but to enrich it, and if you are eager not to hinder the art but to glorify it, turn this munificence to things that are more useful. You will find many things in the art of medicine requiring considerable investigation; become skilful in these, demonstrate your intelligence in them, and from now on hate the distrust of the obvious phenomena.

Have nothing more to do with this in the future, for you will be compelled willingly or unwillingly to use the credible manifestations, or there will be nothing for you to say, even very briefly.

Thus, when the same use follows whether you are investigating it or not, investigating is superfluous. There are, however, many other trainings of the perceptions for people, and also of the reasoning. Be zealous about these, devote your time to them, and trust the appearances apart from investigation. Believe them to be what they appear to be. Agree with me, then, in saying the artery is dilated in length, breadth and depth, **[786K]** and consider after this whether it is possible for the contraction to be perceived or not. But you, so that you might investigate neither this nor anything else, cast doubt on the existence of the obvious phenomena, involving the outpouring of a large number of words without in fact being able to cast down or overturn the use of the apparent phenomena.

We, then, shall do the opposite to you. We shall use all the clearly apparent phenomena without investigation, and we shall believe that such things are what they seem to be. We shall, however, consider closely the more obscure things, trying to grasp them by reason, or by the senses, or by both. I speak more elsewhere about what the path of discovery of these things is; for the present, I shall speak only about what is useful to the matters lying before us.

3. Accordingly, since Agathinus says the contraction of the artery is imperceptible, while Herophilus continually discourses about it being perceptible, it is truly difficult and impracticable to believe one before the other, so much zeal have both brought to the matters of medicine and **[787K]** to increasing the art concerning the pulses, and have practised both reason and perception sufficiently over a long period. So then, it seemed right to me first to train the touch to perceive small degrees of difference, so that, if at any time the contraction should become clearly manifest to us ourselves, we would no longer need another witness. Second, it also seemed right to have read the account of the older men. For if the majority and those who are most trustworthy say it is perceptible, but for us, even after practising for a very long time, it still seems imperceptible, we should not desist and despair of the discovery. On the other hand, if no one else, or only a very few and those who are untrustworthy, say this, and to us having done this it seems unrecognisable, then we should now despair of the discovery of this and agree it is imperceptible.

Having discovered, then, that almost all the Herophileans say it is perceptible, while some of the followers of Erasistratus agree but others do not, and a third group following Athenaeus the Attaleian, one of whom was Agathinus, similarly **[788K]** differ from one another, we recognised there will be nothing further for us from the account of the older doctors, since those who agree are almost equal in number and reputation to those who disagree. Then I considered the teachers who are with us now, who agree they have never had at any time a clear recognition of the contraction. In this, however, it would seem likely some were moved by reason, and to all of them it seemed credible that the touch was not able to perceive the receding artery. For truly, either our perception of the movement itself or of the beat contemporaneous with this does occur in the dilatation, whereas neither the beat nor the movement of the withdrawing and contracting artery is at any time perceptible to us. For recognition is of those things associating with, approaching and affecting the touch, but not of those withdrawing, since it is in this way these things would come to recognition from a long separation, if once and for all it is agreed that there is perception of those things not impinging on the touch.

Saying these and other such things, they readily persuaded me that they also have the touch itself as witness for those things they said. For I was not accustomed to perceive **[789K]** the contraction of the artery, and needed little to set this aside as imperceptible, were it not that at one time, what was like a fire kindled from concentrated rubbing and shining brightly, showed me the truth clearly, so there was no longer need of other witnesses and judges; for me this was enough. However, I shall make known the whole matter, beginning here, and from where the argument led me personally. When bodies touch each other sometimes only the surface is affected while at other times they force, push and disturb, and here the stronger and harder go down into, as it were, the weaker and softer, while the latter yields, is disturbed and is as it were divided by that which is forcing the separation of these from each other, which cannot occur in the same way. Rather, those which touch superficially also separate all at once, whereas with something that is yielding and softer, the other goes further down into this, as it were, and these bodies no longer separate all at once, but only to the extent that they were forced to go further in by the first impact. To such a degree is perception preserved in those things that are withdrawn. For some time **[790K]** will occur in the interval between the first movement of the withdrawing and the last and final movement, after which it is no longer still withdrawing, but will be said now to have stopped. In the whole time in which the one is gradually withdrawing, the other is gradually dispersing the compression, and the movement of each will be perceptible. If this progresses so the one moving downwards is carried upward, while the one touching lies above, in addition to this, when the compression is taken away, the one lying on it will still happen to fall, as it were, on what was previously set against it and pushing upwards, is now carried downwards.

4. It is possible to learn these things through experience by means of bladders, sacs, bellows and all such instruments, for someone to whom the explanation through reason is not very clear. For in all cases it will be found that where the association of those things touching is gentle, and the separation occurs without a lapse of time, but if one [of the bodies] is stronger, the extent to which the weaker is pushed is the extent to which it is moved when the compelling force is withdrawn. If, however, the pressure should come from below upwards, then of course, under these circumstances, **[791K]** the weaker will follow the stronger much more easily and quickly, taking the turning point to the downward as allied with the following. If you fill a sac or bladder or some other such thing with a fluid substance or air, you will be able to touch it, superficially raising and in turn reapplying your fingers, and then pushing and forcing it forward, and here again alternately raising and reapplying [your fingers]. Clearly to vision one of the withdrawals, that from the touch, is without perceptible time, whereas the other, that from the compression, always occurs along with some time. And in relation to the first, the vessel is immobile to vision, whether you are applying or withdrawing the finger, whereas in the second it is moving not only when you are applying the finger but also when you are withdrawing it.

Now assume for me that the artery is analogous to the finger, and the body touching it to the vessel. Clearly the touching body remains one and the same in manner whereas the moving artery advances and withdraws. And let the impact of the artery occur against the body touching it externally like a touch and contact, just as **[792K]** obtains in the case of the weak pulses, and not only some touch but also some thrusting to the depths, as in the case of the strong pulses. Accordingly, it is necessary in the case of the former to say that, when the artery is contracting, it is able to withdraw from the externally touching body, or on the other hand that it is now not able to withdraw. Such a recognition needs time, whereas the withdrawals of the superficially touching bodies occurs all at once and without perceptible time. However, in the other impact of the artery which occurs in the case of the strong pulses, by which our touch happens to be thrust away, the perception of the withdrawal is prior while the recognition of what actually has withdrawn follows, when the perception no longer recognises this. But with the cessation of this the affection on the mind at that time is the inference that what is withdrawing has now withdrawn, while if the withdrawal of what touched superficially is altogether unrecognisable, one must assume by inference what has withdrawn in these cases.

5. That we have no primary perception of going forward or coming back, or pressing, or touching, or thrusting away, without the affections in our own body, **[793K]** escapes no one, I think. So we perceive the upward movement of our own flesh along with the surrounding skin in relation to the impact of the strong pulses. Going along with this is the knowledge of movement and force. Furthermore also, when the artery withdraws, the perception is of the movement in our own body itself. For the compression inwards of either our skin or part of the flesh follows the withdrawing artery, which takes up the original place from which it was pushed upwards, while the knowledge of the separation is conceptually linked. For the primary perceptions are the affections in our own bodies, while the secondary perceptions are of those things present externally bringing these about. Therefore, when the withdrawal of the artery occurs after the pressure, we perceive the primary movement in the case of the strong pulses in the parts that are touching, and concerning this, we say also the contraction of the artery falls upon perception, so that clearly, in respect of the affections in us, there is what is called the coming together of the active perceptions. The inequality of the pressure indicates the return of what is compressed. For we didn't recognise either the strength or the movement of this through anything other than through the **[794K]** inequality of the pressure. When we apply our fingers to the strong pulses, the first impulse thrusts altogether slightly against the touch, whereas from this the pressure always increases more up to the end; this clearly indicates the movement and the strength of what is thrust away. It is certainly remarkable how the weak pulses, even when they happen to be very large, do not seem, when compressed, like those that are very small, but escape detection completely.

And perhaps if someone is unaware that it is necessary to palpate such pulses superficially, he will say the person is pulseless,[554] because he himself palpated badly and made a false statement. The strong pulses are, however, entirely the opposite, needing to be strongly pressed on when touched for the recognition of their strength. For under these circumstances, they seem to you to be pushed away more, when you compress them even more. Anyway, all the strong pulses have their contraction perceptible, whereas none of those that are weak do. Of the strong pulses themselves, this applies more in one case than another, and besides, there is more and less in respect to the strength. This also applies to largeness and smallness, to hardness and softness, **[795K]** and to slowness and rapidity.

For whichever pulse is very large and at the same time very strong and hard, to the extent that the hardness is mixed with largeness and not rapidity, this pulse has a very clear contraction. All the other pulses, to the extent they partake of one of these less or more, will to such a degree be perceived more easily or with greater difficulty. However, I have added to the mixture relating to hardness and largeness, "to the extent it is possible", for the sake of this, so that someone is not able to assume at anytime that the same pulse is simultaneously very large, very strong and in this way also very hard. For it is not possible for largeness to come together with strength as it does with hardness, for you often would find very large and very strong pulses, whereas you would not find those that are very hard and at the same time very large, nor those that are hard and at the same time very large. But the pulse intermediate between hard and soft, which is moderate and in accord with nature, is often found along with the very large. Furthermore also, you will often find hardness along with strength itself, and sometimes a hard [pulse] appears with a very strong [pulse], **[796K]** but you would never find the pulse that is very strong and very hard, just as you would not find one that is very large and at the same time very strong and very rapid, but the very large and very strong are in general also rapid, or if not rapid, at least not slow.

If, then, it were possible for the same pulse at the same time to be very strong, very large, very hard and very slow, the clearest of all the contractions would appear in such pulses. But since the combination is impossible, the pulse that is very strong and very large, but not soft and not rapid will have the clearest contraction. Clear also is the very strong, large and hard pulse. Furthermore also, there are those that are strong and at the same time large and hard, and of these less than those previously mentioned; nevertheless, the contraction is still sufficiently clear. Less than these, but also itself clear, is that of the strong, large and soft pulse. Still clear also is that of the strong pulse which is at the same time moderate in largeness and hardness. The pulse which is large, but moderate in the two remaining *differentiae* has the contraction less clear

554 The Greek term here is ἄσφυκτος (ἀσφυξία, pulselessness; ἀσφυκτέω, to be without a pulse). This is discussed in the appendix on pulse terminology.

than this but not in fact unrecognisable. And whichever pulse is moderate in the three *differentiae*, **[797K]** also has its contraction clear. However, the contraction of the strong and small pulse is difficult to recognise. At all events, such a pulse is also hard and the contraction of the small and hard pulse, which is moderate in terms of *tonus*, is still more unclear. If the pulse is small and not hard, and moderate in terms of *tonus*, such a pulse requires a sufficiently practised perception for recognition. Still more is this the case if the pulse is small and soft but has a moderate *tonus*. The pulse that is weak and is either mixed with largeness or smallness has a clear contraction. If at one time it is not mixed with the very large at the beginning, just as it is not with the very hard, but at another time is mixed with largeness, such a pulse at all events is extremely soft. If, however, it is mixed with hardness, such a pulse will inevitably be very small.

Accordingly, it is necessary for someone who wishes to train his touch to perceive the contraction clearly, to know first that this was left out by our predecessors—that it is not possible to perceive the contraction of every pulse, nor of those we do perceive is it possible to perceive them all with similar clarity—and then also to remember the distinctions **[798K]** I have written about just now, so that, being eager to discover in which pulses it is not impossible to perceive the contraction and then being at a loss, he might in the end abandon every contraction as imperceptible. Or in the case of those that it is not possible to perceive, or not clearly, to attempt these first, then despairing of the discovery of these same pulses, also altogether shrink from the investigation of the others.

Rather, let him begin from those in which the contraction happens to appear very clear, applying the hand, as I said, and remembering the objectives I stated—that it is altogether necessary that the pulse, which is going to make the contraction perceptible to those palpating, is strong, or at least not weak. And he needs not to let go of the fact that the largeness contributes to clearer recognition to no small extent, while there is also some benefit from the hardness and the slowness. As a consequence, the contraction can never be perceptible at any time in the weak pulse, nor imperceptible in the strong pulse. More or less clear along with this is the so-called vital *tonus*, and less and more tension also along with the amount of the dilatation and the quality of the wall of the artery and of the movement.

6. [799K] Therefore, so that those who are fond of learning might have the path of practice complete in all aspects, let us not be satisfied with having stated only the general outline of it. Let us speak in order, beginning from the pulses it is appropriate to train the touch in for one seeking the contraction, but also in which people these exist, both diseased and healthy. If they were to learn this besides, it would be easier, I think, to attain the matter in hand. Those who are healthy, who are moderate in their bodily state and *eukratic* in nature, and particularly in the prime of life, and in the early part of the seasons of the year and settled in a *eukratic* land, who are strong and free of grief, fear, repletion and movement, all have a strong and large pulse. And if they exercise moderately and are warmed in moderate baths, don't drink too much wine, get angry, or behave in too manly a way, they have pulses that are not simply large and strong, but large to the maximum extent and absolutely strong. However, those who are sick with a genuine tertian fever, or with the moderate and simple fever,[555] or have some other strongly hot **[800K]** and, as it were, seething fever, or one of the ephemeral fevers existing in buboes, heatstrokes and moderate fatigues will have very large and hence very strong pulses.

Whether one of the things I spoke of is present in them or not, they may still have strong and large pulses, but less, should it be the case, if they are in their prime, healthy and not moderate in terms of their bodily state, but either fat or thin, or dwelling in a cold place. If you were to palpate their pulses, you would find them to be strong and large, but not similarly to those previously mentioned. But you will not perceive the contraction in all such cases in which the pulse still remains strong and large, if you palpate the artery in the way I spoke of, while [you will] very clearly [do so] in those cases in which it is very large and very strong, as was said. Because it falls more on the touch, if it is very strong, due to this its impact becomes clearer, as was said previously. That also, if it is very large, you would particularly learn in this way, if you were to think about it, that the artery is naturally carried downwards in the contractions to the same extent **[801K]** it is carried upwards in the dilatations. If, however, this is so, it carries each movement of necessity with the greatest distension, upwards in the dilatation, and downwards in the contraction, so the former runs upward to the greatest extent, while the latter withdraws to the greatest extent. We said there is perception of the withdrawing but not of what has become withdrawn. To the degree, then, that the withdrawal is increased in its distension and in the magnitude of the time, so the recognition of this in such a case will become clearer. Because of this, we said the first training of the touch should not be made in the pulses that are the strongest only, but also those that are largest at the same time, and not moving very quickly.

However, it was also said that the hardness contributes to some degree to the recognition of the contraction, but we did not state the cause in that place. Now, on the other hand, nothing prevents us also adding this. Thus, the hardness of the wall of the artery contributes greatly to the force of a beat. You will see this clearly in the case of those things that are external, if using equal strength, you apply pressure to anything with a very hard stone, or wood, or one of the things that is soft. It is inevitable, I think, **[802K]** that what is pressed upon by the stone will be compressed and disturbed much quicker and more, compared to what is pressed upon by the wood. Since in terms of the force, the perception of the contraction is always in proportion of that of the dilatation, the recognition of the contractions will become clearer to the extent that they follow the more forceful dilatations. However, a large pulse is not readily mixed with a hard pulse; I shall speak about the conditions in which

555 The Greek term is καῦσος, which LSJ lists as "a bilious remittent fever (the endemic fever of the Levant)". See Hippocrates, *Aphorismi* III.21 (ed. and Engl. transl. Jones 1931: 128–129) where Jones translates it as "ardent fever".

they come to this in the work *On the Causes of the Pulses* and in that *On Prognosis from the Pulses.*[556] For the present, this much will be enough for beginners by way of an example to learn that, as in those who are strong in respect of the vital (physical) capacity, which some doctors call the vital (physical) *tonus*, they go wrong in relation to the regimen, and some other things, and not least that such pulses occur in relation to drinks of wine that are not timely. For truly, in these cases too, just as those spoken of previously, the contraction of the artery is clear. And it is not difficult to conclude logically that the slowness will also increase the time of withdrawal and to this extent it indicates the movement more clearly. And because of this too, this itself contributes something to the recognition. It has, then, been adequately stated what kind of pulse is necessary to make the perception of the contraction that is going to occur. **[803K]**

7. After this, we must investigate in what way it is appropriate for us to make the application of the hand; whether it is always from above the artery, or from below, or sometimes the former and sometimes the latter, and whether with pressure, or touching, or simply placing, or sometimes in one way and sometimes in another,[557] as the nature of the pulse may require. For nothing about these matters has been determined by our predecessors. But we, having also investigated these matters, will set down those things we discovered about them by reason and those by experience, referring to the general, so that, if this should seem adequate, the discussion might now have everything required from us, whereas if not, someone or other might more easily discover the remainder after many useful discoveries.

Certainly, if we were to apply the touch with pressure, we would discover no perceptible difference were we to place the hand from below or above the artery. And if, on the other hand, we only touch applying the fingers with a light touch (as if raising the fingers), we would not in this way perceive any difference. The third manner of application, which is between pressure and surface touching, cannot be carried out in conjunction **[804K]** with palpations of the artery from below. For it is necessary under the circumstances to free the hand from all voluntary movements, as if you were relaxing it in the application. If we do this, bringing down the touch on the artery by its own weight, it will sink down. In this way, then, the touch is applied only from above and never from below. What often occurs in this is that the contraction also becomes very clear, but is often indistinct, and while not unobserved by those who are practised, it is clear in the case of the pulses that are strong and large but not rapid, whereas in the case of those that are moderate, it is indistinct in all the stated classes. However, it was stated before that by compressing the pulses that are strong and large but not rapid, the contraction will be very clear in all cases, while the causes of these will not be difficult to discover from those things I said about them. For to the degree the touch happens to be overcome by the dilating artery so, when it withdraws, the contraction of necessity becomes perceptible. It is overcome in the compressions to

556 *De causis pulsuum* I–IV (IX.1–204K) and *De praesagitione ex pulsibus* I–IV (IX.205–430K).
557 From "the latter" to "as the nature" is absent from manuscripts F1 and Va1.

the greatest extent in the previously mentioned pulses, whereas in the light touchings, it is equivalent to not at all. In relation to the applications in between these, which **[805K]** I said were like letting the hand fall beside the artery, it is overcome moderately.

Consequently, more of the contraction will be perceived with the compressions than with the simple applications, and no less when touching from below than from above. For the touch follows the contracting artery similarly in each. But it is also necessary for the actual hand that is palpating not only to be sensitive, but also soft. For in this way there will be ready perception of the slightest movements by it. I think it is quite clear to everyone that, when we see and hear, some are naturally better than others and some worse; the same applies when we perceive through the touch. So that if, in addition to this, there is also softness in the composition of the fleshy nature of the fingers, the touch would be more readily overcome in falling on the artery and would be carried to a greater separation in the contraction; I do not think this is unclear to those who recall the things said previously. It is not unclear that, by being carried over a greater separation, the contraction will be perceived more clearly. Consequently, it is reasonable that the softer **[806K]** hand, other things being equal, should perceive the contraction more accurately than the hand that is harder.

Therefore, having already brought this discussion to completion, let us move on to something else. Since there is an enquiry among doctors about the contraction in the arteries, with some saying it is perceptible and some that it is imperceptible, we showed that it is neither at all times perceptible nor at all times imperceptible. First, we stated the pulses in which it is perceptible, and second, the conditions in which we shall find such pulses. After these things, we showed how it is appropriate for someone who wishes to perceive the contraction accurately to apply the touch and of what kind the hand being applied must be.

8. These things being thus, the next matter to be considered is whether we perceive the whole of dilatation and contraction, or whether part of each is hidden to us, and if hidden, what part this is and how much. In this discussion, we shall make a start about the movement of the dilatation. But first we shall remind ourselves that often, in many who are sick and have pulses that are small and weak, especially those who are syncopal, **[807K]** almost all of whom die, some doctors perceive the movement briefly, while some do not perceive it at all. Second, we shall remind ourselves also that the doctor himself often palpates either appropriately and accurately or carelessly and negligently, sometimes perceiving the movement and sometimes not. When, therefore, in respect of the same artery, one doctor says it is without a pulse while another says it is still pulsating, and often the same doctor, on a single occasion comparing different applications of the hand in succession, says the opposite things, it is clear that some part of the movement escapes the touch. Perhaps someone will say the naturally best doctor, applying his hand in the best fashion, and who has trained diligently, perceives the whole dilatation. And how shall we discover this man? Who can assure us that human nature has come to such a precision of perception that nothing, not even the smallest, eludes it. For I am not able to feign a firm recognition in the case of those bodies that are external and which can themselves be touched without there being anything else in front of them, much less in the case of the body which has lying before it fat, membranes and skin, **[808K]** often thicker than the body of the artery itself.

If, however, someone is so confident that he speaks out boldly and says he recognises the first impulse of the movement in the dilatations of the artery, I would gladly question him, asking what kind of application of the hand he recognises this through—whether it is touching the surface, or pressing, or applying the hand without any force. Then, perhaps, if he is one of the great Sophists, who are practised in the actions of the art to such a degree that they do not know if the person has a fever to begin with, but shout down the truth in the discussion, such a person will not know whether there are several applications of the hand, or whether some difference is to be imputed to each of these. And it will seem to him that our [statements] are just long-winded nonsense and he will not remain to hear the beginning of the discussions. But let me say on these matters, as I did on others previously, to the lover of truth, even if there is only one, that the intermediate application which is not forceful will have such a perception of the movement, at least in the pulses that are not weak, as the vision also has. In the weak pulses, it will perceive less than the vision itself. The second application, in which we compress, **[809K]** will make the movement of the artery in the weak pulses still more imperceptible than the former, while in the strong pulses it will provide earlier recognition. The third application, however, in which it is as if we raised the hand, will perceive the movement for a very short time.

Let me attempt to state the demonstrations of these [claims], making a start from those that are themselves clear. What would be clearer than this: all the arteries do not appear to move either to sight or to touch, but only those the skin covers, and this not very thickly, while those concealed by flesh or by certain other bodies have the movement imperceptible, unless it should become perceptible, if at some time the overlying bodies should become severely wasted? And of those arteries having the movement always perceptible, those in the barer parts and under the skin itself reveal the movement more clearly to sight and touch. Also, the largest of the arteries, which runs down along the spine, has the movement imperceptible for a time, unless, if the animal has wasted away, it is apparent in the epigastrium.[558] From this it is clear that the arteries, when they dilate easily lift with themselves the lighter bodies, whereas the heavier bodies are more difficult [to lift]. **[810K]** Those that are sufficiently heavy, they do not move at all. You would understand what I say still more clearly, if, in those arteries having the movement evident, you were to apply externally some other thin body on the skin, such as a leaf, or membrane, or linen, or a strip of cloth, or some other such thing. For it will seem to you to move with the dilatation itself. If, in turn, you apply a second body equal to the first, the movement of both now seems much less, and if you apply a third, it is a lot less, and if a fourth and a fifth [less still], so that those things overlying will seem to you to be completely immobile. And if the pulse is stronger still,

---

558 Galen is again referring to the abdominal aorta.

it will lift up more of the bodies placed on it, whereas if it is less strong, it will lift fewer, while if it is not strong at all, it would perhaps not lift up one body applied. If you think the pulse itself has become weak, not only will it not move one of these bodies placed on the skin externally, but also it will often not move the skin itself. Truly therefore, if this seems to have occurred in this way, you will see the skin being moved to the degree the *tonus* of the artery is stronger. If, however, **[811K]** the *tonus* is sufficiently dissipated, the skin appears altogether immobile.

Therefore, as the stronger movement of the artery is customarily able to move with it not only the skin, but also many of the bodies placed on it externally, while if it is weaker, it moves only the skin, and if it is weaker still than this not even the skin, what is remarkable about there being another movement weaker in turn than this that will not be able to lift up either the membranes covering the arteries or the fat? What touch, then, will perceive such a movement? None obviously. For in addition to the things already said, this is not said—that it is appropriate to compress the skin covering the arteries squeezing it with the fingers, so that we might come to a pulse which rises up deficiently. For truly, in the case of those that have a perceptible movement, but are weak, when we compress the skin, complete *asphyxia* (pulselessness)[559] occurs. What pressure, then, do you think makes the artery unable to move the skin, if besides we should weigh it down with our fingers? Or is it clear also that the preservation of the movement perishes as well? Those who doubt that the first part of the dilatation is perceptible seem to me entirely ignorant of the position of the arteries shown by dissection. For none of them touch the skin. **[812K]** But if also you wish to inspect those arteries that seem to be conspicuous and very near to it by excoriating the skin on either side, it will be apparent to you that there is a considerable space between the skin and the wall of the arteries. For truly, this alone would be an absurd thing for nature to have done, and indeed she has fashioned nothing like it—to prepare an organ capable of dilating in every direction while providing no space for it in which it is moved. But this is unlikely to be so—in fact, far from it. For in the dissections an adequately broad space is apparent around the artery, into which, in those with excessive flesh, this flesh spills over, preventing the vessel dilating to the extent natural to it. And because of this, the pulse of those who are obese is small. For its length becomes shorter and breadth narrower, because it is compressed by the abundance of the flesh, while the rising upward is prevented, because it is weighed down by the thriving condition of the overlying bodies. In those whose bodily state is thin but who are otherwise strong, the pulse has in fact sufficient breadth and length, while also the ascent upwards is significant. Still more, in those in whom the skin becomes clearly relaxed, there is an interval between it **[813K]** and the wall of the artery.

For if the pulse of someone so constituted is strong, or at least not weak, by changing the position of the limb and making it sometimes supine and sometimes prone, you

559 On this term, see the appendix on pulse terminology.

will be able to detect the *differentiae* of the distensions of the artery, which I stated just now, are detected clearly by the compression. For breadth and length, the prone position seems to have the advantage, whereas for height and duration the supine position does. The cause in both cases is the relaxation of the skin: in the supine positions of the hand, what is lying above falls down on to the artery, whereas in the prone positions, it inclines downward and hangs down and is most separated. Because of this, in the supine positions, the artery in its distension flows downward from the higher positions into those that are oblique, and is folded, and as it were wrinkled, and being gathered together there to the greatest extent, particularly hinders the largeness of the pulse in these parts. In the prone positions, the movement of the artery is not weighed down at all, nor is it hindered, so there is no reduction of largeness in the pulse. How then does the pulse not also appear to be dilating here in relation to depth still more, or if not more, **[814K]** at least equal to that in the supine position? It is because the depth is hidden by the increase in breadth and length. And in another way, for the sake of which all these things are said, when the artery is stretched out in the parts of the limb above, it appears to the sight to move over a longer time, and it is imagined also that the ascent upward becomes greater. For what seems to be carried upward over a longer time readily persuades the sight that it has been carried up a greater distance, whereas it has been carried an equal distance, in truth, so whatever position the artery might have when the time increased the prior movement in the supine positions is distributed to the skin, so that to sight, the time of the dilatation increases the illusion of depth in this way. Therefore, these phenomena, and others similar to them, indicate that the first parts of the dilatation escape the sight. As much of this movement as is perceptible to the sight[560] is commensurate with the unforced touch in the strong pulses; in the case of the weak pulses, it is often less. For it is clear that it is not possible in such **[815K]** a manner of touch to perceive the first parts of the dilatation, even if you touch the surface like lifting the hand.

What remains, then, is whether, in the compressing applications of the hand, it is possible to perceive the first parts of the dilatation. But it is unambiguously clear, if we remember the things said, that it is not possible in all the pulses; if it is possible, it is only in the strong ones. And this certainly has advanced to such a point for us that it is agreed neither any other application nor any other pulse makes the beginning of the dilatation possible to perceive, apart from that occurring with compression in the strong pulses. Next, however, let us see if this possibility itself is really true or only plausible. For it seems to me even in these pulses, the first parts of the dilatation still escape detection, although not in fact in a similar way to the other pulses; in these, what is concealed is altogether least. You would understand, if you thought about it, that it is necessary for the initial movement of the artery to be distributed to the first surrounding bodies, then to the skin as a whole, and then finally to our fingers. **[816K]** It is, therefore, not possible for the transference of the movement in this

560 K: τὸ τῇ ὄψει αἰσθητὸν; F2, Va1: τὸ τῇ ὄψει αἰσθήσεως αἰσθητὸν.

way to occur altogether instantaneously. If this is a true statement, it is inevitable that the beginning of the movement is hidden. But also, because the touch perceives a much longer pause occurring after the subsidence of the artery and before the dilatation than it does after the dilation and before the subsidence, it would not be unreasonable for someone to suspect this occurs because some parts of the internal movement escape us. For it is clear that, if the first part of the dilatation is hard to catch, the last part of the contraction would also be unrecognisable. For the touch would not at any time perceive more or greater movement in relation to the contraction, if in fact one must readily accept that its perception is equal to that of the dilatation. Those who say this (i.e. the contraction) is never perceptible show this, and they are not few in number, nor are they to be despised in relation to their practice in the art. We, however, say it is perceptible, and sometimes, particularly as we have already taught beforehand, we demonstrated together also that it is not possible for the touch to perceive as much movement in **[817K]** the contractions as in the dilatations, unless at some time a change should occur in the amount of the dilatation or contraction.

This does, however, happen in the systematic inequalities. Accordingly, let it also be proposed that, due to the magnitude of the internal pause, sometimes the end of the contraction and the beginning of the dilatation escape perception, and let us now put an appropriate end to the other of the investigations. For subsequent to that about the dilatation and whether it is all perceptible, there is a similar consideration about the contraction. However, from those things we have already said about the dilatation, we appear also to have completed the discussion about the contraction. For if it is not possible to perceive more movement in the contractions, but either what is equal to that in the dilatations, or less, which is actually the case, and if we don't perceive all the movement in the dilatations, it is clear that we don't perceive all the movement in the contractions.

9. Two remaining issues exist: to state how much of the dilatation is imperceptible and how much of the contraction. On these issues, we seem **[818K]** to have the general picture already from what has been said. The discussion showed that as much of the dilatation is imperceptible as occurs in the time between the first impulse of the movement in the artery and the beat being distributed to our touch. If, however, we divide this into kinds, we shall find it to be most in pulses that are weak and large at the same time, when we have touched them superficially, and least in those that are strong which we recognise by compressing, and intermediate between both in the intermediate applications of the hand, and in those in which the pulse is unequal besides being either strong or not strong. Certainly, since we perceive the contraction as much as we perceive the dilatation or a little less, the imperceptibility of the contraction would be as much as that of the dilatation, or a little more. These, then, are the primary questions and, as it were, elements of the discussion.

10. The matters concerning the most important parts of the pulse have now almost come to an end. The remaining two issues are more incidental and neither importantly nor primarily parts of the pulse—these are the internal and external pauses of the artery. **[819K]** Since certain things are also prognosticated through these, it is necessary to speak about their recognition. Thus, in the case of the external pause, you would particularly find out how large it is by simply applying the hand, or if you also wish to compress in the obviously strong pulses, apply the force to such a degree that it does not prevent the movement. In the case of the internal pause, if you press to the degree that the present strength allows, you would particularly recognise it in this way. And you will recognise the whole external pause accurately by palpating as you ought, whereas in respect to the internal, as was also said previously, it is with the imperceptible parts of each of the movements—that is, the end of the contraction and the beginning of the dilatation.

It is not difficult to discover the causes of these for someone who happens upon them, if he has not listened in vain to what was said before. Nevertheless, I shall state them no less briefly now. Accordingly, since it is not possible to know if the pulse produces any such thing as the external pause, unless we accept that the artery is moved as much as possible, while we prevent this, when we press on it too immoderately, it is obvious that we would never clearly perceive the amount of the external **[820K]** pause, if we were to press on it over and above its present strength. But perhaps someone will say the largeness of the dilatation is hindered when there is compression, whereas the movement is always made apparent, for the touch is pushed away by the rising up of the artery, and then first stops being pushed away and moved by this, when the movement also stops. However. that this is false in general and requires a distinction, we shall readily know, should we recall what was said previously, when we stated that what sometimes happens is that the artery is moved without those things lying on it being moved with it, it being overcome by the weight of these. Therefore, it is not unreasonable that, when we do not compress it, the perceptible movement of the artery continues in all the pulses for as long a time as they continue to be uncompressed and unforced. So that, whilever the artery is moving by reason of itself, the pulses already often seem to us to pause, and that they begin the pause quicker by as much as they follow the perceptible movement. However, since by as much as the pulse pushes against the touch in the dilatation of the artery, to this extent also it becomes perceptible in the contraction, so it is inevitable that all the first parts of the contraction are not **[821K]** perceptible, when the last parts of the dilatation have become imperceptible.

How, then, is it possible for someone who does not palpate in an acceptable manner to recognise the external pause? For the amount of the compression needs to be measured in relation to the strength of the artery, compressing to such a degree that nothing of the last parts of the movement will be imperceptible. It is clear, then, that it is not possible to press in all pulses, but only in those that are strong, and in these not to such a degree as to interfere with the strength of the artery by the force of the application. It is safest then, whenever you wish to recognise the external pause, to apply the touch without force, whereas, in the case of the internal pause, the opposite applies, for you would not recognise it if you did not use force and compress. It is necessary for the touch to follow the contracting artery throughout, whereas it is possible to compress in the strong pulses only. Therefore, it is clear that in these alone will you recognise the internal pause, although not accurately in them, but often greater than it actually is by a lot, although often by a little. Nevertheless, it is still perceived as greater than it is. For of necessity two times of the imperceptible parts present themselves for each of the movements; the one at the very end of the **[822K]** contraction and the other in the beginning of the dilatation, having different lengths at different times, as I showed previously. The progression in time of the internal pause will not always be equal nor will the diagnosis be accurate.

Book II

1. [823K] I have shown in the book prior to this, how one must distinguish the parts of the pulse. I shall now state what is sufficient for someone to recognise those things that exist in the parts themselves, starting from the discussion of the dilatation. For this is also the most important of the parts of the pulse and has the greatest diversity in terms of diagnosis, while the greatest use for prognosis is through it. Since there are many things existing in the dilatations, the most important of all, and that which relates **[824K]** particularly to the theory of the essence of the pulses, is the movement. In every movement there is some time in which it happens to be either rapid or slow, equal or unequal, and in addition to these, there is the interval during which it lasts. Therefore, let me speak first about these. To most doctors there does not seem to be one recognition of the time of the movement and another of the rapidity and slowness. For if the amount of time is known, they immediately say it is clear whether the pulse has been moved rapidly, slowly or moderately.

To me, however, there seems to be one recognition of the time and another of the slowness and rapidity. For if the time of the movement is shorter, I do not also now say the pulse is more rapid. I think what has misled those who make this assumption has arisen from the definition relating to the rapid and slow pulse, which, if it indicates what they think it does, is erroneous. If, on the other hand, it is as we understand it, the definition is correct, although these people don't agree with it. For the definition states that a rapid pulse is one in which the movement is concluded in a short time, and a slow pulse in a long time. But the terms little and much are said in relation to something. **[825K]** These people think it is the times of the pulses that are being compared with each other. But the truth is not like this. For often the movement of the pulse having the longer time is quicker than that of the pulse having the shorter time. And if someone does not recognise this accurately, he will be ignorant of many important things discovered from the prognosis through the pulses. Therefore, I know that of the things which will now be said, none will seem to be incredible to those of us who have experience in the cases of actual sick people and have often spoken of affected places before exposing the parts, and each of the currently and previously occurring causes and symptoms. For they have the most accurate of all beliefs due to the truth of those things previously said, whereas to others the discussion in respect of the first statement seems incredible and overbold, if we should say that so many and such doctors are mistaken in the diagnoses of rapidity and slowness. If, however, someone should bear up and apply himself to hearing the discussions now stated, I know that, even if I do not completely persuade him to believe, I shall nevertheless keep him from altogether disbelieving those things **[826K]** being said. For it is not possible to easily change someone's opinion regarding those things agreed upon over a long period of time by very many people, and particularly when, in this way, it requires a great amount of precise theory. And if one of them should follow the theory closely, but also that he may judge by experience, he needs a great deal of practice in palpation.

Anyway, the so called formicant pulse requires precise perception for recognition, as it falls to extreme weakness and smallness, so that it is at the beginning altogether hidden from those unpractised in palpation. Consequently, they often say the people are pulseless in whom such a pulse is discovered. Accordingly, if I should contradict those who say it is rapid, and indicate the source of their error, who would be my judge? For how would I teach those who are not able to recognise at the beginning whether this pulse exists in the sick person, what kind of pulse it is? Therefore, I hesitate to say these things about such a theory.[561] It is enough to show its existence by its actions. Since, however, as was also said at the beginning, it is agreed that these writings are not for the many, but **[827K]** await the one person in ten thousand who is worthy, it is for that man one must write the words which ten thousand times before will be treated with contumely as impossible and pretentious, before they come to one who is worthy.

For I need now someone like Archigenes as a listener, so I may speak to him in this way: What then, most worthy sir, are these things you write in your book *On the Pulses*,[562] discoursing on the formicant pulse somewhat according to this statement? It seems to me to be a compound of smallness, weakness, frequency, and necessarily also rapidity. For I do not know what the last thing of all those said means to you. As a consequence, I understand and accept everything else said before this as being particularly true and fitting for you. But may I enquire how "rapidity is necessarily joined to the formicant pulse" is said? For it is more than what would simply be said, apart from the addition of "necessarily". Furthermore, this is also false. But even without the addition to the statement, in the error of the touch alone what is misleading would be revealed, while the added "necessarily" displays a certain logical suspicion. Therefore, I would need to learn what this is. For I don't see a single necessity as to why **[828K]** I should think a weak, frequent and small movement is also immediately rapid. Indeed, the ant itself, from which the pulse is named derivatively, is not something we see that is moving rapidly; rather we see its movements are slow, weak and frequent, but in no way rapid.

561 K: ὀκνῶ περὶ τῆς τοιαύτης θεωρίας; F2, Va1: ὀκνῶ πολλάκις περὶ τῆς τοιαύτης θεωρίας.
562 As noted earlier, Archigenes of Apameia wrote a work on pulses, now lost and known mainly through the writings of Galen.

Which of these does rapidity follow of necessity? Is it the weakness? But I would be astonished if, all other things being equal, something will be moved quickly due to weakness of the capacity, for the complete opposite holds in other similar cases; the strength of the capacity when associated creates rapid movement, as we demonstrated in the work *On the Causes of the Pulses.*[563] But the frequency is not a cause of the rapidity. Who, travelling from Athens to Piraeus of necessity frequently, also immediately moves rapidly? For there are some reasons for doing something frequently and others for doing something rapidly. But perhaps he says that due to the smallness the formicant pulse is necessarily rapid. For what else would someone think? Such a thing is in fact then reasonable, if also we are now seeking the truth. For it is necessary, when the artery is carried over a very short distance (interval), that it uses up less time **[829K]** than when carried over a very great distance (interval). That movement which is limited to a short time is not necessarily rapid. Do you then say that the movement occurring over a long time is rapid? Perhaps someone might raise this question, but I would not say so. For in general, I don't think the amount of the time alone should be considered, but also the amount of the distance (interval). Therefore, when you wish to question me with respect to the movement occurring in a short time—what kind I say it is —immediately add the amount of the distance, so that I may answer you. If you don't, I will add it and make the distinction, saying that the movements occurring in less time are not in every case more rapid than those brought about over a long time. But if the same ratio exists in time to time and in distance to distance, they are concurrent. If, however, either the ratio of time to time or distance to distance becomes greater, I say the movement will be quicker in the one instance and slower in the other. I have explained the whole matter. If you have understood, perhaps the discussion may now have an appropriate end. If you have not understood, I shall explain to you **[830K]** at greater length what was said about rapidity.

Those movements that pass through equal lengths in equal times, we call *isochronous* and *isodromous*,[564] whereas those which cover a greater distance in an equal time, we call quicker. Similarly, also with those that cover an equal distance in less time. For example, if two men cover fifty *stadia* in one hour, we would say they were *isodromous.* If, however, one travels fifty *stadia* in one hour while the other covers thirty *stadia*, it is clear that the former is moving quicker, covering a greater distance in an equal time. If one travels fifty *stadia* in one hour while the other happened to take three hours, it is clear that the former is quicker, if he completed the equal distance quicker. There is certainly nothing obscure in these examples. Perhaps, however, there would be in what follows for those who don't pay close attention. For those who

---

563 *De causis pulsuum* I–IV (IX.1–204K).

564 The two Greek terms romanised here are ἰσόχρονος used in a technical sense in sphygmology (LSJ has "even, regular" referrring to this passage), and ἰσόδρομος which is found several times in Plato (e.g., *Timaeus* 38d, Engl. transl. Jowett in Hamilton and Cairns 1989: 1167–68) meaning "keeping pace with." In the present context, the argument makes the meanings clear.

do pay close attention to those [statements], there will be similar clarity to these [that follow].

Let us make two movements, the one completing fifty *stadia* in one hour and the other one hundred and fifty *stadia* in three hours. Should we not then say these are *isodromous?* Very definitely. So if the one covering fifty *stadia* in one hour **[831K]** happened to move for three hours, he would I suppose cover the hundred *stadia* in addition to the fifty. And if the one that happened to cover the hundred and fifty *stadia* in three hours were to move for only one hour, then altogether he would only cover fifty *stadia.* Consequently, they would be *isodromous*, having completed equal distances in equal times. Therefore, it is necessary to pay attention to the lengths of the distances and the lengths of the times, so that, if there is the same relationship of distance to distance as there is of time to time, we may say the movements are *isodromous.* If, however, either the relationship of distance to distance is more than that of time to time, or the relationship of time to time is more than that of distance to distance, we no longer consider these movements to be *isodromous.* So let the relationship of one distance to the other be four to one while that of one time to the other is two to one. It is clear that the ratio of one distance to the other will be greater than that of one time to the other, and in this way one of the movements would be quicker. Suppose, then, one of the distances is four hundred cubits and the other a hundred, while **[832K]** the time of the former is two hours and that of the latter is one hour. Accordingly, if one of the movements covered four hundred cubits in two hours, while the other covered one hundred cubits in one hour, it is clear that, if we observe the former moving only one hundred cubits, it will be moved in half the time. And in this way, it will be quicker than the second, being moved at equal distance in less time. If, however, we observe the second movement over four hundred *cubits* itself requires four hours, in this way it will be slower than the first, since in fact it has covered an equal distance in a greater time. It is clear, then, that whenever the comparison of distance to distance has a greater ratio than that of the measurement of time to time, of necessity such a movement will be quicker. If, contrariwise, the ratio of the movement of the first distance to that of the second is less than that of time to time, of necessity the first movement would be slower.

I shall also give an example of this. Suppose the distance of the first movement is double that of the second movement, while the ratio of time to time **[833K]** is four to one, so that if it should happen that the first movement is four cubits in four hours and the second movement two cubits in one hour, it is clear that, if we suppose the first movement to cover only two cubits, which would take two hours, this would be slower than the second, which covers the same two cubits in one hour. Should the second cover the four cubits, it will be moved for two hours and in this way will be quicker than the movement of over four cubits in four hours, for it completes the equal distance quicker.

Accordingly, since this is assumed, let us now consider two pulses unequal in terms of rising up to a height, for we need this dimension particularly for the discussion, and let us make, if you wish, for the sake of an example, the depth of one ten times that of the other and the time of one twice that of the other. Would it not be clear to us, if we recall something of those things previously stated, that the pulse having ten times the height will be much quicker than the other? For if, while maintaining the same movement, the low pulse is increased ten times in the dimension of depth, it will become similarly as high as the other, but in a much longer time than the other. Moreover, a certain shallow pulse is found to be shorter in time than the high pulse and slower than the pulse of long duration, and this is what **[834K]** was said from the beginning—that the recognition of the time of the movement is one thing, while that of the time of the rapidity is another.[565]

But perhaps these things are conceived as far as the argument is concerned, but do not seem to be among the truths. For if you compare the pulse of a healthy person in his prime with the formicant pulse of a child, you will discover the rise in height to be ten times greater in the former. However, if in fact you were to heat moderately this same person in his prime, either with baths, massages or exercises and then throw him into a state of anger and irritation, you would find the height of his pulse not to be ten times but fifteen times that of the shallow distension of the formicant pulse in the child. But the formicant pulse in the child is of benefit to the stage of life, as long as it is not altogether slow, whereas with the formicant pulse in an old person, would someone experienced in palpating the pulse say this is not more rapid than that which is vermicular? For I do not know whether any other pulse appears clearly slower than this. The formicant pulse escapes clear recognition due to brevity, whereas the vermicular pulse is not obscurely slow, but if anything is very clear.

Why, then, do the majority number this pulse among the rapid pulses, unless **[835K]** wishing to preserve the homology with the formicant pulse, they declare it to be among those that are imperceptible? For since these pulses readily change into one another, and the final collapse of the vermicular pulse readily generates the formicant pulse, while the first increase of the formicant pulse generates the vermicular pulse, it seemed to them clearly irrational to say that one of these is rapid while the other is slow.

Accordingly, since they happen to have said incorrectly that the formicant pulse is rapid, in order to maintain consistency, they agreed falsely and clearly contentiously that the vermicular pulse is also rapid, and yet they liken it to the progression of the worm, just as they liken that of the formicant to the progression of the ant, being accustomed to see them somehow as swift creatures, then likening them in this way, while at the same time saying they are not swift and not perceiving the contradiction among themselves. But when they have once and for all fallen into so great an error, they also say that all the pulses in the stomachical conditions[566] are rapid, al-

565 Galen's laborious argument is essentially aimed at establishing his claim that the rapid pulse must be distinguished from the frequent pulse.
566 In LSJ the term στομαχικός is listed "as of the stomach" and "disordered in the stomach" but given the range of meaning of στόμαχος, it may include the oesophagus and cardiac orifice of the stomach.

though not all of them are rapid, and there are countless other errors, which it is more fitting to speak about elsewhere. **[836K]**

All such errors followed with them due to the disjointed and wholly confused nature of their concept of rapidity. For determining the rapid pulse by the shortness of the time only, without defining in addition how it has magnitude, they conjectured falsely from these things, and it is from these things that they err in the diagnosis. But we showed clearly, I think, that one pulse is of short duration—someone might name it in this way—while another is rapid. Similarly too, there are the opposites to these, the one being of long duration and the other slow.

Perhaps you may think the whole discussion has already been completed, and we know clearly that the rapid pulse differs from the slow pulse. I, however, professed to steer clear of such a long discussion on the issue, but the truth is not in fact like this. Rather, the things now being said are a proem to the recognition of the rapid and slow pulses. In this whole previous discussion when comparing two movements with each other, I have demonstrated the slower and the more rapid, but have not said anything about slowness and rapidity. However, **[837K]** there is a great deal to say about how much a slower movement differs from slowness, and in turn, a more rapid movement from rapidity. Immediately someone says this particular movement is slower, we might question him as to what he says it is slower than, for slower is always said of something that is slower than something else. If, however, he says this particular movement is slow, we do not still question him as to what it is slower than, for we would then seem to be ridiculous. And if we were to say a pulse is slower or more rapid, someone will invariably ask us what we say it is slower or more rapid than. However, when we say slow or rapid, no one would still ask us than what.

Therefore, are the rapid and slow among things said absolutely, while more rapid and slower are among things said referring to something else? Perhaps someone will readily regard this as true, but to me it doesn't seem to be so. Rather, it requires considerable investigation because such a theory relates to other things and is common to very many matters. For we must certainly not define rapid and slow, or more rapid and slower, while **[838K]** omitting large and small, and larger and smaller and likewise those other things said in this way. We must consider whether the same pulse can be called larger and large, and smaller and small, or whether different matters underlie such terms. And we always think of the larger as larger than the smaller, and the smaller as smaller than the larger, while we don't think of the large as large compared to anything, but as absolutely (simply) large, just as we don't think of the small as small compared to anything, but as absolutely (simply) small. The same also applies to shorter and short, longer and long, broader and broad, narrower and narrow, and by Zeus, to softer and soft, and harder and hard, and all the other things said in this way that are worth distinguishing. The issue extends to almost the whole medical art and all the other arts. For we also say hotter and hot, colder and cold, drier and dry, and moister and moist, and in fact in all these, more and much, and in turn less and little. Therefore, we put the discussion to the test in the case of something as an example; if we should happen to find something more, **[839K]** we shall attempt now to transfer the method in this way to all other matters, while if we are at a loss, we shall at least transfer the same difficulty to everything. One or other of two things will result: either we shall discover a method of very great use for many things, or we shall set aside the foolishness of seeming to know those things we don't know.

Well now, let us take up the investigation of the large and the small, and in what way they differ from the larger and smaller. And first of all, the things that we say in relation to the life, to which we are accustomed and with which we have been brought up, we shall call to mind for ourselves what we think the large and the small are. Then there is the matter of larger and the smaller.[567] Do we not cease from saying larger absolutely, unless saying more than something is added, whereas we ourselves say large absolutely and we understand compared to other things without the addition? Are then large and small said in relation to themselves? But if this is so, it is of course altogether indicative of substance, or quality, or one of those things said in relation to the other categories.[568] It is not then indicative of substance, just as Socrates or Plato or horse or man, nor of any quality nor then of form, **[840K]** nor of colour, position or conformation, nor of any of the other things at all said in this way. But neither do they indicate naturally where or when, or being, or having, or doing, or suffering. Rather, it is necessary to seek these things only in what is said in relation to quantity or relatively. If then, as it seemed a little earlier, it does not seem to hold for the category relatively, we must seek these in relation to the category of quantity. Furthermore, it is very difficult to define how large we must think each of these is—whether the *stadion*[569] is large but the *cubit* small. And further again, we say the mountain is small, even if it is a *stadion*, while the ant is large, although it is clearly much less than a *cubit*.

But perhaps someone might say, whatever is sufficient for one of our comrades, to indicate to them what the amount is, although this is not in fact definitive, as a *podiaion* is, but as that person said, in breadth and indeterminately. Do we not, then, have to encompass with boundaries the width of these, just as of those other things we say in this way, like a chorus, a city, a mountain, an army and a marshy lake? For each of these **[841K]** is said of a quantity that is indeterminate. It is possible to say that we do not already think of a chorus as three or four dancers, nor yet of a thousand, just as we do not think of a city as composed of three houses or extending for a thousand *stadia*, or a mountain not to be only a *cubit* high, but somewhat larger, although not in turn extending up to the moon, or a marshy lake to be composed of three jars of water, nor such that it floods the whole earth. For of each of these, even if it is not possible to define the amount precisely, it is possible to discover many bodies which deviate entirely from the concept of themselves in each direction, those that are not yet such and such, and those that are no longer still such and such.

567 Kühn's Greek text has μικρῶν as do mss F1, F2, Va1, whereas Kühn's Latin translation has *minus*.
568 On the categories, see Aristotle, *Metaphysica* XI.12 and elsewhere (Engl. transl. Ross in Barnes 1984: 1687–1688).
569 The three Greek measures are: στάδιον, 606.75 feet; πῆχυς (*cubit*), 18.25 inches; πούς, approx. a foot long.

2. We are not, then, able to say what limits there are to the large and the small, just as we are not in each of the things mentioned. But the concept of the large extends so broadly that it also refers to the whole cosmos itself, and if someone is able to think of something larger, this would also receive the predicate (category) of largeness. But also the ant, the gnat and the millet seed, and if there is anything **[842K]** smaller than this, whether it be animal or seed, it may receive the predicate of largeness. For not only an ant, but also a millet seed and a gnat is called large.

This much at least is clear then—that when we compare each of the members of a single genus we say either small or large, but not absolutely. And in the same way as the larger differs from the large, so the larger can be said to be larger than everything smaller, whereas the large is not larger than everything, or of something, but only of those of its own class. In this way too, it is necessary to make a distinction regarding smaller and small, and regarding the quick and the quicker, and the slow and the slower in the same way, if in fact some horse is said to be slow and an ant quick. Similarly also, you will find each of the other things named in all the classes of the pulses and in relation to the whole medical art. And already, in the whole life of people, there are both hard and harder, soft and softer, frequent and more frequent, and all other similar things, both those that are at some time compared with those things in another genus (class) and those that at some time only with those things in their own genus (class). Have we, then, already covered the whole matter, or is some **[843K]** distinction left out in this discussion?

We do not use the term great comparing in this way all those things from the same class (genus). Thus, Achilles and Ajax would not customarily be called great in the same breath, and nor would Polydamas and Milo, if in fact Achilles were greater than Ajax or Polydamas greater than Milo.[570] Rather, it is clear that by saying the median, the moderate and the proportional in relation to each class, we are comparing other things by a rule and criterion; those that are above this, we say are all great, and those that are opposite, small. And this is so in every aspect of life, and especially in the technical arts. For all these are concerned with the pursuit of due proportion, and are eager to provide this for the essences (substances) with which they are concerned. The doctor says someone has eaten too much of a particular food, when it is beyond the nature of the body being nourished, and in turn too little, when it falls short of a balanced nourishment. Anyway, in this manner, we say two *minas* of meat is a small amount for an athlete, while with one who is sick, we often forbid him to gulp down a plate of barley juice, as being more than a due proportion.

---

570 Achilles, son of Peleus and Thetis, was the greatest of the Greek heroes, as recorded in Homer, *Iliados* II.681–685, while Ajax was the next greatest warrior, and said to be of great size (see, e.g., *Iliados* II.768–769 and III.226–229). The Polydamas referred to here is presumably Polydamas of Skotoussa (5th to 4th century B.C.) and not the Polydamas of the *Iliados*. The former was the victor in the 93rd Olympiad (408 B.C.). He was immense in size and reputedly of great strength. Milo of Croton (6th century B.C.) was an athlete of legendary strength who enjoyed multiple victories at the Olympian and Pythian games.

Therefore, we shall say a large pulse and a small pulse **[844K]** are set apart from the moderate; there will be many of each of these (i. e. the large and the small), but only one that is the moderate, which is the one we must pursue, storing it away in the memory to apply to all the others like a true measure defined by Nature herself. Let us now consider how we might find this. The whole discussion hastens towards this, and all the things previously said are for the sake of this. We might find it, if we are able, by method. For it is useful to have one rationale concerning the matter relating to all the proportions, and not to seek in turn how we must discover that relating to proportional (fitting) movement, that relating to the pause, and that relating to the body of the artery. And first let us see if it is necessary to seek the one pulse of due proportion in each class, or in many, and another which is of due proportion for the age (stage of life), and another as in states of bodies, and another as in *kraseis*, and another as in the seasons, or places, or climatic conditions, or pursuits, or psychical (mental) conditions. For in all these, we say the pulses are large in some cases, or small, or moderate (of due proportion), **[845K]** and we say some are rapid, some slow, and some moderate, and similarly in relation to the other *differentiae*. And perhaps it will seem to someone more reasonable and more consistent with use to establish one moderate in each class of pulses. Therefore, let us consider this more precisely, and taking up one class of pulses as an example—that of the large and small, about which we spoke a little earlier—let us attempt to discover the median and moderate of these, separately in the stages of life, in seasons and in places, natures, *kraseis*, and all the other such things. If we should discover this, it will be sufficient for use, both to know what lies before us in respect of the defining characteristic of each of the pulses mentioned, and to recognise the deviations of this towards a contrariety to nature (abnormality). The teaching, as now stated, appears to provide this for us.

Let us, then, examine precisely if it is possible to discover this. For it seems to me very difficult, perhaps even impossible. The difficulty lies in **[846K]** not being able to take anything of the things said separately and in relation to itself. For when is it possible, separately and in relation to itself, to see the difference of the pulses in the seasons without a body, and not how it is in terms of *krasis*, bodily state (*hexis*), age, and the other things mentioned? Consequently, you would not be able to say such and such a pulse is characteristic of spring, or such and such of summer. Nor, if you were asked, would you be able to say in a certain body whether its *krasis* is hotter or colder, whether it is thin or fat, in the prime of life or old, in a cold place or a hot, in a man or a woman, and similarly in relation to the other *differentiae*. As a consequence, it seems necessary to consider, in relation to the whole nature of the supposed body, the characteristic and non-characteristic pulse, so that we might see, should it so happen, that the pulse in Dion is such and such in nature, and that in Theon is such and such. But this too is paradoxical, for not only in terms of use in relation to each person would we need to have a specific memory of the characteristic pulses and would not be able to recognise them in anyone we had not previously encountered, but also in that from Dion himself, should it so happen, would we be able to take this one characteristic pulse in each class separately, **[847K]** and apart from the prevailing circumstances of the external matters.

At what time will you take Dion's pulse? In winter or in summer, when he is in Rome or Libya; when he has become thinner or fatter; when he is relatively idle or working harder; when he is deficient in food or has eaten abundantly; and when he has taken wine or water or is thirsty? For it is impossible to take the body of Dion as specific and the same at any time without these and countless other circumstances. Accordingly, the difficulty [we had] from the beginning still remains now, for we are unable to discover in the whole nature of the underlying body the characteristic pulse which we also called moderate in equality to it.

Perhaps, then, it would be better not to approach the diagnosis of the pulse in this way, but, as I wrote in the work for beginners,[571] to learn thoroughly the change in the case of each of the stated causes. For there is the possibility of discovering a certain method of such a thing, which now having written about before, we shall examine in order whether it is sufficient for the whole or is lacking in some respect. The method would be as follows, if we take up by way of an example **[848K]** one of the natural causes which change the pulses, like exercise: to consider what is set before us—that is, in what way will exercise make a change of the pulses. Taking, then, a particular person, let us examine all the *differentiae* of his pulses first, and then, after he has exercised, examine again which of them has been intensified and which relaxed and up to what amount. In this way you will discover to what extent exercise naturally changes the pulses. Similarly also regarding baths, drinks, foods and all other things, from the experience in relation to each person, you will readily discover the changes. In this way the experience is very easy, but the conclusions drawn from such experience are no longer easy, for there is a danger, as in every experience, of erring on account of the indefiniteness even here. And this same indefiniteness occurs in two ways:

- Not dividing the general into all the *differentiae* in it.
- Not considering, in the causes of change, those that are *proegoumenic*, or primary, or however someone might wish to call them, and those that have followed *per accidens* (contingently).

The failure to divide the general into all the *differentiae* in it is as follows: **[849K]** The explanation needs an example so it becomes clear. If someone should declare that the pulses of those who have exercised become such and such, without distinguishing who he says has exercised, or how he has exercised, or when, or how much, he errs on account of not dividing the general into the *differentiae.* For I say of someone who has exercised, how old he is, how strong or weak, how he is in terms of humors, bodily state (*hexis*) and *krasis,* how he is in the construction of the parts of the body, and how he is in terms of habit with regard to exercise. For the matter in a particular person is divided into so many parts.

571 *De pulsibus ad tirones* (VIII.453–492K).

In turn, how he has exercised is divided into other *differentiae:* whether intensely or not; whether quickly or not; and whether evenly or not. For he will not be changed similarly in relation to each of these. Likewise also, it is necessary for the change to be differentiated according to what time [he exercised]. For it is not the same thing to exercise having eaten or fasted, when grieving or joyful, or when eager or sluggish, or having been cooled or heated, or to have slept more than necessary or less, or moderately. Similarly, too, the change in the amount of the exercises will create differences. For moderate exercises will not **[850K]** change the pulses in like manner to more or less exercise. If the change is *per accidens* or primary it must be distinguished in this way: whether or not the person was full of superfluities, for if he were when he began to exercise, a large amount of humors in a bad condition would flow to the stomach and the orifice of the stomach would happen to be bitten by the badness of the humors, or weighed down by the amount, or he would be upset and made nauseous, and because of these things the person would either be exhausted, enfeebled, cooled or choked up. For each of these things will change the pulses and someone will perhaps think they have been changed by reason of the exercise. For surely, if someone is plethoric and more sickly in the structures in relation to thorax and lung, when he begins to exercise rapidly, he will become dyspnoeic, and the pulses will be changed by reason of the dyspnoea, whereas perhaps it will be thought to be related to the exercises. Therefore, how one must make a distinction so as not to be mistaken in the experience was stated at greater length elsewhere, but will be said here again briefly. We must reduce all the stated differences to these three in general:

- The body assumed to be undergoing the experience should be healthy and **[851K]** not be prevailed upon by any external change.
- The exercise applied to the person should be kept moderate in quantity.
- Consideration should be given to the kind of exercise, lest the person encounter something accidental and not according to the actual nature of the exercise.

If you look at these three things, you will easily discover how and to what extent an adventitious cause changes [the pulses], and it will be possible for you in this way, having learned thoroughly by experience the change in relation to each kind of cause, to use this in the diagnoses of the diseases, discovering easily how much the pulses are changed by external causes and how much by the disease. As a consequence, it will be no longer necessary for you to know beforehand what kind the moderate is in nature. Instead, knowing what is characteristic in each of the underlying things, you will be able to recognise in this way how much the pulse is changed by other causes and by the disease itself. Furthermore, if we do not need the knowledge of the moderate for these things, we shall be seeking it superfluously.

How then shall we say something is large or rapid before discovering the moderate? For it seems the original enquiry is still preserved. But why do we need to say large or rapid? It will be enough for us **[852K]** to call to mind the changes alone, so we may speak as follows:[572] "moderate exercises change the pulses towards large, rapid, strong and frequent." We shall also say this, and it is what others say when they wish to indicate that in exercises the pulses become large, rapid, strong and frequent. But first we shall not simply say "in exercises", but we shall add "in the moderate [exercises]". For the exercises beyond the nature of the exercising body alter the pulses to another kind of change. Then, however, we shall not say in this way absolutely large and rapid; we shall also say larger, more rapid, stronger and more frequent, and all the other things we need to say in this manner, like for example, that in relation to the ages (stages of life), the largest are those of people in their prime, while the smallest are those of old people, or those of people past their prime are larger than those of old people, but less than those of people in their prime. The same applies in the case of the seasons, and we shall make the same teaching in all the other cases, comparing the pulses with each other in each of the classes, and not speaking about them absolutely.

Such a teaching will perhaps suffice for the Empiric. And **[853K]** we shall consider this a little more precisely. However, for someone who wishes to discover the primary and so-called *synektic* causes of each of the pulses and to practise the art with reason and method, it is essential to know the moderate in each of the classes. Suppose, then, we are investigating why the pulse of the child is small and are at a loss about reasonable causes. For surely, we would be at a loss, were we not beforehand to have carefully enquired for ourselves about the causes of the genesis of the large and small pulses. For we shall not discover in the case of children those things that generate the small pulses, and furthermore, if the reckoning of the cause is in error in any one whatsoever, then it will be rightly suspected in the others in which it seems to find a way. It is appropriate in this for it to be agreed everywhere, if it is true. Because of this, then, it is necessary to know the moderate. For perhaps the pulse of the child is above the moderate.

Anyway, Herophilus once termed this pulse "good-sized".[573] Why in the world, you might ask,[574] does Archigenes say it is small? Perhaps I could state for you the reason for the disagreement, should you first know that what is proposed has been shown—that it is necessary, regarding the **[854K]** theory concerning the causes in the pulses, for the man who wishes to be logical to know the moderate and to prefer to know everything by the reckoning of the cause. And this is perhaps less, (for someone will say he does not wish to be logical), whereas let us now look at what was deferred a little earlier, as being worthy of investigation, which is that none of the Empirics is able to make the observation accurately and articulately before discovering the absolutely moderate. For although he sets himself the task of not only knowing if it changes the pulses to

572 Kühn's Greek text has λέγουσι while the Latin translation has *dicamus*. The latter is followed on the basis of context.

573 The Greek term is εὐμεγέθη. Kühn's Latin translation has *magnum*.

574 Again, there is a discrepancy between Kühn's Greek and the Latin translation, the former having φησί τις and the latter *quaeres*. Again, Kühn's Latin text is followed.

rapidity or largeness or any of the other parameters, but also of knowing by how much it changes them, for the whole use of the prognosis through the pulses is encompassed in this particularly, it is impossible for someone, ignorant of what accords with nature, to diagnose how much it changes [them]. It is necessary to know beforehand what accords with nature of each of the diseases, if the amount of the present change is going to be recognised. But he has not encountered all those who are diseased when they are healthy, so that he might learn from the experience, and then, having memorised this, compare the nature in them when diseased.

How, then, will he still be able to gain his objective through experience? Or again, **[855K]** even here, will the observation be according to certain common factors? How and in what way? For it is not still similarly possible here, as it is in the case of external causes, to keep the other factors unchanged, while only changing that on which we made the observation. For truly, colours, forms, smells, lengths, breadths and depths of bodies, and thinness and thickness, and what relates to classes, ages (stages of life), places and seasons, and the differences in the prevailing climatic conditions exist of necessity in all people, so that, since you would not find anyone without these, it is difficult to conduct the observation in them by those things mentioned a little earlier. So we shall need some other method in the case of the kinds of external causes. You would learn it is difficult to make the observation in this case, not only by reason, but also through the actions themselves, should you wish to find two people differing from each other in one respect only, who are similar in all other respects. For you will not easily find them, in addition to which very many conjunctions of the changes also occur, and a long time is needed for the observation, as well as an extraordinary and **[856K]** remarkable memory. It is necessary for the Empirics to observe closely the other things and those that make no difference to the change of the pulses for a very long time, so that in this particular thing itself, they may know for sure that it does not differ at all. And it is this particularly illogical and unmethodical activity which extends for a limitless length of time, and because of this, we always say, just as I showed also through other things, that the art consisting of observation alone, without rational thought is one of those that is impossible.

And even if all these things are in fact agreed upon, anyway the amount of their change in each case will become impossible to remember, because we are unable to refer either the excesses or the deficiencies to any defined measure. For how would someone remember all these, when of necessity they exist in the bodies combined with each other, along with the amount of them in each case? For all in all what will such a combination be, with white up to so far, broad chested up to so far, and similarly thickness up to so much, and breadth and depth in respect of the body up to so much, and to go beyond this, being male and a child, and in the season of summer, and in a place cold in nature, **[857K]** and in a dry condition the pulse is such and such.

It will be necessary to combine certain such combinations, which are limitless in number and cannot be observed or memorised by any one person, but as far as discussion is concerned they are only able to involve themselves in the diatribes of the Sophists and are unable to refute the actual actions of the art. If, what is more, such conjunctions are going to differ from one another in terms of more and less in all the things spoken of, and that if they do differ, they will lead the observer to such a degree of confusion, is I think, perfectly clear. Therefore, it is perhaps better not to waste more time, nor to delay in demonstrating the difficulties and limitlessness of the Empiric observation, but to show a certain logical method itself. Using this, along with experience, we shall readily grasp what is useful.

What might the method be? The following: when we have discovered the very best constitution of the body,[575] it is necessary first to consider what kind of pulses there are in this, and we must do this not simply or casually, but taking the person in such a condition in which the pulses preserve the characteristic movement, **[858K]** not having been changed by any of the external causes. This would happen if he has neither held himself in prolonged idleness nor has recently exercised, but also has not fasted for a long period nor has recently now eaten. It is also necessary, obviously, that the ambient air must be *eukratic* and the person free from psychical affections, having a soul that is calm and pure. After these things, having discovered the abnormal constitutions of the body, first the *differentiae* of these will be enumerated, then the excesses and deficiencies in these compared with the moderate and the best, first reckoning up how much they are likely to change the pulses in such natures compared to those in the best constitution, and then to put the reasoning to the test through experience itself, if it agrees with the appearances. By doing this often and through prolonged practice training the reasoning and the touch, then at that time a person may now think he has secure knowledge of the moderate in all the classes of the pulses and whatever kind of excesses and deficiencies there are of this.

I have stated the whole matter. Whoever does not know the nature of our body or the abnormalities, **[859K]** or has not considered them by reason, or has not taken care to recognise them when they are encountered, would gain nothing much from the present discussion. The right course, then, is to learn those things beforehand and in this way come to these. For truly, he will not find any chance disagreement occurring in these in the doctors of ancient times with each other, or in the doctors of recent times, not only with each other, but also with the doctors of ancient times. For neither is it agreed upon by all what the best nature is, nor is the number of the abnormalities discovered in some way. And very few doctors have written about the signs through which someone might recognise either the best constitutions or the abnormalities. And even if all these things had been agreed upon by our predecessors and properly discovered, to recognise the natures from the actions themselves would need prior practice, and in this way to undertake the art concerning the pulses.

Perhaps some people will be vexed by the large amount of introductory theory, but we shall not be vexed, if the matter is treated with contumely. For we never expected this to please or benefit all. This we know very clearly and precisely—apart from the

575 *De optima corporis nostri constitutione* (IV.737–749K; Engl. transl. Penella and Hall 1973, and also Johnston 2020: 431–451).

one person who loves the truth, **[860K]** such discussions will seem to be longwinded nonsense to everyone else. I was diverted to the description of these, esteeming the one over the many. And certainly, if someone has observed us stating beforehand things concerning the sick that have previously occurred and are going to occur in the future, I also think this person, if he is not altogether naïve, will be enamoured of the theory. Even if it does seem overly long, it does in this way promise a useful and good outcome.

3. Therefore, if someone has first learned how he must recognise the best nature of our body—the signs of this have been stated elsewhere[576]—let him know the pulse in this is the median of all the others, and as it were some kind of measure and rule in all the matters concerning the pulses, and let him recall what we say is the due proportion and moderate, compared to which we say all the others are rapid and slow, and large and small. The pulse that is in due proportion and in the best nature in relation to everything, we rightly do not call either large or small, or rapid or slow; nor do we call it hard or soft, or any of the other names signifying the disproportions. But just as in the work *On Kraseis*,[577] **[861K]** we said the best *krasis* of all is neither hot nor cold, neither dry nor moist, neither soft nor hard, (for all these consist of some excesses and deficiencies of that one *krasis* that is *eukratic* and in due proportion), so too we shall now neither designate nor think of the pulse that deviates from the best nature as large or small, or rapid or slow, or anything else deviating from moderation. Nor do we call the person himself, in whom we find such a pulse, by any of the names designating the disproportions. We call him neither fat nor thin, neither hot nor cold, neither moist nor dry, nor any other such thing, but *eukratic* and in due proportion in all the parts of the body. Accordingly, in this person, when you learn to recognise him, considering all the *differentiae* of the pulses, remember and judge all the other pulses in relation to them, and call them in this way either hard or soft, or infrequent or frequent, or any of the other terms analogously. Therefore, when now we seem to understand the whole matter, **[862K]** it would be no bad thing to stir up some necessary doubt, so that nothing seems to be left still undiscovered. Let us attempt, then, to add some explanation to the discussion, which will persuade us and at the same time teach us, first that we already understand the whole matter, and next that we demonstrated the most complete recognition of the matters before us.

Therefore, we must now begin with the doubt, and we must say that it is not possible for all the arteries naturally in the body either to be equal to one another or to dilate equally. For how would it be possible for the artery adjacent to the spine to be equal to those in the neck, or those in the axillae and groins to be equal to this, or for their dilatation to be equal to the dilatations of those? Much more certainly, does this apply to the arteries at the elbows and knees and more still to those at the wrists and

576 In the work referred to in the previous note and in *De bono habitu* (IV.750–756K; Engl. transl. Johnston 2020: 452–461).
577 *De temperamentis* I–III (I.509–694K; Engl. transl. Johnston 2020: 2–279).

ankles, and even more still to those at the temples and in the fingers.[578] Thus, it is not possible for one pulse to be the measure of the amount of the dilatation, not even in the best natures themselves, although it is likely that there should happen to be one origin in the movement and *tonus*, and the other classes in relation to all the arteries. For it is not possible, when a person is in accord with nature, and particularly **[863K]** when he has acquired the best constitution, for some of the arteries to pulsate quicker and some slower, or some more strongly and some more weakly, or some harder and some softer, or some more frequently and some more infrequently. Rather, it is necessary for all the arteries to be moved similarly to each other and to the heart, although just as their bodies differ in the mass, so too are the dilatations likely to be different. Perhaps, therefore, someone will think the resolution of the difficulty is at hand. For surely, there is nothing to prevent someone who always touches only the arteries at the wrists from always knowing how large the pulse is in the best natures in these arteries, or by comparison with this how large each pulse is of those in the other arteries, or also, by Zeus, should someone so wish, of those in the ankles, or any other part of the body, it being necessary to have remembered the amount of the dilatation in the case of the best nature, so that having a measure from these for the pulses occurring in these same parts in the other natures, will be neither strange, nor difficult, nor hard to recognise.

But he who says these things doesn't know **[864K]** it has been demonstrated in the work *On the Best Constitution of the Body*,[579] that such a nature naturally exists similarly in large and small bodies. For if he knew this, he would not still think that the arteries themselves, neither in each part nor in those dilating, would be equal in mass in all the best natures, but would be quite different in the large bodies and the small, so that the initial uncertainty remains as we are not able to discover one mean in relation to the amount of the dilatation. Is it, therefore, worthwhile, since we are ultimately at a loss, to admire all those doctors who always term pulses large and small, as may happen, but do not teach how large the mean is, which we shall discover to compare them with? For they do not seem to have come to any distinct concept of such a pulse; rather, they predicate relative to it, large or small and equally larger and smaller.

Anyway, I am often accustomed to enquire of the best doctors among us, when they have palpated someone and said the pulse is large or small, what they say it is larger **[865K]** or smaller than. They say, than what accords with nature. Should I question them again on whose accord with nature, not only do they think the question is not worth asking, but they are also amazed that I think to say it is of anyone else apart from the sick person himself, whom they happen to be palpating at that moment, since they say that children also already know this—that in health itself one person

578 The arteries referred to are, in order: abdominal aorta, carotid, axillary, femoral, brachial, popliteal, radial, posterior tibial, superficial temporal and digital.

579 See note 575 above.

has a pulse that is larger than another, and one a pulse that is smaller than another. Do these people then assume this, whereas others better than them, who have left us their commentaries on the pulses assume otherwise, or similar to that? There are certainly those who say the middle of spring and the middle of autumn produce the largest pulses, while the middle of winter and the middle of summer produce the smallest, whereas the times between those mentioned produce the medians of the large and small. These people do not seem to me to say either large or median absolutely, but in relation to something. Anyway, because of this some of them also add in the statement "as in seasons", writing as follows: "The middle of spring and the middle of autumn, as in seasons, make the pulses very large." And the same people, in the **[866K]** stages of life (ages), do not simply write, as another might, that the pulses of those in their prime are large, whereas those of children and the aged are small, but even here they write in addition, "as in stages of life." As a result, they speak in this way of the large and small, as if they also said larger and smaller, clearly comparing them with one another, but not to the mean. That they compare them with one another is, I think, clear from the actual nature of the matters about which they are discoursing. For certainly the pulses of all people do not happen to be very large in the same way in the middle of spring, but each according to the specific nature [of the person]. And of course, this itself, by virtue of thick-witted reasoning, is altogether disjointed and confused. For it would be no small degree of foolishness, if someone were to think that the largest pulses occur in the middle of spring in someone who was bilious, hot and burning up in nature, or the opposite to this in someone who was phlegmatous and cold. For in the discourses and as far as argument goes, it would perhaps seem to be like this, whereas if you were to put it to the test, it would not be found to be so. You would see that those who are hotter and dryer in *krasis* do not await the middle of spring, but have the largest pulses very much earlier than you yourselves, whereas those who are more moist **[867K]** and colder are the opposites to them being much beyond the middle of spring when summer is already beginning. In respect of the best constitution of the body it is possible to find you yourselves having the largest pulses only in the middle of spring. The same applies also in relation to the stages of life (ages). For those who are *eukratic* have their pulses largest in the middle of spring, whereas those who are *dyskratic* have them earlier or later. But none of the doctors made a distinction of these for us; rather they would reply that this refers to all people similarly disposed in each of the seasons.

And yet I do not think the many kinds of things said by Hippocrates about the seasons[580] would escape anyone, for he clearly showed that neither natures nor stages of life (ages) are all similarly disposed in the same seasons. Thus, the winter makes some people stronger and healthier and altogether better, while the summer does the same to others. How, then, is it possible for one who is benefitted to have the same change of

---

580 See for example Hippocrates, *De aeribus, aquis, locis* 10 and 11 (ed. and Engl. transl. Jones 1923: 65–138).

the pulses as one who is harmed? How are they not ashamed to write so simply for us in this way and so vaguely, that the middle of spring creates the largest pulses? For if it is not added in whom they are largest, we have learned nothing. How, on the other hand, **[868K]** when they say the pulses become small in summer and winter, do they not venture to write as well how small, or which of the seasons makes the pulses smaller? For what they say is, of course, no small error, if someone were to think that the summer makes the pulses smaller than the winter does, when as may happen, it makes them larger, or if the winter makes them larger than the summer does, when the converse happens to be true. For if someone were to say any one of these three things—either the summer compared to the winter, or the winter compared to the summer, or the pulses to be equal in both, he will not be speaking the truth, unless he adds, in whom. However, if he should add and distinguish in relation to the natures of the bodies and the stages of life, each of the three things will become true, just as we have distinguished in the specific discussions about these. But each person has not been able to refute either himself or others about these things, so they make many false and undefined statements. Nor are they affected in such a manner for any other reason than that they are entirely ignorant of the mean in each class of pulse. And why, since they do indeed contend bitterly with each other about all these things in this way, do they not on the whole enquire about them, **[869K]** but distressingly neglect them? It is for no other reason than that they have a mixed-up and disjointed mind about these things, having all followed the one first person who erred, and then, since they introduce nothing worthwhile, either harmful or beneficial, to these same actions of the art, they not only contend bitterly with each other, but also with Herophilus.

And of course, with respect to these things, since contention exists about obscure matters and many and detailed discussions are required for this to be demonstrated, with regard to things that are clear in this way, like things it is possible to grasp through perception alone, they themselves neither write distinctly nor refute those who are in error, nor do they mention whether one of the ancients has written something different from what seems to be the case to them. But they have said the pulse of the child is small, and similarly also, that the pulse of the old person is small, but they still do not write which of the two of these is smaller. And indeed, they clearly have not encountered Herophilus, who has written about both of these, and not cursorily, in his books. But perhaps I also make myself a laughing stock, wishing to distinguish such things for them—they, who have not stated the primary and most simple things properly. **[870K]** Thus, they say the pulse of the child is small, whereas Herophilus has never at any point said it is small, but at one time that it is sufficient in magnitude, and at another that it is noteworthy, or terms it in some way thus. And then, when they enquire as to what the cause moving the arteries is, they omit none of the things written on this point by Herophilus, while of the different theories pertaining to the actions of the art, regardless of whether Herophilus wrote correctly or incorrectly, nothing is mentioned.

By the gods, would it not certainly be much better not to seek the four powers (capacities) governing animals, as stated by Herophilus, nor to contend bitterly and gainsay him on these. But if they wished to bring him down and refute him utterly as rashly foolish, were they to call to mind those ideas of his which are clearly at odds with the obvious phenomena, would they then be able to gather these together as witnesses, not few in number, against him? For they will have as evidence those things elaborated on by the followers of Herophilus concerning the pulses, and almost all those now elaborated on, to the effect that the pulses of children is smaller than that of old people, and in fact much smaller. So that someone very recently **[871K]** said to me, ridiculing the whole question, that the matter is not worthy of investigation. For the pulses of children are much smaller, and because of this it would be as well not to have written about them. Having, therefore, a myriad of such witnesses and other things always contentious towards Herophilus, I do not know how they have omitted such things in the first place. Nor do they know if he has even written them. For they don't devote much attention to the art of medicine itself, but only to those doctrines which are able to increase the nonsenses in their diatribes. Therefore, they choose only the chief points of the books of Herophilus, passing over the other things altogether, and because of this, do not on the whole know what Herophilus wrote, that carries across greatly to the actions of the art, but is opposite to what they themselves say.

Anyway, because of this and what they had to learn about the rhythms while they were still children at their music lessons, they do not on the whole carry this across into their writings on the medical art, suffering the misfortune of late learners, not being able to remain silent, even if the art is different. But none of them has yet written on how it is necessary to take the significations from rhythm. However, Herophilus travelling the opposite road to them, **[872K]** leaves aside those things which someone suitably trained in the art must have learned from the musicians, conversing with them as if they knew, and takes from them what is useful for the medical art. And yet, when they have gone through this remarkable *technologia* (systematic treatment),[581] as they themselves call it, concerning the rhythms, they are no longer at pains to show them how someone might prognosticate adequately through these, or how they might become signs. Therefore, they say the same thing—that the pulse of children is small—while Herophilus writes the opposite, without being zealous to the point of considering how small, as if it were possible to conceive what kind of dilatation is called large or small among people.

For truly, in the case of bladders, coppersmith's bellows, wine-skins, stomachs, uteruses, and absolutely everything that has acquired some cavity, we are accustomed to term dilatations large and small. And often, in fact, we say the size of the stomach is small, or as may happen, the size of the bladder is large, or that of the wine-skin is

581 The Greek term, τεχνολογία is retained in Kühn's Latin translation, and is used by Cicero, *Epistulae ad Atticum* IV16 § 3 (ed. and Engl. transl. Winstedt 1902: 314). It is applied to particular branches of learning, such as grammar.

small, while that of the stomach is large. Sometimes, we might also say the bladder has come to a large size and a very large dilatation, or the wine-skin and sack have, **[873K]** although often they are able to accommodate twenty such bladders within themselves and don't yet have a very large dilatation. For we conclude, I think, that we must compare its distension to the hollowness of each of the bodies, and not simply claim to use one measure in relation to all. For it is possible to have a certain measure in each, in which the walls containing their internal cavities happen to be neither stretched nor relaxed. Therefore, when there is either tension in them, or the hollowness is apparent, we say at that time the dilatations are large and small. When, however, it is no longer possible to still discern a perceptible cavity, but the walls are not suffering under the tension due to the dilatation, we say such a distension in the body is moderate, in due proportion and according to nature. Why, then, is it difficult, starting from these instances, to understand about the artery, as this also exists as a hollow body, necessarily having boundaries of the movement in each direction, and not to be distended immoderately (for it would break), nor to contract to the smallest? Is it the nature of the walls being hard and difficult to contract[582] that prevents this?

4. [874K] There is, then, an adequate dilatation of the artery, such that no part of it is either unduly stretched or lacking in relaxation; we shall say this is the mean and in accord with nature. If not, however, we shall be compelled to say, in relation to one and the same person, that the pulses are large, small and moderate. And yet, in the best nature, it is strange for them not all to be moderate, but much more strange that the same pulse should be large and at the same time small and moderate. In comparing large to the smaller, small to the larger, and moderate to both together, what is right is to call it larger when it is compared to the smaller, smaller when it is compared to the larger, and intermediate when compared to both. In this way, it is possible for us to call the pulses in the best natures larger and smaller than each other, or all moderate, if we remember anything of what was said above, when we said it was necessary to look to the nature of the proposed body, whenever the matter before us is to discover the moderate, the large and small, while whenever we simply compare **[875K]** the sizes of the bodies to each other, to use the terms larger and lesser and intermediate between both, and not as yet large, small and moderate. For the larger is what is more than the lesser, whereas, as the large is larger than the moderate of the same class, so that both terms are used relative to something, but we may compare the larger with all those that are smaller, whereas the large is compared only to the one median of the same class, for if it is larger than that, it is called large.

And we need to conceive of the counterparts to these in the case of smaller and small. It must also be said that the moderate and the median are in accord with the proper nature, just as the immoderate and non-median are contrary to nature. Accord-

---

582 The term δυσσύμπτωτος is listed in LSJ as meaning "not coalescing easily" citing this passage only. Kühn's Latin translation has *contrahitur.* The structure under consideration and the context suggest the translation given above.

ingly, since the artery is a hollow body, it is inevitable that sometimes its distension is proper and moderate for the sake of the use for which it was created by nature, and besides this, of the walls of the artery, while sometimes the distension is accomplished immoderately. However, it is shown in other instances, as in the best natures, it is natural to be dilated to such a degree that is sufficient to fulfil the use without distressing the walls. Therefore, it is necessary for someone who wishes to direct his attention to one standard in the **[876K]** diagnoses of the large and small pulses, having discovered the best nature, and knowing any one artery whatsoever, to remember in this the size of the body and of the dilatation, and to calculate by what amount the artery is likely to be larger than this, and the dilatation in the best natures. Then, if you discover that it will be as large as estimated, call the pulse like this moderate. If, however, it is either larger than this or less, call it large or small respectively. You will know the largeness of the artery in the case of the best nature, either through anatomy (dissection), whether this should occur by chance or on purpose, or by calculating from those things said in the work, *On the Best Constitution of the Body*.[583] And yet, even if you don't know what the artery is in those having the best nature, but have remembered their pulses, then again, in encountering the other, you would be able to calculate whether the arteries are likely to be larger or smaller than that. In this way, it will now be possible to compare the size in the dilatation and to calculate whether the pulse is moderate, large or small. For if **[877K]** the person being compared has arteries that appear to be larger to the extent that their dilatation is larger, we shall say the pulse is moderate. If, however, the largeness of the artery seems to exceed that of the dilatation by more, or is less, we shall call it in this case large or small. If either the largeness of the artery seems larger by much or the excess of the dilatation, or is less, we shall in this way call it large or small. If, however, the largeness of the artey is less, while that of the pulses is greater, we shall call it large. And this also applies to the remaining classes of the pulses, in relation to the movement itself, in which we recognise rapidity and slowness, both in relation to the pause, in which we recognise frequency and infrequency, and in relation to the *tonus*, in which we recognise strength and weakness, and besides these, in relation to the body of the artery, in which we recognise hardness and softness. For in all these, we shall discover the moderate pulse in the case of the best nature and the immoderate pulse, either through excess or deficiency, and we shall examine them by comparison not with each other, but with the moderate.

Book III

1. That the concept and diagnosis of the moderate pulse in each class is derived from that of those which are immoderate, and that in the bodies that are best constituted, all the arteries pulsate moderately, we went over in the book before this one.

583 Galen, *De optima corporis nostri constitutione* (IV.737–749K; Engl. transl. Johnston 2020:434–451).

What signs we must give consideration to, and in what manner we must apply our fingers to the pulses to train the touch for ready recognition of which of these are moderate and which immoderate, we shall speak of next, starting from the class relating to the movement. **[879K]** We have shown adequately through what has gone before, that the amount of time alone is not a sufficient sign of moderate and immoderate movement; rather we must look at the distensions (dimensions) through which the moving body is carried. Then we also utterly refuted those who explained the definition of the rapid pulse badly, although we ourselves did not say how it must be understood better, but now we shall. We shall not need a long discussion. Those things that contribute to the concept of this we explained beforehand through what was previously stated; the discussion will be useful to us in another way for the present matters. Moreover, those people were in all likelihood mistaken who defined for us the movement of the pulse in a short time as rapid, and conversely, in a long time as slow, before the significations of the names were differentiated. They thought the same thing is indicated by much and more, and in turn by little and less. However, since we have shown there are those things said regarding whatever happens on each occasion, and those things said only regarding the mean of the same class, there is not now any longer a problem when we say how the movement of the rapid pulse is circumscribed in a short time, **[880K]** whereas that of the slow pulse is circumscribed in a long time. For if we were to know beforehand how much distension the moderate pulse has achieved in how much time, and then we find that another, having an equal time, has less distension, we shall say that such a pulse has been moved in a long time.[584] So let no one be astonished, if of two movements having the same quantity of time, we say only one has occurred in a long time and not both equally. For surely, of two children who have been sent to different places at the same time, then have returned at the same time, we praise the one as accomplishing the journey in a short time, while we blame the other for accomplishing it in a long time. We say the time is short or long not absolutely, but looking at the length of the journey. It is in this respect, then, that we shall also speak of a pulse occurring in a short time as being rapid and one occurring in a long time as being slow, and we shall accept the stated definitions, explaining them in the manner we have spoken of.

If, however, such a mode seems to be forced by this, remember that the definitions were shown to be false in relation to the other signification of the statement we heard, so let us attempt to make other **[881K]** definitions. For how it should be defined is not now the matter before us to consider; our task is to articulate accurately the concepts of the matters, lest erring in some way in these, we also err regarding the diagnosis of the pulses themselves. Now we have stated what is necessary for the diagnosis. However, if someone should wish to learn to construct some definitions for himself, it would not perhaps be disagreeable, if he were to practise the exercise, although it is not in fact necessary for the matters before us. Wherefore, we shall not say how we

584 There are some minor differences between Kühn's Greek text and F1, Va1 in this sentence.

must define such pulses, for it is not on the whole necessary, even if it should appear to be so, being characteristic of the previously mentioned matter—enough has been said about these matters in the fourth book of *On the Differentiae of the Pulses.*[585] Since I think it worthwhile for the present matters to make mention of the concept only, let us pass on to the continuation of the discussion. For now is an appropriate time to say what we find in them, which contributes greatly to their diagnosis. Rather, if truth must be told, the whole matter will now be stated.

We do not think it worthwhile to consider the amount of either the time or the distension (interval), or even the **[882K]** proportions in the pulses. We have given a detailed account of these in the book prior to this one, when we taught about the more rapid and the slower pulse. There we spoke about the error that inevitably attends those who judge the pulses by the amount of time alone. To me, this seemed to be problematic for the diagnosis—time-consuming, hard to work out, difficult and almost impossible to judge. We recognise the rapidity and the slowness in the first impulse of the artery, but we need a considerable time, if we are to discover the proportion of one time to the other and of one distension (interval) to the other (interval). As a result, we do not await these to draw a conclusion on the diagnosis in the case of those who are sick, but just as if we happen to be observing things that are present, the actual rushing motion of the impulse is sufficient to make the rapidity obvious, as it is in the case of the artery. For it is also possible to find its movement either hastening forward or falling away. We recognise how each of the moving bodies is moved from the places in which they are in each case. When it spends a long time in the places, we say the movement is slow, whereas when it is ceased readily, **[883K]** we say it is rapid.

I know the discussion will bring up an issue of physics—whether the moving body also remains in some way. It is shown, then, in the writings on motion that this is for the most part true. But we must now be on guard against letting ourselves get caught up in such great issues. It is enough for this to be said: that perception alone makes apparent the diagnosis of the rapid and slow pulses; that what is moving appears readily to leave behind the places in which it is each time, but for the most part remains in this; and the rapid is named in this way, while the slow is named in relation to the transitions from the places, since perception makes the distinction not in relation to the time of the whole movement, nor specifically and alone by comparison with the amount. And in this way, the recognition of rapidity and slowness becomes easy for us, which it would not have been, if we waited to consider the amount of times and intervals, taking the proportions in these to consider. But obviously, we do not wait. In relation to the first impact alone on vision, taking and availing ourselves of an inexpressible shortness of time, we recognise the stopping of the moving body, from which we compute the **[884K]** transition. Anyway, those things we see from afar, sometimes seem to us to remain for a longer time, even if they are moving as quickly as possible, from which clearly we recognise the movement by reason and not by percep-

585 *De pulsuum differentiis* IV (VIII.695–765K).

tion. For when each of the first perceptible parts of the moving body remains for the first perceptible time in the first perceptible place, then the thing being observed seems to be immobile. However, when the first perceptible part remains for a shorter time in the first perceptible place compared to the first perceptible time, then it seems to be moving.

Each of the things spoken of can be divided into an infinite number of times according to the nature, while, when measured by perception, some first parts have also acquired indivisibility, by which it is also clear that every movement is recognised by reason and not by perception. But because reason is closely associated with perception, while the transition of the mind is very quick, it often seems not to be recognised by reason but by perception. For of those things in the first time regarding perception, the first perceptible part remains in the first perceptible place regarding perception, of these alone, **[885K]** we agree the movements are clearly grasped, not by sensation, but by reason, just like the shadow in the case of a *gnomon,* moon, sun and all the stars, and all those other things most remote from us. Of those things, however, as was said, the time of the first movement regarding perception is shorter, and only the change of the places seems instantaneous, for everything which occurs in a time less than the first time regarding perception seems to be instantaneous.

These things, then, were better to have been pursued as knowledge privately by one who prefers to learn well those things now before us and the many other things of the medical art. Such things will, however, suffice for us regarding demonstration that not the amount of time and of the distensions (intervals) are to be considered, but the quality itself of the movement, which is hard to name. Beginning from these things, we said, unless you were altogether incapable and entirely without diligence, you would have no difficulty attaining the truth, as far as human capacity allows. Still more, if you were to study the diagnosis of the unequal pulses in relation to rapidity and slowness, you would discover that the genesis in these is in the quality of movement. Anyway, the beginning of the movement of one dilatation often appears clearly more rapid, whereas **[886K]** the next in succession appears slower, but in turn the end appears faster, when we are not counting any quantities either of times, or capacities, or intervals, but are considering only the actual quality of the movement. We must not, therefore, palpate the surface, when we are considering this class of the pulses. In this way, we would perceive only the external limit of the movement and not the whole of it. From these things, it would be already obvious that such pulses need to be, in all instances, neither small nor weak. It is not possible to divide the movement in three ways, either in the small distensions (intervals) or in the weak capacities. In the first, the smallness prevents this, not allowing a division into perceptible parts, whereas in the second, the weakness of the nature sinks down in response to the pressure. Anyway, because of this it is not possible to detect any inequality in the lethargic pulses, although it may perhaps exist, because at all events it needs pressure for the diagnosis, but does not stay behind due to weakness. Therefore, someone who is going to discover the appropriate palpation for each of the pulses needs to be sufficiently intelligent and at the same time practised in their diagnoses. **[887K]**

If the pulse is very large and at the same time very strong, it holds up against strong pressure, whereas each of the others, to the extent that it is defective in either strength or largeness, does so less, to that extent. Then, it is necessary first, to diagnose precisely how much it is lacking of the very large and the very strong, and second, to calculate how much pressure is sufficient to allot to those in this state. For inevitably there is error in each direction, if we either exceed the proportion or fall short. In those cases in which we now compress too little, something of the internal movement will escape us, whereas in those cases where we compress too much, it will be something of the external movement. As a consequence, we sometimes don't recognise the first movement as being unequal; then too, we often give a false opinion about the whole from those things we diagnosed badly in the parts. For if either the first part of this is more rapid than the second part is, while only one of these is recognised by us, we are inevitably ignorant first that the whole movement is unequal, so then we shall think it is either rapid or slow, but not that it is one of these absolutely. For if the first part of the movement is rapid, while the second is slow, **[888K]** someone would not be right if he were to call this either absolutely rapid or absolutely slow.

Since I have brought completion to the discussion and must pass on to other matters, it is not then appropriate to examine closely the moderate pulse in relation to the movement in the case of the best constitution, nor in relation to the quantity of distension and times, but also the quality itself, as we said, bearing this in mind too: that we said how the moving away from the places needs to be considered, and that which seems to move away more rapidly, we call rapid, while that which seems to move away more slowly, we call slow. And this is apparent at the very first application of the touch, before an artery has passed through the whole distension. As a consequence, if the pulse is equal, we do not wait for the whole dilatation to form a judgement about the whole from a part of this. If, however, the pulse is unequal, we said it is necessary to follow the whole dilatation, and to apply the touch differently, as the pulse may have largeness and strength. These things, then, are sufficient for the present regarding the class pertaining to movement. For it will be necessary for us to mention this again in the diagnoses concerning the pulses.

2. Let us speak next about the quantity of the dilatation. Now since there are three distentions (dimensions) of the body **[889K]** of the artery—length, breadth and depth—it is not possible to perceive any one of them as a whole. This is absolutely clear in relation to length, for the artery pulsates up to the extreme limit on each side, while we apply our touch to a short part of it, which is as much of the mass of the presenting body as comes to the surface. However, this itself is what makes the pulse either long or short, since we are not able to touch the length of the whole artery. Rather, even here again, it is clear that in each nature of a body, there is something under the skin, while it is obvious that in the best nature this is the moderate. What is more than this makes the pulse long; what is less makes it short. It cannot be defined by the number of fingers, as some say, if it falls on four fingers, it is long; if on three, it is intermediate; and if on two, it is short. For there is not one measure of the best nature of a body, but in each case you will discover the amount without difficulty, starting from one, only if you are able to calculate what the proportion is. If, in turn, you compare other people to that, those that are equal in relation to the length of the whole body, you will say in this way have pulses that are either long or short. Similarly, **[890K]** you will say they are either broad or narrow, or high or low and in relation to three dimensions at the same time, large or small, for you always judge the others by reference to the mean (moderate). It was said before that, in relation to each of the other classes in the pulses, the mean is not simply one in relation to the quality[586] alone, but in relation to each part of the body specifically.

However, I don't propose to speak of this now, but calling to mind those things said, it is already appropriate to return to the actual matter before us. What is proposed, I think, is to say it is not possible to recognise through perception the amount of any distension precisely. The cause is in part common to all instances and in part specific to each. The common will be spoken about first. Accordingly, since none of the arteries is bare (lies exposed), but fewer and thinner bodies surround them all in many places, while more and thicker bodies surround them in many other places, it is somehow clear to everyone that the perception of the movement will not be equal in all, but more in those which fewer and thinner bodies surround, and less in those which more and thicker bodies surround. So then, this is common. What is specific to each of the distensions (dimensions) [is as follows]: **[891K]** in relation to length, there is a very great difference between what truly exists and what is apparent; in relation to breadth the difference is small; in relation to depth it is least. For in length, although the whole artery often extends for two or three *cubits*,[587] the movement of the artery appears to three or four fingers only, all the rest being concealed by the number of the bodies lying in front of the artery. In relation to the breadth, when it is not possible to apply the touch to the actual body of the artery due to the many or thick bodies lying around it on either side, due either to their thickness or their abundance, the movement does not penetrate to the outside, in many places the movement is lost altogether, while in many other places it is straitened to the minimum, while the whole movement may not be apparent at all due to the artery not being exposed in any way. Nor in depth is the artery in any way completely exposed so that in this the whole movement of it appears clear. Rather, if there is nothing else, skin at any rate lies around it, and we perceive the pulse through this, and it is altogether concealed.

Therefore, whensoever these things are like this, we must attempt to show the method of the best diagnosis in each **[892K]** distension (dimension). Let us speak, then, about the first distension relating to the length. It is very clear that we must consider as the best that in which the most movement is perceived. So, if it were possible

---

586 K: ποιὸν; F2, Va1: ποσόν.

587 A *cubit* is defined as the length of the arm from the elbow to the tip of the middle finger (approx. 18–22 inches or 46–56 cm).

to perceive the whole movement, this would be best. Since this is impossible, when none of the available parts of it escape being recognised, we must give an account of the best palpation. We must attempt to state how someone might do this, remembering first that all the differences of sensory recognition lie in these three things: nature, practice, and manner of application. Thus, one who has naturally very keen perception, and in whom the organ of sense has been trained sufficiently, along with the way he must apply this to the objects being perceived, will be best at recognising those things lying within. On the other hand, one who is lacking in any one of these things, would, to the extent he is deficient, be worse when it comes to accurate diagnosis. What will be spoken about, then, is how someone who is the best in nature and has been trained carefully would apply the touch precisely, so that he might perceive the movement in the artery to the full extent of human capacity.

And since **[893K]** we propose to speak about the first dimension (distension), that of length, we shall state the most suitable manner of palpation for this. Let it be entirely without pressure, touching the surface in many places, and like lifting the hand. Such a method is particularly useful in the pulses that are small and at the same time weak. For if we press on these to some degree, it destroys the movement for the most part, and often altogether. Depending on the quantity of the compression, and in addition the quality of the *tonus* in the artery, sometimes the movement is destroyed to a great extent, but sometimes less, and sometimes altogether. And because of these things, it is safest, for the recognition of the length, to apply the touch without pressure. The strong pulses do bear moderate compression without distress, but there are no pulses at all that are harmed when palpated without pressure. When, therefore, such a manner of palpation is most useful in the weak pulses and harmless in the strong pulses, no one would choose any other method instead of this.

However, it is also necessary to specify this phenomenon clearly, as in the compressions in different parts of the length of the artery, the **[894K]** movement does not seem[588] to be harmed similarly in all. In the parts above, the harm is greater, while in the parts below it is slight, and in fact is often hidden in the strong pulses, although not in those that are weak. In the latter, if we compress the part below, it is quite clear that the harm which occurs involves those parts above. How we must understand the parts above and the parts below here is quite obvious; the parts more distant from the heart are lower, while the parts nearer the heart, we call upper. In the case of the limbs, this is very familiar. However, in the case of the parts in the head and neck, it is not similarly familiar. What is common in all cases, as was said, is that the compression of the parts nearer the heart brings about great harm to those parts immediately adjacent, whereas often with those further away this is not altogether perceptible whenever the pulse is sufficiently strong, although sometimes it is apparent in those pulses that are not strong.

---

588 K: φαίνεται; F1, Va1: βρίσκεται.

These things, then, will be sufficient concerning the distension in length; let us speak now about the remaining two. It is not necessary to think of these in this way as being two, as is seen in the case of four-sided and cuboid bodies, where breadth and depth are distinguished separately, **[895K]** but as in a circle and sphere where all is commingled, and one circumference is apparent, whether drawn or visualised. The same form must be thought of as existing in the case of the artery, for in the dissections, the form appears in many places cylindrical and in many places conical. How, then, did it come to doctors that it was necessary to speak separately about breadth and separately about depth in arteries, as it seems also possible in the divisions to term one dimension the length of it and the other the circumference? Never mind that the majority often also name in this way; it is not always possible either in these, or in any other detailed description of the things which are clearly apparent, to name length and circumference only, but often they will be compelled to say a certain dilatation of this is high but narrow, and in turn another is low but broad. These things are clearly apparent in this way, as no one will dispute. And because of such appearance, we were compelled to have the different concepts of breadth and depth in the case of the arteries. Just as we recognised such differences by perception, so too we often acknowledge **[896K]** that the movement appears to us to be precisely more circular.

Now we shall speak about each of the *differentiae* in common—how someone might make the best application of the fingers and be able to recognise these. For surely there would seem to be no small difficulty contained in this recognition, since we are not able to perceive all of the distension in any one application, not even when we touch the surface, for such an application perceives only the last stages of the movement; everything before this will be unknown; and not when we press sufficiently, for in addition to the fact that it is possible to do this only in the strong pulses, contrariwise such an application will not be perceptive of the last parts of the distension, perceiving only the first parts. In relation to these, it is sufficient if we use some other, intermediate application among those spoken of, that we will be ignorant of the first and the final parts of the distension. That what I say is true, someone who wishes to do so will learn by putting it to the test. For if he presses the strong pulses, and it is permissible to press only these, he will not know how large some are. For in most of the compressions, some of the movement itself is lost, as was also said previously. However, the distension is no longer small, but **[897K]** is almost wholly destroyed. If, on the other hand, you touch superficially, you will not be able to surmise how large a distension the movement was carried to. Such therefore is the difficulty. The resolution of this is not to seek to recognise the amount of the dilatation altogether by perception, but if we are also able to do so by some reason, not to consider this to have been harmful. For what lies before us is not to know altogether by perception those things that have happened to the pulses, but simply to know them. Therefore, all such knowledge requires perception, but not in fact only this, as it is not always possible to discover the whole, although it is often sufficient to create any one sign of the perceptible affections of the whole essence of the matter. Anyway, this is how it is in the case of the matters now before us. For if you were to touch superficially, it is inevitable that you either perceive the impinging artery in more parts or less, after which reason closely associated with perception brings us to a recognition of whether the dilatation is large when many parts move the touch, or small when only a few do so.

For suppose the impact of this to be circular. Would it not be quite clear, if the circumference of the circle is substantial in magnitude, that the whole dilatation **[898K]** of the artery is thick (large), whereas if the circumference is small, the whole dilatation is thin? For the thin dilatation cannot be bounded by large circles, nor the thick by small; rather, it is necessary that the amount of the circle bounding the dilatation be proportional to the amount of dilatation. In the same way, even if the boundary of the dilatation is not precisely circular, but is, as it were, curtailed in part, which is found in the high and narrow, or low and broad pulses, the computation of the amount of the dilatation would become similar. For the limits of the dilatation and those of the ascent to a height will at all events be segments of a circle, and these will either be greater or less on each side, appearing either more concave than the circumference itself or altogether more convex, rendering the line now appearing either longer or shorter than a precise circle.[589] And in this way, we calculate the inequality of the distension in breadth and depth in respect of the amount of the whole dilatation.

For suppose the impacts on the sides of the artery appear more convex and more inclined forward, while somehow more concave **[899K]** and lower in relation to depth, so that it does not still seem to be a circumference of a circle, but like a straight line, it is clear that such a pulse will be broad and low. For the movement in breadth, when extended still further, would in all probability create breadth, while that in height, when impeded, would create lowness. Conversely, if the parts of the movement in the transverse planes fall far short of the mean, the middle itself would appear markedly convex, it is clear that such a pulse will be high but narrow. When the impact of this in relation to its breadth becomes altogether narrow, so that the movement not only seems to have diminished in the transverse plane, but also the whole movement is completely lost, it is difficult, or rather impossible, to know how much there is in these, what kind of pulse it is. For unless we take it upon ourselves to make certain distinctions, we shall not know whether a pulse is high, low or moderate in terms of depth. All are possible. For surely, if a pulse is low and at the same time narrow, it will appear thus; and if it is high and at the same time narrow, it will appear thus; and if it is moderate in terms of depth and at the same time narrow, [it will appear thus][590]. For what is in such a narrow pulse is at all events due to the **[900K]** movement in breadth (in a transverse plane) having been prevented. However, the problem is that concerning the depth; a pulse which has the movement beginning from a large distension, and from a slight distension, and from a moderate distension

589 The Latin text has: "longer, narrower or broader" rather than "longer or shorter". The manuscripts F1, F2, and Va1 have the same as Kühn.

590 Τοιοῦτος φανήσεται, present in Kühn after στενὸς εἴη is absent from F1, F2, and Va1.

will produce the same impression in those palpating superficially. For they will perceive the movement as narrow.

How, then, will it be distinguished? Often by the affections of the touch but always by reason. Things were said previously about the affections of the touch, from which it is possible, due to this, to take something clear. It is possible in the case of the strong pulses, about which we have spoken often, that they are the only ones not destroyed when we compress the artery. It is very easy, then, in such pulses, to know the depth of the pulse. Perhaps to you, when you press, through the touch itself, if that has been trained and is gentle and of an accurately perceptive nature, there will be some recognition of the distension of the movement. However, if not, at least the duration of the movement will indicate the amount of the distension. If it is possible for you to judge specifically the quality of the movement, in which we find the *differentia* relating to rapidity and that relating to slowness, and specifically that relating to the time, in which we recognise movements of short duration, **[901K]** none of these would escape your attention in relation to the compression. If, however, you were to detect both, there would be no difficulty drawing an inference about the distension.

Suppose the movement is rapid but of long duration. It should somehow be clear to everyone that it has been brought to this by the very large distension. Contrariwise, let the pulse be slow and of short duration; this would, I think, be in relation to being brought to the least distension. If the pulse is strong, recognise the amount of the depth in the very same way, whereas if it is weak, perception is no longer able to help us at all, for when compressed, the movement in such pulses is altogether destroyed. Under these circumstances, the whole must be tracked down out by reason. It will be said in the treatise, *On the Causes of the Pulses* that not all *differentiae* are able to be mixed with all others, and it will be distinguished which are and when and how.[591] As a consequence, it will be taught in these to know clearly from the other phenomena also what concerns those that are not apparent, and why a pulse that is narrow and at the same time weak doesn't ever become high. Such things will certainly be sufficient about these matters.

3. We shall speak next about the rhythms, keeping to the same order of the teaching in relation to **[902K]** what is proper for the diagnosis, since surely we must first go over the matters pertaining to one part of a pulse according to their nature and particularly the classes constituting the dilatation, as we did in the treatise *On the Differentiae of the Pulses*.[592] But the classes existing in one part that are left over from what was said require a common teaching, whereas the class of the rhythms is linked to the two classes in the diagnosis previously mentioned, as the actual discussion about these will show. We shall on this account speak next about rhythms, putting off the discussion about hardness and strength to the book connected with this, so that each of the

591 *De causis pulsuum* I–IV (IX.1–204K).

592 *De pulsuum differentiis* I–IV (VIII.493–765K).

treatises has the complete teaching pertaining to the recognition of the classes lying alongside each other.

Accordingly, since the rhythm arises in the comparison of the time of the dilatation with the time of the contraction, it is clear that we must first be able to distinguish each time specifically. But this is not very easy. For if we recall those things said in the first book, when we looked into whether it was possible to perceive the whole of the dilatation and contraction, or some part **[903K]** of each escapes us, we shall know how difficult it is, or rather impossible, to make the diagnosis of the whole time in these. For it has been shown there that the first parts of the dilatation and the last parts of the contraction are unrecognisable. To the same extent that some of the movement would escape notice, I think some of the time would also escape notice. First, this same thing must again be recalled and distinguished—that it is not the same as was sought in the first book and now a little earlier. There the enquiry was simply about the dilatation; here it is about the amount of time in this. Therefore, in investigating simply whether it is possible to perceive the whole dilatation, we shall say what would be unrecognisable is as much of the movement as would escape the touch. However, in the recognition of the amount of time in this, we consider how great the interval is in which it has been moved. And it has been shown a little earlier that, if we have accurate recognition of the end of the dilatation only, by our combining computation with perception, the indication of the amount in relation to this will be made. Therefore, it is possible, although perceiving no other part of the movement except the external limit, being ignorant of how much this is in terms of time, **[904K]** we shall recognise how much of an interval it has carried. One of these, then, had its investigation in the first book, while the other we finished going through in detail not long ago.

Therefore, let us call to mind again, that in the strong pulses, if we compress the artery moderately, we shall recognise the first parts of the movement, but will be ignorant of the amount of the interval. Further, let us also recall this beforehand, which was shown right at the beginning of the second book—it is not the same thing to examine closely a quantity of time of a movement and to recognise rapidity or slowness. For our present purposes, we need all these so that we may know accurately and definitively, if the recognition of the interval in which the movement happens to have been carried differs from the time, considering which one of each of these is either rapid or slow. For the first one stated measures the amount of the interval, while the second measures the amount of time of the movement, and the third the nature of the movement. Therefore, for the diagnosis of the rhythms, we need neither the first nor the third, but only the second itself. For that very reason, therefore **[905K]** it is necessary for the application of the touch to be such that it observes most precisely the whole movement of the artery. As those things in the first book stated, when we were investigating whether it is possible to perceive the whole movement, it is appropriate to have called to mind our knowing that and this are one and the same problem in importance, but are put forward differently in speech. Certainly, if we consider all these things beforehand, we shall know the difficulty of the diagnosis before us.

Let a particular pulse be large but weak. It is not, therefore, possible to detect the movement of the whole distension, but only that of the outer limit. Similarly too, of the contraction, if that is in question, only the beginning and the withdrawal [can be detected] but nothing else. Therefore, it is necessary to palpate such pulses superficially, as has been said often. With such an application of the touch, it is possible to recognise the amount of the distension of the dilatation, but not possible to recognise the time of the movement. In the case of the contraction, it is not only impossible to discover the amount of the time of the movement, but also that of the distension itself. Since this is completely imperceptible, inevitably there would be loss of all the things existing at the same time with it. **[906K]** As a result, not only the amount of the distension and of the time of the movement, but also the nature, in which the diagnosis of rapidity and slowness has arisen for us, will be altogether imperceptible to us. For how would the amount of the movement still be able to be recognised, when there is no perception of movement at all? The essence of the matter must exist before the quality. Therefore, since this altogether escapes perception, it is not possible for the quality to be recognised. How then, in such pulses, is it possible to recognise the rhythm, when the contraction is utterly lost in them, while only the outer limit of the dilatation is preserved?

But not in the case of any other of the weak pulses is it possible to perceive the contraction. Accordingly, since we only recognise the movement of those pulses that are not weak, we shall recognise the artery in the contraction and dilatation of these only, and the rhythm of those only, obviously making the application of the fingers moderate in this way, according to what we said earlier, so as not to hinder any of the external movement. For if we compress the artery more than we need, we shall compel its movement to stop.[593] Otherwise, in these, any rhythm is unrecognisable. **[907K]** If the first parts of the dilatation and the last parts of the contraction escape [our detection] throughout, unless what escapes is altogether short, the rhythm will not be greatly harmed. Perhaps someone might say this too. Nevertheless, there is need of a distinction in the discussion. For what is said would be true in relation to the very strong pulses, whereas it is no longer true in relation to the others, and particularly those that partake only a little of strength. For what escapes perception in these pulses would already also be sufficient to be able to harm the rhythm.

Therefore, from all the things said, it has already become clear that only in the very strong pulses is there recognition of the rhythms. In the case of all the others, it is either altogether unrecognisable, as in the case of those that are weak, or far from accurately, as in the case of those that are less strong but not yet weak. Such then is the difficulty. We must however attempt to remedy this as far as possible. First, then, since the same harm does not occur in the dilatations and contractions in the case of the weak pulses and in those that are insufficiently strong, it is possible for the difficulty to be relieved here in some way. For **[908K]** it is clear that the ratio which the whole time of the dilatation has to the whole time of the contraction is the

593 K: τὴν ἀρτερίαν παύσασθαι; F2, Va1: τὴν μᾶλλον ἤδη ἀρτερίαν παύσασθαι.

same as those parts of these which escape the touch have to each other, and those that are preserved will have the same ratio to each other. As a consequence, we shall not have greatly harmed the prognosis from these. If the whole time of the contraction compared to the whole time of the dilatation is double, as may happen, and is similarly double when what is preserved is compared, we shall have a similar prognosis from the rhythms as we have from the perceptible parts of the movement, and as we would have if the whole of this were preserved. However, in the case of the weak pulses, it is not similarly possible to explain away the difficulty. For truly, if people seek to find resolution of the difficulty by comparing the time of the dilatation with the whole remaining time, it is false. First, we must attempt to state this clearly, and then we must prove in what way we find fault with it. Accordingly, since there are four parts of the pulse—dilatation and contraction, and two pauses, one after the dilatation and before the contraction, and the other after the contraction and before the dilatation—those who say the contraction is unrecognisable to perception, **[909K]** establish separately the time of the dilatation, which they call the time of movement, and separately the time of the two remaining parts, which they call pause and contraction, and the ratio of these three times to each other, they say is a rhythm.

However, of those who say the contraction is perceptible, some compare the time of the dilatation alone with that of the contraction, saying this is the rhythm, whereas some assign to each of the movements the time of the pause after this. Therefore, there are these three opinions regarding the rhythms in the pulses:

- There are those who concede that, from the second movement, they have nothing to say in regard to the stated difficulties.
- There are those who seem to find a solution equally from the first and third movements, because they compare the time of the apparent movement with the whole remaining time of what they call the pause.
- There are those who, because they compare the time of the apparent movement along with the pause associated with this to the remainder, make the beginning of the contraction a boundary of the combination of the two prior times.

In each case they are worthy of wonder. In the case of those of the first group, they only remember the things said about the contraction, **[910K]** forgetting about those things said in the case of the dilatation. [As regards those of the second group,][594] we pointed out that in the weak pulses the amount of the distension in the dilatations is known by inference, while the time of the movement is unrecognisable, since only the external extremes of this come to perception. I don't, then, know how they say they compare the time of the dilatation to the whole remaining time, for the perception of the time of this is altogether small, unless they wish to compare this itself alone with all

594 Galen does not indicate the group being referred to here in relation to his preceding list, as he does with the first and third groups, hence the addition.

the remainder. In this way the rhythm would become for them the ratio of the apparent time of the movement in the dilatation to all the remaining time. However, if they say this, they are saying something which is possible but not in fact also useful, since nothing can be prognosticated from such a rhythm, in which the time of the apparent part of the dilatation is compared to all the remaining time, which is a combination of the time of the external pause and of the contraction, and third, of the internal pause, and fourth, of the non-apparent part of the dilatation. In addition to not being able to take anything useful from such a teaching, **[911K]** these men further confuse the concept of the rhythm, as if someone claiming to be a musician said the rhythm is not the ratio of the time of the *arsis* (rise) to the *thesis* (fall),[595] but the time of the apparent part of the *arsis* to all the remainder. Similar to these, those of the third group err, thinking better than those men only in assigning the time of the external pause to that of the dilatation; inasmuch as they forget that the time of the contraction is unrecognisable to perception in the weak pulses, they err similarly. On the other hand, these men in turn add on a third choice, which is that the first impulse of the contraction is always perceptible, although this is not perceptible in the case of the weak pulses.

Therefore, in relation to all these things, it is inevitable that many difficulties arise in the sects on the diagnoses of the rhythms, and because of this, I think, the followers of Herophilus did not attempt to write anything on the prognosis from these. However, Herophilus himself mentions rhythms in regard to the prognoses in many places, but it is difficult to discover just what he means by rhythm; whether it is in fact the ratio of the time of the dilatation alone to the **[912K]** time of contraction alone, or whether he assigns to each of the movements the pause which follows it. And because of this, it is not agreed by those called Herophileans by him what they ought to think about these matters, for neither does his own statement show one of the two clearly, nor is the nature of the matters sufficiently credible. Therefore, if we preserve the other of the opinions, following the musicians, we shall add the times of the pauses to those of the preceding movements. If, however, it is for the use in regard to the prognoses, if we compare the one movement to the other, we shall consider the time of the pauses individually. Therefore, when either one of the two is necessary, either to destroy the concept of the name or to overturn the use, we shall find ourselves in a difficulty that is hard to manage. Nevertheless, it is necessary, since it is appropriate for one of these to be chosen, to prefer the one that is useful for the prognoses, disregarding the name used among the musicians. For what harm will we do, if we judge beforehand from reason the time in relation to the movements, although we have no specific name for it? The art is harmed, not because we are at a loss with the names, but because we are unable to prognosticate with them. **[913K]** Accordingly, for the sake of the use, let the rhythm lie in the ratio that exists between the time of the movements, since we shall be investigating separately at some time the doctrine of Herophilus. But even if we were to establish this, the difficulty would still seem to remain, since

595 On these terms, see von Staden 1989: 276–279.

neither the whole time of the dilatation can be known, except in the very strong pulses, nor does the contraction have a perceptible beginning in all cases.

How, then, did Herophilus first lay down a certain time regarding perception, and using this to measure the others, say they are either two or three or more, or are either complete, and as they call them *aparauxōn* (not subject to increase)[596] or also diminished by a little or a lot or a great amount? For he seems to write these things as if accurately recognising the times in the case of all the pulses, either of the movements alone or also of the pauses after them, for it makes no difference regarding our present difficulty, unless we must hold that he writes such things not for all cases but for those where it is possible and for those alone. Perhaps we shall consider this specifically at some later time. For the present it is not our task to examine the opinion put forward by Herophilus, but to discover what is at the same time true and useful. **[914K]** And we must attempt to demonstrate this clearly, as was found by us through experience in the actions themselves. And this is as follows: If we compare the nature of the movement in the dilatation with the nature of the movement in the contraction, we find the nine primary *differentiae* shown in the diagram; in each of these, however, there are many others unstated in the discussion, but recognisable in terms of more and less in the actions themselves. What kind of thing each *differentia* naturally indicates, we shall state in the writings on prognosis. But now we shall first subjoin the diagram,[597] which will be readily clear to those practised in the first book of *On the Differentiae of the Pulses.*[598] Next, we shall mention how in turn many differences in quantity exist in each *differentia*, although they are of the same class.

Table 7: Tabulation of the nine possible combinations of speeds in dilatation and contraction.

Dilatation	Contraction
Rapid	Rapid
Rapid	Moderate
Rapid	Slow
Moderate	Rapid
Moderate	Moderate
Moderate	Slow
Slow	Rapid
Slow	Moderate
Slow	Slow

596 Kühn's Latin translation retains the transliterated term here (ἀπάραυξος). LSJ gives the meaning in parentheses above, citing only the present passage and adding "dubious".
597 The table is set out slightly differently from Kühn's text for greater clarity. It is in fact the form used in all three manuscripts followed here.
598 The work on prognosis from the pulses is the fourth of the long treatises. The tables relate to those given in the first of the long treatises (*De pulsuum differentiis* I [VIII.493–565K]).

Accordingly, since the first *differentiae* listed in the diagram have each of the movements rapid, it is possible that the movement in dilatation is extremely hastened, while that in the contraction exceeds the moderate and median by a small amount, or conversely, that in the dilatation exceeds the median and moderate by a small amount, while that in the contraction does so by a very large amount, or both do so by a small amount, or both by a large amount, or one by a little, and the other by a lot, or by a little less, or much less, or otherwise in any way whatsoever. For the variation is obvious. And it is not possible to say either that this does not appear to happen or that it naturally signifies nothing. For it does happen and it signifies a very great deal. It is customarily sufficient for us to prognosticate from such a diagnosis without requiring the primary movements to be complete and augmented. That such a diagnosis does not need the whole **[916K]** movement, but might also be possible through a part of the movement is clear to everyone, unless at sometime it should be unequal. Under these circumstances, sometimes the part is not equal to the whole. Nevertheless, in respect to the extremely weak pulses, it is not possible to use such a movement. For it is not possible in those either to recognise the first beginning of the contraction, or in fact, any worthwhile part of it.

Book IV

1. [917K] Those things existing in the pulses by virtue of the specific reason of their essence have all been stated previously. We shall now speak about the other things, which cannot be recognised apart from the movement of the arteries, but are not specific to this. Certainly, when we palpate the artery in the dilatation something like a blow to the touch occurs in us, which, whenever it is forceful, will seem to you right to call a beat, although this is not so, if it is without force. We must consider carefully whether what we say is a forceful impact is truly a beat. **[918K]** For it seems to follow not a few kinds of pulses; hard and strong, large and rapid, high, spasmodic and clonic. Therefore, so that no one, being deceived by the common symptom, should err in the diagnosis of these, we must consider all of them carefully. If it is possible to determine them by other signs, having learned those, we would already have everything. If, however, it is impossible, it will at least suffice to know this is common to many pulses. The best principle of the speculation is to learn the cause by which one common affection seems to follow many pulses in this way—that is, the forcefulness of the beat. So then, since it is necessary for the touch to be thrust away by the pulse, there is going to be perception of the impact of this. However, it is not thrust away similarly in every case, but sometimes more, sometimes less, and sometimes such that, if you only touch it, these movements thrust against and upset the touch very much. These create an impression of force of the beat. However, when the movements are able to thrust very greatly, by the hardness of the artery bruising the touch, as it were, and by the strength of the capacity moving the arteries, causing upset, as it were, and by the magnitude **[919K]** of the dilatation reaching to the maximum extent, and by the rapidity of the motion, thrusting briefly, and due to the height, they sink down inwardly, as it were, and due to the tension are difficult to move, and due to the *clonus*, along with tension, hardness and rapidity of the impact, they have this appearance.[599] It is, then, possible to encounter every day many doctors who say that all the pulses mentioned are strong.

If, then, they were aware of the differences in these, but nevertheless they all wished to call them by a common term, we would not mind at all. Since, however, as they apply a common term, and in this way think there is one nature of all, we must attempt to show them their error, beginning from those things that are most obvious; for example, from the rapidity. It is easy here to detect the error for someone who is able to understand that the rapidity of the impact is not the same thing as the forcefulness of the beat. For, among other things and for the most part, the rapidity is not able to thrust against the touch if the pulse is not also large. From this it is clear that it does not have this symptom by virtue of its own nature. For whenever it is rapid alone, not due to largeness existing, but by the presence of largeness being added to it, **[920K]** how would this be called specific? Anyway, the pulse that is rapid and at the same time also small does not thrust against the touch very much; certainly this is very much more characteristic of largeness than occurring *per accidens.* For if we were to call to mind those things we have said about the recognition of the largeness, it was not recognised by how much it withdraws inward, but by the length and at the same time also breadth and depth contacting the touch in many parts.

If we recognise by perception that the magnitude is sufficient in relation to the circumference of the pulse, it now remains after this that we have inferred the ascent from a great depth, which is not followed by the touch for the whole interval; rather, we only perceive the outer termination. Anyway, if we should wish to perceive the movement from the first impulse of the dilatation in the artery, which is very large, as was also said before, we shall be prevented [from doing so]. As a consequence, the pulse no longer seems large to us. But this itself is not possible to do in all the large pulses, apart from those that partake of strength with the largeness. For in those that are large only and not also strong, it is not possible to encounter them in their ascent, nor **[921K]** to compress the artery, for they are wholly lost immediately.

It is clear, then, that the largeness is not recognised by how much the pulse pushes against the touch inward, but by its falling upon the touch in many parts of the length and circumference of the artery. Is, then, the high pulse forced inward still more? And further, what the large pulse has acquired in the three dimensions, has itself been acquired in one. If not that, then not this. But the large pulse was shown not to be able to thrust against the touch still more without strength. Therefore, the high pulse would not be able to. However, if with the large pulse there is often weakness, while with the deep pulse but not with the large pulse there is in many instances strength and

599 This is one of Galen's particularly long and convoluted sentences extending over half a Kühn page. It seems to be a list of factors which create or contribute to a forceful pulse beat, but the translation is somewhat tentative.

never weakness, this is another discussion. For the deep pulse will have, as I said, greater thrust, not by reason of the height specifically, but because it is altogether strong, or at least is not weak. Therefore, all the pulses mentioned seem very much to require strength in addition to thrust against the touch. And because of this, those of the pulses that have strength throughout always beat strongly, whereas those that do not, don't always do so, as for the most part these also appear strong in many instances to those **[922K]** who consider such things more carelessly, believing that such things always exist. However, other pulses in turn beat forcibly due to their inseparable hardness, just as the clonic and spasmodic do. For these have the forcible beat for no other reason than their inseparable hardness. And these two seem to be the only kinds of pulses that beat forcibly: [that is] those that are strong and those that are hard. However, there are those who have said that even in these there is need of strength in addition to the hardness, for hardness alone cannot make a strong beat, just as it cannot if you bring in a stone gently, for there is also the need of a certain *tonus* in this, if it is going to beat strongly.

But it is legitimate to ask these people what it is they mean—whether there is no body that is altogether hard, but this term hard is said by people without reason and there is no matter underlying it, or there is a matter in relation to what is said, but nothing of this is recognisable by perception, either by some other sense or through the touch. All these things are paradoxical: that there is no body that is hard; that [hardness] **[923K]** is recognised more by reason than by perception; and that it is recognised by some different sense of perception more than by touch. And if in fact it is agreed there is something hard and it is recognised through the touch, what remains to be discovered by arguing about the name, saying that without *tonus*, is it not possible for the beat from this to become forcible? For if they don't wish to call it forcible, they will be compelled at any rate to say something like bruising, or thrusting, or overturning. What is being disputed is in fact the *tonus*. To begin with, they are not able to agree what the *tonus* in the bodies of animals is, in respect to those things that are unbreakable in their whole substance, whether least particles, atoms or *anarmoi*.[600] For it is agreed only by those who say this is a unity. As a result, the *tonus* is not distinguished by perception, but if it were, through what sign or signs or in relation to what indication or demonstration would it be credible? All of us, as people, recognise a hard and soft body by perception. And it is legitimate to enquire not about the existence of these, but about the recognition; and more, not about this itself, but about how a person might explain to someone else the affection occurring to himself in relation to the touch in the **[924K]** impact of the hard bodies, and in what way this differs from those that are strong. But since this has been neglected by those before us, let us attempt to do this, establishing as a beginning to the discussion, those things agreed by all people.

---

600 Galen here raises the question of what constitutes *tonus* in structures if they are constituted by particles and void, as envisaged by atomist theories in their various forms.

Accordingly, since we call certain things strong that are not only animals, like lions and bulls, but also some inanimate things, like the downward flow of rivers and large winds, while other things we call hard, like bronze, iron and stones, it must be remembered, what it is at anytime from each of these, when different affections occur in our bodies from our associations with them. For the force of the impact is common to both —for example, from strong winds and from hard bodies—while what prevails, I think, in relation to the winds and the flow of rivers is the disturbing effect on ourselves. On the other hand, in relation to stones and wood, it is compression and bruising. In the case of the former, even if we are very willing to walk against them, we are altogether constrained and disturbed and are thrust forcibly backwards apart from the compression, bruising and pressure on the parts of the body. In the case of the second, however, it is this **[925K]** compression, bruising and pressure that prevails. You would especially learn this, if, on one of the parts of the body, you placed a stone that was not very heavy, but such that, if you wished, you could easily throw it off. You will recognise how, when it lies there, it weighs down, compresses and bruises and as it were goes down into the skin, while when it departs, it leaves behind some footprint and sign of the compression involving the part. For the part appears hollower than it was and, in most instances, the resting place is preserved, and in addition to this, it is also livid. The wind in fact, even if it happens to be very strong and disturbs someone,[601] and even knocks him down, does not somehow condense and create a hollow in the body. For being soft, it readily divides and flows around. Because of this, it often easily drives forward and overturns the largest boats but does not produce a greater hollowness in any of the wood in these, as it would if it were a stone or iron falling on them. These are able to bruise and shiver into pieces, but are not able to thrust away and overturn like winds. Therefore, those things which would on no account yield to the nature of flesh we have, we call hard, whereas in the case of those things which would yield to the flesh itself, like water, have we not **[926K]** called these soft? And this is what Plato has to say on the matter: hard things are those to which our flesh yields; soft things are those which yield to our flesh.[602] Suppose then our touch falls upon something moist. It will somehow be clear to everyone that it will sink down into this, whereas the moist thing will yield and give way. On the other hand, if the touch were to fall on something hard, it is clear that it itself will sooner yield and become hollowed.

Therefore, so that we now complete the discussion, let us think of our hand in encountering a swift flow of water at sometime or a strong wind blowing from the opposite direction, while at the same time a stone is applied gently to the other hand. Would it not be clear from the former that the hand will be upset but not compressed, whereas with respect to the stone, the hand will be upset and compressed by it. For the hard object in itself is not enough to prevent us going forward, but due either to the weight

601 Following K: ὢν τύχῃ: κἂν ἀνατρέψῃ τινὰ; F2 and Va1: καὶ ἀνατρεπτικότατος
602 Plato, *Timaeus*, 62b (Engl. transl. Jowett in Hamilton and Cairns 1989:1167).

or the external movement acquired by it, it is able at times to constrain us. On the other hand, the strong object in itself is conceived of as constraining, preventing, thrusting against and upsetting. It is clear, then, that one sensation occurs in us in the association with strong objects and another in the association with hard objects. For the latter only compress the flesh while the former restrains us somehow from going forward. **[927K]** Often, however, both come together in the same object, so that, if what is hard falls on us at the same time as being strongly moving, inevitably our perception becomes twofold from these, the one as compressing and the other as upsetting. If this is so, as indeed it may be, one must posit that the recognition of the hard pulse is in the touch being compressed, so that the force of the beat is common to both, while specific to each is the compressing of the hard and upsetting of the strong. These certainly are the affections of the touch and the customary names for them among the Greeks.

2.[603] Someone might also be able to indicate these matters through other names, just as of course those who divide the continuity of the whole substance by the void easily do;[604] that is, all who say these same bodies, which are small, undivided, indivisible and without joints, sometimes come together and sometimes separate, create the generations and destructions. For these people, as they know nothing of vital *tonus*, say that when these small bodies are crowded together and many, and are swiftly carried, then the strong beat arises from them, just as **[928K]** in the case of the winds and the flows that are carried downwards and impact. What they call "crowded together, many and dense" they say is the same as what is dispersed to some extent in empty space and is the solid substance by which the empty space of the arteries is filled, for the emptying of this, as the name itself also indicates, is named after the empty space, as this is always contained in all the comings-together. In this, there is less or more and sometimes the bodies are hard and sometimes soft, and now the internal spaces of these are full, and in turn, should it so happen, they are empty.

Therefore, they say the wall of the artery becomes either hard or soft due to the amount of the empty space which has been intermixed with what has come together, while the internal space is full or empty due also here to the amount of the interspersed emptiness in it. They say a certain pulse is large, but does not in fact have a firm impact, when the empty space in the cavity (lumen) of the artery is very great, while the primary bodies are few. However, it is large and at the same time also upsetting, when the empty space is little but the bodies are many. It has, then, already become clear **[929K]** that it will make no difference for those people to say the pulse is full or strong. For truly, the affection of the touch is one and the same, and the nature of the pulse is one and the same. Thus, when there is a small amount of empty space present in the lumina of the arteries, due to the abundance of the masses (*ogkoi*), the

603 Kühn's text erroneously numbers this section Γ'.

604 This is a further reference to atomist theories. Galen goes on to give what he considers would be an Atomist's account of the pulse in terms of strength and fullness.

pulse is called full, while due to the strength of the beat it is called strong, so that among the names, one is indicative of the actual essence of the pulse, while the other is indicative of the nature of the beat. It is not, therefore, that there is one pulse that is full and another that is strong, for by the two names, one matter is signified relating to different concepts.

But we, on the other hand, who say there is a certain vital *tonus*, will necessarily agree that there is a twofold nature of the pulses, if we also consider the diagnosis of these more carefully. However, the consideration becomes particularly difficult by virtue of the fact that none of our predecessors has thought it worthwhile to write anything about diagnosis of pulses. They have left out some aspects altogether and touched on others only slightly. Thus Herophilus, when writing of the affections of the perception, never mentioned fullness, which he never perceived. However, those who have followed him and are called Herophileans, **[930K]** wish to say anything rather than grant this, almost all of them being sophistical and loquacious, and not eager to put into practice any other of the Herophilean theories in the actual actions of the art. And in addition to the many other outrages they have inflicted on the art, there is also the full pulse, which they introduced into our diagnoses, so although the art is clearly small, we do now have something we should do. For it is neither possible to pass over a pulse given credibility in this way by men of repute, nor must we establish what we have never yet readily found to diagnose in practice, following these strange assertions. But first it is necessary to show that this pulse proceeds as far as the concept, and then that Herophilus himself did not recognise it by perception. Therefore, if they themselves did not confuse the signification of the name, I too would now have something to say about those I denounce. However, since they are variously confused on this matter, neither agreeing with each other, nor each with himself,[605] it is first necessary for us to give consideration to the signification, lest in some way we conceal these things from ourselves and, like ravens or jackdaws, present only sounds, **[931K]** while being entirely ignorant of the matters in relation to which these are said.

Therefore, since there are many who write the term "full pulse", I am resolved to pass over the others, unless somehow I shall mention them briefly in relation to the necessity of the discussion. It will be enough for me to have the writings of Archigenes and Agathinus to hand for consideration. Archigenes speaks about it in this manner: "There is a full pulse which displays a firmer artery and the collapse of this when stuffed juicily, whereas the empty pulse makes the rising of the artery bubbly, so that, in response to the pressure of the fingers, it collapses, lacking a solid foundation."[606] Certainly, in this description, first what the term "firmer"[607] indicates is not very clear, I

605 Following πρὸς ἀλλήλους, the manuscripts F1 and Va1 have τε καὶ πρὸς ἄλλους.

606 Here, and in subsequent instances, the quotation marks follow Kühn's Latin translation as indicating statements by Archigenes.

607 The two terms are ναστός (close-pressed, firm, filled full) and κενός (empty). Both are somewhat exceptionable to Galen in relation to the pulse, but particularly the former, which he considers in what follows.

know, because it is not a term customarily used by the Greeks to speak about such a matter. For they called some bread "firm", but I know of no other body named in this way by them. However, Archigenes himself, for it is most proper to learn the custom in the names of this from him, seems to me to use the term "firm" (*nastos*) instead of "full" (*plēres*). Anyway, in the discussion about a strong pulse,[608] **[932K]** when describing in full what was said by Magnus in the statement, he writes in this way: "For in general," he says "the strong beat does not occur, unless what makes the impact is firm, large and carried quickly."[609] A little later, he says this: "And the strong pulse is made up of rapidity, largeness, strength and fullness." The "firmness" above, he has now called "fullness", as if there is no difference speaking this way or that. But also, he seems to me to use the term "firm" everywhere in the same way. Therefore, let us now change "firmer" to "fuller", so that we may understand what it is he says. The statement is as follows: "There is a full pulse which indicates the artery is fuller." Although this is clear in itself, it teaches us nothing more.

Therefore, it would be better for us to go on from the whole matter of "firmer" to consider what "juicily" refers to.[610] For the statement seems to indicate something like this: the full pulse indicates that, in the impact, the artery is stuffed full juicily (succulently). Whether this "stuffed full" is in turn the same as "filled full", or **[933K]** indicates something else, I don't have a clear opinion on. And certainly, when next he set (contraposed) the definition of fullness and firmness[611] against that of emptiness, he didn't take the latter to be the opposite to firmer or being stuffed full, which I think he ought to take, if he were defining an opposite matter. Now he took nothing; but what does he say? "An empty pulse makes the rising of the artery bubble-like." For this "bubble- like" is opposite neither to the "firmer" nor to the "stuffed full". But if, therefore, it is only opposite to the juicy (succulent), the juiciness (succulence) is somehow like being more moist, while the bubble-like is more air-like. Contrariwise, it is attached to the discussion of the empty pulse, "so that, in response to the pressure of the fingers, it collapses, lacking a solid foundation." You would not find this is the opposite for any of those things from the full pulse. For to the term "lacking a solid foundation" the term "resistant" is the most antithetical.[612] But this is altogether left out in the definition of "full". And yet, it was possible for this to be said in the same way—that is, in relation to the pressure of the fingers being resisted (having a firm foundation), it collapsed. But it was not said unless in some way this is to indicate

---

608 In Kühn's Latin translation this is taken to refer to a book *De vehementia pulsus.*

609 These two statements given as quotes are exceptions to the comment in note 606 above.

610 Following τὸ ἐπιφερόμενον, the manuscripts F1, F2, and Va1 all have ἐν τῷ λόγῳ τὸ τῆν ὑπόπτωσιν αὐτοῦ διασεσαγμένην, absent from K.

611 "Firmness" (τοῦ ναστοτέρου) is omitted in the manuscripts F1, F2, and Va1.

612 The two Greek terms, in the genitive case, are κενεμβάτησις and ἀντίβασις. The former is from the verb κενεμβατέω meaning "to step on emptiness, reach a cavity, lack a solid foundation (metaphorical)". LSJ also lists it as "emptiness of the pulse" citing this passage and *De pulsuum differentiis* I.6 (VIII.509K).

the artery being firmer, **[934K]** so that in each definition, it is not made clear at anytime what the full pulse is, but only how it is diagnosed.

If this is the case, perhaps I may now be able to follow. For if the discussion is an explanation of our affections relating to touch, in no way engaging with the actual nature of the matter itself, I am able to be reminded of all those things the touch senses in the dilatation of the artery. Perhaps, then, we shall discover in this way what it is that Archigenes wishes to make clear. The things he experiences are these. What is perceived is the impact of a moving body on the touch in a certain time, and this sometimes appears to impinge on it in relation to many parts and sometimes not, and whether to compress and again not to compress, and when we compress, whether it resists or yields.

These are the primary factors, and as it were elements, of the whole discussion about the pulses and the affections of the touch. All the other things are either compounded from these or are conceived relative to something in this, or are the parts of those things mentioned compared with each other or the whole with the whole. In relation to which of these, then, is the term "full" said? Is it to the movement itself? But in that there is rapidity and slowness and the origins of the rhythms. But when in relation **[935K]** to many parts does the touch perceive the dilatation? At the time when the pulses are long, broad, high and large. But when in relation to few parts? Here obviously the opposites of those things just mentioned apply. But when is it compressed? When it is a hard pulse. But when it is upset? When it is a strong pulse. What need is there to talk at length about the opposites to these or the means? When, therefore, will it become full? What affection of the touch is experienced? Will it be when more heat or more sharpness is perceived? But such a diagnosis is not from the *differentiae* relating to the pulses. When is it moist or dry? But these affections are not among the *differentiae* of the pulses. For surely, even apart from the pulses, it is possible to diagnose them by placing the hand on the body of the sick person. Does he not, then, term the kind of resistance one thing in the strong pulses but another in the full pulses? Perhaps Agathinus also means some such thing. But if this is the case, he ought to have made this very thing clear to us and not put forward sphinx-like riddles. Or does he not only wish us to enquire into matters that are difficult in this way, but **[936K]** to consider into the bargain what they say, and spend a great deal of time on this, or on the discovery of the matters themselves?

Therefore, let us now look also at the statement of Agathinus and enquire as to whether he means there is one kind of resistance in the strong pulses and another in the full pulses. He himself writes in this way: "What follows in the considerations of the full and empty pulse is this: on the one hand, the tension and straining of the full pulse is due to the *pneuma* being present throughout, while on the other hand, the falling away of the empty pulse and the apparent disappearance of the resistances is like some watery bubble seeming to burst."[613] By this we know we should be grateful now to the extent that he does not put forward riddles like the sphinx, as Archigenes and many of the others do. For he is right in saying it is not possible to investigate this at leisure. Therefore, what he says is now clear, for he understands the tense and straining *pneuma* throughout the whole is seen in the full pulses, while in the empty pulses, there is weakness, readiness to collapse and no resistance. Whether the innate and vital *pneuma*, which is throughout the bodies, as they themselves understand, is tense, or the airy and material [substance] in **[937K]** their lumina is under tension, the statement does not make very clear. The universal opinion of men is evidence that what he is talking about is the vital *pneuma*. But whether it is this or that, we must call upon him and those who follow him so far as present purposes are concerned, not to mix doctrines with the affections of the senses, but as an Empiric would do, not to value experience lightly. First, when the actual things said in relation to touch are discussed with us, then in turn he should teach us how, from these, we must, through reason, move to the doctrines. Therefore, what affections of the senses will there be in relation to this, if we were to change the name of *pneuma* and say beat? The argument will be of this sort: the full pulse presents throughout a beat that is tense and resistant, whereas the empty pulse is like flowing and giving the appearance in the resistances which seems like some watery bubble bursting.

Having grasped these things properly in this way, let us turn the pages of this little book of his and see what he says about the strong and weak pulses. The statement is as follows: "On this account, then, it is immediately clear that the strong pulse is one which strikes the touch forcibly and disturbs it by leaping against it strongly; this is at once clear, **[938K]** and is agreed upon by the majority." Then, moving on, he says: "Clearly, if the opposites are thought of, the weak pulse approaches the touch feebly and lightly."[614] Agathinus is right on these matters, but he doesn't differentiate those concerning the full and the empty pulse. For the beat being "captious"[615] throughout is the same as striking the touch forcefully and disturbing it strongly, one of these being said of the full pulse and the other of the strong. Moreover, dying away in the resistances is the same thing as appearing to the touch feebly and lightly. The first of these is said about the empty pulse and the second about the weak pulse. Thus, Agathinus explained the affections of the touch in this way, in relation to the strong pulses and the weak, but Archigenes in another way, writing thus: "So that if one beat is estimated in relation to itself, the strength that exists is, as I said, the *tonus* of the movement of the arteries." Then next: "The firmness of the impulse appears in relation to

613 This is assumed to be Agathinus' statement although it is not indicated as such in Kühn's Latin translation.

614 This is taken to be a quote from Agathinus' book referred to earlier.

615 The Greek term used here, ἐξεριστικός, is somewhat unusual. LSJ lists two meanings: "captious" and "stubborn" while Kühn's Latin translation has *contentus*. Perhaps "resistant" would be a suitable English rendering here.

the whole dilatation, and if we should press our fingers against it, **[939K]** the beat falls more strongly."[616]

These things are, in this way, clear concerning the affections of the touch. Among these, we shall have nothing specific to say on the affection of the fullness, except what some of those of the present time said—that whenever the pulse rises up still further, it is resistant. However, there are two kinds of pulses, not one. Rising up to the greatest extent occurring in the three dimensions makes the pulse large, whereas, when it is in one dimension only, it makes the pulse either long, broad or high. However, resistance is, in turn, the same thing as strength. If someone should wish to call the pulse that is large and strong at the same time by one name, let it be called the full. Let us not differ about a name. However, let him know that, if he introduces from without something now besides the stated *differentiae,* taking two or three of these to apply as specific names should he so wish, there is no objection. But our investigation is not about names but about matters. If someone has something more to say on these, we shall gladly learn. To conceive of some difference of the pulses according to the content of the artery, for they call in this way the fluid contained in it, I fully agree with, and am able to think of it as sometimes being thicker, **[940K]** and sometimes thinner. And in turn, it is sometimes more viscous, or watery, or more or less in amount, or more airy, or however otherwise it might be. I also conceive of *pneuma* alone being contained in the arteries, and of this sometimes being pure and ethereal, and sometimes hazy and vaporous. But what lies before us now is not to take on the concept of what is possible, but to enquire into the reality of these things. For to conceive of opposites is possible, whereas to conceive of what exists at the same time is impossible. For example, we can certainly think of the arteries as containing blood and as not containing blood. We would not be enquiring about what we had not conceived at all. But what lies before us in those things we are seeking is not this—that is, to come to a concept of them—but to demonstrate their existence. For there is no one who would not conceive of blood being contained in the arteries; what in truth we are seeking is whether this is so in a state that accords with nature.

I know many things were said about this by the followers of Archigenes. And yet, if this is recognised by perception, they talk a lot of nonsense in vain, for it is possible to say outright that Erasistratus speaks falsely about the things we learned through the senses, **[941K]** in attempting to persuade us to change our position on these by reason. But no one would dare to say that the fluid in the arteries is perceived by the touch. How I wished this was so above all, so I didn't have to deal with matters, when I was seeking to discover the truth by reason. I do not understand, then, what it is they wish to say: "The pulse is full when it displays collapse of the artery that is stuffed full juicily (succulently)."[617] Is the touch itself able to recognise this, or is the touch being affected what affects, and is able to provide the starting point of the inferred af-

616 This second statement is not given as a quote in Kühn's Latin translation.
617 See the definition given above at 932K.

fection towards an indication of the condition of the arteries? But if they say this—that it is possible to draw inferences on the conditions in the arteries from the affections involving the touch—I would not gainsay them. For Praxagoras does this and Herophilus and all but a few, some more, some less, some worse, and some better. Perhaps that Praxagoras does so may seem surprising to you. For he does not say the humors are contained in arteries. Nevertheless, from the pulses, he attempts to make calculations about certain kinds of these. But nowhere does he introduce a common affection of sensation. For he would, in this way, **[942K]** make the discussion enigmatic, so that someone reading it would either not understand, and think something profound and remarkable is hidden in the discussion, or would obviously be worn out by seeking continuously but finding nothing. For how would he find what didn't appear at all? So, my friend, diagnose not only kinds of humors from the *differentiae* of the pulses, if you wish, but also the characteristics of the soul, just as Hippocrates attempts to do. Only preserve for me the affections of the senses, but neither dare to speak falsely about what was not experienced at all, nor claim to recognise from one affection one pulse or another.[618]

There is a certain pulse which has so much resistance that, even if you compress the artery, it is not moved by the hand. Watch for this one, calling it what you wish. You may call it now strong and now full, or in turn perhaps something else, and by the multitude of the names, think you are doing many things very extraordinary. Even I have signs through the pulses of kinds of humors and of affections of the soul, but from those things I spoke of before, I attempt to diagnose all the perceptible affections, without ever at anytime **[943K]** calling a pulse precipitate, light, heavy, of moist appearance, airlike, *diēgkōnismenon*, strained, *mogulon*, tight, nodding, loosened or *ropalōdē*, or any other of those things they fashion with their logical concepts, but do not recognise by sensory affections.[619] So now also the perceptions normally confined to hearing alone he transfers to the touch. For to speak of shrill, noisy and rumbling pulses is nothing other than to discourse about touch as if it were hearing. Clearly, then, in this way they make truth itself untrustworthy.

Some man of the present time, and by no means the most foolish, said that the theory of the pulses is gossip, and what he says is very true, because these people speak many words but teach nothing useful. But even if not in their discourses, at least in their writings, they ought to feel ashamed to suffer what the Comic poet said:

618 For Praxagoras' fragments on the pulses and related matters, see the sections on Physiology and Dietetics, and on Pathology, Diagnosis and Prognosis in Steckerl 1958: 55–90. Also Lewis 2017.

619 These are taken to be terms invented by Archigenes; where no attested translation exists, the term is simply transliterated, as in Kühn's Latin translation. The penultimate and antepenultimate terms are taken to be respectively from the verbs περιλύω (to loosen) and κατανεύω (to nod). The term ῥοπαλώδης is listed in LSJ as "term for the pulse invented by Archigenes," citing the present passage; no translation is given.

Excellence in prattling but entirely unable to speak.[620]

So, you write for me "full pulse" but are not able to explain with sufficient clarity what you are trying to show—whether **[944K]** you are indicating the quality of the wall of the artery, or the amount of the actual substances poured in, or the kind, or of the *pneuma*, or also of a certain capacity of the heat, or quality. For you seem to me to say all these things inarticulately and in a confused manner. What do you wish to call to mind by using the fullness of wine as an example? Why, in the discussion about fullness, do you sometimes speak with these terms in this way: "So that also, if someone should wish to look at fullness not in relation to substance but in relation to capacity (potency)."[621] And a little later: "Perhaps the torpor of the capacity in such cases characterises the full pulse, these being names for the *dyskrasia* of the *pneuma*. And in the same way, wine-tasters, when they are tasting wine, recognise the fullnes of the wine, so too those experienced in the pulses observe closely the fullness in relation of the torpor of the capacity, even if it is 'semi-empty'."[622]

In these words referring to the capacity, he seems to me to be thinking about the full pulse, whereas in the same words about fullness, he mixes everything up and brings confusion at the same time. Someone might learn from the actual statement, which is as follows: "In the way the consistency of wools is apprehended on the spot, in relation to which **[945K]** they are said to be empty or full, the quality of wine is also recognised not by being prescribed but by being applied straight to the tongue, in regard to which we separate the full and empty wines, and we apprehend the body of the wine as empty or full, and as one thing in the accord with nature, but another in the conditions contrary to nature, and in the case of inflammation, oedema and inflation (*emphysema*),[623] we also distinguish in this way the unexercised tissues from those that have been exercised."

Conversely, in the same statement, the things said about wine would seem to be consistent with the fullness in relation to the capacity, whereas those said about inflammation and oedema would seem to be consistent with the fullness relating to the actual body of the artery. However, the example of the wools, I do not even begin to understand. I have never heard anyone speak of full and empty wool, so I might ask what he means. And indeed, I have never heard from anyone apart from Archigenes those things said about inflammation and oedema, but in these I am able to

620 According to Plutarch (*Alcibiades* 13.2 [ed. and Engl. transl. Perrin 1959: 28]) this belongs to Eupolis, son of Sosipolis, Athenian comic poet of the 5th cent. B.C.

621 This and the subsequent passage are given as quotes, presumably from Archigenes' work, following Kühn's Latin translation.

622 The term "semi-empty" is the translation of μεσόκενος as listed in LSJ. It is listed as "of the pulse, dubious sense", citing this passage.

623 The term "inflation" (ἐμφύσημα/*inflatione*) is present in both the Greek text and the Latin translation of Kühn, but not in the manuscripts F1, F2, and Va1. The definition given in LSJ is, "an inflation of the stomach, peritoneum or cellular tissues ..."

conceive what someone might wish to indicate by the actual term "full". For the opposite to empty (porous)[624] is called full. **[946K]** I have, however, heard slave dealers and gymnastic trainers speak of certain bodies as empty or full, but this is not analogous to oedemas and inflammations. Rather, just as these terms are said most authoritatively and customarily, they are similarly used in naming vessels full and empty, when taking up the skin and drawing on it, diagnosing whether it is loose or tight. In this way they find evidence of emptiness from the looseness, and fullness from the tightness. And often in fact, from such a diagnosis, they called the thinner more full than the thicker. We are compelled to speak such nonsense because of those who first spoke nonsense, saying nothing to the purpose. For like this, Archigenes, writing an explanation of names but making no attempt to teach the matters, brought into this many examples and names, which were neither all able to suit the one matter, nor were customarily used as names among the Greeks. For among some of the Greeks, it is possible to find wine or wool having been written about as full and empty. Accordingly, it is necessary for us to go to wine merchants and women, and ask them what they wish to indicate, and how this is useful. Certainly, if someone is wholly desirous of using the term "fullness", displaying **[947K]** clearly the matter to which he applies the term, there is no longer any need for examples. But concerning the use of names, enough is said in the third book of the work *On the Differentiae of the Pulses*,[625] and we call this to mind now for the sake of demonstrating the confusion in the teaching of Archigenes. For it was not necessary to complicate matters by undertaking a long and endless account, since it is possible in a few words to proceed with the teaching easily in this manner. Those things he wrote of obscurely in many words, I shall now state briefly and clearly.

3. There is one class of pulses which indicates what the composition of the actual body of the artery is; there is another which indicates the quality of the substance contained within the artery; and third, there is one that makes known the capacity of the innate *pneuma* of this. The name for these three classes is "fullness". In writing these things, if he wrote clearly and quickly, he would not now be delaying us, while we seek first what he says, and second if he speaks badly. But if we were to know directly what he says, we might move on to the enquiry as to whether he speaks badly, and so in turn also **[948K]** make the argument ourselves. Of these three classes, O Archigenes, the first is diagnostic of the composition of the artery and is not in fact called full by the Greeks, as we showed in the third book of the work *On the Differentiae of the Pulses*.[626] There our discussion was about names, whereas here it is no longer about names; rather we consider the diagnosis of the pulses themselves. You may call this what you will, but allow us to use the terms of the Greeks. They term a certain pulse "hard", and opposite to this, they term another pulse "soft", and the third between both they term "moder-

---

624 The Greek word translated as "empty" here is χαῦνος rather than κενός, as used previously.
625 *De pulsuum differentiis* III (VIII.636–694K).
626 See previous note.

ate" and in accord with nature in the composition of the body of the artery. Therefore, in respect of this pulse, which you call full but I call hard, with what perceptible affections of the touch do we recognise it? That the body of the artery is of necessity either condensed or spongy, no one disputes. But this is not now the matter before us; rather, it is how we diagnose this. Therefore, teach us this clearly without complicating [things]. But you don't do this, and we don't in fact blame you for this, nor **[949K]** do we exult, because we clearly explained the affection of the touch occurring in one way in the strong pulses and in another in these pulses.

Through you, elders, who carefully investigated these matters, something was discovered that is useful to us. This would not otherwise have been discovered had we not been trained by you, who had toiled and laboured before in these matters. Let us now leave aside what is diagnosed and distinguished of the strength by the compressive affection of the touch. We assigned the disturbing and resisting to that. Next, through that, let us investigate the remaining two classes of fullness, the first being that conceived of in relation to the *pneuma* itself. It is not possible to conceive of this as what you say it is; we also showed this in the third book of the work *On the Differentiae of the Pulses.*[627] But since I now wish to cut short the discussion of those things because I have already spoken at too great a length, you say it is either the fieriness of the heat or its *tonus*. These are things we have been able to discover from your statements, as from the oracles of Apollo. Therefore, now is already an appropriate time to teach the signs of these. Certainly, we have learned of the *tonus* and are pleased we know. For the pulse resists **[950K]** when compressed, as you say, and disturbs the touch. However, of the heat and the fieriness of this, we only need to learn what we would acquire through the sense of touch. For I think after touching, we shall not be unaware of any difference in heat. And we wonder at that, if you number such things among the *differentiae* of the pulses but not among those of the fevers.

But let that be, for we have no differences now about this, and let us consider the diagnosis of the remaining class of fullness. In this, I have nothing to conclude through the touch as to whether *pneuma* alone is contained in the arteries or also some humors. But you were not able, as I also said previously, when you were speaking against Erasistratus or Praxagoras about blood being contained in the arteries, to call on sense perception as a witness, although you would undoubtedly do this, if it were clear. But no one had recourse to sense perception as witness to those same things—that the arteries were clear of humors, or were organs needing *pneuma* only. For touch is not able to make known whether *pneuma* alone is contained by the arteries or **[951K]** if some humors also are. Therefore, setting aside now their vain nonsense, let us ourselves alone say it is not possible to diagnose through any of the senses the amount of blood contained in the arteries, or what kind it is. First, let us remind ourselves that in relation to the moisture in those with dropsy, we are not able to diagnose by

---

627 *De pulsuum differentiis* III (VIII.636–694K).

touch whether *pneuma* is contained in the spaces of the abdomen, unless there is pitting of the skin when pressed.[628]

Therefore, as pitting does not often appear in the true ascites and the so-called tympanitic dropsies, we are compelled, for accurate diagnosis, first to percuss the epigastrium, so we know if it resounds like a drum, while second to both change the posture and turn the person quickly onto his sides for the perception of a certain splashing.[629] For us the tympanitic sound is a sign of *pneuma*, whereas the splashing is a sign of fluid. It is not, however, possible, by touch alone, when applied to the distended epigastrium, to diagnose whether this has been affected by air or fluid. And further, in these cases, the skin having wasted away, as well as the whole of what is fleshy overlying the peritoneum, and **[952K]** the several layers of the abdominal wall, due to the thinning of the surrounding structures and the abundance of the contained substance, we are able to form a judgement through the touch as to whether fluid or *pneuma* is contained. Nevertheless, it is not safe in these cases to trust touch alone. How, then, shall we know anything precisely through the touch in the case of an artery which has a thick wall in this way, that is six times the thickness of the peritoneum, while the skin overlying it is often significantly thick, for we do not consider all those sick with consumption, and there is a third accession of the movement which has no small power to confound the diagnosis? And further, when it comes to being confounded, it is enough that what is underlying the artery is continually in motion. From what source, then, is it possible to know either the quantity of the humor contained in the arteries or the quality? It is by the magnitude of the pulse that we know the amount of the contained substance, but we by no means recognise here that this is fluid humor. On the other hand, it is by the hardness that we know the consistency of the body of the artery.

However, the resistance **[953K]** of a beat is not a sign of thick substance but of the strength of the vital *tonus*. Surely it would be laughable to suppose at anytime that the fluid contained within the lumina of the arteries is thick in such a way as to resist when pressed, for if it were mud, it would yield easily when pressed, unless they think it is stones or sand contained in the arteries. Only in this way would the thickness of the substance resist pressure; otherwise, it would not do so at all. I would be willing to fill a wine-skin with fluid substances for those who declare the touch to be accurate in this way, such that it produces a louder voice than the truth itself in recognition of what kind of substance is contained in it. For not only would they not distinguish thin water from thick wine, but if one of the sacs were filled with liquid pitch and the other with air, they wouldn't recognise it in this way. Only the splashings and sounds are able to distinguish what kind of thing is contained; perception through touch is the same in

628 In this sentence γαστήρ is taken as referring to the abdomen or belly generally rather than the stomach, while the verb βοθρόω is translated as "pitting", which is the descriptive term still applied to oedema when the impression of the pressing finger remains after the finger is withdrawn (see also LSJ).

629 These same examinations and signs are still important today.

every case. But not even the splashings are able to indicate the qualities of the fluids; rather this itself alone **[954K]** is common—that fluid is what is contained and not *pneuma.*

And how indeed would someone trust those who are not able to recognise the quality of what is contained in the sac through a single wall to be able to recognise the quality of what is within the artery through the skin, which is no less thick than the sac, through at least two membranes, and of necessity two arterial walls? Therefore, let they themselves say these things, without giving a false account of Herophilus, or with a revered name put to shame those ignorant of the Herophilean writings, or from this furnish belief with reason. For it is shameful to dispute with witnesses as in a law court. If you have something to say regarding demonstration, I shall gladly hear you out. However, to call on Herophilus and the Herophileans as witnesses is to shun direct conflict and to discover reasons, evasions and devices through fear of refutation. It is clearly not for the sake of further enquiring about the matter that they introduce recorded statements as false witnesses. Does Herophilus say this? He does not. But do you not speak falsely? Show how I speak falsely; show how he speaks. **[955K]** And then a statement, should it so happen, that is ambiguous is put forward and conflict exists around this—what the statement says and what it means, and now, in relation to the statement and concept, dialectical arguments and rhetoric are rolled out, and in relation to the ambiguity other things, and in relation to syllogisms, by Zeus. But also they bring across in this way the whole rhetoric, as if pursuing an hypothesis without bringing the matter under consideration to a conclusion. For what has not been said by more recent doctors regarding the issue, for those who are equipped to understand Herophilus on this *differentia* of the pulse, and by those trying to show that he did not know? Each of these, then, are reluctant to make the effort and are deserving of censure; the former for their ignorance and the latter for their love of contention. We are also not prepared to put in the effort—at least not enough for us to practise the specific theory of the art, but we wish to know what Herophilus said, and what Heracleides, Chrysermus and Hegetor did not explain correctly, and what Apollonius, Bacchaeus, and Aristoxenus would say. And if we do not understand, we are altogether compelled to "enjoy the benefit" of two evils: that we talk nonsense **[956K]** when we should not and that we do not wish to do this, just as those men do not. Anyway, now I need to do one of two things: either to seem to say the opposite to Archigenes, Herophilus and countless others, or to show that according to Herophilus there is no full pulse.

If, therefore, I undertook this as a task, I would be like those men worthy of blame for my vain effort. Now, having refuted their opinion in relation to the actual truth of the matters, I shall release those hastening towards the actual activities of the art regarding those things, having myself already completed the proposed enquiry. However, for those who also wish to learn ancient history thoroughly, and have the time for this, I shall add all the things showing that nowhere does Herophilus make any use of the full pulse. But first I shall begin from the actual statement which those men put forward in the first book by Herophilus on the pulses where he has written what seems to me to be the only thing they have read. It is as follows: "In general, then, pulse seems to differ from pulse in quantity (πλῆθος), magnitude (μέγεθος), rapidity (τάχος), strength (σφοδρότης) and rhythm (ῥυθμός)."[630] For when they bring this forward, they ask what **[957K]** the quantity is, as if we do not know what kind of thing he is indicating by the quantity, since fullness is necessarily signified by this.

Therefore I, as I think, am glad to answer. For I say quantity indicates frequency. Again, however, I would reply to others that the quantity indicates strength, and I might say anything more than fullness, so they may know how much licence there is to those who wish to talk nonsense, for why is the quantity indicated more by fullness than by frequency? Is it because we say the first syllable in fullness (*plērotēs*) and quantity (*plēthos*) with the same elements? But if this is so, *plēgē* (a blow), *plēsion* (near), *plēsmonē* (repletion) and *plēktron* (a plectrum), and countless other words have the first syllable compounded from these same elements. But they say the significations of these are far removed from that of *plēthos* (quantity). Also, why does the signification of *plērotēs* (fullness) differ so very greatly from that of *plēthos* (quantity)? But it is, they say, reasonable that the *differentia* relating to fullness exists in the pulses and is now said to have been known to **[958K]** Herophilus. Why is it not also reasonable for the (*differentia*) of hardness to be known and stated? Surely these are also *differentiae* of the pulses and Herophilus was altogether acquainted with them. Therefore, on the question of the source from which he speaks of the *differentia* relating to fullness, it would be better, I think, to read all those things written by him on the pulses long beforehand, and certainly for me, there is this law of exegesis: each man should find clarity for himself and not chatter at random using empty notions and unprovable utterances, which someone understands. Where else, then, do you discover the term fullness besides in Herophilus? For you find the term frequency countless times. Therefore, when the *differentia* relating to fullness does not come to recognition, in the manner that has been shown, nor does Herophilus mention it elsewhere, whereas the *differentia* relating to frequency is accepted by all, and is discovered countless times in the writings of Herophilus, it would be more reasonable to think that frequency is meant more than fullness. I do not now say that Herophilus has written fullness (*plēthos*) instead of frequency, for it would be a similarly laughable **[959K]** explanation to that given by those men, if I were to say this, but what I do say is that it is easier to show what is signified by frequency and hardness and all the other things more than it is with fullness. However, I think what fullness indicates to Herophilus has been explained clearly by others, so how do we now show that he does not signify fullness?

First having introduced directly what he says on this, I shall write out the whole passage, so you may particularly marvel at the stupidity of people who think that by fullness he means quantity:

630 The quoted statement appears only to be preserved here.

> On the whole, then, pulse seems to differ from pulse in quantity, size, rapidity, strength and rhythm. From the difference in these aspects it becomes clear that sometimes it is characteristic and sometimes not. It seems, however, that one pulse differs from another and in general is recognised as I said by rhythm, magnitude, rapidity and strength.

If it appears that in the same rhythm, one pulse differs from another in rapidity, magnitude and strength, would someone wish to take a more credible witness of the opinion of Herophilus than Herophilus himself? For it seems, he says, that one pulse differs from another pulse, as was said, in rhythm, magnitude, rapidity and strength. **[960K]** How, then, if in general the quantity indicates a certain difference of pulses, does he overlook this now, repeating the statement neither casually nor carelessly, but along with the addition of "as was said"? How does he say the pulses in the same rhythm differ in rapidity, magnitude and strength? How next, setting out the *differentiae* in relation to age, does he distinguish the pulses in other aspects, while omitting the fullness? And distinguishing the pulses in relation to the affections, seasons and activities and all the other things, he always mentions all the other *differentiae*, but never fullness. What is more, to almost the exact same statement in each of those things he subsequently wrote, he never added "the quantity", except in the first book only. He would not have omitted this in all the statements, if it were a name of a *differentia* of pulses.

Let this be a summary of what I said in the works about the difference in the pulses as stated by Herophilus. Through those works, I have explained to the fullest extent the **[961K]** customary usage of the names relating to pulses. Therefore, I shall now put an end to the discourse about the diagnosis of the pulses. Those things about either the distorted position of the arteries or anything else of the individual matters that I put off in this discussion, I shall speak about in the two following treatises—that about the causes in pulses and that about the prognosis through them, each of which will be in four books.

V *On the Causes of the Pulses* (*De causis pulsuum, libri IV* IX.1–204K)

Book I comprises 12 sections as follows:

1. Causes in general are divided into three types: *synektic* causes of genesis, *proegoumenic* (internal antecedent) causes of change, and *prokatarktic* (external antecedent) causes of change.

2–3. The particular case of heart and arteries is considered. A number of factors could and do contribute to movement: innate heat, *tonus*, specific *krasis*, *pneuma* and a bodiless power (*dunamis*) of essentially unknown nature. Another cause is the use or need (*chreia*), which is threefold:

- preservation and cooling of the innate heat;
- expulsion of the residues of breakdown of humors;
- generation of psychic *pneuma*.

4. Another *synektic* cause is the body of the artery which dilates to draw in what cools the innate heat and contracts to expel sooty superfluities. The four key factors in determining largeness/smallness, rapidity/slowness and frequency/infrequency of the pulses are strength/weakness of the capacity, status of the innate heat, hardness/softness of the arterial wall and use/need.

5–6. Combinations of causes which give rise to large and small pulses.

7. Criticism of the views of Archigenes and Magnus.

8–10. A consideration of the pulses in various combinations of these causative factors.

11. Galen lists eight combinations of the three factors—strength/weakness of the capacity, increased/reduced need and hardness/softness of the arterial wall—and discusses all eight.

12. This section takes one pulse—the rapid—and examines it in relation to the eight combinations.

Book II comprises 15 sections as follows:

1. First there is a statement of intention. Having dealt with the causes related to the five basic pairings of strong/weak, hard/soft, large/small, rapid/slow and frequent/infrequent, he turns to inequality or unevenness, both in a single pulse and in a series of pulses, the latter being termed "systematic" inequality. The basic causes are non-uniform *dyskrasias* of the heart and arteries and vitiations of the capacity and the arteries.

2–3. These sections deal with systematic inequalities. In 2, he considers causes of damage to the arteries, and in 3, weakness of the capacity and the production of six types of abnormal pulses: tapering, deficient, imperceptible (*asphyxia*), recurrent tapering, intermittent and intercident.

4. On inequality in a single pulse, of which there are nine basic *differentiae* determined by the nine possible combinations of rapid, moderate and slow in respect of the two components of dilatation, which may or may not be interrupted by a pause. He gives the causes and significance of the various changes, the latter pertaining mainly to prognosis.

5–10. Consideration of some of the variations: when there is no interruption (5); the dicrotic pulse, which is clonic (6); the bounding pulse (7); the undulating and vermicular pulses, regarded as one kind differing in quantity (8); the tapering pulse (9); a brief statement about the nodding and inclining pulses (10).

11. Inequalities due to position of the artery.

12. Spasmodic/convulsive pulses associated with spasms or convulsions.

13. A summary of the natural causes of all the unequal pulses: variations in the three dimensions (length, breadth, height); non-uniform *dyskrasias* of the arteries; distortion of the arteries; changes in the bodily state.

14–15. In these final two sections, he gives brief consideration to irregular and regular pulses, all of which are unequal. There is a basic division in that some preserve an equality over certain periods and some are completely in confusion (in modern parlance, regularly irregular and irregularly irregular). Changes of rhythm remain to be considered, but two important factors are changes in the need of

https://doi.org/10.1515/9783110612677-011

dilatation and contraction, and age or stage of life.

Books III and IV describe specific factors which cause change in the pulses, following the sequence of On the Pulses for Beginners, but adding detail particularly on causation, which was deemed unnecessary for novices.

Book III comprises 17 sections as follows:
1. An introductory section.
2–12. Sections devoted to natural causes of changes in the pulses, as follows:
  (2) gender;
  (3) a hot nature;
  (4) thinness;
  (5) age/stage of life;
  (6) seasons;
  (7) places;
  (8) pregnancy;
  (9) sleep;
  (10) change from sleeping to waking;
  (11) acquired bodily states;
  (12) *kraseis.*
13–17. The remaining five sections are devoted to non-natural causes of change, as follows:
  (13) exercise;
  (14) hot baths;
  (15) cold baths;
  (16) large amounts of food;
  (17) wine.

Book IV comprises 27 sections as follows:
1. An introductory section on causes of changes to the pulses which fall under the heading of "contrary to nature".
2–6. Five psychical causes. as follows:
  (2) anger;
  (3) joy;
  (4) grief;
  (5) fear;
  (6) pain.
7–27. Twenty-one physical causes, as follows:
  (7) inflammation;
  (8) pleurisy/pleuritis;
  (9) abscesses;
  (10) wasting/*marasmus*;
  (11) consumption/*phthoe*;
  (12) peripneumonia;
  (13) lethargy;
  (14) phrenitis;
  (15) lethargy and phrenitis;
  (16) *katoche*;
  (17) convulsions;
  (18) paralysis;
  (19) epilepsy/apoplexy,

(20) *synanche*;
(21) orthopnoea;
(22) hysterical choking/*pnix*;
(23) stomachical affections;
(24) dropsies;
(25) elephantiasis;
(26) jaundice;
(27) hellebore.

Book I

1. [1K] Of the causes changing the pulses, there are the causes of their genesis, and the causes of their change only. Those of genesis are the use (need) for which they arise, the capacity by which they arise, and the organs (arteries)[631] through which they extend. All the rest are causes of change—both those called *proegoumenic* (internal antecedent) and those called *prokatarktic* (external antecedent).[632] For the class of causes is threefold, and not in pulses alone but also in all other things. One class, which is primary and most significant, and which **[2K]** people call *synektic* (containing),[633] is named from its holding together of the essence of the pulses which, as was said before, is a cause of genesis. The other two classes are not causative of the coming into being of the pulses but causes of change in those that have come into being. At any rate, thickness of humors, or abundance, viscidity or acridity are not able to create pulses, but are able to change them. In the same way too, a cold or hot bath, winter and summer, and in general cold and heat are causes of change of pulses, but not of their genesis. These causes are, then, the same as the so-called *prokatarktic* causes spoken of just now, while those spoken of previously, as relating to the humors, are *proegoumenic*. To speak generally, those causes that are outside the body but change something in it are called *prokatarktic*, being forerunners of the conditions in the body. These same conditions, whenever they change the *synektic* [causes], become *proegoumenic* causes of these [conditions].

Suppose then, the skin is condensed by cold falling [on it] externally, and due to this condensation, the natural transpiration is held back, and being held back is collected together, and then in this way kindles a fever, following which the use (need) **[3K]** of the pulses is changed, and because of this also the pulses. Here, the external impact of the cold is a *prokatarktic* cause, while all the others, right up to the use (need) of the pulses, are *proegoumenic*. Thus, the *prokatarktic* cause, through the mediation of *proegoumenic* causes, changes the use (need) of the pulses, which is one of the *synektic*

631 The Greek term ὄργανον is variably translated organ, instrument or artery, in the latter case when this is obvious from the context.

632 On this Stoic-based classification of causes which Galen employs in certain instances, see Johnston 2006: 31–37, 110–112.

633 On this term, see von Arnim 1903: 2.121, in addition to the references above.

causes, and in this way makes a change in the pulses themselves. For it is not possible for any *synektic* cause to be changed; it remains unchangeable, being complete in itself. In the case of one of the *synektic* causes, before the change is handed on, it is not possible for the pulses to be changed. And this is why these are the most important, major and primary causes of the pulses, while all the others [act] through them. Because the external change that has occurred is transmitted to the *synektic* causes, it is called a cause of the pulse; in fact, only in relation to kind, essence and nature, and in no other way can they change them.

Accordingly, it is necessary for someone who wishes to become adequate in the art concerning the pulses to be variously practised in all the classes of causes, so that in regard to whatever is presented to him, he is able to describe the change that has occurred in the pulses. An excellent beginning of the practice is to weigh up the *synektic* causes, and in what manner **[4K]** each is able to generate a pulse naturally. For it is by the reason they lead to the generation of the pulses, that they also lead to their change. The differences in quantity and quality in the causes bring about the differences in quantity and quality in the pulses.

2. Now the heart, along with all the arteries, continuously displays the so-called pulsatile movement, which is active whilever the animal lives. However, when the animal dies, there is no pulsation to see, either of the arteries or in the heart itself. From this it is clear they were moved by some cause, although it is difficult to discover what this is. For one person says it is the innate heat, another the *tonus*, another the specificity of the *krasis*, another the whole constitution of the bodies, another the *pneuma* alone, another some of these things, and another all of them together. Already some have introduced a bodiless power (capacity) which uses the proper organs of movement previously mentioned—either all, some or one of them. This same cause, then, is what creates the pulses, **[5K]** whatever it may be, even though we are ignorant of its essence. From its being able to create pulses, we called this a power (capacity), just as, I think, we are accustomed to call every other power (capacity) from its ability to do what it is able to do. For the power (capacity) of something is also the concept of it we have acquired in its relation to something, and because of this, we name it in this way when we are ignorant of its essence.[634] And when we say strength or excellence of a power (capacity), or conversely weakness and badness, we must understand these terms in regard to the function (action).[635] For the active, productive and creative cause of the function (action) is sometimes better disposed in regard to this [power] and sometimes worse, and because of this, the function (action) is now accomplished better, or again, should it so happen, worse. Thus, to have the power (capacity) better with regard to its action is excellence and strength, while to have it worse is weakness

634 On the term *dunamis* in Galen's usage, see *De naturalibus facultatibus* I.4 (ed. and Engl. transl. Brock 1916: 16–17) in addition to Brock 1916: xxix-xxxiv.

635 Throughout ἐνέργεια is predominantly translated as "function" sometimes with "action" added in parentheses and sometimes used solely depending on the context. This is to make the contrast between ἐνέργεια and ἔργον as "action" clear. See Brock 1916: xxix-xxx.

and badness. Now about the power (capacity) bringing about the pulses, it is sufficient to distinguish these things for our present purposes.

3. Concerning the use (need) of the pulses, it is necessary to call to mind those things we demonstrated in other [writings].[636] They are for preservation **[6K]** of the heat produced in the whole body, for cooling this, and if something smoky should arise from the combustion of the humors, for expelling this immediately. And it is said in those [other writings] that the pulses also contribute to the generation of the psychical *pneuma.*

4. What remains is the third *synektic* cause of genesis of pulses, which is the body of the artery itself. For every power (capacity) functioning requires suitable organs for the sake of something in regard to the action. What constitutes a suitable organ in the present instance is impermeability in the walls and a hollowness within, as is the case in the arteries. Therefore, as the craftsmen in the cities—shoemakers, smiths and painters—accomplish their work through certain instruments, so too does the power (capacity) bringing about the pulses do its own work through the arteries as instruments, achieving the preservation of the innate heat. It differs from the other craftsmen only to the extent that it is bound in the organs, is spread everywhere through all these, and touches nothing external, as those other things do. And accordingly, when it draws in the cooling substance, the body of the **[7K]** arteries dilates, whereas when it expels what is sooty, it contracts.

Sometimes the need of the dilatation predominates and sometimes the need of contraction, but at all events the rest of the movement follows the other part. However, since the pulses invariably arise in relation to these causes, it is necessary that all the kinds of them follow the changes of the causes. As a result, there are as many primary and generic *differentiae* of pulses as there are of the causes. The *differentiae* of the capacity are strength and weakness; of the innate heat, heating and cooling; of the wall of the arteries, hardness and softness; and before all these are the due proportions in relation to each class of those mentioned. But the due proportions bring about the pulses in accord with nature, being the means of the extremes, whereas the disproportions bring about the pulses contrary to nature. The other *differentiae* are clear, although that in relation to the strength of the capacity has something worthy of investigation, which the discussion will consider as it goes forward. Thus, strength of capacity necessarily brings about a strong pulse, weakness an indistinct pulse, hardness of arteries a hard pulse, and softness **[8K]** a soft pulse.

One and the same kind of pulse does not always necessarily follow the increases and decreases of the heat that accords with nature (natural heat). For neither the rapid pulse, as it seemed to some, nor the large, nor the frequent always occurs when the nature is febrile, just as the slow, small and infrequent does not when the fever is quenched. But those who dispute about these matters are similarly mistaken, like Ar-

636 See Galen, *De usu pulsuum* (IV.149–180K). Also, Greek text and English translation in Furley and Wilkie 1984, and English translation in the present volume.

chigenes when he differed from Magnus about a rapid pulse—that it does not occur more from a strong capacity than from a weak one. Neither one of the two follows of necessity. A rapid pulse is more characteristic of strength of capacity and abundance of heat, and besides, softness of the arteries, whereas a slow pulse is characteristic of the opposites. Similarly also, a large pulse is characteristic of a strong capacity, abundant heat and soft arteries, while a small pulse is characteristic of the opposites. In like manner too, a frequent pulse does not necessarily follow any of those things previously mentioned, but is more characteristic of a weakness of capacity, abundance of heat or hardness of arteries. And an infrequent pulse in the same way is more characteristic of the opposites, but does not always **[9K]** follow. I shall attempt to demonstrate each of the things mentioned, resuming the discussion afresh.

5. If more heat has collected in the body, the animal invariably needs greater respiration and larger pulses, for which reason these become larger, and with this, both more rapid and more frequent. For as those who are thirsty not only take a lot to drink, but also take the drink rapidly and frequently, so too do those in need of the cooling substance draw in a large amount of this and do so rapidly and frequently. This, then, is why I said pulses characteristic of excessive heat are large, rapid and frequent, although they do not necessarily follow this heat. For such pulses to be generated, there is need of a strong capacity and soft arteries. If the capacity is weak or the artery hard, it is not possible for pulses to become large or rapid, since the artery becomes refractory to the moving capacity due to hardness, while due to weakness of the capacity itself, it is impossible to raise the arteries up to the appropriate degree.

Accordingly, it is necessary for the three classes of causes to work together with each other so that the pulse **[10K]** is made rapid or large; there is need for the genesis of these to be stimulated, for the capacity to be in a good state, and for the walls of the arteries to be soft, or at all events not hard. For, just as neither those who are very old nor small children are able to thrust forward or advance very much, even if they have a particular need of hurried movement, so too the weak capacity is not able to raise up the arteries rapidly or very much, even if it needs to. And again, just as the soft bladder is readily distended to a great degree, while the hard bladder is not, even if you were to force it very strongly, in the same way too, the wall of arteries that is soft is easily raised up still more, even if the capacity is weak, whereas the wall that is hard resists and withstands the moving capacity, even if this should happen to function very strongly around it.

And there is one method relating to all these when only the cause under scrutiny is changed while the others remain the same, of making the determination of the change of the pulses. For if you were to consider many changes at the same time, first it would be unclear whether the change occurred due the whole collection of all the causes acting together at the same time, **[11K]** or through any one of these. And then, if it is also agreed that it is due to one, what this one of them is may remain as yet unclear. Anyway, should the heat alone be increased, while the capacity is preserved and the walls of the arteries remain unchanged, the pulses will, of necessity, become larger and more rapid, and will be increased together with the excess of heat until they reach as far as the very large dilatation. They will not now appear very rapid before the heat is increased beyond the very large dilatation, while when it is increased beyond the very large dilatation, it is cooled insufficiently by this. Accordingly, under those circumstances, the deficiency will be balanced by the rapidity of the movement, and in this way the pulses become very rapid. However, if these are not sufficient for what is burning up greatly, the time of the pause between the movements, which previously existed, is already curtailed. Often it appears so short that the movements seem to be continuous with each other, and not divided by a pause at all. This occurs when the burning up is extreme.

Since there are three modes of relief in all excesses, it uses all at the same time. For it has the benefit of most of the cooling substance and is very rapid and very continuous. **[12K]** The first of these creates the largest pulses; the second, the most rapid; and the third, the most frequent. For it does not, when it is sufficiently cooled by the magnitude of the dilatation, stir up the rapidity, nor when it is adequately benefitted by both, does it stir up frequency. However, if either one of these is deficient, it is always equalised by the other—largeness by rapidity and rapidity by frequency. To speak generally, one cause of frequency of pulses is the deficiency of the previous function, and this also arises due to other causes, which I shall speak about in order, but also due to abundance of heat, as now demonstrated. The matter before us is already almost at an end. For it was shown that no large, rapid and frequent pulse inevitably follows an abundance of heat; largeness and rapidity are particularly characteristic of it, and at times also frequency contingently.

6. It would, I think, already be clear about the pulses that follow in the deficiencies of the innate heat. For truly, nothing follows these inevitably, but the small and slow pulses are particularly characteristic, and now also **[13K]** the infrequent. For just as the nature that is cooled beforehand needs neither much nor rapid cooling, so too it does not due to brevity. This "through brevity" is through frequency and "through long duration" is through infrequency.

7. Since enough has been said about these matters, let us complete what was put off a little earlier, considering the continuation of the argument as far as it is useful for our present purposes. For what is in between of the large and small pulses is the pulse which is in accord with nature and moderate. When we compare and measure all the others with this, those that are larger than it we term large, and those that are less than it, small. It is the same for the median of the rapid and slow, which is moderate and also has an accord with nature, and those in the middle of the frequent and infrequent, and besides these, those in the middle of the hard and soft are, by the same reasoning, in accord with nature.

But the pulse in the middle of the weak and strong does not seem to be in accord with nature. For a strong pulse is an action of a strong capacity, just as a weak pulse is of a weak capacity. To be strong in respect of capacity is not, of course, contrary to nature, but is excellence and accord with nature, **[14K]** while the weakness of the capacity is badness and contrary to nature. For as far as it is in between strong and weak, the road is somehow placed in the border region to both from one to the other; to the extent it is better than the weak, so it is worse than the strong, and conversely, to the extent it is worse than the strong, it is better than the weak. Here they say it is not therefore necessary to seek accord with nature, other than in the highest and strong, and it is from this the strong pulse is "an offspring". There are those who agree this is true, but dispute it as a generalisation, and say the pulse in accord with nature in this class arises through the optimum and best state [of the body—*hexis*], therefore just as also in all the other classes it is not the utmost and extreme, but is one of those not in accord with nature; one other and stronger than this. For they say that something specific has happened to this class of the pulses compared to all the others, and for this reason, they have encountered the difficulty and errors in those not able to see this.

For although it is the strong capacity that gives rise to the strong pulse, there is a certainly also **[15K]** of that a mean of the strength, and what moves further away departs from an accord with nature. At any rate, they say that in rages, and often in exercises, and in the drinking of wine a greater strength of the movement of the arteries than accords with nature is found, and it is agreed by all that one must consider, in the case of an animal, the pulse having a precise accord with nature in one who is at rest, free of all large and strong changes. A strong change is one that is from anger, exercise and wine; this is not therefore one in accord with nature, but occurs when the nature is strong, and in relation to none of the external causes stirring it up forcibly. In this manner, the pulse is in some way median, but in some way not in the class of accord with nature. Inasmuch as there are some other pulses stronger than it, and, by Zeus, others less strong, to the extent it is median, insofar as there are three *differentiae* in this class —strong, weak and the median of both—it is not from the median but from one of the extremes, and to this extent it is not still median. All the rest, then, **[16K]** about this is, I think, well and truly stated—that the pulse in the class of accord with nature is to be found occupying a median place between the extreme *differentiae*.

There are those who argue against this, saying the strong pulses do not occur in the times mentioned, but only appear to do so, and that those who palpate them too carelessly are deceived by the associated occurrence of rapidity and largeness in them, and sometimes also depth, as in the rages. It is clear that the judgement on such a disagreement pertains to the diagnosis. What, then, is the true diagnosis? It is necessary for us to make this apparent. After drinks of wine, at least after those that are moderate, the pulses are clearly larger and more rapid, and in this way also become stronger. However, after rages and moderate exercises, it is not still similarly clear, but if some strength is associated with them, this is slight and difficult to recognise. Whatever the truth may be, we are not prevented from completing the matter before us, since we seek this alone—whether largeness and rapidity of pulses is characteristic of a strong capacity, which it seems to be **[17K]** to those who do not consider it cursorily outside the actual nature of the matters. For since it is agreed that the arteries are dilated by a capacity, it is reasonable [to assume] they are distended more when the capacity is strong, and less when it is weak, faster when it is strong, and slower when it is weak, just as it is with ourselves when we are walking, running, rowing,[637] digging or doing anything else whatsoever.

In these things, the stronger man not only displays greater action but also swifter, other things being equal in both. For if the weaker should be stirred to action by the need, while the stronger man has no need at all, or only very little, the weaker man will act more rapidly in this case. We must not, however, compare them in this manner, but as was said before, when all the other things are similar for both, and only in relation to the one thing we happen to be investigating on each occasion when the change is occurring. For example, if there is need to flee when robbers are approaching, people do not of course stride equally—a strong young man and a weak old man do not run with equal speed. If, however, you compare movement **[18K]** of an old man, who fears the approach of robbers, with the movement of a young man, walking in a relaxed manner in a town, the old man seems to you to move faster and more, but this is not because he is able to move faster and activate his limbs more than a young man, but because he has need of hastened movement due to the fear. It is this by which he is compelled to use all his capacity for the action. There is greater strength of the capacity in the young man, but there is nothing that needs this for the action. If, in fact, it is needed, similar to the old man, he will act faster and more by as much as the strength of the capacity is better in him.

Here, then, it is worth censuring both Archigenes and Magnus—and Archigenes much more. For if Magnus did not perceive the whole truth, he did perceive part of it well, whereas Archigenes, in attempting to add the needed remainder, overturned what was properly discovered, writing thus (for it is right to set down the actual statement): "Therefore, in the case of the newborn, the pulse is entirely small and not strong, and very frequent and rapid."[638] Magnus, however, says it is not rapid, wishing to establish definitively **[19K]** that the rapid pulse is not characteristic of weakness, but the frequent pulse is. For those of such an age are weak. And proceeding a little further, Archigenes says: "It seems to me rapidity does not lie in relation to strength, whenever I shall find, in the cases of those who suffer from cholera and heart disease,[639] the contracted movement of the arteries." Such, then, is the statement of Archigenes—very credible and easily able to mislead the hearer, as may be concluded from the statements the majority of doctors approve. For they almost all agree that the rapid pulse exists in those with cholera or heart disease.

But I have said enough about rapidity and slowness in the first book of *On the Diagnosis of the Pulses*,[640] when I show the rapid pulse is one thing, and the short-duration pulse another. It is similar with slowness; the slow pulse is one thing and the long-

637 Kühn's Greek text has ἐρεσσόντων (rowing) while Kühn's Latin translation has *metimus* (reaping). The three manuscripts Fi, F2, and Va1 all have ἐρεσσόντων.

638 As elsewhere, the quotations are as indicated in Kühn's Latin translation.

639 The Greek term καρδιακός LSJ lists as "of persons suffering from heart diseases" citing this passage but adding "probably equal to καρδιαλγής" (heartburn), citing Dioscorides I.112.

640 *De dignoscendis pulsibus* I (VIII.766–822K).

duration pulse another. It is not, therefore, appropriate to still speak about these here, but only to mention the conclusion—that in the previously mentioned affections the pulses are of brief duration and not rapid—and in this way to return to the subject before us. For nothing further will be accomplished **[20K]** by Archigenes, if the pulses are not accepted as being rapid in the choleric and cardiac affections. And yet, even if this one thing is agreed, the rapidity did not become a reliable sign of a strong capacity, although it did not prevent it being characteristic of this, since we have already clearly determined that some pulses necessarily follow the *synektic* causes, while some are characteristic in these, but do not necessarily follow. For the genesis of these pulses needs one cause, and they will follow this constantly. And the strong pulse will be the consummate sign of the cause of the strong capacity, as a weak pulse will be of the weak capacity. Of these, however, unless many causes combine together, it is not possible to establish the genesis. We shall say these are characteristic of the causes, but do not also indicate some one of them as necessary. Similarly, if a pulse sometimes arises in relation to one cause, but not constantly from this, it will be characteristic of all the effecting causes and will not in fact show with certainty any one of them.

For if there is rapidity of pulses, there is not necessarily also strength of capacity, but there is sometimes the greater **[21K]** need of the generation of the psychical *pneuma*, or sometimes a larger share of heat, or softness of the arteries, or some or all of these together at the same time. It is the same too, if there is largeness of pulses, or strength of capacity, or consumption of much psychical *pneuma*, or a larger share of heat in the heart, or softness of the arteries, or some of these, or all of them. Therefore, Magnus was not mistaken when he said the rapidity is characteristic of a capacity that is strong. For sometimes also the pulse does become faster through this. Archigenes, however, errs in gainsaying this; first, because he speaks falsely about what is apparent, thinking the short-duration pulses are all rapid, and then, even if they are rapid, thinking that none of them is itself characteristic of the accomplishment of a strong capacity, which is obviously false. For if pulses always become rapid through strength of capacity, but not constantly through that alone, this would follow as a conclusion, if we shall also agree that the previously mentioned pulses are rapid. If, however, **[22K]** Magnus thought the rapid pulses are not characteristic of a strong capacity alone, but that they are also constantly used as signs, he erred.

For if there is not rapidity of pulses, and there is also already strength of capacity, but as was said previously, either a stirring up of the need, or a greater softness of the arteries occurs, the arteries are moved faster. Therefore, since the rapidity sometimes follows strength of capacity, sometimes abundance of heat, or consumption of psychical *pneuma*, or a softness of arteries, Archigenes would have found it better to add Magnus' statement. For in this way it would ultimately have become sufficient for all the diagnoses.

Accordingly, since Archigenes left this out, and yet it had previously been set in motion by Herophilus, we shall make the complete teaching of the diagnosis of the causes bringing about rapidity, when we add as a remainder those things Magnus properly wrote, and still more, Herophilus. However, first it is necessary to call to mind those things stated in the second book of *On the Diagnosis of the Pulses*, so there is no misunderstanding in the discussion when we follow the signs by means of the names.[641] For we certainly say there are rapid pulses, and similarly slow pulses, not only in those who are diseased, but also **[23K]** in those who are among the healthy. And we say this is normal in them, just as in others who have the best constitution of the body, where the pulses are moderate in nature, and neither slow nor rapid. And in these same people, they are neither large nor small, but moderate in between both, and neither frequent nor infrequent, but also moderate in this class. Accordingly, from this, let the discussion first proceed from the person who has the best constitution. Certainly, such a person is, as it were, a standard and measure for all other people. For we call pulses large and small, and rapid and slow, and the terms analogous to these in relation to each class, signifying the excesses in respect of these, without referring to anything else but this.

Now if such a person becomes hotter than he should be, and is changed in this respect alone, the strength of the pulses would not change, either in relation to the class of hardness or that of softness; nevertheless, in all cases there will be a change towards largeness and rapidity, and sometimes also to frequency, depending on whether the heat is increased more or less. For if **[24K]** it should become clearly hotter than it was to some small degree, the pulses will be larger, but not also clearly more rapid. If, however, the heat is increased still more, and the magnitude of the pulses increased in proportion to this, it is quite clear that the rapidity will be also. If the heat is increased to the greatest extent, the pulses will become very large, although not yet very rapid, but as was also said before, first at that time, when the dilatation is increased beyond the very large. Under these circumstances, the frequency will also already be clearly recognised. We have stated beforehand the reasons for these phenomena, while it is not difficult to discover them from what has been said.

If, however, such a person should become colder than he was, what first clearly appears is infrequency; second, slowness; and third, smallness. And perhaps it would seem to someone that the phenomena contend both with each other and with the assumed causes. For just as it is necessary, in the case of the heat, for there to be first largeness, second rapidity and third frequency, so also, in the case of coldness, it is necessary for there to be first smallness, second slowness and third infrequency. What, then, is the resolution of the problem? **[25K]**

It is that, when the capacity is strong, even if the use becomes a little less, the primary action is not necessarily either small or slow, although of course it is reasonable for the secondary action to be so after a longer time, if the use is not stirred up. For when it was stirred up, it spurred on and roused the capacity towards action, so in this way, when it is not stirred up, it allows a longer pause. However, should the cooling become greater, and because of this the use is sufficiently lessened, so the rapidity of the movement will also be lessened with it, while the dilatation will be immediately

641 Galen, *De dignoscendis pulsibus* II (VIII.823–877K).

smaller, whereas, if the cooling advances still further, the pulses become very infrequent, but not necessarily also very slow and very small; it is enough always for them to become slower and smaller than those in accord with nature. They would never become thoroughly slow or small while the capacity is strong, even if the use were sufficiently lessened. These things appear to be similar in the case of the other actions. For example, at any rate, in the case of walking, when the use is increased —that is, if there is a need to traverse many *stadia* every day—we progress faster and more, and **[26K]** we make the times spent in the lodgings of short duration. When it is reduced to such a degree that, if it should so happen, we need to traverse only fifteen *stadia* each day, we spend still more time in the lodgings, but we don't reduce the action to the level of extreme slowness and smallness. It is similar in the case of respiration. But I have said enough about this in the work *On Difficulty of Breathing* (*On Dyspnoea*).[642]

8. Taking up again the matter of the pulses, let me say that when they are changed, either through heat alone, or through cold, if none of the other *synektic* causes depart from their original nature, of necessity the change of the pulses is as I have described. However, when there are not only matters of need, but aspects of the capacity are already changed as well, there are in this way four different conjunctions in all, and in relation to each of these, the arteries are manifestly moved as follows: in relation to the weaker capacity along with heat, they are overcome to the greatest extent by both the causes; when the capacity is sufficiently weak and the heat burning, the pulses become small and slow, and very frequent. If, however, **[27K]** the harmful effect is moderate, they are similarly frequent but moderate in respect of the distension and movement. Why, then, are they similarly frequent but not similarly small or slow? Because there is nothing more frequent than extreme frequency, so they are similarly both. However, if in fact another were able to become more frequent than this, those that are very weak would possess it altogether to such a degree, and there would be nothing by which to ever fulfil the use. Following the moderate harms of the capacity along with the heat, for what reason do the pulses become frequent, although still not small or slow but moderate? Because, when the capacity is weak, but not yet reduced to an extreme degree, a moderate dilatation and movement would sometimes occur, as in the case of weaker people, moderate walking stirs them to use. For what in the strong occurs without great necessity, does so in the weak when they are compelled. In these same people also the continuous action happens whenever they are stirred up by need, and as it were, forced [to act]. But these people, even if there is also difficulty, nevertheless fulfil the need. However, those who are extremely weak, even if they exert themselves as much as possible, and add as much as they can **[28K]** to the magnitude, rapidity and frequency in the actions, do not, in this way, fulfil the need. That all those who are weak in the capacity have a weak pulse was stated before, and the pulse is

642 *De difficultate respirationis* (VII.753–960K).

weak to the degree that the capacity has been affected. These things are enough about the *kakosphyxias*[643] due to an abundance of heat and a weakness of capacity.

Let us pass on to the other conjunction composed of weakness and cooling. In this case, the pulse becomes weak to a degree determined by how bad the capacity is, and small and slow, just as the pulses are also in the previous conjunction. But they do not, like those, come to extreme frequency, unless at some time the capacity has suffered to an extreme degree. It is clear that in such cases, the pulse will be very small and there will be a great deal of intermittency and failing.[644] However, when the capacity is moderately weak but the cooling is severe, the pulse does not become frequent at all, as when the need is fulfilled adequately and befits such a capacity.

A third conjunction of a change of pulses is when heat and capacity are increased at the same time. In this, **[29K]** the pulses become very strong and very large, but not similarly very rapid, while for the most part they are not clearly very frequent, although sometimes they are clearly beyond what accords with nature. And this occurs when the heat is increased to the maximum extent. Indeed, under these circumstances, the pulse is very large and rapid, although it does not serve its use due to the great amount of heat. The remaining and fourth conjunction of the change relating to this class of pulses is when strength of capacity and deficiency of heat come together at the same time. These pulses are moderate in magnitude, but of course slower and fairly infrequent, and particularly when the cooling prevails to a greater degree. For under these circumstances, the need of the pulses is brief, and everything is fulfilled easily due to the magnitude of the dilatation, so it is reasonable that the arteries rest to a greater extent. For as when the need was stirred up, the movement of these was continuous, in this way now, when it is lessened, a long interval occurs. Nevertheless, the amount of the distension of these is not clearly or obviously hindered due to the strength of the capacity. Nor does the slowness prevail to any great degree for the same reason. The strong capacity preserves the measures of the movement in accord with nature, even if the **[30K]** need does not call for this.

For the need of the pulse requires to be greatly lessened, when strong cooling prevails, so the pulse obviously becomes smaller. As for the most part this preserves the amount of the distension in accord with nature, it becomes slightly and not very clearly smaller, although it appears to a greater degree slower or smaller in the coolings. This closely resembles those walking, when they don't need the movement to be stirred up. These people take away less of the magnitude of the walking than of the rapidity. And if they sometimes need to walk quicker, they add more to the magnitude than they do to the rapidity. Here we must pay precise attention to what is being said, lest we should

643 This is translated as a technical term meaning a bad pulse in general. See also Galen, *De sanitate tuenda* IV (VI.238K; Engl. trans. Johnston 2018: 347), and Aretaeus, *De causis et signis acutorum morborum* II.8 (ed. Hude 1958: 29; Engl. transl. Adams 1856: 281).

644 The two terms are διαλείπω meaning to leave an interval between, intermit, discontinuous and may refer to pulses of various kinds in heart block; εκλείπω means to leave off or cease. It is not clear that these represent different phenomena.

take this lightly and think, in the case of every condition, the change in relation to largeness and smallness is greater than that in relation to rapidity and slowness. For it does not occur more continually but more readily, and because of this also, sooner. It is easier to see the change occurring from moderation to largeness and smallness, than that to rapidity and slowness. Nevertheless, when the change reaches a significant magnitude, if the **[31K]** increase in heat is still happening, then the rapidity increases suddenly. And it is reasonable for this to happen, as we also said previously. For to fulfil the need, it suffices for the heart to increase the magnitude of the dilatation, there being no need for it to also increase the rapidity in proportion to this. However, when the magnitude is already significant, then also it stirs up use, and at that time also everything remaining adds to the rapidity. If, on the other hand, this is not sufficient, it makes the movements continuous, which is the same thing as frequency. But this is enough about these things.

9. Passing in turn to the differences of the arteries, let us examine first the characteristic pulses of each of these, and then directly also the changes associated with the other causes, just as we did a little earlier. It was stated before that, when the body of the artery becomes harder or softer than it was, and the pulses themselves are made hard or soft, such a thing does not need more lengthy considerations, it being attested by all the palpable phenomena that the quality of the pulse beat is determined by the nature of what is beating. **[32K]** It has also been stated before that largeness is characteristic of those arteries that are soft, and smallness of those that are hard, and this will be stated again now through a continuation of the discussion.

The soft artery is readily stretched in all directions and is easily dilated, but the hard artery not without considerable force. The largeness follows the readiness of the dilatation, while the softness follows the resistance and *dyspatheia* (difficulty of being affected). Similarly also, rapidity follows the first class and slowness the second, but not in the same proportion to largeness and smallness. For a pulse due to softness of the arteries is a little more rapid but much larger than that due to hardness. The membranous substance of the body of the artery, even if it should sometimes become harder, is ready for and compliant towards rapidity of the movement, but is the most naturally unsuited of all things to the largeness of the dilatation, if in fact it must be stretched easily in all directions, when making the very large dilatation. Softness is such a thing. And whenever the body of the artery has lost to the greatest extent the accord with nature, either towards softness or hardness, the soft pulse becomes slower than the hard. **[33K]** For the body that is a little softer than accords with nature is more compliant to the largeness of the dilatation than a body that is a little harder, and because of this also, towards the rapidity of the movement, while the body that is relaxed into softness to the greatest extent and, as it were, collapses and sinks down into itself due to excessive moistness, is not similarly ready for the largeness of the dilatation, or for the rapidity of the movement. Because of this, it falls short of the hard body in rapidity, but now still exceeds in largeness, not only the hard body, but also the body that is in accord with nature.

Nevertheless, such conditions in which the arteries come to an immoderate excess of hardness or softness are not without the change of the other *synektic* causes, for it is not possible for either the strong capacity or the heat to still remain within the limits of an accord with nature. This, then, is why, when we speak of hardness and softness of the [artery] alone without the other causes, in respect of the characteristic pulses, we shall say they become larger and more rapid in the soft arteries, and smaller and slower in the hard arteries, and even larger or more rapid **[34K]** through the cause we spoke of previously. On the other hand, in frequency and infrequency, they vary as much from each other and from the moderate as they deviate from an accord with nature in largeness, smallness, rapidity or slowness.

It was also said before that there is one cause in general of frequency of all actions whenever something is more deficient in accomplishing each use. Thus, the dilatation of the hard arteries is smaller, and because of this, the action of the use is more deficient. However, the dilatation of soft arteries is to a significant degree the opposite, while the use is adequately fulfilled. It is reasonable, then, that frequent movements follow in hard pulses and infrequent movements in soft pulses. On the other hand, if at some time cooling coincides with hardness or heating with softness, the hard pulses would in this way become more infrequent than the soft. In relation to combination, such a change of the pulses arises when there are already some differences of the *synektic* causes. But since, in relation to the actual continuation of the discussion, we first mentioned the conjunction involving hardness and cooling, let us now say everything about this. For when the body of the arteries becomes harder, along with cooling of the innate **[35K]** heat, it will make the pulses smaller and slower than those in accord with nature. If the hardness prevails, they will be smaller more than slower; if the cooling prevails, they will, conversely, be slower more than smaller. In the same way too, they are more frequent if the hardness predominates over the cooling, but more infrequent if the cooling prevails. If the deviation towards contrariety to nature becomes equal in each of these, they will be smaller to the same extent as they are slower, while in infrequency or frequency, they are not more different than those in accord with nature. For as much as the need of the generation of these declines due to the cooling, so the amount of the dilatation and of the time of movement are reduced together.

Nature does not, therefore, need to move the arteries more continuously, as in the majority of the small pulses in which the dilatation is more deficient than the need, if indeed the hardness is sufficient, while the cooling is altogether slight. This itself will then be what was previously mentioned, when the dilatation that occurs being less than the need, makes the movements continuous, just as if, contrariwise, the cooling becomes very great, but the hardness less than the least which accords with nature. For then, **[36K]** it is even more necessary for the pulses to become more infrequent or smaller, because the need is diminished even more than the amount of the dilatation. Such, indeed, is the first conjunction of the causes.

The second conjunction is that in which softness prevails along with heat. But here too, of necessity, there are three primary differences: sometimes the change occurring due to softness and heat is similar, while sometimes either the softness or the heat is prevailing. If, then, the degree of softness deviates from an accord with nature as much as the heat, of necessity the pulses become large and rapid, although not immediately frequent, but only when the need of their generation exceeds the magnitude of the dilatation. If, however, the softness prevails, largeness will predominate over rapidity. If, on the other hand, the heat prevails, it is no longer possible to make a simple declaration; a distinction must be made. For when the heat is very great, so as not to fulfil the need by the action, of necessity the pulses become still more rapid or large, but clearly also immediately frequent. Whenever the nature sufficiently enjoys the benefit of the need, the pulses will become large and rapid to an equal extent.

The third conjunction is of the **[37K]** causes changing the pulses in relation to these classes, whenever hardness should come to the same degree as heat. Thus, when a large deviation towards contrariety to nature occurs in each, the pulses will be very rapid and very frequent, but not yet very small, unless the aspects of the capacity are in a bad state; this is not assumed at the moment, but they are less than those which accord with nature, although they have not in fact come to the extreme of smallness. If, however, the aspects of the heat should be immoderate, while the artery is not very hard, but deviates a little from the moderate, the dilatation of the arteries is not so much less than accords with nature, but might also become more. For since the pulses characteristic of the hardness of the arteries are small, whereas those of an immoderate heat are large, it stands to reason that in such conditions the pulses become large, being similar to what prevails, just as when hardness prevails, the pulses become small. However, it is at all events rapid, whichever of the causes gains ascendancy, and much more so when it is the heat. Under these circumstances also, the frequency is more apparent.

The remaining conjunction, in relation to the classes of the pulses now before us, is whenever cooling comes to the same degree as softness. And since here too, the large [pulse] **[38K]** is characteristic of the softnesses, while the small is characteristic of the coolings, the pulses are necessarily similar to whichever of these two departs more from the normal. Consequently, the pulses will be large when the softness of the arteries prevails, but small in the coolings. They are moderate, however, when they are increased towards largeness by the softness to the same extent as they are decreased towards smallness by the cooling. For to speak generally, the pulses become moderate not only in the case of the causes now before us, but also in the case of all the other things contrary to nature, which are similarly able to change each of them to opposites. And the majority of doctors hold a false opinion that those diseased in this way are in a state of accord with nature. But still more will be said about these matters in the treatise, *On Prognosis from Pulses.*[645] For the present, it is sufficient to have made this alone clear, to the degree that it is pertinent to the matter before us—that all the pulses in accord with nature are medians of the extremes and moderate, but not all those that

645 *De praesagitione ex pulsibus* (IX.206 – 430K).

are moderate are in fact in accord with nature. **[39K]** Moreover, all such pulses are infrequent and slow, and more so when the cooling prevails.

10. Since the things that are likely concerning these have been spoken about, let us pass next from the *differentiae* of the arteries to those from the capacity, first calling to mind that as many affections of pulses follow a preponderance of heat, as there are needs of generation of psychical *pneuma*. Thus, hardness of arteries, along with weakness of capacity produces pulses that are small, slow and frequent, but along with strength of capacity, it produces pulses that are small and frequent, but not also slow; rather, sometimes they are very much more rapid than accords with nature. For to the extent that deficiency in moderation of the dilatation is due to the hardness of the arteries, so it is compensated for by the rapidity and frequency. If in fact the artery is soft, but the capacity is strong, the pulses obviously become larger but to a small extent slower and more infrequent. If, however, weakness of capacity comes together with softness of the arteries, when the deviation to contrariety to nature of both becomes equal and moderate at the same time, **[40K]** the pulses appear moderate and have a close similarity to those in accord with nature, except those that are changed to softness, whereas if they are sufficiently immoderate, they become small, slow and frequent. And it is similar if also the aspects of weakness prevail by much. For at that time, the pulses also become small, slow and frequent. But actually, when softness prevails, the pulses seem close to those that accord with nature. For the capacity, being slightly short of an accord with nature, does not hinder the preservation of either the quantity of the dilatation or the quality of the movement, as it does when it occurs together with hardness of arteries; under these circumstances, the dilatation is hindered. However, of necessity, the pulses become frequent to the extent that each of the actions is deficient. Whenever the capacity is somewhat weak, while the walls of the arteries are soft, the movement doesn't become slower and the dilatation doesn't become smaller, and because of this the need is adequately fulfilled, so no change of the pulses towards frequency occurs.

11. What remains is to put together at the same time all the *differentiae* of the *synektic* causes, which are **[41K]** eight in number. There are two, when the capacity and the need are reduced at the same time, and sometimes the artery becomes softer than it was and sometimes harder. There are two others, when the capacity is reduced but the need is increased, while the arteries become either soft or hard. And there are another two, when the capacity is strong, while the need has been reduced, along with the changed *differentiae* of the arteries. The remaining two are when the capacity is strengthened and the need increased, along with the previously mentioned *differentiae* of the arteries. Thus, when the capacity and the need have been reduced at the same time, the pulses change altogether to small and slow in relation to each *differentia* of the arteries, while they are not necessarily either frequent or infrequent, but each of these happens to them in turn, depending on the *differentia* of the arteries and the amount of damage to each of the *synektic* causes. For when the damage in the capacity is slight and the arteries are not very hard, no change occurs in relation to frequency of pulses.

As was also said before, what happens is that not only the pulses, but also absolutely all the actions happen to be more frequent, **[42K]** due to the deficiency in each of them. However, that the deficiency is measured by the need was also said before. Therefore, whenever this is reduced, while the damage of the capacity is slight, then the action would not become deficient, unless due to immoderate hardness of arteries. Moreover, if the capacity is greatly damaged while the need is moderately reduced, and the arteries are made sufficiently hard, of necessity the pulses become smaller than the need, and due to this the action is deficient; and the more deficient, so the more frequent. But it is not possible for the actual smallness or slowness of such pulses to be the same continually. For depending on whether the capacity or the need has been damaged more or less, of necessity also the pulses themselves are changed more or less to smallness and slowness. Thus, they become smaller due to weakness of the capacity, due to reduced need, and due to the hardness of the arteries, but they suffer this less or more, depending on the amount of change of each of the causes, either one alone, or two together, or all of them having been damaged at the same time. **[43K]**

Alone, when the pulse becomes very small, it is either when the arteries are made very hard or the capacity is weakened to an extreme degree, for either one of these, as was said, when damaged, is sufficient to make the pulse smaller. However, in two together, as when smallness becomes extreme, it is not that one of the mentioned causes comes to an extreme of damage, but that each of them falls short of extreme damage to such a degree that the pulse produced from both is changed similarly to that which has suffered extreme damage from either. In this way too, if the three classes of causes come together for the generation of one pulse, they will sometimes make as great a change as would occur from one of them that is damaged to the extreme. And it is necessary to bear this in mind throughout the whole discussion, so that someone is not deceived that the change occurring from all the causes having been damaged to an extreme degree is thought to be greater than that from one. For it is sufficient, as was said, due to any one cause whatsoever having been damaged completely, for the pulses to come to an extreme change. At any rate, the collapse of the capacity to an extreme degree **[44K]** makes the pulses very small, even if the walls of the arteries are soft, while the need stirs them up, so that it is also useful in turn to call to mind in general, that often several causes are required for the generation of the extreme pulses, whilst often one is sufficient.

For a very large pulse would not occur due to extreme strength of capacity, unless other factors were also to contribute. Nor at any time would a very small pulse arise due to a weak capacity. And what is more remarkable than this is that the most opposite pulse is often characteristic of the remaining causes, and the change occurs from one of these that prevails. One thing or another prevails, should it so happen—weakness of capacity due to which the pulse became very small, or again hardness of arteries, or again in turn dissolution of need. For just as, when the capacity is extremely weak, neither the softness of the arteries nor the urgency of need will increase the magnitude, so too, when the arteries are made hard, the strong capacity contributes nothing to the magnitude, nor does the urgency of the need. Just as due to one cause having been extremely damaged, an extreme change of pulses **[45K]** often occurs, so sometimes, when all the causes contribute at the same time to the generation of one pulse, what is created from them does not yet show the extreme change.

If at the same time the capacity is weak, there is hardness of the arteries, and dissolution of the need, the pulse arising in this way doesn't inevitably become very small, although there are all the causes mentioned that naturally generate a small pulse. But if neither any one of these whatsoever comes to the extreme of injury, nor the sum of all of them collected together is equal to it in one very great badness, it is not possible for the pulse to become very small. What I said, then, is that there is nothing so necessary to consider in relation to the changes of the pulses as the amount of the damage of the causes. But I shall also say this in the work *On Prognosis from the Pulses.*[646] In the present discussion, let me move on to the third and fourth conjunctions of the eight previously mentioned, here leaving behind to some degree now the discussion about the first two. For if something were to be left out in relation to these, it would be readily discovered from those things previously said.

Thus, whenever the capacity becomes weak, or the need is stirred up, **[46K]** or the artery is hard, you will find the change of the pulses arising most to frequency, second to smallness, and third to slowness, but not in fact of necessity, nor continually, nor in relation to each of these things whatsoever, nor all occurring at the same time. For if the damage is great in relation to each of the causes of the damage, the pulses become very small, very frequent and slow. They are slow because the capacity is also weak; they are very small due to this and the hardness of the arteries. Because they are such, they are also, for that reason, very frequent, and much more, as when the need is stirred up. And moreover also, if apart from this, weakness of capacity were to come together with hardness of arteries, enough to make the pulses very small and very frequent, while into the bargain, if the force from the need adds to these things, it contributes not a little to approaching destruction, but would not make the pulses still more frequent. For apart from this, they were already very frequent. If, however, the damage of the arteries and that of the capacity are slight, while the need of their generation is considerable, they would not become smaller than those which accord with nature, nor would they become slower, but sometimes **[47K]** clearly more rapid and even if not clearly larger, at least clearly equal to those that accord with nature.

For when the need is stirred up sufficiently, even if the capacity is weaker than accords with nature, it [the artery] is moved rapidly. However, the magnitude will have been increased much more than the rapidity, unless hindered through the hardness. But now, due to this, the rapidity becomes more manifest than the magnitude, because nothing impedes the artery from becoming a little harder than it was, which opposes the magnitude most of all, while in such cases the pulses become fre-

646 *De praesagitione ex pulsibus* (IX.205–430K).

quent because the need is not fulfilled. However, if a slight change of this towards a contrariety to nature occurs, the frequency will be altogether slight and difficult to discern. Under these circumstances, the need is fulfilled by the rapidity itself and the largeness. To speak generally, it is necessary to compare, in the case of such a conjunction, the action performed by the capacity through the arteries with the amount of the need. By as much as the action falls short of the need, so the pulses advance towards frequency. And in turn, they become smaller and slower in those so disposed, to the extent that either the factors pertaining to the capacity or those pertaining to the arteries **[48K]** depart from an accord with nature. But these things are sufficient about the third change of the causes.

Now I must speak about the fourth conjunction,[647] in which there is a weak capacity, while the need is increased, along with the softness of the arteries. It is clear, then, even in this case, that if the increase in the need becomes very great, while both the capacity and the arteries fall a little short of an accord with nature, the pulses will increase much more towards largeness than towards rapidity. In the combination mentioned before this, the hardness of the arteries acted in opposition to the largeness of the dilatation, whereas in the combination now before us, not only does the softness of the arteries not act against the dilatation, but also acts with it most of all. The pulses, then, will clearly be larger than accords with nature, but clearly not also similarly more rapid. Such pulses will also be more frequent than accords with nature, although not in like manner, nor as in the third conjunction, because now the need is fulfilled more than in that. If, however, the matters from the need and the capacity should change a little from an accord with nature, or the change of the arteries towards softness is great, **[49K]** the pulses in such conditions will become close to those in accord with nature, and in the quantity of the dilatation, the quality of the movement and the duration of the pause; they will change from those in accord with nature only in the softness of the impact, for this is characteristically inseparable from these. If, however, those things deriving from the need and the arteries do not have a great deviation, should the capacity labor strongly, such pulses would be excessively small and frequent, and moderately slow. Such things are sufficient about this combination.

Let us move on to the next two in turn, which have in common the strength of the capacity and the reduced need of the generation, but differ in the arteries. Certainly, the one of these in which the artery is hard, while there is strength of capacity, but a need that is less, makes the pulses altogether smaller, but not necessarily slower or very frequent. But if the matters of the need are sufficiently reduced when the arteries are moderately hardened, not only will the pulses become smaller, but also clearly slower and more infrequent. If, however, the badness of the arteries is more, or **[50K]** that of the need, such pulses become notably small, and because of this, frequent, whereas due to that same thing, also less slow than that in the previous condition. For

647 Galen uses the two terms, συζυγία and συμπλοκή apparently interchangeably. The former has been rendered "conjugation" and the latter "combination".

if the need is not fulfilled and the capacity is strong, they do not then become significantly slow. However, the need is not fulfilled due to the smallness of dilatation. This customarily follows the hardness of the arteries. If they were soft, the pulse would not be clearly less than that in accord with nature, unless the matters of the need were not greatly lessened. And altogether such pulses would seem to be more infrequent to the extent the matters of the need had been reduced. And furthermore, they would also clearly be slow after significant changes of the need, although not clearly so after those that are lesser.

But since we have already discussed these combinations sufficiently, let us now make mention of the remaining two, in which, when the need is increased and the capacity strengthened, the walls of the arteries are sometimes soft and sometimes hard. When they are soft, the pulses will become in all instances large and rapid, but not always frequent; this will only be when the increase of the need is excessive. For under these circumstances, neither the largeness of the dilatation, nor the rapidity of the movement **[51K]** is sufficient to fulfil the need. However, when the excess is small, the largeness alone serves sufficiently. If, on the other hand, the walls of the arteries are made to become hard, the pulses in such conditions change in various ways. Thus, if the hardness is slight, while the change in need is substantial, the pulses will be larger, more rapid and more frequent than those in accord with nature. If, however, the change towards contrariety to nature is equal in relation to each of the causes, and neither one nor the other is very strong, the pulses become equal to those in accord with nature, but more rapid and more frequent. If the damage of both the need and the arteries becomes severe, the latter being hardened to a greater degree, while the former is stirred up more immoderately, the pulses are now even more rapid than those in accord with nature, but smaller and more frequent.

12. Since the likely things about these have been stated, what remains is to choose one particular pulse and to teach, in this case, the use of the practice in it. Let it be the rapid pulse, which we said is characteristic of a strong capacity, of a need that is stirred up, and softness of arteries. How this **[52K]** is discovered in one sick person will be determined, when it occurs by one of the causes mentioned, when by two, and also by all. The method is as follows.

Since the strong pulse is inseparable from the strong capacity, while the soft pulse is inseparable from soft arteries, whenever there is rapidity apart from both of these, it only occurs when stirred up by need. When, however, it occurs with either one, and is wholly due to that, providing this sign, of necessity it is also due to the increase of the need. For it was shown before that it is not possible for rapid pulses to occur, just as it is not for large pulses, without their being called forth by need. It is, then, possible to recognise how much the capacity is changed, either to strength or weakness, from the pulses particular to it. The same also applies to the quality of the arteries. For we certainly also have pulses indicative of this. How much the need of the generation of these is, must be judged by logical conclusion from the change in terms of magnitude, frequency and rapidity. For if the pulse is both very large and very frequent, and at the same time also very rapid, the need is stirred up to an extreme degree. If, however, it is simply large and **[53K]** slightly more rapid, but not yet very large and frequent, the change of the need would not be great, and besides this also, from the other signs apart from the pulses, it is possible to know beforehand how the matters pertaining to need are. For it is not difficult to determine whether there is a large amount of heat or an expenditure of psychical *pneuma*. The heat would not escape the touch, and for the most part, by the already very clear recognition of the dyspnoea, and by the thirst also through the internal perception of the one suffering [it]. The expenditure of the psychical *pneuma* occurs in those exercising, and in those perspiring and in the inward fluxes, none of which escape the doctor.

It is not necessary for us now to say still more about these, because it is not a secondary aspect of the matter before us presently; such a discussion is part of prognostics. I mentioned this now, wishing to show the followers of Archigenes, in addition to those who have spoken wrongly about the generation of the rapid pulse, that it is not difficult to discover the effecting cause on each occasion. This will be even clearer through those things that follow this, **[54K]** if we go through beforehand the genesis of the remaining pulses—for example, those called irregular, arrhythmic and unequal[648]—and particularly if such pulses also relate to a single dilatation of the artery.

Book II

1. [55K] Regarding the strong pulse, the hard, the large, and besides these, the rapid, the frequent and the opposites of these, what the generation of each is, how many causes it needs, how they act together with each other, which of them follow the generating causes of necessity, and what characteristics are in these causes that do not have a necessary consequence, have been made known in the first (i. e. the previous) book. The next task is to go over the generation of the remaining pulses, and first, of those that are unequal. **[56K]**

Since of these, some weaken the equality in one movement of the artery, while some do so in a number; it is customary for doctors to call the latter a systematic inequality. And as this is much clearer than that occurring in one pulse alone, it would be better to make the discussion about this in the first place. Now to speak generally, all such pulses follow the unequal (non-uniform) *dyskrasias* of the heart and the vitiations of the arteries or capacities. For they supervene in obstructions, compressions, hardenings and excesses, the last now being understood as the measure regarding the capacity. But it is not enough to know only about these; it is also necessary to distinguish all of them individually, and to add the appropriate demonstration for each.

There are three *synektic* causes of the generation of pulses—artery (instrument), capacity and use (need). It is clearly not possible to establish a systematic inequality in the pulses through the change of the need. For it is impossible for the need to change

648 See the Appendix on pulse terms.

suddenly or all at once to such a degree at any time, that the first pulse, should it so happen, is large, or rapid, or more frequent, while the second **[57K]** is small, or slow, or infrequent. For whether there is an increase of the natural heat, or a diminution, or greater or lesser consumption of the psychical *pneuma*, none of these would effect a sudden change, such that the pulses were made unequal.

You would learn this accurately from the paroxysms occurring in all the mild fevers, such as those from buboes and the genuine tertians, and those due to heatstroke, cooling or any other such thing. For the whole time from the beginning of these to the peak is proportional to the increase in heat, and in changing the pulses to largeness, rapidity and frequency, there is never at any time a second pulse which shows a clear difference from the first. But if you were to leave some long time, they would seem clearly larger, more rapid and more frequent, but not immediately more different, unless at some time one of the things previously mentioned should join in. I speak of excess in relation to the diseased capacity, or obstruction, or compression, or hardness of arteries, or some non-uniform *dyskrasia* of the heart itself, which is not still a mild disease. For certainly also in those who are healthy, **[58K]** whenever they should happen to have bathed or exercised, the *pneuma* is consumed, while the heat is increased, and the pulses appears larger, more rapid and more frequent than those which are in accord with nature. But the second clearly doesn't yet have any difference from the first. Whoever says such pulses are unequal, even if they don't seem so, overlooks what is customary in names, but teaches nothing wiser. For in the case of those pulses which we are accustomed to call unequal, reason does not discover them by some inference from premisses; rather we clearly recognise them as being so by perception. Let such pulses, then, also be called *anomalous* (unequal, uneven), should someone so wish, but know that none of the perceptible inequalities arise through the change of the need.

Perhaps this much is enough for a demonstration that the so-called systematic inequality never follows a change of need. Let us further add the phenomenon sometimes appearing in rages, since these also seem to be a kind of seething of the innate heat. Accordingly, the pulses in these sometimes become unequal. This never relates to the specific reason of themselves, for they were always so, but when a certain fear **[59K]** is associated with them. Reasonably, under such circumstances, they appear varied and unequal, as they would be changing not only due to different causes, but also to those things opposite in nature. Due to the desire for vengeance, they become deep, large, rapid and strong, while due to the expectation of being acted upon more than acting, they become small, shallow, weak and slow. They seem to change alternately with each other, according to the nature of the predominant cause. Sometimes, the anger prevails more and sometimes the fear. But also, the fear itself, at least when it is strong, creates unequal pulses, due to affliction of the capacity, not change of need.

How, then, does the capacity that is afflicted make the pulses unequal? For this is the next thing to speak about. It is by the capacity not altogether prevailing over the arteries, as when it does prevail, it makes all the pulses similar and equal. If, however, it errs in some way concerning the action, like when abating, it destroys the equality by this. It is necessary to pay attention precisely to what is being said, so it is not thought that the systematic inequality follows necessarily every affliction of the capacity. For this is not so. Whenever it becomes weak by reason of itself, it rarely **[60K]** at any time makes the pulses unequal. But when it is strong in itself, but is weighed down by excess, or when the arteries are altogether refractory, then it produces the inequality mentioned.

It is possible to learn more clearly what I mean by way of an example. Let us call to mind two men walking at leisure. Let one of them be very strong and at the same time also very young, but carrying a very great burden. Let the other not be carrying anything at all, but let him be very weak due to disease or old age. These men progress neither very much nor quickly, and don't differ from each other in respect of action in this way. But they will differ in that the weak one walks similarly for the most part, while the strong one walks differently and unequally, sometimes in one way and sometimes in another. For the weak one, even if he should exert himself, is not able at any time to make the action large or rapid, while the strong one is himself similar for the most part, but sometimes, using greater eagerness, will act more rapidly and more, for a little while at any rate, and then, again afflicted by the burden, is brought back to the initial slowness and smallness of the action.

Now this same thing also befalls **[61K]** the action of the arteries. For when they have fallen to a sufficient weakness of the capacity, the pulses become weak, small and slow. However, if the capacity is strong by reason of itself, but is weighed down by some excess, it produces in this way pulses that are varied and unequal. In most cases they are weak, slow and small, whenever it is severely burdened, and conversely in most cases, strong, rapid and large, when only burdened a little. But whensoever neither a very strong nor a very mild affliction arises, such that in magnitude it is near to the opposite, you are not able to say either more or less, or the strong compared to weak, or large to small, or rapid to slow. In this way the harms of the capacity produce pulses that are unequal.

2. The damages of the arteries are of the following kind: whenever the surrounding regions of the artery, which are for the purpose of receiving the arteries when they dilate, are taken over by some other body, of necessity, when the arteries are confined by this, they are unable to dilate sufficiently, Therefore, in summary, the deficiency in relation to each action sometimes becomes significantly concentrated, so that the action falls a long way short of the need. **[62K]** Already, at that time also, the capacity is compelled to function forcibly and strongly, striving to shake off and push away those things surrounding the arteries and hindering their movements. And in this, the pulse necessarily becomes larger than before, and stronger and more rapid. However, it would not now for the time first become more frequent, but became so, immediately from the beginning. For it has been shown, in the book before this one, that frequency of pulses is a necessary consequence of the more deficient functions of the need.

If, then, the capacity were to have preserved the equal impulse in the case of these, in all subsequent dilatations, the pulses would always have been similarly large, strong and rapid. Now, however, it does not preserve [this], but comes again to the movement from the beginning; in this, the inequality occurs in various forms and continuously. This failure to preserve is the reason it achieves nothing and labours greatly in vain. If there is some moisture that has been poured around the arteries, impelling the strong and forcible movements, it is compelled at another time to flow differently to other places, and is either carried down some channel capable of drawing it off and expelled from the body, or is also thinned over time, and ultimately, **[63K]** being resolved into vapour, is transpired. And in this it is inevitable that the inequality becomes variable and continuous, since the abundance of fluids is neither preserved equal over the whole intervening time, nor [do the fluids] obstruct the arteries themselves, but often change place from the unimportant parts to those that are important, and then, from these back again to the unimportant parts.

If the affection of the bodies around the arteries is either some inflammation, or an abscess, or induration (scirrhosity), by which they are compressed and constricted, inevitably in this way over a longer period of time, unequal pulses are intercident in the equal pulses. For always the actual parts of the arteries are compressed by themselves, or by things equal in magnitude, or by similar masses. And sometimes also inequality of pulses arises due to an abundance of blood poured in, either by the veins or by the arteries themselves. But this is very quickly settled by phlebotomies. For the abundance in the veins, as the arteries lie alongside them, compresses and constricts them. When the mass contrary to nature is evacuated by phlebotomies, [the procedure] immediately reduces the mass of the veins, and immediately also provides ample space for the movement by the arteries **[64K]** in which the natural action unimpeded takes hold and they reacquire the equality of the pulses and with this, an accord with nature.

The blood in the arteries themselves, and how this occupies the paths of the *pneuma*, does not need discussion, and particularly if it is also somewhat thick, or viscous, or even impacted in the extreme ends of these. For under these circumstances, since it is neither able to draw the *pneuma*, nor send it forth, it creates a varied and at the same time very severe inequality, and especially when the obstruction of the blocking humors occurs either near the heart itself or further on, but in important parts. To speak generally then, when the arteries are severely compressed or obstructed, some inequality arises involving the pulses, even if the capacity has not been affected. If, however, the capacity has a problem together with such a condition of the arteries, it should somehow be clear to everyone that of necessity there will be a more varied inequality, just as, in fact, if also there is a non-uniform *dyskrasia* in the heart itself, concerning which we shall investigate the whole argument in the prognostic treatise.[649] If, however, the aspects of the arteries are good, whereas the capacity is greatly weakened, **[65K]** it would not be as previously stated, but would become the so-called tapering (mouse-tailed, curtailed) inequality. In relation to the impact of the fingers, the capacity being overcome always creates pulses that are smaller and weaker. Further

649 *De praesagitione ex pulsibus* I–IV (IX.206–430K).

demonstration is not needed for the argument. But what is useful for the present matters must and will be added.

3. The capacity which creates the pulses is not necessarily weak in equal measure in all the conditions, but is very much more or less. In fact, it first makes the pulses slightly smaller and weaker, but not tapering (curtailed, mouse-tailed). However, if it is greater than this to such a degree that the applied fingers are already a burden to the capacity, there is production of the tapering pulses. There is a twofold difference in these: some of them return to the initial movement (people call these recurring tapering), while others somehow come to an end on the spot. One must know that the former **[66K]** indicate a smaller weakness. For somehow the capacity still fights back and, as if collecting and gathering itself together, often increases the magnitude of the pulses in this way, and increases the *tonus* also. However, those that remain extremely small, and are not yet returning, show the capacity has been dissipated more than in relation to these, just as if ultimately they fail, they indicate an extreme weakness.

Of course, that the diagnosis of all these occurs, when the fingers are applied to the arteries and not lifted at all, is obvious to everyone. For if we do lift them and then reapply them a little later, we discover the movement has become deficient in each impact; we call this complete *asphyxia*[650] and not a remittent pulse. I think *asphyxia* dffers greatly from a remittent pulse, for the one is like a deprivation of all action, at least in relation to perception, while the other is a partial loss. Therefore, whenever the movement is apparent to you when you apply your fingers, but dissipates as you keep them applied and little by little ceases altogether, we call such pulses remittent and regard the condition as intermediate between that of the tapering (curtailed) **[67K]** pulses and complete *asphyxia*, which is as more severe than the tapering pulses, as it is better than *asphyxia.*

These three conditions differ, then, in the amount of the damage. However, a fourth, which is that of the recurring tapering, is the mildest of all. Let a fifth be added to these, which is that of the intermittent pulse; this is obviously antecedent to that of the remittent. For what was in one or two, or to speak generally, in a certain number of pulses, creates this state of the remittent in all cases. The *differentiae* of the intermittent pulses themselves are also clear. For the harms of the capacity are greater, whether more and less so, when fewer movements are wanting. The recognition of the so-called intercident pulses is opposite to that of the intermittent pulses. For clearly, these would be in the systematic of the inequalities and originate from the strong capacity, but are disturbed by an excess of humors or obstruction of arteries. So then, in order that the disturbing factors are repelled, the pulses are often made larger, stronger and more rapid, and so also in such conditions, **[68K]** the pulses become greater particularly and deep. And because of this, there are signs of critical excretion. The in-

650 The particular meaning of this term, which has changed over time, is absence of palpable pulsation —see the Appendix on pulse terms.

termittent pulses arise not only due to a weak capacity[651] but also due to a severe obstruction of arteries, or some compression and constriction, whenever they are rendered entirely unable to draw to themselves. This, then, is enough about the inequality of pulses in aggregation (systematic inequality).

4. Next we must speak about the inequality in one pulse. There are many individual *differentiae* of this. We must attempt to reduce all these to generalities, and when these become clear, even if one of the individual *differentia* is omitted, it will be easily discovered from those previously mentioned and the rest. And let the beginning of the discussion be the same as that in the discussion about *differentiae* of these. In that, we began by distinguishing first the inequality of the movement itself,[652] which is either interrupted, remains continuous, or is recurring. And accordingly, we had taught that there are nine *differentiae*, since, whenever the artery dilates, a pause occurs in the middle, dividing the whole dilatation into three time periods. Next, let us consider here each of these, **[69K]** seeking their causes, and first of the first mentioned of these, when the artery begins the movement rapidly, and then after a short pause, the rest is moved slowly in the added dilatation.

This occurs when need is stirred up, while the capacity supports to the extent it is able, hindered either by damage of arteries or a significant excess, since if the dilatation were never interrupted by the pause, it would be large throughout. Such would not occur apart from the need calling and the capacity supporting. You may learn that it is not possible for a small dilatation to be interrupted by a pause, if you realise that none of the small pulses can be divided into two perceptible parts. What in fact this present discussion is about is the pulse being divided into two clear parts—one before the pause which is rapid and one after the pause which is slow. If each of these should have significant largeness, this whole would be of the largest. If not, it would not be altogether small, even if perceived as being from two small ones. This is clear from the strength of the capacity and the rapidity of the first movement, just as in fact it also **[70K]** indicates the stimulus of the need by the movement after the pause, for without this, the capacity would not at any time complete its initial impulse, but where the movement was first stopped, there would stop the dilatation, creating this one small pulse. Therefore, just as also the small pulse manifestly occurs countless times in the systematic inequalities, it is not because the capacity cannot be moved, but by its being impeded. So it often makes the pulse in succession to it very large, gathering itself up, one might say, as if collecting and recovering itself, and using all the *tonus* forcibly. Thus, when it is possible for such a thing to now be made by it and it is not made, this shows the need to be stirred up. From what source is it, then, that immoderate excess, or compression of arteries, or obstruction impedes the movement?

651 From the phrase δι' ἀρρωστίαν δυνάμεως ὁ διαλείπων in Kühn, the manuscripts F1, F2, and Va1 omit δυνάμεως.

652 Following αὐτῶν, Kühn's Greek text has ἠρξάμεθα δ'ἐν ἐκείνῳ διαιρούμενοι τῆς κινήσεως αὐτῆς; the manuscripts F1, F2, and Va1 have διαιρούμενοι τῆς κινήσεως.

First, it is from the capacity being strengthened and stirring up the need, for there would be nothing still remaining as a cause of the damage outside those things mentioned, and then also from the actual nature of these. For either the excess is the burden which is not incapable of stopping the capacity in the middle of the dilatation and, as it were, **[71K]** stopping in this way to add what remains, whether it be severe obstruction of the arteries, or some forcible compression, there is good reason for the movement to be restrained to the extent that the filling *pneuma* flows away from the place of dilatation. But when this is completely restrained, the total closing off of the channels of the *pneuma* is indicative, and because of this, we say the compression and obstruction are severe.

This is certainly common to all the dilatations divided by a pause; what is specific to each of them will now be stated. Of the first, there is significant weighing down of the capacity by the excess, when the pulse becomes like this due to excess. For the later movement after the pause would be anything other than slow, the previous movement being rapid, unless the capacity had not been overcome and severely burdened by the excess. Conversely, it exists in the case of the second *differentia*, in which both parts of the movement are rapid. For the capacity manifestly restores and maintains the *tonus* and contends with the excess. However, there is obviously a condition in between both, in which the second movement is midway between rapid and **[72K]** slow, which is moderate. These, then, are the three *differentiae* of the dilatations divided by a pause.[653]

There are three others, whenever the first movement being slow, the second changes in three ways, being either slow, rapid or moderate. Obviously, the worst in these is that in which the movements are both slow; much more suitable than this is that in which the second movement is rapid; and in between both is that in which it is moderate. In this way too, if the first movement were to be moderate, the second would change in three ways. The rapid movement indicates a more suitable condition, the slow a more destructive condition and the moderate movement an intermediate condition.

Of all nine *differentiae*, arising due to whatever cause, the worst is when both movements are slow, the best when both movements are rapid, and intermediate when both are moderate, while the others are in proportion between the moderate and each of the extremes. But also, the duration of the pause, to the extent that it should become greater in relation to each of the nine *differentiae* mentioned, is that much more problematic. Conversely, the duration of the movements when it is increased, signifies a better condition. In like manner **[73K]** too, the large size of both the dilatations is best. Therefore, just as the small is worst, so too the strong is best

---

653 The Greek here is somewhat difficult (or Galen's explanation is simply obscure). What he is describing are nine *differentiae* of pulses in which the dilatation itself is interrupted by a pause: there are three in which the first movement is rapid and the second rapid, slow or moderate, three when the first movement is slow and the second rapid, slow or moderate, and three when the first movement is moderate and the second rapid, slow or moderate. In this and the following 2 paragraphs the causes and consequences of these are considered but not exhaustively and systematically.

while the weak is worst. However, if all the bad things mentioned should come together in one pulse, this would be the worst of these, just as if all the best should come together, it would be the best. For if someone were to say that either excess, weighing down the capacity, had overcome this significantly, or there was some obstruction in the important arteries not far from the origin, this would indicate largeness in the actual obstruction, or if compression and constriction, it would show that almost all the lumen of the arteries was occupied. The best in these is obviously that in which both the movements are rapid, large and strong, and the duration of the pause short, indicating the capacity is overcoming the excess, or that the obstructions and compressions of the arteries are either small, or not in important vessels, or far from the origin. Although these things properly belong to so-called prognostic theory, they are recorded at this time due to the commonality of the discussion.

5. [74K] Moving on now, let us consider the pulses having continuous movement for the whole dilatation itself, but not equally rapid throughout. A lot was also said about these in the treatise *On the Differentiae of the Pulses*—how many there are of this class in all and also how many pulses in each of these [*differentiae*].[654] As a result, it is not still necessary to go over the number of these now, but we must go on to their causes, as far as possible going through them in general, to make the whole discussion shorter.

Now those of them which distribute the rapidity from beginning to end, show the capacity to be less burdened, whereas those that are the opposite show it to be more burdened. When the capacity is not dissipated, none of these occur. All are large, and particularly as far as the movement is as much as three intervals and three times, or four, which can be perceived. It is characteristic of those that make a sudden change to the opposites that not only is the capacity burdened, but there is also obstruction or compression of arteries. I speak of suddenly making a change to the opposite if, **[75K]** when moving rapidly, they have suddenly become slow, or conversely, from slowness of the prior movement, rapid. These occur through some obstructions and constrictions, but are less severe than those that are intermittent, and when the capacity itself is burdened by excess, but less than in those that are intermittent, for it is not of course equally bad for the movement to be changed and to be hindered. And yet individually the particular pulses of this class of inequality sometimes and in some way differ from each other, and some of them are more deadly, while others are better, and it is easy for someone encountering these to discover a distinction beginning from these things I have said regarding the intermittent pulses. And indeed, I shall say more about these in the work *On Prognosis from the Pulses.*[655] For certainly the present discussion requires nothing more of these than relate to the commonality of the class and the common causes of the generation, which I shall now stop speaking about.

654 *De pulsuum differentiis* I–IV (VIII.493–765K).

655 *De praesagitione ex pulsibus* I–IV (IX.206–430K).

6. But now we must pass to the still remaining class of inequality in relation to one part, which we call dicrotic pulses. I said these are, of necessity, clonic. **[76K]** It is not, therefore, possible to discover the causes of them accurately before carefully considering those of the clonic pulse. Accordingly, this occurs when the capacity is strong, the artery hard and the need pressing. For if one of these is lacking, the pulse would not become clonic. For certainly the constant factors in respect of the genesis of this are forcible dilatation and hard arteries. The pulse would not otherwise vibrate and shake in every direction, unless both these factors were to come together. If the artery alone is hardened, but the capacity does not compel it to distend to the maximum extent, there would certainly not be any clonus; this would only be the so-called hard pulse. If, however, the capacity should wish to dilate the arteries to the maximum extent, and these were to respond with moderate softness, [the pulse] would in this way be large, but not what is in fact clonic, nor will it be hard at all. If, therefore, it is going to be clonic, it is clear that the capacity is naturally compelled to dilate the arteries to the maximum extent. But not unless compelled by the pressing need would the capacity urge on action strongly at any time. Consequently, it is reasonable that we said the capacity must be strengthened, the arteries hardened, and the **[77K]** need pressing.

But it must be said that the urgency itself arises due to some cause—this must not be omitted. There are, someone might say, one proximate (immediate) cause, but many agents of this. What, then, is the one, and what are the many? The one is when the need remains unfulfilled in respect of function (action); the many effecting this are, to speak generally, weakness of capacity, *dyspatheia* of arteries, and increase of the need. But in the case of the class of pulses now before us, we must discount the weakness of the capacity, for this was shown to be strong. What remains, then, is either the need being increased, or the arteries being refractory, or also both coming together at the same time, when the dilatation will be less than the need in the clonic pulses. But the need is increased by the predominance of the heat, or by the consumption of the psychical *pneuma*, whereas *dyspathic* arteries arise due to obstructions, compressions or hardnesses. These same hardnesses arise due to inflammations, indurations, hardenings, dryings, coolings or stretchings of the arteries. And in this way, we already discover all the causes of the genesis of the clonic pulses.

For certainly, when the *dyskratic* inequality advances, **[78K]** it will increase the clonus, but is not in fact a cause of the genesis of these. But the inequality of the pulses in relation to different parts of the artery follows such *dyskrasias* proximately, for where the part of the artery is more moist, and at the same time hotter, there the pulse becomes larger and more rapid, whereas where the part is either colder or drier, there it becomes less and slower. Furthermore, it urges on the part first mentioned to movement and initiates the second part later. Consequently, in this, it will also jointly increase the clonic pulses and will become a proximate cause of unequal movement in different parts of the arteries. Accordingly, the so-called dicrotic pulse, for let me now speak about this, arises as follows. When the part of the artery to which we apply the touch is harder and colder, and because of this those parts on either side which we do not palpate become more difficult to move, and then excites these before the dilatation, drawing the middle back again inward, and again raising it, of necessity it twice strikes the touch. Such a pulse is then dicrotic only, **[79K]** and not yet also clearly clonic.

However, some other pulse is at the same time sufficiently clonic and becomes dicrotic in the following manner: when the artery is carried up clonically, to the extent it is bare under the skin alone, it is carried up readily, while those parts on either side of this are weighed down by the thickness or excess of the bodies lying on them, so the movement of these does not fall on those palpating, of necessity under these circumstances, when the evident part is already completely carried up, the non-evident parts on either side are still being carried up, and in this way the middle, which is brought down beforehand, is again raised up along with those parts on either side. As a result, they return and fall for a second time on those palpating externally. It is clear, then, that such a pulse will be dicrotic, which has some advantage over the clonic in this way only—it has acquired the inequality of movement in relation to the large parts of the artery. For when the whole movement falls on the fingers similarly, apart from what is clonic, while the parts on either side of it are different, the dicrotic pulse arises, whereas in the pulse that is clonic alone, there is no more quivering and shaking a little in all the parts, **[80K]** but the difference of the movement is in this way in large parts, which the dicrotic does not have. It is clear, then, from what has been said, that the dicrotic pulse sometimes occurs following a major *dyskrasia* and slight hardness of the moving artery in this way, and sometimes with major hardness along with some *dyskrasia*, or thickness or abundance of humors and flesh.

7. Certainly, the so-called bounding (like a gazelle) pulse has in common with the dicrotic pulse the double beat but not a common cause. For it does not follow of necessity the hardness of the arteries or the *dyskrasias*, but whenever the capacity to the extent it preserves the *tonus* is itself in accord with nature, this constrains the movement, due either to an immoderate excess, or obstruction, or compression of the arteries, then a pause divides the dilatation. As it ought, the second part of the dilatation after the pause seems stronger than that before the pause. For, as someone might say, the capacity, having been for a time overcome, collects itself and composes, strengthens and recovers itself, and resets itself **[81K]** more strongly, striving to overcome and thrust aside those things that are distressing. Furthermore, it does this often. However, if it were at any time overcome in relation to the second impulse,[656] such a pulse would in this way be worse than the bounding, but is without a name. And in fact some call all such pulses in which a pause interrupts the continuity of the dilatation, bounding. Names, however, hold no interest for us, other than those for the purpose of clarity of teaching.

8. Since the matter has already been clearly explained, we must go on to another kind of unequal pulses, which people sometimes call undulating and vermicular. For it was shown regarding these that they differ only in quantity, both being subsumed

656 The manuscripts F1, F2, and Va1 all omit ὁρμὴν following τὴν δευτέραν.

under one kind. And obviously they are of those pulses having the inequality in different parts of the artery in one dilatation. Causes of the generation of these are weakness of the capacity, excess of humors, or extreme softness of the arteries. For whenever they go through from the beginning to the arteries individually, the *tonus* is unable with equal power[657] to distend the whole vessel; in this [situation] **[82K]** it begins the movement first nearer the beginning, and the more distant parts later.

If, therefore, such a thing should occur equally through all the arteries, always the first part (of the artery) would seem to be moved first, as if something happened to be flowing unimpeded through these—either *pneuma* or fluid—and happened to bring about the pulses. And this is what Erasistratus says always occurs, being deceived by the appearance, although none of the other doctors, nor any of the philosophers or laymen are. For all the parts of the arteries clearly appear to everyone to be dilating at the same time. It is rare and not in accord with nature for the parts nearer the heart to seem to be moving first and those further away second.

Now such a pulse is not in any way undulant or vermicular, but when there is added to it the appearance of something flowing through the arteries which seems to be held back somehow, and to remain and rest, and then to flow again, and to be held back again, and this occurs through the whole artery, such a pulse is either undulant or vermicular. For always if the movements individually are circumscribed by pauses, **[83K]** what happens is something like the appearance of the succession of waves or the creeping of a worm. Therefore, if the capacity is burdened by excess, this holds up, cuts through and separates the *tonus*, and here the major part is gathered together, and in this way many dilatations occur like waves succeeding one another, in relation to which the *tonus* is divided and here the immobility circumscibes the wave. If, however, the capacity itself alone is fatigued without moisture, being unable to dilate the whole artery all at once, and carrying this out in sections, what would also occur is similar to many terminations of waves, when dilatations succeed one another.

In fact, what is different between these pulses and those occurring due to excess is magnitude. And because of this, these are vermicular due to the smallness, while those are called wave-like due to the magnitude. The weakness of the capacity is a cause for the former, whilst for the latter it is the preponderance of the moisture. And furthermore, such pulses occur due to extreme softness of the arteries, like what happens to soft young twigs taken from the highest level, and particularly those **[84K]** raised up near the point of taking, whereas some of those that are twisted still remain low-lying and even on the ground. And such a thing is not seen in the hard and dry twigs, but if you wish to lift them up, all that is taken up is lifted together uniformly. The same sort of thing is what also befalls the arteries whenever they are imbued with much moisture. For they do not still receive the *tonus* similarly in all the parts from the beginning, but as if they are bent and twisted, they are broken through in many places, and in this case the pulse will be, in many instances, undulant, and sometimes also ver-

657 K: ὁ τόνος ἰσοσθενῶς ὅλον ἐξαίρειν; F1, F2, and Va1: ὁ τόνος ὅλων ἰσοσθενῶς ἐξαίρειν.

micular. For with a strong capacity it is undulant, in this way being able to raise up large parts of the arteries, but with a weaker capacity it is vermicular, for under the circumstances it is obvious from a short interval the *tonus* is overcome. But of necessity this immediately becomes unequal, and in other respects also intermittent, as also obtains in the case of the dropsies. In these, although the *tonus* still holds out, just as in the discussions about those we shall speak of need, it is broad and undulant, but when overcome it becomes unequal in many ways. **[85K]**

9. But since enough has been said about these also, let us mention next the tapering (curtailed, mouse-tailed) pulses which follow a weakness of the capacity creating the pulses. For whenever the *tonus*, which always extends from the heart to the individual arteries and is more dissipated in the parts further away, in this way, of necessity, the movements in these become smaller and weaker. How, then, do some say they sometimes find lesser movements in the parts nearer the origin?[658] For it is not possible to say in the case of these, that the *tonus* in the parts nearer[659] the origin is dissipated more than that in the more distant parts. They seem to me either to have created a picture for themselves in this way due to the inequality of the position of the arteries, or if what they are saying is really true, it is necessary to refer the generation of all such pulses to an inequality of *dyskrasias*, so that the capacity that has extended from the origin (heart) to the arteries is with good reason[660] better utilised in the *eukratic* parts of these and worse in the *dyskratic* parts. For due to *tonus* that is equal and the same, the organ (heart) that has been mixed with moderate moisture and heat is moved more readily and more, **[86K]** whereas that which is dried out or excessively moist, or cold, is moved more sluggishly and less.

10. How, then, shall we investigate the natural causes of the so-called nodding and inclining pulses similarly, or can they in equal manner sometimes also arise in this way? At any rate many have clearly appeared to me to arise through other causes. For when the capacity is weak but there is colliquescence of the bodies surrounding the arteries, and as a result some of these are in a particularly flabby condition, and because of this are very light, the movement is enabled to rise up and because of this particularly its magnitude is made apparent. For when the dilatation in this is compared to those on either side which are smaller, the parts display the movement that is tapering in relation to both. It is not possible for this to occur in the very strong capacities, as it is necessary for this to be immobile and not to bear the weight of the overlying bodies, so that it may make perceptible the prominence of the distension in the barer part compared to those in the parts that have been weighed down. **[87K]**

11. How, then, does the inequality of the pulses arise from the position of the artery? For this still seems to be wanting in the discussion. How otherwise than just as Hippocrates said, when teaching us in general about every distortion? For surely, it is

658 K: ἐν τοῖς ἐγγυτέρω μέρεσι τῆς ἀρχῆς; F1, F2, and Va1: ἐν τοῖς ἀνωτέρω μέρεσι τοῖς ἐγγυτέρω τῆς ἀρχῆς.
659 K: ἐγγυτέρω; F1, F2, and Va1: πέρασι.
660 F1, F2, and Va1 omit εὐλόγως.

just like that of the uterus and those of the spine and the muscles occur when they are extended into those parts of the bodies in which they are distorted. It is because there are many causes of tension, and in general detection was difficult for the majority of the doctors of old. For inflammations, indurations (scirrhosities), abscesses, and the generation of certain unconcocted *phymata* (tubercles), and altogether all the tumors (swellings) contrary to nature are able to stretch the contiguous parts towards themselves, and besides, hardness, dryness and excessive cold, acting like some congelation and spasmodic tension—one or other of these—distort the arteries. And we must particularly consider the non-uniform *dyskrasias* causing distortion.

12. But since I have also stated the probable things about these, I must add the discussion about the spasmodic pulse, in which the name is fittingly derived from the actual form, **[88K]** and naturally signifies the affection. For you will think the artery is being stretched towards both ends, and a little later a spasm seizes the person. What, then, is the problem here and for whose sake is it being put forward? If the heart is an origin of arteries as it also is of nerves, there is no problem, and it is put forward needlessly. If, however, there is one origin of nerves and another of arteries, and a pulse is an action of the latter, while perception and voluntary movement are functions of the former, so there is no association of the functions in these, in this way there will not be any commonality of the affections. Or do those who think there is no relationship between them speak wrongly and falsely? Certainly, it is not as the father of the doctrine, the divine Hippocrates, says: "There is one confluence (flowing together), one union (breathing together), all things are jointly affected with all, and there is one common nature. For if this were not so, there would be two organisms and not one of each of us."[661]

Now the heart sends great arteries to the brain bound in association, while the brain sends nerves to the heart, and even if they are not large, they are at least able to express the command of reason. Also, in this way fibres from the nerves come to an association with the individual arteries, through which there are not **[89K]** only actions, but also a sharing of affections with each other. And there is nothing still remarkable if the whole class of nerves being connected with the arteries, or a commonality of the affection is apparent, when the branches individually are mixed with each other and the origins are, as I said, connected.

13. Since also the matters pertaining to a spasmodic pulse have been spoken of, what still remains that we must add is the investigation of the natural cause of all the unequal pulses.[662] What is left, as I think, is how deep (high) and shallow (low), broad and narrow, and long and short pulses occur. For certainly, these are very numerous and vary especially in respect of all the inequalities, both those in relation to a single pulse, and much more, those in a collection [of beats]. Now what happens

---

661 Hippocrates, *De alimento* 23 (ed. and Engl. transl. Jones 1923: 355).

662 What is implied here is the contrast between physiological and pathological causes in modern terms—that is, between the Greek φυσιολογία and παθολογία.

to the arteries while dilating in all those who are healthy and the majority of those who are diseased is that we perceive the upper part of the circumference [of the artery] moving more than the lateral parts. For the former is nearer our touch and is in contact through fewer bodies, whereas those parts on either side of this are further away **[90K]** and are surrounded by more intervening bodies.

Accordingly, the actual pulse that accords with nature, should someone examine it accurately, is higher, since the movement in it is always deficient in breadth, but more substantial in height. But since this is for us a measure and standard of all the others, we call it neither high nor narrow, but as seems right and proper, moderate and a median of all the extremes. Since we are accustomed to this and are brought up with it, and often find it not only in those who are healthy, but also in those who are diseased, we easily recognise those that fall short of it or go beyond it in relation to any one of the dimensions whatsoever. For certainly, the long and the short are also spoken of with reference to this. And if anyone is not accustomed to this, not only would he not recognise accurately high and low, and broad and narrow, but also long and short. So then, we say some person is average in every dimension, not comparing in relation to perception the length with what comes to hand equally in breadth and height, but with that in relation to potential. For none would be found to be short in this way when the length is deficient compared to the breadth. In this way too, **[91K]** we term a pulse moderate, when we make the analogy to an accord with nature and a moderate standard of equality, and not to perception.

Always in this, the length exceeds the depth, and the depth in turn exceeds the breadth. For if we were to conceive of the artery itself, bare of all the surrounding bodies, the breadth will be equal to the depth. In fact, in the case of such [an artery], we shall not be in a position to say which is more—the width or the depth—but as with a body that is precisely circular, every direction is like itself, so too there is no difference of the dimensions. And indeed, when it is placed in relation to some other body, one might already say "below" of those parts it supports and "above" of what is opposite. But in respect of what is in the middle of the dimensions, they think of height from below upwards and depth from above downwards. Here there are height and breadth and there the other of the dimensions, depth and breadth. But if it is not only fixed at one point, but also from the sides themselves, and besides these, is surrounded by certain bodies above, it should somehow be clear to everyone that those parts we contact through the fewest coverings **[92K]** will show the movement more clearly than those we contact through more.

But it is in fact the position of the arteries under the skin—such are the arteries we palpate when diagnosing the pulse. Of these, it is the upper part we contact through the skin only; those parts on either side deviate as much toward the side as determined by the longer dimension and the more bodies intervening. And because of this necessity, every pulse which accords with nature is always higher, if you compare the dimensions with each other. And it is called moderate and a standard, like the Canon of Polyclitus.[663] For as when we compare the other statues of a man to that, we call them stout or slender, broad or narrow, or some term of the other excesses, so too, when we compare the pulses to the standard of accord with nature, we term them high or low, broad or narrow, long or short, stout or slender, or large or small.

What, then, are the causes of the generation of these? For this is what lies before us to state. What was at hand was either to inculpate a non-uniform *dyskrasia* of the arteries or distortion in the position. For it has been shown how, in such conditions, **[93K]** the movements appear unequal. But I find the high pulses occurring in the rages and before the crises, but the broad ones in general after the critical excretions. I am not still able to inculpate *dyskrasia* of arteries or distortion of position only, for sometimes they do arise due to one of these, but the majority of conditions in which such pulses appear are outside the causes mentioned. For neither a non-uniformity of *krasis*, that is to say one of such a degree that it changes the nature of the arteries so as to bring about unequal pulses, nor the change of position, are ever able to exist apart from severe damage or a long time, neither of which exist in those who are angry, nor those who have had a crisis, nor in those who are not yet come to such a critical point. But such a thing is what occurs in them. It makes the pulses larger in those who are angry and those who will be brought to a crisis, but do not become larger to such a degree as it appears.

For in the movement towards height and in the dilatation of the artery, as much as comes forward is all apparent, whereas in the movement towards the side, the greater part is hidden, for the same reason, I think, in relation to which also in those **[94K]** who are healthy without some condition, the movement to height was shown always to be predominant. For that very reason, when we call to mind the state of the pulses at that time which we term moderate, then now find them substantial in height, although not substantial in breadth, but either altogether small, or the movement not to have increased at all, we would reasonably term such pulses high. And if also something is clearly added to breadth, we would call them large and at the same time high, but if nothing at all is perceived, only high. If, however, they were increased in both height and breadth, but the prominence of height were not notable compared to the breadth, we would term such pulses large only. For when, by the same token, the increase of height exceeds that of breadth, by which it also exceeds an accord with nature, it is necessary for what was called moderate to preserve the same form that now exists. It is not then high, just as that is not either, but only seems more than this. For those preserving the same form, it is not possible to call one moderate and the other high. But if the one is increased in all directions in the same proportions, [we call it] only larger.

---

663 Polyclitus was an Argive sculptor active during the middle to late 3rd century B.C. His most famous work was the Doryphorus. He is said to have written a book (his *Canon*), detailing the principles of his art, as exemplified by that work. Galen refers to Polyclitus and his *Canon* in a number of places, including *Ars Medica* 14 (I.343K; Engl. transl. Johnston 2016: 211).

For surely then, it is similar concerning length. **[95K]** When there is an increase together in the same proportion in depth and breadth, it makes the whole pulse seem larger, as has already been said. There is not, therefore, need of simple equality of the increases in the three dimensions of movements, but equality in proportion, so that the pulse becomes larger. To those accustomed to the analogy of accord with nature, the extent to which what is added is outside the proportion is easily recognised. Such are some of the generations of the pulses under consideration.

The next thing is to state the causes by which the natural proportion in these is either preserved or destroyed naturally. It is preserved, then, when there is no great change from an accord with nature towards plumpness or thinness of the bodily state, but not immoderate dryness, or moistness, or some great necessity of extreme dilatation. It is destroyed and demolished through one of these things being present —for example, let all the other things be good, but let the capacity be hastened in this way to make a great dilatation, and that the spaces around the arteries have no more room for its mass. For it is necessary, in such a state, for the pulse to have appeared high, **[96K]** even if you have none of the other factors to blame. For until the dilatation of the artery is increased to such a magnitude so as not yet to be constricted anywhere, it preserves the proportion in accord with nature in the three dimensions. When it is constricted to such a degree, it takes away from that dimension, and does so to the extent that it is henceforth hindered. Because of this, in those who are angry, even if they should have a faultless bodily state (*hexis*), the pulse becomes high. For such a bodily state does not naturally tolerate the severe excess of the dilatation, but already the artery is compressed from below and also from the sides. Therefore, only the bare part of the artery under the skin alone receives all the heat, whenever the dilatation is hindered in all the other parts, whilst only the part of it above is preserved free from interference. It is necessary for the movement in that part to predominate over the other parts, which is nothing other than a high pulse.

Since in all those who are healthy and the majority of those who are sick, in the largest dilatations of the superficial arteries which we palpate, all the other parts are compressed and constricted, and prevent their movements in these, the parts rising up to a height, **[97K]** receive the whole dilatation entirely free and without interference. Because of this also, it has been observed and written by doctors that a pulse of those who are angry is high. For they either did not see the rarity, or looked at it carelessly, or thought this pulse characteristic of each of the affections, since they often saw this pulse accompanying them. Seldom are there those who choose to remain silent, when they are at a loss about the cause. Someone might blame them for this, considering them worthy of forgiveness on all other things. But in this way, countless things worth knowing will not be spoken of. It is necessary for one who loves the truth to describe in detail every phenomenon accurately, even if he is at a loss about the cause.

Whatever, then, is the phenomenon? In many of those who are thin and not very strong in the capacity, the pulses are large when they are angry, although they don't yet also seem high. Anyway, what the cause of this is seems not to be far from those things assumed initially, if in fact, under these circumstances, we said the strength of the capacity was needed in the high pulses so as to be able to bring about a very large dilatation and not be overpowered by the **[98K]** regions surrounding the arteries. If this is so, it would somehow be clear to everyone, that in the previously mentioned conditions, it would not be possible for the pulse to become high, since the capacity is not strong and the spaces around the arteries have become very wide due to the attenuation. But since it is also found to be more rapid than expected, it is, I think, also clear from this why the broad pulses have arisen. For whenever the capacity somehow falls short of the strength that accords with nature, while there is significant width of spaces, and some weight is added to the parts above the arteries, such a pulse seems broader when compared to the moderate. This weight which is added is often the wrinkling of the skin itself, as occurs in those who are old and those who are thin, but not in those who have been dried out in respect of the bodily state (*hexis*), from which class are those who have recently had a crisis. However, it is often due to moisture, and particularly when the actual bodies of the arteries have become imbued with moisture, as in the dropsies. For the parts of these that are placed at the peak of the height flow in and are carried downwards, in some cases to the actual lumina of the arteries and in others to the sides. **[99K]** Of necessity, then, there is prevention of the pulses rising to a height to such an extent that the back of the artery is cast downward. And in particular, it is already inevitable that the breadth gains ascendancy from both sides—from the compressed parts of the artery flowing away to the sides widening the parts below and from these the height itself beforehand having been cut off. However, I think either of these is sufficient to render the pulse broad, for either the height is diminished or the breadth is increased. When both occur together, this is why an obvious and significant breadth is displayed in such conditions.

How, then, does the so-called long pulse arise? For this is found sometimes when the movement falling under the four fingers is indeed narrow and faint to such a degree that there is an impression of a thin string, or also there is no difficulty discovering from which things we said they are set in motion. For when the impact seems either like some thin string or like a hair, some great hardness is being displayed by the actual walls of the arteries. But not only that: for in this way, the pulse would simply be hard. Since it is also narrow and low, of necessity **[100K]** the capacity is weaker, so that it dilates forcibly, the artery being refractory due to hardness.

Why, then, if this is true, does such a pulse not also immediately become short? Or shall we also discover the cause of this by recalling the conditions from which such pulses arise? What are the conditions then? The thin state of the body and skin that is somehow stretched all around and drier. If what is relaxed and at the same time thick conceals much of the movement, what is thin and stretched preserves it all absolutely. And it has already become clear that there must be a coming together of thinness of the whole bodily state (*hexis*), dryness of the skin, hardness of the arteries, and weakness of the capacity, but not extreme. If this were so, the person would be pulseless and not long-pulsed,[664] for the completely dissipated capacity altogether fails to raise a perceptible movement—not in a vessel that is soft, much less in one that is hard.

In this way, then, the long pulse is named on the basis of some excess, but the specific name for it is thin and slight, although it is absolutely and properly long, by which it doesn't take into account the amount of the other two dimensions. **[101K]** Such a pulse is brought about either through the thinness of the body or through its own particular magnitude. However, the pulse is short when, in any one small part whatsoever, the artery is covered by skin only, whereas everything on either side of it collectively goes down into the depths. For then, necessarily, the superficial part of it under the skin clearly shows the movement in its entirety, whereas the parts that are deep are concealed by the many bodies lying in front of them. This exists naturally in many people, and indeed in a few disease conditions, whenever, in the condition, some distortion of the artery occurs, and only a small part of it is manifest, while all the other parts are drawn inward towards the depths. There is also such a pulse of those called short *par excellence*, since the specific name for this is thick and swollen. The absolutely short pulse is engendered by plumpness or smallness. I have now spoken about almost all the causes producing inequality; if some small matter has been omitted, there should be no difficulty discovering it from what has been said.

14. We must now speak about the irregular (disorderly) and regular (orderly) pulses. Accordingly, since all such pulses are unequal, **[102K]** as has been shown in the books on the *differentiae* of these,[665] they differ in that some preserve the equality in certain periods, while some are altogether in confusion, it is necessary, in brief, to recognise this one thing common to them all—that the cause, which is what brings about the inequality in each case, is stable in those that are regular but unstable in those that are irregular. And yet, order is characteristic of Nature, if anything is, just as equality also is. If, then, one of these is in confusion, it indicates lesser damage, whereas if both are in confusion, the damage is greater. Both are in confusion in the irregular pulses, but one or other in the regular pulses, and are clearly so in the unequal pulses.

15. The causes in the changes of the rhythms (for these remain to go over), someone might discover without difficulty, if in possession of those things proceeding from the beginning. For if we have demonstrated correctly that, in the dilatations of the arteries, some air-like substance is drawn in, whereas in the contractions, the residue and smoke, as it were, of the burning up of the humors is expelled, it will be necessary that when there is a pressing need of the dilatation, it becomes larger and more rapid, and when of **[103K]** contraction, the contraction [does likewise].

---

664 The contrast is between the terms ἄσφυκτος and μακρόσφυκτος; for the former LSJ has "without pulsation (Galen, VII.194K)" and for the latter, "having a long (i.e., spatially extended) pulse" citing this passage.

665 *De pulsuum differentiis* I–IV (VIII.493–765K).

Therefore, in the stages of life (ages) in which there is the greatest expenditure of humors, and these are the stages up to the prime, the dilatation falls a little short of the contraction, whereas in those stages when expenditure is least, which occurs in the elderly, the time of contraction is comparatively much greater than that of dilatation, if, as surely is the case, transpiration in relation to the whole body is the work of the contraction of the arteries. For in the dilatations, they draw into themselves via those passing out through to the skin, just as through all other openings. Therefore, under these circumstances, nothing is excreted from the body, but altogether more is drawn inward. Conversely, in the contractions themselves it is excreted, while nothing is drawn in.

It is clear that in the stages of life (ages) in which Nature works on the humors, not only for nourishment but also for growth, it is inevitable that more superfluities arise in these. For the amount of the superfluity is always in proportion to the amount of the humors worked on. Here then, when there is greater working up [of humors], there is the greatest superfluity, whilst when there is the greatest superfluity, continuous at the same time with this is the need of much **[104K]** evacuation, just as, I think, is the case in the superfluities in the stomach and bladder. But we recognise the expulsions of these through the senses, whereas reason discovers those in the arteries. And I do not know whether someone will agree that the arteries are evacuated through what appears externally or in the way I spoke of. I have demonstrated as much about these as was necessary in the work *On the Use of the Pulses.*[666] The things now said are sufficient for the present discussion.

Book III

1. [105K] The previous book adequately explained the changes of the pulses from the primary and most important causes. We shall now go over the changes from all the other causes—those that are not outright [causes] by themselves, but are called the primary causes for the change of pulses. However, since of these, there are those that are contrary to nature, those in accord with nature, and those in between these, which people also call non-natural,[667] it is reasonable to start from those that are in accord with nature, and apply to the discussions the same order as the one I have also used in the work written for beginners.[668] **[106K]** For there would be an explanation of both other things, and at the same time an indication in it of those things correctly stated and those things reasonably omitted; there are however those that have been omitted completely and those not differentiated in it.

---

666 *De usu pulsuum* (V.149–180K).

667 For an overview of causation in Galenic medicine, see Johnston 2006: 81–125. On the specific matter of non-natural causes in the pulses, see Bylebyl 1979.

668 *De pulsibus ad tirones* (VIII.453–492K).

It seemed necessary to omit every aspect of the theory that was going to be obscure for beginners. Not least also the consideration of the cause has been omitted. For it were better for beginners to be practised empirically in the actions themselves first, and then later to have been taught by reason (theory) the whole essence of the matter. Accordingly, this work is composed for those who are practised and able to learn the more complete arguments. I have discoursed sufficiently on the so-called *synektic* causes through the first two books; through the remaining two in turn—this one and the one after it—I have taught about all the others. We must now begin the discussion, taking in hand each statement of those written in that book,[669] and then explaining it. And the following is the first of those suitable for the present discussion. **[107K]**

2.* *Men have a pulse that is generally much larger and similarly much stronger, and a little slower and significantly more infrequent than women.*

This is the statement. And why this statement is more fitting to speak about first has been shown by that book. Consequently, although it is not my custom to write about the same things often, we must now go to the causes and must prove the necessity of the nature of the matter—that is, for what reason such aspects of the pulses are so in men and so in women. This is as follows. Since the male is hotter and drier than the female, as was also said by the best of the doctors and philosophers of old, and as I have shown elsewhere, it is inevitable now that the need of the generation of the pulses and the state of the organs (arteries) is different in them; if this is so, it also applies to the actual pulses. For it was shown in the previous books that the large pulse is characterised by an abundance of heat. If, then, the male is hotter than the female, he would immediately have larger pulses, and if he were much hotter, **[108K]** they would be much larger. But he is also hotter and much hotter, so he will have pulses that are larger and much larger. But that he is hotter has been adequately demonstrated by those things I have already written about *kraseis*[670] and those things previously stated by the ancients. However, that he is much hotter, neither you nor anyone else has yet shown. So let me now demonstrate this, or the whole argument is not yet credible.

We must first call to mind what was set before us in the statement, and we must read all of it, paying attention to each term. What does it say? That in general men have a much larger pulse than women. What we must consider carefully is what this "in general" is. For it is not, certainly, a needless attachment—the whole authority of the argument is in it. The generality of the statement is destroyed by the fact that we do not have similar *kraseis* (are not similarly blended/tempered) right from the beginning, nor have we used similar practices and regimens; sometimes it is possible for a

669 The passages from *De pulsibus ad tirones* are in italics. Those in which there are minor differences in the Greek text are indicated by an asterisk; more major differences are indicated by a footnote.
670 Galen, *De temperamentis* (I.509–694K).

woman to be found whose pulse is not much smaller, or is not small at all, nor equal, but is clearly larger than in some men. **[109K]**

Suppose for me a man more abounding in phlegm by nature, in whom the *krasis* would be cold and moist compared to a woman abounding in bile by nature, and is thus hotter and drier. Do not, however, let this alone be sufficient, but let him be settled in Pontus and her be settled in Egypt; and let the man live indoors and in idleness, and let him have affected habits in respect of diet, whereas let the woman live outdoors and exercise a lot, and let her eat moderately. The pulse of such a woman will be greater than that of the man. But if you keep each living the same life in nature and they have been blended in the characteristic *krasis* from the beginning, and are in one place and using the same [regimen], not only would the pulse of such a man become greater, but also much greater, But to the degree it depends on the natural *krasis* it would simply be greater and not much greater. However, since the difference comes to each from the characteristic practices, if she were to live a life free of toil and he a life of much labour, the man would, in this way, become much hotter, and because of this would have much larger pulses than the woman.

"In general" is added **[110K]** in the statement for the more naturally clever of the beginners as a kind of goad, exciting and stimulating them to the investigation of the whole nature of the matter. For they will immediately also consider the difference in nature in a man and a woman, and will see the male is stronger than the female throughout the whole animal class, and that in the female, it is natural to bear children and suckle them, and altogether to nuture those born in accord with nature, while for the male it is to provide food for them and to fight on their behalf against those who scheme against them. From these things they will know there is always the difference of male and female which simply brings about the larger pulses. In this way too, it is in the nature of males for the pulses to be absolutely stronger, and of course the pulses in males are much stronger compared to those in females, because in the former the natural *tonus* is necessarily increased in their labours, whereas in the latter, who are idle it is diminished. And if someone should not clearly discover any of these things himself, but should come to a notion of them through enquiry, or at the end, thirsting as it were for those things he desires to discover, but has not yet discovered, **[111K]** should he come upon someone able to teach him, he would in this way very quickly learn and especially remember.

But these things are like this. However, the strength of the capacity existing in males would also itself give something of largeness to the pulses, so that not only through the amount of the heat but also through the strength, the pulses of males will become larger. This, then, has also been shown in the first book—large pulses are characteristic of strong capacities. Do men, then, differ from women only in this way with respect to causes of pulses, or does the matter of the state of the body (*hexis*) come into play significantly for each? For in men the whole body breathes freely and is pure; it is without superfluity and is not compressed, whereas in women the opposites [apply]. In them, the spaces between the vessels are narrow, since fat, soft flesh and certain phlegmatous superfluities occupy them, the body is condensed and short of breath, and in many places of the body accumulations of stagnant, thick and unconcocted fluids exist. By all of these things their arteries are inevitably compressed and constricted, since they are naturally much **[112K]** less in terms of mass, so that everywhere and in every way the bodily state is prepared by these things for smallness of pulses. They have, however, one thing alone suitable for largeness, which is the softness of the arteries, but they derive no benefit from this. First, this is because the causes of smallness become greater and stronger. Second, it is because, while the difference between men and women is greater in those things, it is least in that due to softness of the arteries.

It is said the bodies of men are not drier in this way when compared to those of women, as for example, a stone is drier than mud. Nor would you find much excess in the actual walls of the arteries. But also, the arteries of men have such a constitution that they are responsive in terms of action. If, however, the excess were large and the arteries of the woman much softer, this itself would make the pulses smaller. For where there is nothing weighing down, compressing or constricting, then the softer artery is more suitable for a large dilatation, whereas where, as was said before of the female bodily state (*hexis*), the compressing, **[113K]** weighing down and constricting is considerable, there the softer artery is more naturally unsuited to the dilatation, so it is more easily overcome by all the disturbing factors.

It is clear from what I have already said that all these things are stated about the bodies which are constituted best and following regimens proper for their constitutions. Of course, the whole and definitive discussion about these, brought to completion with demonstration, has been written in the second book of *On the Diagnosis of the Pulses.*[671] Consequently, one must also read that treatise, if one intends to pursue correctly those things pertaining to the pulse. And furthermore, it has been shown in the third of those books, how greatly the majority of doctors are mistaken about the rapid pulse, thinking, when it comes to the recognition of this, that the short-duration pulse is immediately also rapid. And never mind the pulse of women, when we recall what has been pointed out in those [books] for the sake of the matters now before us, which they think is sufficiently rapid. It is more rapid, but does not exceed the pulse of men by a significant and large difference.

This, then, is the truth and has been rightly written for beginners—the **[114K]** pulse of men is a little slower than that of women, but the cause is not in fact added, just as it is not in anything else written there. But I shall state it now. To the extent that the capacity is weaker and the heat less in women, the pulse must also be slower; such a pulse was shown to be characteristic of both the causes mentioned. But if it is much smaller, and because of this also the need is not significantly fulfilled, it adds as much as possible to the rapidity. The first book showed this to be inevitable, and anyone who does not recall what was written in that book, encounters these [statements] in vain. If, then, the addition of the rapidity is of such a degree as to fulfil the

671 *De dignoscendis pulsibus* II (VIII.832–877K).

whole need adequately, there would be no need of frequency. However, if the action falls a long way short of the need due to the smallness of the dilatation, and there is little benefit from the rapidity, it is already inevitable that the pulse will become frequent to a significant degree. For surely, this was also shown in the first book.

Thus, women differ from men in the pulse in this way, and the male infant differs from the female infant analogously, although not similarly in every respect, but as I said before. **[115K]** For the male has a pulse that is absolutely larger, stronger and more infrequent than the female, but not yet by much. These things are enough about these matters. Let me move on in turn to the statement teaching about the difference in respect of the *krasis*.

3. *Those who are hot[ter]*[672] *in nature [have a pulse that is] much larger, more rapid and more frequent but not much stronger.*

This discussion also contains within itself a major question to which someone is unable to apply the understanding of beginners, unless he is very intelligent. For of those who are hot in nature, they say there are those who are moist and those who are dry. It is necessary, then, not to say simply the pulses of those who are hot in nature, but first of those with moistness and second of those with dryness. For not all those who are hot in nature have the same pulses; there is also a third group comprising those who are *eukratic* in terms of moist and dry, but *dyskratic* in terms of heat. That there are those before me who have wrongly overlooked these [factors] when they divided the *kraseis* has been shown by the treatise I wrote on *kraseis.*[673]

Indeed, it is customary to call those **[116K]** hot who, of the four powers,[674] have this one alone above the others. These [people], then, to the extent that they are hotter than those who are *eukratic*, have pulses that are to this extent, larger, more rapid and more frequent. And the cause is clear beforehand to those who have recalled the things written in the first book. But why did we say such pulses are not much stronger? And indeed, in the first book of this particular treatise, we were uncertain as to whether the pulse could become stronger without that by which the capacity is helped towards strength, because to beginners all such pulses seem stronger. For how does the pulse that is rapid and large at the same time, according to some ill-conceived phantasy, lie beside that which is strong, so that some of those doctors who have left us commentaries said that a pulse compounded from largeness and rapidity cannot be otherwise than strong. It is not, then, necessary to set in motion so great an enquiry for beginners. Of benefit to them in many cases are the indefinite and unresolved rather than the distinct, whenever the latter requires a profound theory, prolonged discussions and a lot of time. But **[117K]** in the first book of this particular treatise, we discoursed about this sufficiently, intending in those books following, when the discussion

672 Θερμοί here; θερμότερος in *De pulsibus ad tirones* VIII.465K and present volume..

673 *De temperamentis* (I.509–694K), in three books (Engl. transl. Johnston 2020).

674 The powers (δυνάμεις) here are taken to refer to the four elemental qualities: earth, air, water and fire.

comes to the matter of rhythms, to say something again. But here, for the present, we shall put an end to the explanation and turn our hand now to another statement.

4. *Those who are thinner [have a pulse that is] larger and more infrequent by a lot but stronger by a little.*[675] *In nature, then, they differ in this way.*

Here again one must understand those who are thin without any other condition. For if there is either a change of the *krasis*, or of the capacity, or greater thinness arises due to some affection, the change of the pulses will be compound and no longer due to the thinness alone. Thus, thinness itself alone, by reason of itself, makes the places occupied by the arteries wider and the bodies lying on them lighter, and contributes a large enough magnitude to the pulses. Then, if the need remains equal, the magnitude increases, and then inevitably the pulses become more infrequent to the extent that they are larger. If they were larger by a lot, they would in this way also become more infrequent by a lot. They seem to such a degree stronger to the extent that **[118K]** they are in contact with the things touching them through thinner bodies. Therefore, such things appear not by acquiring some *tonus* but by displaying that which they have to a greater extent. Concerning rapidity and slowness, nothing is said beyond these things, for nothing is taken on in addition of either one of them from the thinness, but they are similar to the thicker, whenever they possess both strength and need similarly. Next in succession the following has been written about ages (stages of life).

5. *[The pulses] change in relation to the stages of life (ages) in the following way. The pulse of the newborn child is very frequent, while that of the old person is very infrequent. Those in the stages between are all in proportion to the extent that people happen to be nearer to a child or an old person. Similarly, the pulse of the child is very rapid, while that of the old person is very slow. Those of the other ages are in between in this. However, the difference in respect of the infrequency between the old person and the child is much greater than that in respect of rapidity. In the difference pertaining to strength and magnitude, the largest, as in the ages, is that of those in their prime and the smallest is that of the aged, while in the middle or a little larger is that of children.* **[119K]** *And the pulse of those in their prime is very strong, while that of old people is very weak; in between these is that of children.*[676]

That frequency and infrequency of pulses is conceived of in a twofold manner has been shown in one way by those speaking of the perceptibility of contraction and in another by those saying it is imperceptible, which is more clearly stated in the books, *On the Differentiae of the Pulses,*[677] but not also in this same book I have written for beginners. I thought it worthwhile for them to be practised first in the imperceptibility of the contraction. It is then more accurate, and at the same time also much

675 Possibly σφοδρότερον. Here, δ' ὀλίγῳ. *De pulsibus ad tirones:* δὲ οὐ πολλῷ.

676 There are the following notable changes from the statement in *De pulsibus ad tirones:* (i) sentence 2—πυκνότατος and ἀραιότατος instead of πυκνότερος and ἀραιότερος (ii) sentence 3 (end)—ἐν τῷ before μεταξύ (iii) sentence 5—μέσος δὲ βραχεῖ instead of μέσος δὲ αὐτῶν βραχύ.

677 *De pulsuum differentiis* I–IV (VIII.493–765K).

more useful, to think this class of the pulses exists in the quantity of the pauses. For in this way we shall reach a precise diagnosis in relation to the rhythms. But since in general the conditions producing the frequent or infrequent pulse make it appear to be such to each of the sects of these, and the conjecture of the cause would be common to both. But it would be better, reducing this to what is truer, to go forward in the discussion in relation to that, no longer now in fact making out that the matter before us is for beginners, but teaching thoroughly the whole truth. **[120K]**

Accordingly, children, since they have not a little heat, need not a little of the substance cooling it, and because of this need large pulses. And since they have the greatest consumption of the humors, as serving not only for nutrition but also for growth, in the workings up of these they inevitably collect in the whole body an abundance of vaporous and, as one might say, smoky superfluities, so that also because of these things, they need a very large contraction of the arteries, and indeed also the substance of the psychical *pneuma*, not only for nourishment, as in those who are fully grown, but also as a necessity for the associated increase in the organs. As a result, and because of this, they will require larger dilatations, and in infants the need for the arteries to contract almost comes to an equal level to that of the dilatation. For there are very many superfluities, and because of this, the need for very large contraction, while the heat, if this is not also simply very large, but certainly also due to the increase of the psychical *pneuma*, comes by the need of the great heat of the dilatation in them.[678]

To the extent that in youths there is a falling short of the excess in acquired heat, this **[121K]** is brought forward from the need through the *pneuma*. Therefore, children need movements that are both very large and at the same time also very rapid, and because of this, the action in them falls short of the need to a greater extent. For surely, if they are as rapid and large as possible, they are not yet adequate for the need. It is necessary in this, then, for the movements to become frequent. In the case of old people, however, everything is the opposite; the heat is slight, and in proportion to this also, there is the generation of sooty, as it were, superfluities and the nourishment of the psychical *pneuma*, so that it is reasonable for the pulses to be opposites in them also—that is, small, slow and infrequent.

What must be remembered here are those things said in the second book of *On the Diagnosis of the Pulses*,[679] in order that we may see what many doctors were deceived by when they supposed the pulses of children to be small; they are large, if you compare them to the masses of the arteries, but small if you make your assessment from the immediate impression. This is the case, if you consider only the amount of the distension in them, comparing them to those of people in their prime. Children do not in fact have very large pulses, if you don't **[122K]** examine them in relation to the masses of the arteries. For the capacity doesn't come to a peak of strength in them, and the

678 The meaning, and therefore the translation, of this complex sentence is rather uncertain; there is a textual variation as follows: K: τοῦ πλείστου χρείᾳ θερμοῦ τῆς διαστολῆς αὐτοῖς; F1, F2, and Va1: τοῦ πλείστου χρείᾳ τὰ τῆς διαστολῆς αὐτοῦ.

679 *De dignoscendis pulsibus* II (VIII.823–877K).

constriction and compression involving the walls of the arteries, which are still soft, takes a lot away from the largest dilatation. Nevertheless, the pulse in them is still so great that it is above the median, and in rapidity is similar in relation to the dilatation and contraction, so that also the need for both movements is equal and not as it is in the case of the aged, for to the extent that the need of the contraction falls short of that of dilatation, so to such an extent the contraction is slower than the dilatation.

It is reasonable for it to fall short in the aged because altogether little or no superfluity arises from the burning up of the humors, due to the weakness of the heat, and because there is no significant nourishing and no growth at all. However, the use of the dilatation, even if not great, at any rate still remains moderate in the aged, for it needs to fan the innate heat and nourish the psychical *pneuma.* And in old people, the change to infrequency is much more than that to slowness, due to the need of the pulses not only to be fulfilled completely by the **[123K]** action, but also to be entirely released by reason of itself. There is good reason then for the pause to become very long in this, just as also in fact that it was not fulfilling along with increasing greatly, forcing there to be continuous movements.

In the case of newborn babies, it was said previously that not only is the need of the generation of the pulses increased to an extreme, but also that of the very large dilatation which alone perhaps would fulfil the need, having said before that not a little is taken away. We said that what is taken away is as much as the capacity falls short of the peak strength, and the arteries are weighed down and constricted, being soft due to moisture in those of such an age. For truly, we also said this before—arteries that are soft without compression, weighing down and constriction contribute to largeness of pulses, and with these things, to smallness, and particularly when the capacity is more deficient in *tonus.* To the degree to which the pulse of old people has undergone a change from the frequency of children to infrequency, almost as much has the movement of the contraction changed from rapidity to slowness. And the cause is clear beforehand, if in fact **[124K]** we correctly said the need of contraction has been completely released in such stages of life (ages), but not in fact, if need of the dilatation has completely gone—and I have also shown this. From this, the movement in the dilatation is less slow than that in the contraction. The beginner is able to diagnose this alone. Consequently, it was reasonably said that the difference between an old person and a child is greater in relation to infrequency than it is in relation to slowness.

But now, in fact, everything remaining about this actual difference has already been specified. Over and above this, [there is] the change relating to the rhythms in the stages of life, about which I said nothing, thinking the discussion to be too obscure for beginners. For as with the frequency, so too with the rhythms, there are two schools of thought:

- those who say the contraction is not perceived at all, comparing the time of the dilatation with the whole remaining time, which is compounded not only from the contraction but also from the two pauses.

- those who say the contraction is perceived in the times of the movements making up the rhythms.

How much more accurate this latter school of thought is, and how much more useful for prognoses, will be made very much clearer in the treatise that is going to be written.[680] **[125K]**

However, it is not unclear now, unless someone has listened cursorily to what has been said. From these same things, there will no longer be any difficulty discovering the differences of the rhythms in the stages of life, and the cause from which they arise. Since the greatest need of each of the movements of the artery is in newborn babies, and the least in those who are old, while the whole time between these stages diminishes both the needs, but more that of contraction, it is reasonable that, by as much as people move away from infancy towards old age, they will always have both the movements slower, but more that of the contraction. It is from these that the differences of the rhythms in relation to the stages of life arise; I say more about these in the treatise, *On the Diagnosis of the Pulses.*[681] But it is also quite clear why the pulse of those in their prime is strongest, while that of those who are old is weaker. For the capacity is strongest in those in their prime and weakest in those who are old. And so too in the other stages of life—as they are strong or weak, so the pulses will differ in weakness and strength.

6.[682] *In relation to the seasons, pulses are* **[126K]** *largest and strongest in the middle of spring, as in times of the year, but moderate in rapidity and frequency, and similarly in the middle of autumn. As spring advances, it takes away some of the largeness and strength but adds to the rapidity and frequency. And finally, when summer takes hold, the pulses become weak, small, rapid and frequent. As autumn advances, it takes away from all [aspects]—largeness, strength, rapidity and frequency. As a result, when winter comes on, they have been changed to small, weak, slow and infrequent. The first stages of spring are like the last stages of autumn, and the last of spring like the first of autumn. And the first stages of summer are like the last stages of summer, while the first of winter are like the last of winter, as those of the middle of summer and the middle of winter deviate equally from each, changing similarly. Midsummer is in one respect similar and in one respect opposite to midwinter. For the pulses are small and weak in each, but rapid and frequent in summer and slow and infrequent in winter. They are not as small in summer as they are in winter, but less in summer. They are not as weak in winter as they are in summer, but less in winter.* **[127K]**

Those things relating to the seasons are referred to these three causes: the need, the capacity and the arteries (organs). For it is on account of each of these things

---

680 *De praesagitione ex pulsibus* I–IV (IX.205–430K).

681 *De dignoscendis pulsibus* I–IV (VIII.766–961K).

682 There are a number of minor differences from the *De pulsibus ad tirones* text that don't affect meaning. However, the final sentence in that text is not included in the present text (see start of VIII.467K).

being different in the seasons that the changes of the pulses arise. Thus, the most *eukratic* condition of the ambient air is in the middle of spring, wherefore bodies are also healthy. It is necessary to be always mindful of this: spring is the best of the seasons for a person to be alive. There is the greatest strength in the capacities, the greatest *eukrasia* in the heats, and the greatest moderations exist in the arteries, as they would neither be soaked by excess moisture, nor parched by dryness, nor as it were made brittle and stiffened by cold, nor as it were released and weakened by heat. Therefore, the largest and strongest pulses clearly occur in bodies that are best constituted. For in the case of such things, the changes of the pulses in relation to the stages of life (ages), seasons (times), places and other similar things were written out sufficiently for doctors in other places and have been shown in the second book of *On the Diagnosis of the Pulses.*[683]

But also, for those so disposed, that the pulses will become large and strong has been shown in the first **[128K]** book.[684] For these are characteritic of a strong capacity, a moderate need and arteries in accord with nature. If, in fact, the matters pertaining to need were more immoderate, not only would they become large and strong, but also rapid. And if it were even more immoderate, so as not to fulfil the need in terms of the action, some change to frequency would also immediately be added. But if the matters pertaining to need are moderate, there would not be any change to either rapidity or frequency.

But we need to add this to the discussion—"as in seasons". The pulses do not become very large absolutely and very strong under these circumstances, for others are found much stronger and larger in their natures than these, but compared to the other seasons, such pulses occur in the middle of spring. In summer, matters pertaining to the need are increased, while matters pertaining to the capacity are reduced. Consequently, as a result of this, they take away some of the strength mentioned, and likewise also some of the magnitude, but less. The urgency of the need adds something. And because of this, something noteworthy comes to the rapidity and the frequency. In winter, in turn, things are the opposite of summer; the capacity is stronger, although not to the same extent as in **[129K]** spring, while the need of the generation of the pulses is reduced. Wherefore, the pulses in winter are weak more than those in spring, but strong more than those in summer, being somehow in the middle of both the seasons, because the capacity is also in the middle, being in fact significantly small, because the capacity is lacking something in respect of its peak strength and the need is sufficiently reduced. Because of this same reduction, [the pulses] are slow and infrequent. It must again be remembered even here what was said by Hippocrates, which I explained elsewhere.[685] Need of these is, in the case of the changes, related to stages of life, seasons, places, climatic conditions and all such things, so that someone must not think I am

683 *De dignoscendis pulsibus* II (VIII.823–877K).
684 *De dignoscendis pulsibus* I (VIII.766–822K).
685 Hippocrates, *Aphorismi* 1.14 and 15 (ed. and Engl. transl. Jones 1931: 104–107).

speaking in opposition to him. For that man did not simply say, "those things that are growing have the greatest heat," but that the innate heat is added as another kind of acquired heat, of which those in their prime have the greatest share. For the innate heat is *eukratic*, whereas the acquired heat, which is also *akratic*[686] and already somewhat fiery. In this way too he says, "bellies are hotter in nature in winter and spring." These, then, are not simply hottest themselves **[130K]** but with the innate heat. In the summer they partake of this least; in them the acquired and dry heat prevails, which melts the body more than it nourishes it. However, the functions are not thereby made stronger, but are made unlike those occurring in the case of moderation. For those readers who remember these things, there will not seem to be any disagreement with Hippocrates, either in the present discussions or in those hereafter.

Next, I have written for beginners about places and climatic conditions. And the discussion is clear from those things already said previously, and having a most recognisable cause. Consequently, I shall subjoin this statement only, which is as follows.

7.[687] *About the places it is like the seasons. In the very hot places, it is like the middle of summer; in the very cold places, it is like winter; in the eukratic places, it is like spring; and analogously also in the places in between. And of the other [climatic] conditions of the surrounding (ambient) air, the hot are like the hot seasons, the cold are like the cold seasons, and the intermediate are like the middle of spring.* **[131K]**

8. *In pregnancy the pulses become larger, more frequent and more rapid. The other features are preserved [in accord with nature].*

In the case of those who are pregnant, nothing is changed from before, apart from aspects of the need. Whatever the situation already was, it is altogether extended and increased, since the woman is no longer still breathing and pulsating for herself alone, but also needs to do so for the foetus. Consequently, it is reasonable [that the pulses] will be more rapid, larger and more frequent, but not in fact stronger or weaker, or harder or softer by the actual reason of the pregnancy. Nor in this do the aspects of the capacity and the arteries change in those who are pregnant.

9. *Sleeps*[688] *should be natural if anything is. However, they also change the pulses. When they begin, they make them smaller, weaker, slower and more infrequent. As they proceed, they increase the slowness and infrequency, and particularly after food.*

---

686 The main meanings of the term ἄκρατος listed in LSJ are (i) applied to liquids, especially wine: "unmixed, pure, neat", and (ii) to persons: "intemperate, violent". Kühn's Latin translation has *intemperatus*. Here it presumably means the heat is not tempered or moderated by anything.

687 There are three differences between the two texts:
(i) sentences 2 and 3: *De causis pulsuum* περὶ; *De pulsibus ad tirones* παρὰ;
(ii) sentences 2 and 3: *De causis pulsuum* οἷοι before μέσου; *De pulsibus ad tirones* οἷον before μέσου;
(iii) sentence 4: *De causis pulsuum* οἷος μέσου is added after εὐκράτοις.

688 The distinction Galen makes between the singular and plural of ὕπνος is retained in the translation, although in English it sounds somewhat awkward.

They become larger and stronger, and over time change in turn to weakness [and smallness],[689] *but preserve slowness and infrequency.* **[132K]**

The present is not the proper time to go over what the cause is through which, in sleeps, concoctions are better, not only in the stomach, but also in the veins and arteries, and the whole bodily state (*hexis*) of the animal. There is no agreement among all doctors about the cause, or what it might be. Therefore, I now speak about this in general, [and say] that the capacity, which brings about the concoctions, turns itself more inward and towards the depths during sleeps. In fact, because there is the movement inward and it acts on the material, at first it becomes as though weaker and weighed down, but a little later enjoys the benefit of the nutriment it works on, and itself grows stronger and increases the heat in the animal, just as also those who partake of cold foods immediately perceive themselves as colder, and many of them[690] in fact shiver and have obvious rigors, then a little later they become hotter. Certainly, if the innate heat is what concocts the nutriments and the movements controlling this, as Hippocrates assumed,[691] those outwards produce wakefulness while those inwards produce sleep, and it would be very easy for us to explain on natural principles the factors relating to the **[133K]** changes of the pulses in the sleeps.

If, however, there is some other capacity which is the artificer of this, but all the capacities governing the animal have in fact been taught to be affected in common with each other,[692] it is hardly remarkable that this physical (natural) capacity is somehow sympathetically affected with it. But once this is agreed—that the capacity producing the pulses is strengthened and again becomes worse—the conception of the cause is easily followed. For in the beginnings of the sleeps, when the capacity itself is weighed down and worse, it is reasonable for the pulses to become smaller and weaker, and yet after a little while, being strengthened, they become on the other hand larger and stronger. Why, then, do they at some time also become slower and more infrequent? Erasistratus says it is due to the pause of the voluntary movements, which also concoct this well, but are not held responsible for the sleeps themselves. Hippocrates, however, and those who say the heat tends inwards in the sleeps, and are almost all the notable doctors and philosophers, will ascribe the causes of this to the diminished, increased and different movements. For **[134K]** they say [the heat] becomes less than it was, and is, as it were, smothered, having fallen all of a sudden on the large amount of material within, as in fact a fire does, should it fall on a large amount of wood piled up. For just as this fire itself, when it has prevailed over the material, is strengthened and increases to a large size, so too the innate heat does, when it has enjoyed the benefit of the nutriment, and worked up the humors on which it fell, becoming now larger and stronger than it was.

689 Added in *De pulsibus ad tirones.*

690 The manuscripts F1, F2, and Va1 omit αὐτῶν after πολλοί γε.

691 Hippocrates, *Aphorismi* 1.14 and 15 (ed. and Engl. transl. Jones 1931: 104–107).

692 The verb here is συμπάσχειν. See Aristotle, *Analytica Priora* 70b16 (Engl. transl. Jenkinson in Barnes 1984: 113).

But we have not yet spoken about causes of slowness or infrequency. In fact, the truth is otherwise, at least in my judgement. When, in those who are sleeping, the outward movement of the heat is diminished, while the inward movement prevails, it is reasonable for the contraction to become more rapid but the dilatation slower. Yet what also obviously occurs is this: those who do not follow closely the contraction of the arteries, but make the diagnoses from the dilatation alone, declare that the pulse of sleepers is absolutely slow. However, it is not absolutely slow, but slow in relation to another part of the movements, whereas the whole is not more slow than fast. For the dilatation of itself is slow, whereas the contraction is fast, but it is necessary not to say either of these is absolutely fast or slow; **[135K]** rather, they must be distinguished by the parts, just as Hippocrates did in the case of respiration.[693] Furthermore, the pulse in those who are sleeping is small in another of its own parts. For it contracts to the maximum extent whereas it does not dilate to the maximum extent, but as much as is necessary to restore the opposite movement by the contraction. Thus, the artery is moved continuously at equal intervals in the contraction and the dilatation, unless at some time it should suddenly change the rhythm. But it does differ in the limits from which it begins and with which ends.

For when the movement inwards takes the greatest share, coming to an extreme contraction, and here again begins the outward movement, it no longer makes the dilatation maximal. However, when in turn again the outward movement prevails, the opposite being weak, it does not still begin from the extreme contraction in the dilatation or end in extreme largeness. However, from here in turn, when beginning the inward movement, it makes the end of the contraction from when it began to dilate, and in this way it is always making a kind of ebb and flow of movement, sometimes being carried more inward, **[136K]** and sometimes outward. To those perceiving only the dilatation, it is inevitable that the pulse in the sleeps seems small and slow. However, it is inevitable for these same people, as it would also be for the more unpractised regarding the diagnoses, that the movement in the dilatation seems slight, the major part escaping notice. For in the case of the artery moving in this way, those who are not very practised do not follow closely the limits of the movements inwards, but both the end of the contraction and the beginning of the dilatation escape them.

Consequently, it is probable that a large part of the movement escapes those who are unpractised—almost all the contraction and not the least part of the dilatation. And as much of the movement as would escape, being imperceptible to those palpating, being added to the time of the pause, makes the pulse seem infrequent, since, in relation to the truth itself, in no way is the pulse of those who are sleeping more infrequent, just as respiration is not. But clearly in this also, the expiration takes the greater share, occurring along with exhalation, while the time of the pause is not increased. I have now said almost everything about sleeps, adding the remainder for purposes of clarity.

693 It is not clear what Hippocratic work is being referred to here.

The pulse of those who are sleeping, **[137K]** in relation to the external pause is like that of people who are awake, whereas in relation to the internal pause, it seems considerably different, since our perception does not follow the limits of the movements inward. But it is also smaller and weaker than the pulse of those who are awake, and particularly right at the beginning, for something of magnitude and strength is added to it later. But it is also slow in the dilatations while it is rapid in the contractions, and on the whole is moved more inward. As a consequence also, in saying it is strong in the contractions, you would not be mistaken. For the artery also sets out more strongly to expel the residues of the processed humors, which are, as it were, smoky in the sleeps, as these arise more at that time, so that also, because of this it takes a greater share of the movement inward. And yet, in fact, when the work on the humors has ended, but the person remains sleeping, such a sleep would clearly be immoderate, while each of the two movements becomes smaller and slow, weak and infrequent in due proportion. For sleep, as you hear the poets say, is a brother of death,[694] and in it, there is only one thing in common with the living, which is the **[138K]** processing of food; all the others are like those who are dead. There is no sight, no hearing, no understanding, no thinking, no talking, and [the person is] without sensation and movement, and is tossed about without thought. Moreover, if this, which is the one thing they have in common with the living, is destroyed, what else might sleep rightly be called but a road to death? For surely, we have often already said beforehand that what is first and in accord with nature is the movement of the heat outward and from itself, while second is the movement inward and to itself, occurring due to the food. Therefore, when it is not still exerting itself at all concerning the food, nor is there any recall of the characteristic movement outwards, but each one of the two is inactive, clearly it will be brought close to death, and the *tonus* of it is dissipated; because of this, the pulse becomes weak, while the movement is ended, and for this reason [the pulse becomes] small, slow and infrequent. And in fact, if smallness, slowness and infrequency advance to the greatest extent, it would come near to not being moved at all, and this would be death. In this way, there are all the things that relate to sleeps.

If, however, Archigenes says, regarding the things stated, that the pulses seem very full, but Apollonides says they are **[139K]** very empty, there seems no need for us to speak at length about such things, having adequately demonstrated in the previous treatises how needlessly the followers of Herophilus introduced in addition this class of the pulses, as also many other things. And it is not at all surprising for Archigenes and Apollonides to say opposite things on the matter of recognition, for each wrote, not about the affection commonly perceived, but of specific opinions.

10. *The pulses of those who have undergone the change from sleep to wakefulness are in the immediate term large, strong, rapid and frequent, and have some clonus. After a short time, they return to moderation.*

---

694 Hypnos and Thanatos were twin brothers, born of Nyx and Erebos.

We do not undergo the change from wakefulness to sleep in a similar fashion to that from sleep again to wakefulness; the former occurs altogether gradually while the latter for the most part occurs abruptly. For truly troubling dreams and someone shouting, making a noise or touching the sleeper produces a rapid and sudden change. For that very reason also, the previously described pulses appear very clearly in those undergoing a change in this way, **[140K]** but not in the case of the others, in whom the wakefulness has occurred due to the actual movement of the innate heat outward, these escaping the notice of those who are practised. For sleep occurs when the innate heat, due either to some fatigue or greater dryness, is turned towards the nutriment, or due to immoderate moisture is unable to extend outward. The first is healthy and in accord with nature, whereas the second is said to be like that in deep sleeps (*kōma*) and lethargies, to which in turn it is set against the wakefulness in the *phrenitides* and all the insomnias contrary to nature. Under these circumstances, the innate heat is excessively dried and, as it were, burnt up, and because of this is also moved immoderately outwards.

But for the present, I must leave aside the conditions contrary to nature, which it is better to term insomnia and deep sleep (*kōma*), rather than wakefulness and sleep. The generators of sleeps according to nature, just as also of wakefulnesses, are the kinds of things stated before. At the same time, when the abundant moisture that is needed for the natural heat in animals is wanting, and because of this comes together to the internal organs and the stomach when the sleeps occur, whilst **[141K]** this is adequately enjoyed, it already has the quality in accord with nature, when wakefulness is brought about. At the time, it is necessary for the changes that are occurring to have some force. For in the working up of the food, when a large amount of vaporous moisture grows up in relation to the heat itself and in the channels and all the bodies in between, up to the external surface, the movement to the outside is not yet similarly easy for the heat. Accordingly, in order that this is shaken off and thrust through, and that the pores are cleaned out, and that the body provides a passage for itself in all directions, strong and forceful initial movements are needed. When the body is already breathing well and is pure, it will recall the natural moderation. Because of these things, therefore, immediately in those awakenings from sleeps, the pulses become large, strong and rapid, and besides this, they take on a little clonus, due to the force of the movement, while a little later they return to the moderation that accords with nature.

Next after these, I have written for beginners about the acquired states (*schesis*) of the body, having termed the states thus, just as Hippocrates also did in his work, *On Joints:* "And certainly **[142K]** a state (*schesis*) of the body also differs somewhat."[695] I have written on the pulses in each of these, the causes of which it would not now be difficult to add for someone following what has been said. For what will be said in the case of the natural (physical) differences of the body will also be said of the ac-

695 Hippocrates, *De articulis* 8 (ed. and Engl. transl. Withington 1928: 212–219).

quired differences. Consequently, there is no need to speak at length, but only to write the statement that is next, as a reminder of the matters, the statement being as follows.

11.[696] *The acquired states of the body (schesis) change the pulses similarly to the natural states. For the person thin in nature who becomes well-fleshed*[697] *has a pulse in proportion to such a nature, while the well-fleshed person who becomes thin has a pulse like those who are thin in nature. It is clear that, apart from the change of the capacity, it is necessary to examine the difference relating to thin and well-fleshed, and similarly in the case of all the other [factors], so that, in relation to the one change that has occurred alone, we make the argument about this in each case. Those things said in the case of the well-fleshed, it seems necessary to extend further to the case of those who are obese.* **[143K]**

A distinction has been made in this way in those works on the states of the body.[698] It was similarly added also in *On Kraseis.*[699] It will be sufficient alone in the case of these to copy the statement as a reminder; it has the following form:

12. *Also the kraseis of the body which are acquired change the pulses analogously to the natural kraseis.*

And it is clear even here, just as we also said about the seasons, that the hot *kraseis* make pulses like the hot seasons do, and cold like the cold. The *differentia* of the *kraseis* in relation to moistness and dryness is encompassed in the states (*schesis*) of the body. The thinness is characteristic of the dry and fatness of the moist. These things have been shown in the treatise *On Kraseis,*[700] just as also that hardness of the body follows the dry *kraseis* and softness the moist *kraseis.* And because of this also, the pulse in the former is hard and that in the latter soft.

13.[701] *Next, it is [time] to speak about the other changes* **[144K]** *occurring from the non-natural causes. Exercises at the beginning [and practised] up to the point of moderation, make the pulses strong[er], large, rapid and frequent. Much exercise, exceeding the capacity of the one exercising, makes the pulses small, weak, rapid and [very] frequent in the extreme. Great excess, such that the person can hardly still move, and only with long rests or not at all, and is very enfeebled, [makes the pulses] very small, weak, slow and infrequent. If he is overwhelmed to the point of dissolution of the capacity, the pulses are specific for that. I shall speak a little later about what a dissipated capacity makes the pulses like.*

---

696 Two additions from the *De pulsibus ad tirones* text are made to second sentence and included in the translation. These are: (i) φύσει after ἰσχυὸς and (ii) ἔχει τὸν σφυγμὸν after παραπλήσιον.

697 On the term εὔσαρκος, rendered here "well-fleshed", see Papavramidou 2004.

698 Presumably Galen is referring here to *De optima corporis nostri constitutione* (IV.737–749K) and *De bono habitu* (IV.750–756K; Engl transl. Johnston 2020: 431–462).

699 *De temperamentis* I–III (I.503–695K; Engl. transl. Johnston 2020: 2–280).

700 See previous note.

701 Changes made on the basis of the *De pulsibus ad tirones* text are indicated by parentheses.

This particular statement has been written in that book.[702] Now the cause is also added. In the exercises the capacity uses all the *tonus* and because of this the pulse becomes strong. The factors pertaining to need are increased, and because of this, the pulses are large and rapid. And since they are increased in two ways—by the consumption of the *pneuma* and by the increase of the heat—they become, due to this, frequent also, not sufficiently fulfilling the **[145K]** need resulting from their largeness and rapidity. However, when the exercises are beyond the capacity of the one exercising, the pulses are small and weak, due to the weakness of the capacity, frequent due to the urgency of the need, not very rapid, due to the weakness of the capacity, but instead of these, very frequent, and to the extent the action falls far short of the need. Whenever someone is compelled to labour to such an extent that not only is the labour beyond his capacity, but also he continually needs periods of rest in between, the whole body is cooled in such a situation, while the pulses are weak and small due to the weakness of the capacity, but neither rapid nor frequent due to the cooling. The reasoning is clear to someone remembering those things that are primary.

14.[703] *Hot baths, whilever they are moderate, make the pulses large, rapid, frequent and strong. Those that are immoderate make the pulses small and weak, although they are still rapid and frequent. If, however, someone were to stop at this, they would be small, weak, slow and infrequent.* **[146K]**

In baths, when taken at the appropriate time, something altogether beneficial occurs to the body. For we should have recourse to baths when we are tired, heated, cooled, dried, have a sense of some distress or sense a non-uniform *dyskrasia*, or need evacuation. But in all the times mentioned, a hot bath restores the body to *eukrasia*, while the capacity becomes better, there is some opportune increase of heat, and some dispersion of *pneuma* together with the sweats. As a result it is reasonable for moderate hot baths to make such a change in pulses, of the kind I showed to occur when the capacity is benefitted and the need is increased. It is clear beforehand that immoderate hot baths dissipate the capacity, while additionally increasing the need immoderately, and because of this they create very frequent pulses that are rapid, similar to those beforehand, whenever the capacity comes to extreme collapse. For in this way, the so-called formicant pulses would occur. If people who use immoderate baths stop and no longer bathe, it is clear that the bodily state (*hexis*) is cooled and the capacity is dissipated. **[147K]** Consequently also, such a change of the pulses occurs like that we showed in the first book to follow weakness of capacity and dissipation of need. For the present, this alone will suffice to show how the change from each goes through the *proegoumenic* and *prokatarktic* causes to the primary, most important and so-called *synektic* causes. How the pulses are necessarily changed in those has been shown by the first book.

---

702 That is, *De pulsibus ad tirones* (VIII.453–492K). See particularly section IX (VIII.467–468K; Engl. transl. in present work).

703 There are two differences between the texts in the final sentence; these are indicated in the notes to the translation of *De pulsibus ad tirones*.

15. *Cold baths immediately make the pulses small, weaker, slow and infrequent.*[704] *Afterwards what they are made like depends on chance. At all events, there is either torpor or strength. In someone who is made torpid and cooled, the pulses are small, weak, slow, and infrequent, but when he is warmed up and strengthened, the pulses are large and strong, but moderate in speed and frequency.*[705]

Cold baths immediately cool altogether, whereas later they either strengthen or dissipate the capacity, along with cooling. This is to make torpid. As a consequence also, the changes **[148K]** of the pulses are in proportion to the causes mentioned. I shall now speak about what was not possible to specify besides for beginners. The movement inward of the arteries in the cold baths predominates over the outward movement, so the heat would immediately move more in the same direction in them. And if it were possible to palpate the arteries from within, all the phenomena hitherto apparent in the dilatations would now be displayed in the contractions—largeness, rapidity and strength. However, since this is impossible, we name all the pulses in this way from the phenomena in the dilatations. Small, because at such times the dilatation is much less than before. The same also applies to slow, because the movement has clearly also been reduced. But infrequent too, because all the other features in which the inward movement prevails are also displayed as such to those who don't perceive the contraction. And the causes were stated previously. I said weak not absolutely, but wrote weaker, indicating by the term the brevity of the change to this.

16.* *A large amount of food, such that it weighs down the* **[149K]** *capacity, makes the pulses unequal and irregular. Archigenes says they are more faster than more frequent. Food in a moderate amount makes the pulses large, rapid, frequent and strong. Too little food, so the person is insufficiently nourished, is not like the moderate, but creates less change and of short duration.*

In relation to each heading of those set out in this book, although it would have been possible to state first the opinions of others on these matters, and then to refute what was incorrectly assumed, we do not do this, but immediately go to what is approved and true, hoping to cut short the size of the work as far as possible. For since this is now long in itself, what must we think, if it also took on besides the previous opinions along with the required refutations? Would the length not be extended limitlessly? However, for someone who has the leisure time for the teachings and the enthusiasm to read all the opinions of the doctors, I know this particular treatise will clearly be a stepping stone for him to discover for himself on what matters they speak the truth and on what matters they are mistaken. **[150K]**

And yet I have also written about these in other places, completing many things previously mentioned in this way. In fact, certainly concerning the question now before

704 The text of *De causis pulsuum* is followed here on the grounds of sense. The text in *De pulsibus ad tirones* reads "small, slow, infrequent and more infrequent."
705 K: πυκνότητι; F1, F2, and Va1: σφοδρότητι.

us, in those places where I enquire of Erasistratus into the art concerning the pulses, I think I have clearly made known the difficulty of the theory.[706] But also, as far as we differ from those considered to know best on matters relating to the pulses, it seems to me on these an intelligent man, starting from what has been written here, would decide without difficulty. For he will discover most of them; in some cases, these exist in the statement alone, and in other cases, in the accuracy of the definitions, like those now before us. For Archigenes wrote in this way about the nutriment weighing down the capacity: "By the compression [the pulses are made] weaker and smaller, and more rapid than more frequent." We didn't write in this way, so how did we write, "A large amount of food, such that it weighs down the capacity makes the pulses unequal and irregular." And for someone concerned about the actions of the art, and not about food alone, but about all the other conditions in which the capacity is weighed down, **[151K]** will discover unequal pulses. And this in fact is itself what distinguishes the weak capacity from one that is weighed down. In this, the amount of damage is immediately indicated at the same time. For if, in many large and strong pulses, they discover a few that are small and weak, the damage of the capacity is slight. Contrariwise, if there are few that are large and strong, but many that are small and weak, in such conditions the burdened capacity is suffering greatly. If, however, small and weak pulses are occurring continuously, with none of the opposites being intercident, this is not in the burdened capacities, but in those that are dissipated. Should they be strong in their specific substance, but weighed down by something and compressed, the strong and weak pulses are always intercident, even if they are in the extreme stages of damage.

Thus, Archigenes writes carelessly [in stating that] the pulses become weaker and smaller in capacities that are weighed down by the foods. For these are signs of a capacity that is dissipated and not of one that is compressed. However, just as in the *differentiae* relating to size and strength, inequalities happen to exist in this way, so too do they do so in the *differentiae* relating to frequency and rapidity. **[152K]** Therefore, I do not have to conclude how Archigenes wrote that the pulses are faster more than they are more frequent, unless he intended to make this clear—of those that are unequal, the rapid are nevertheless more than the slow, or the frequent more than the infrequent. And yet, this is not simply true in this way, when stated without qualification. But the damages that are less make the pulses rapid more than frequent, while conversely those that are greater make them more frequent than rapid. The cause is clear beforehand from those things that have been shown in the first book—if in fact rapidity is characteristic of a strong capacity, then frequency is characteristic of a weak one. And of the other things said subsequently, the calculation of the cause is clear beforehand, if someone calls to mind those things written in the first book.

17. *For the rest, wine changes the pulses similarly to foods, but differs in making the change immediately—and the change from wine stops before that from food—and in*

---

706 K: δόξης; F1, F2, and Va1: τέχνης.

this rapidity and magnitude increase more than strength and frequency. For **[153K]** *by almost as much as moderate food provides more and more lasting strength [of the body],*[707] *does wine increase the magnitude.*

This very statement explains itself, reminding us of the things said by all doctors about the power of wine. For the increase from this is rapid because wine is both moist and hot. He says: "For those who need a rapid increase, water is a cure. Whenever, then, the fluid is led in by the heat or moisture, this is even more so. But also, when people are nourished all at once and rapidly, the excretion is also rapid." Because of this then, just as it is rapid, so too is the increase from wine of short duration for bodies, for it is easily dispersed as vapour, That due to its nourishing similarly to foods, it will change the pulses, while due to heating particularly,[708] it will add more evident magnitude and rapidity, is entirely clear to those who remember the things I stated in the first examples. And the cause of none of the things stated subsequently through its introduction will be unclear to someone who recalls what I have written in the first book. **[154K]** Therefore, passing over most of these, let me call to mind only those having more obscure causes, one of which is the following.

Why, then, do the vermicular pulses sometimes follow choleras, strong fluxes of the belly, menstruation (the so-called female flux) and all those things from evacuating the body dissipating the capacity, while the formicant pulses follow the dissolutions of the capacity due to a *kakoethical* fever? Because it is necessary for the pulse to be slow and altogether infrequent for it to be vermicular, it is neither possible for the first part of the artery to manifestly be moved first, nor for many "circumscribed components"[709] to occur in one dilatation, unless the movement is slow, and at the same time also infrequent. However, in the acute fevers—for of necessity the *kakoethias* are acute—the abundance of the heat doesn't allow the pulses to become either extremely slow or altogether infrequent. But it is slow, or at least not altogether rapid, in the dissipated capacities, even if some err in the diagnosis of this; it is not extremely slow and not altogether infrequent, but always frequent, and when the factors of the heat **[155K]** urge it on, very strongly frequent.

It seems to me to be an appropriate time to end this third book here. In the fourth book, I shall add those things that remain, starting from the psychical affections.

Book IV

1. [156K] Of the causes changing the pulses, I have written, by way of the previous two books of those that effect the change through themselves, containing the genesis of these, and because of this have been called *synektic.* The causes that precede those,

707 "of the body" only in *De pulsibus ad tirones.*
708 K: θερμαίνειν; F1, and Va1: τρέφειν.
709 The Greek term is περιγραφή; Kühn's Latin translation has *circumscriptiones.* Presumably Galen is referring to palpable separate components which can be recognised.

some of which exist in the actual bodies of the animals and some of which befall them externally, I set out to explain in two works previously. Although also, in the book I have written for beginners about the pulses, **[157K]** I have spoken about these beforehand under headings and without demonstration, I judged it better to make mention of them now through their actual names at the same time as defining them more precisely, and giving their causes. Therefore, I made the discussion of the third book about the stages of life (ages), places, seasons, natures, and besides these, about the so-called pursuits, foods and drinks. In this, being the fourth and last of the proposed books, my discussion will be about the causes contrary to nature, making a start from the psychical affections and going through in succession also the somatic affections. And here, in relation to each subject, I have first written the statements from the book I wrote for beginners.

2. *[The pulse] of anger is high, large, strong, rapid and frequent.*

The things I say about anger are clear to someone who remembers them from before. Now it is already necessary to add about this what I did not make clear—whether the pulse of those who are angry only seems stronger, or really is so. For I left this alone **[158K]** as almost indeterminable, although I did go over it more in the first book, almost unjudged. For someone would not in this way have been persuaded at that time, if we had said that it doesn't only appear so, but is stronger, as now will be added, previously having heard in the intervening discussion many things contributing to the belief, and especially those said about the unequal pulses. For the pulses seems clearly in these to be alternately weak and strong. This is not, of course, because the capacity is rapidly undergoing a change in this way to weakness and strength, but because it is contending with the distressing causes and not always overcoming them. Very clearly, in the good resolutions, the pulses become strong, the capacity rousing itself under these circumstances over getting rid of the distressing factors. This also seems to occur in the case of the psychical capacity, and not only of the vital (physical) capacity. Anyway, some have been dissipated to an extreme degree when, due either to age, or disease, or some antecedent fatigue, or suddenly seeing the approach of robbers, enemies or some wild animal, they run swiftly, they flee from the danger no less than the strong. In this way too, due to an earthquake, some spring from **[159K]** their houses, as they do due to a blaze, and run very swiftly, when before they hardly went forward due to the weakness. Therefore, in all such instances, it is clear that the capacities governing the animal, when seized by some necessity, function more strongly, and one must not be amazed, if this occurs in the angers. And it appears and makes sense, as also in those exercising, and in acting like a man in any way at all,[710] to see the pulses change, not only to largeness but also to strength.

3. *[The pulse] of joy/pleasure is large, infrequent and slow,[711] but not in fact different in strength.*

710 K: καὶ ὁπωσοῦν ἀνδριζομένων; omitted in F1, F2, Va1.

711 βραδὺς (slow) is not included in *De pulsibus ad tirones.*

The heat in those who are joyous is dispersed to the whole body and its movement outward prevails, just as the movement inward does in those who are grieving. It is reasonable, therefore, for their pulses to become larger, but not yet stronger. For some acquired need is not added due to the affection or stronger actions. It is clear that these things are said in the case of moderate joy,[712] changing the pulses according to the specific reason. For often, contingently the opposite pulses occur **[160K]** when excessive joy occurs, and because of this dissipates the capacity. But at that time in fact the pulses of a dissipated capacity arise. It was said before that the stronger affections effect the whole change of the pulses on their own account.

4.* *[The pulse] of grief is small, weak, slow and infrequent.*

Grief occurs along with cooling and the inward movement of the heat. It is reasonable, then, that in grief the opposite pulses to those previously mentioned as following the joys are brought about.

5. *[The pulse] of fear that is sudden and severe is rapid, clonic, irregular and unequal, while if already long-lasting, is like that of grief.*

In the majority, fear immediately bringing about disorder of thought changes the pulses to inequality, as one would expect. I have spoken about this before. In time, however, fear makes the pulses like those of grief, for fear that is long-lasting is not much different from grief. In all those [fears] that are long-lasting or **[161K]** become very strong, pulses follow like those of a dissipated capacity. For surely all these dissipate the capacity—immediately those that are strong, but chronically the opposites. This was also made clear a little earlier from the things said to be due to the joys. And now this has been written in addition, because of those things said in relation to the introduction subsequently, and because of what was said about all the similar things is in general fitting. What has been said has no need of more superfluous explanation; let us move on now to the continuation.

6.[713] *Pain is something that changes the pulses. Severe pain or pain in important parts changes them, as inflammation also does. [Pain] that is still mild and just beginning makes the pulses larger, stronger, more rapid and more frequent, whereas when it is increased and becomes very strong, so as to already damage the vital tonus, the pulses are smaller, weaker, rapid and frequent.*[714] *And the more the pain lasts or becomes stronger, the more it increases each of the aspects mentioned. Pain which already also dissipates the capacity* **[162K]** *effects the change to weakness, smallness, a false impression of rapidity and excessive frequency.*

Just as in the case of all the other distresses, so too in the case of pains, the capacity naturally urges itself on to contend with and repel everything troubling, wherefore it is reasonable for the pulses to change to large, rapid and strong. If, however, that accom-

---

712 K: τῆς συμμέτρου; omitted in F1, F2, and Va1.

713 There are minor changes plus the omission of καὶ ἀμυδρότερον, present in *De pulsibus ad tirones* and several manuscripts, and in parentheses above.

714 Post τόνον: Kühn's Greek text has μικρότερον καὶ ταχὺν καὶ πυκνόν; F2, and Va1 have μικρότερον καὶ ἀμυδρότερν καὶ ταχὺν.

plishes nothing, it is clear that in it, from excessive fatigue, the actual dissipation of *tonus* occurs, and in this way the movements undergo a change to weakness and are made worse by the distresses of the capacity. Why I said a false impression of rapidity is stated in the third book of *On the Diagnosis of the Pulses.* And indeed, why frequency is linked with almost all the small pulses has been stated in the first book of this treatise.[715]

7.* *The pulse common to every inflammation is saw-like (serrated), as one part of the artery seems to be dilated but another does not, and this latter clearly appears to be harder. This pulse has some clonus and is rapid* **[163K]** *and frequent, although not always large. The specific characteristic of each is that at the beginning the pulse is larger than accords with nature, and stronger, more rapid and more frequent. However, when the inflammation increases still further, all these features are in fact increased, and the pulse is clearly now harder and more clonic. As the inflammation reaches its peak, the pulse is more clearly harder and more clonic, but smaller than before, although not in fact weaker, unless the affection exceeds the capacity. But also the pulse becomes very frequent and rapid. If, however, the inflammation is sufficiently chronic and is already hardened into a scirrhosity, then in addition to the features mentioned, there is thinness and hardness of the pulse. These things come about when the inflammation changes the pulse in the animal as a whole, through either the magnitude or the importance of the part in which it is situated. When the whole animal is not involved, the pulse in the inflamed part is as described. Also, each of the things mentioned either increases or decreases according to the amount of the inflammation or according to the nature of the inflamed organ itself. Thus, the more sinewy parts make the pulses harder, more saw-like (serrated) and smaller, while the more vein-like and artery-like parts do the opposite.* **[164K]** *The greater of these themselves is in the artery-like parts, and this pulse readily becomes unequal and irregular.*

It should, then, already be clear what the pulse would be like in inflammations involving the liver, spleen, kidneys, bladder or colon, or the stomach, and of the pleuritides and peripneumonias, and in short, all those parts in which a fever follows the inflammation of the part, except for those due to the nature of the symptoms following of necessity in them, and of those chance accompanying features that would be able to change each pulse. And this will happen to change the pulse to such a degree when a mixed change occurs in it by reason of the inflammation, and what the nature of the place is, and what the present symptom produces. For in those with inflammations involving the brain,[716] *convulsions will readily be produced, chokings in inflammations of the lung, syncopal episodes in those with inflammations of the opening of the stomach (cardiac orifice), wastings in those with inflammations involving the liver, failures of concoction (apep-*

---

715 *De dignoscendis pulsibus* I (VIII.766–822K).

716 There is again the issue here of whether the brain or the diaphragm is being referred to (τὰς φρένας). Kühn's Latin translation has *septum transversum* (i.e., diaphragm). The modern reader would find the association of convulsions with the diaphragm odd, but this may have been in keeping with Galen's ideas on pathophysiology.

sias) in those with inflammations involving the stomach, and retentions of urine in those with inflammations involving the kidneys. And the more sensitive parts change the pulses due to the pains, whereas the more insensitive parts do so in relation to **[165K]** *the condition alone. From all these then, various kinds of changes of the pulses arise following the inflammations. And that it is necessary to distinguish these changes was described comprehensively in other works. What is stated now is as much as is useful for beginners.*

The statement itself from the introduction clearly teaches thoroughly the change occurring in the pulses due to the inflammations. How the primary and containing causes effect the genesis of these and make the stated change with regard to the inflammation is the next thing to speak of, making the beginning from this point.

In every inflamed part these two things necessarily happen: the part is hotter than accords with nature, and it is under tension. Both these things are shared by the whole body, whenever either the part is important or the inflammation is sufficiently severe. But the increased share of the heat in relation to the whole animal is a fever, and it is clear that this will change the need of the pulses. And tension from what is inflamed is distributed also to the arteries. Because of this, the pulses will be harder and the increased hardness will in fact be more in the sinewy parts, because these are also stretched more than the others **[166K]** in the inflammations, as they would have already been stretched in nature, and the contiguous parts would naturally be stretched more with them, as they would be stronger.

The gain of heat is customarily displayed, making the pulses large and rapid by reason of itself. When, however, there is some deficiency in the need, then it is also already frequent. It is to be expected, then, that in all the fevers following inflammation, the pulses become harder, and besides, due to the *dyskrasia* in the arteries, there is inequality in a single impact. And it is clear that, when both the things mentioned concur, due to these the form of the pulse will be of the kind that is saw-like to palpation. For in this way, it is inevitable that the person palpating will get the impression of an artery that is hard, there being an unequal impact (pulse-beat). It is, then, as I said, obviously more rapid, unless in fact the hardness resists, and clearly larger. If, however, it does resist somewhat, it will take on a clonic movement by as much as the largeness was taken away. That in this also it will be more frequent to the extent that the dilatation does not adequately fulfil the need, is clear beforehand[717] to someone who recalls the things demonstrated in the previous writings, as it will also be that **[167K]** an equal change of the pulses does not follow equal magnitudes of the inflammations, but those in the vein-like parts produce larger pulses, while those in the sinewy parts produce lesser pulses, just as they are also harder and more saw-like.

None of these things is obscure, if we recall that it is inevitable for the arteries to be stretched more by sinewy parts that are inflamed. And it is clear regarding the artery-like parts, inasmuch as there is an abundance of heat in the arteries, that being

717 K: πρόδηλον; F2 and Va1: ἄδηλον.

hotter will increase still more, and because of this, compel the pulses to become very large. And if some inequality should coexist for whatever reason, it will show this more clearly. If we recall, we showed there are two primary reasons for inequality of the pulses—a weakness of the capacity moving the arteries due to excess weighing down, and besides this, the conditions of the arteries themselves, which are certain obstructions, impactions, stretchings and compressions. It is, then, in no way remarkable that the pulses in inflamed artery-like parts are made unequal, since the capacity producing them is more easily fatigued in such parts, and the arteries themselves are in a bad state. The **[168K]** other things said in the book I have written for beginners on the changes of the pulses existing following an inflammation require, I think, no explanation. Thus, it is now time to move on to some of the subsequent statements.

8.[718] *[The pulse] of those with pleurisy is rapid, frequent and not very large. It would also seem strong. It is not weak, but is not already strong to an extent commensurate with the affection. It is necessary to bear in mind in all cases that we need to examine the change in each of the matters, distinguishing how much it is due to the affection and how much to something else contingently, and not due to the affection. Accordingly, the pulse of those with pleurisy, since it somehow makes the artery more sinewy and harder, so as to tend towards strength, deceives the unpractised, who are unable to distinguish a hard beat from a strong one. In this way also, the majority [of doctors] being unable to distinguish the other differences of pulses are perhaps quick to find fault with those things written here and when they hear them, not being acquainted with the things correctly stated, they are prejudiced [against them]. But it is not necessary to spend time on these in the present discussion. I have written specifically about these in* **[169K]** *the diagnosis of the pulses.*[719]

Therefore, I advise training the reasoning power and at the same time also the touch, so as to be able to recognise the pulses in respect of the touch in their very actions, and not to distinguish them by reason alone. However, the starting point of the practice in the actions is the teaching through reason. And for sure it is not possible to explain the quantity of the frequency, although there is a great difference certainly between what exceeds the customary measure of the pleuritides and what falls short. For if there are the excesses there will inevitably be a change to peripneumonia or threat of collapse (synkopē), while the deficiences end in lethargy (kataphora) or damage of sinews (nerves). In this way too, the kind of the inequality is, as it were, saw-like (serrated), which is no less specific for those with pleurisy. When reduced, it is a sign of pleurisy that is of the soft type and will be [readily][720] *brought to concoction; when increased, it is a sign of a pleurisy that is troublesome and difficult to concoct. Such inequalities, with weakness of the capacity, are acutely dangerous. If however the capacity is strong, they will be concocted over time or changed to empyema or a consumptive marasmus*

---

718 There are several minor variations in the two texts, not affecting the translation, plus the three changes made on the basis of *De pulsibus ad tirones* and indicated by parentheses.

719 *De dignoscendis pulsibus* I–IV (VIII.766–961K).

720 *De pulsibus ad tirones:* ῥαδίως; *De causis pulsuum:* οὐχ ἥκιστα.

succeeds them. Therefore, the pulses of what is concocting **[170K]** *gradually is devoid of every change contrary to nature. When there is a change to empyema, the pulses become specific for those with empyema. The pulses in those with marasmuses are the same as in those with consumptive wastings.*

The actual statement from the introduction teaches thoroughly the phenomena themselves that befall those with pleuritis (pleurisy) but doesn't yet state the causes in detail. Let these be stated now. Pleuritis is an inflammation of the membrane joined under the ribs, and on account of this, the patients are acutely febrile, as the heart lies nearby. Because they are febrile in this way, the pulse is rapid in them; because the affected part is sinewy, the pulse is manifestly hard. Since it is hard, it is also, on account of this, smaller than is sufficient for the fever. Therefore, it necessarily also becomes frequent, since the heat requires largeness of pulses whereas the hardness doesn't allow this. It was shown in what has gone before that, if the magnitude of the pulse falls short of the need, this is made up by the frequency. However, the frequency of these cases is unequal, because also the humors producing the inflammation are not of one kind. Thus, if they should happen to be more bile-like, the fever **[171K]** would become more acute and some damage of the capacity would follow. If, however, they are more phlegm-like, the fever is milder but more readily comes to lethargy (*kataphora*).[721] For always in the pleuritides and the peripneumonias, the head is adversely affected, receiving the smoky fumes of the putrefying humors in the inflammations. But when the putrefying humor is bile-like, the vapour is acrid, just like some soot, and this bites, consumes and irritates the structures in relation to the brain, and in this way leads to delirium.

When the humor is phlegm-like, and not like acrid smoke but like some thick and foul mist, as if some cloud is being sent to the head, this moistens and weighs down the brain and compels the person to become weighed down and stuporous. It is reasonable, then, for a change of the pulses to greater frequency to proclaim either peripneumonia or collapse (syncope). In the case of peripneumonia, the inflammation always attacks and harms those structures contiguous in the side, as it has been generated by a hot, seething and readily moved humor. Syncope, on the other hand, is due to a badness of the capacity, or through injury of nerves, or lethargy (*kataphora*) when the brain **[172K]** is filled full of the vaporous phlegm-like humors. For also in turn, if this is able to drive away the harmful agent to the branches, the affliction of the nerves follows; if it is not able to, lethargy (*kataphora*) follows. Thus, with such humors, the more infrequent pulses occur, while with the opposites—that is the bilious—the pulses that are more frequent than the customary measure of the pleuritides coincide. The median of both, which is the measure of infrequency and frequency in the pleuritides, occurs whenever the bloody humor is what brings about the affection. In the

721 On this somewhat difficult term, LSJ lists a range of meanings, including two rather different medical meanings: diarrhoea (Hippocrates, *Aphorismi* 7.86 [ed. and Engl. transl. Jones 1931:217]) and lethargic attack (Hippocrates, *Epidemiae III* 6 (Engl. transl. Jones 1923: 244–247). The basic meaning is 'a downward stroke' or 'descent and fall'. For Galen's definition see *De comate* 1 (VII.643K).

same way too, the median of hardness exists in such a humor and occurs in the inflammation that is moderate in terms of magnitude. It is increased due either to an excessive magnitude of inflammation or bilious humors, just as it is reduced by the opposites It is also clear that the saw-like (serrated) kind of impact will more clearly follow the hard pulses.

9. *The pulse of those suffering from empyema*[722] *is, at the beginning, like that of inflammation at its peak. For this is the beginning of the actual empyemata. Sometimes, it is unequal and irregular; in all cases it is hectic. When the pus is already present, the other aspects are similar, but the pulse is more equal (even).* **[173K]** *In the ruptures [of the empyemata], it is weaker, broader, slower and more infrequent.*

There are those who have a large amount of pus contained in the open space between the chest wall and the lung; some term this *empuos*, some *empuikos.*[723] These arise due to the magnitude of the inflammation in the pleural cavities being changed into pus. Sometimes they also follow peripneumonias. Now [doctors] also call *empuos* all cases in whom an internal organ that is inflamed suppurates. Accordingly, the statement previously written about the empyemas from the introduction, particularly in the case of the first things said, and now in the case of all the others, is true. For certainly when the inflammation changes to pus, inevitably there is some inequality and irregularity in the pulses, due to the bad state of the arteries and the weakness of the capacity. Into the bargain, a kind of struggle of the nature with the symptom comes into play. For the fluid from the putrefaction, flowing chaotically one way or another, at one and the same time stands as an obstruction to the outflow channels of the *pneuma* and urges on the nature to expel **[174K]** the distressing agents. In this situation, the pulses become large, [strong],[724] high and rapid. A little later, when either the capacity is worn out in the struggle or a stronger obstruction arises in some part, they change to the opposites until, when the vital (physical) capacity rouses itself, it turns towards the expulsion of the distressing agents. And even in this in turn the pulses are made higher and stronger than they were before.

Thus, when such a change becomes continuous, of necessity some inequality exists in the pulses and irregularity, and these remain throughout until rupture occurs. Because of this also the pulse of these cases is called 'hectic'.[725] For this term is indicative to the younger doctors that it remains the same throughout and does not create some

722 Although the term *empyema* has come to be particularly associated with collections of pus in the pleural cavity, it still retains the original, more general meaning—for example, *Stedman's Medical Dictionary* 2000: 585 has: "...pus in a body cavity; when used without qualification, refers specifically to pyothorax."

723 The Greek terms are ἔμπυος and ἐμπυικός. They are simply transliterated here, as in Kühn's Latin translation; both would be rendered empyema in English (see previous note). The verbs of derivation are: ἐμπυόομαι (to suppurate) and ἐμπυΐσκω (to cause to suppurate).

724 After μεγάλοι, F1, F2, and Va1 add καὶ σφόδροι.

725 See Galen, *De pulsibus ad tirones* 8 (VIII 460K), *De febrium differentiis* I 10 (VII 315K), and Aristotle, *Problemata* 920b27 (Engl. transl. Forster in Barrnes 1984: 1436).

kind of periodical fits, and in turn does not have intervals, as in the other fevers. For the previously mentioned agitation in the movement of the arteries remains until the rupture occurs. However, when the suppurating part does rupture, it is clear that everything will become the opposite. And it is necessary to recall what was said in the second book about the broad and high pulse, and not to wish **[175K]** to hear about everything often.

For certainly, with the ruptures of the collections of pus (*empyemata*), the tension of the vessels already stops, the capacity is wearied by the conflict, and the person has inevitably become weak. When these things come together, the pulse is wont to become broad. The rising of the arteries to a height in this will increase particularly as regards the sense impression, before it bursts. Still bearing that in mind and comparing the present appearance to it, we accept the pulse becoming significantly broad.

10. Next is the discussion of those who are wasting away (those with marasmus). On this itself, I shall first write those things that are useful for the discovery of the cause. After this, I shall subjoin the statement from the introduction. The affection was named from the fires of those who are wasting. These are quenched by lack of nutriment. On the same basis, whenever the innate heat in animals dies away, the pulses are inevitably weak, since of necessity the capacity is made weak in those so affected. And for the same reasons, the pulses also become small. Furthermore, they seem to have some rapidity, because this is inseparable from those with fever. And since they are febrile, they are weak **[176K]** in respect of the capacity, and because of this, they have a pulse that is small, and is of necessity also frequent in them. This was shown in the previous books. It was also shown that the nodding pulse follows weakness of capacity, being tapering (mouse-tailed, curtailed) in each part.

It is clear that such people consistently have the same pulse, since the condition is not started from the fluids or the *pneuma*, but arises in the solid bodies. For in these cases the whole state of the body has become like *anthrax*.[726] And because of this, the younger doctors term such pulses "hectic". In these, the heat is already completely quenched—this particularly happens in old people—and the pulse becomes more infrequent. Phillipus calls such wasting (*marasmus*) "old age from disease."[727] This particularly befalls those past their prime, but sometimes also those still in their prime. And it has been properly written about these by Phillipus that their condition is not exactly like the fiery *anthraces*, but those that have been quenched in some way, so as to already be changed to ashes. These things are enough for the **[177K]** man of intelligence to follow what comes next. Thus, I have now subjoined the passage from the introduction.

**The pulse of those who are wasting doesn't change in one way. As far as possible it is necessary to distinguish these through clearly apparent differences. Those with inflam-*

---

726 The basic meaning of ἄνθραξ is "charcoal" or "coal", the term later becoming applied to a malignant pustule or carbuncle. The general meaning is perhaps intended here. See Hippocrates, *Epidemiae III* 7 (ed. and Engl. transl. Jones 1923: 269), and Galen, *De tumoribus praeter naturam* 6 (VII.719K).

727 See Galen's *De methodo medendi* VII.6 (X.495K; Engl. transl. Johnston and Horsley 2011: 2.292–293).

mations that are not being resolved, when they are gradually fading away and dying, have pulses that are weak, more rapid, very frequent and tapering (mouse-tailed) in size in a single beat. Archigenes calls them "nodding" and "inclining", wishing to make abundantly clear the brevity of the dilatation along with the kind of "nodding" of the limits on either side (i. e. starting and finishing). For it is not that they cease all of a sudden, but contract as if making an angled return of the parts on either side gradually, there being a tapering away in magnitude in relation to each of the parts. This is not in these cases alone, but also exists in the majority of those who are wasting away in any way whatsoever—in all those [wasting away] due to inflammation, but also in many of the others who are wasting away, if not also in those who are wasting away due to some occult inflammations. And this would be specific for those wasted away because of inflammations **[178K]** *and would not be present in any of those wasted otherwise. The pulse is hectic in all those wasting away and this is very common in them. Secondly, there is a tapering inequality in the magniude of the dilatation. Furthermore, this also exists in the majority. Thirdly, there is frequency. And this exists without exception in all those wasting away from inflammations. But it is also present in those acutely endangered from cardiac or gastric syncopes, if they seek to escape the acidity by a drink of wine, and are wasted away over time, unless you were to say these people perish due to small inflammations not apparent to us. And some of these people have the "nodding" pulse, unless again it is those with inflammations, although they say others waste away without inflammation. This is hard to envisage. These people have a hectic pulse or one that is weak and very frequent; and some call this 'nodding'. This is the second difference of the pulses in those wasting away.*

The other and third difference is having an infrequent pulse. But in these also at any rate the preceding fever **[179K]** *made them frequent, and the extreme dissipation of the capacity made them very frequent. However, in the intervening stage, when all fevers are cooled, if those affected are not yet dead, the pulse makes the change to infrequent. This form of marasmus is characteristic of old age, and particularly when those parts related to the thorax and lungs happen to have been affected. Such people preserve the feverish hardness of the pulse, even if it is infrequent. In very few of those who are wasting away is the pulse changed to another inequality, apart from that mentioned in relation to magnitude.*

11. *The pulse of those termed phthisical (consumptive) is small, weak and soft, and moderately rapid and hectic.*

Next the discussion is about those termed consumptive (*phthisical*).[728] The pulse of those who [suffer from *phthisis*] is small and weak, due to weakness of the capacity, but not overly rapid, like the majority of those with fever. This is because they are not febrile with a burning-hot fever. They also have the hectic pulse that relates to the fever. Anyway, from its taking over the whole state of their body (*hexis*), the

728 The manuscripts F1, F2, and Va1 combine and modify the opening two sentences of this section; this has no effect on meaning. Kühn's Greek text is followed.

pulse of the fever is termed thus. **[180K]** Next to these in the book written for beginners, there is a discussion about those with peripneumonia. It seems to me to be better to write the statement here beforehand; it is this:

12. *The pulse of those with peripneumonia is large, [and it is somewhat undulant (wave-like)],*[729] *weak and soft, similar to the pulse of those with lethargy, except those which go beyond bounds in the inequality in a single pulse-beat and the so-called systematic [inequality]. In relation to a single pulse-beat, the pulse sometimes becomes interrupted, as it were, turbulent and dicrotic. In the systematic inequality, there are other differences, the pulse sometimes being intermittent and sometimes intercident.*

This same statement teaches thoroughly and precisely what happens in those with peripneumonia. For one who remembers those things which have been demonstrated in the first commentaries, there is no difficulty discovering their causes. There is nothing remarkable in their pulses being large, or weak, or unequal due to inflammations of the lung, it being at once a spongy and soft internal organ and full of innumerable empty spaces, **[181K]** lying so near to the heart. For, on account of being spongy, soft and porous, it draws the blood, by its being more phlegmatous and thicker in consistency, to inflammation, whereas the thin and bilious blood for the most part flows through and can in no way be plugged up in the internal organ. I call such a flux 'phlegmonic' (inflammatory); it soaks the walls of the vessels, dissipates the *tonus* of the capacity in them, and in this way renders the pulses large and weak at the same time.

Also, because the organ is near the heart, the inequality becomes multiform. For we showed the inequality to follow compressions and constrictions of arteries, and on the whole plethoric conditions, and particularly, when they are near the heart. That in the peripneumonias the arteries of the lung are compressed, weighed down, constricted and weakened has no need of discussion. Because of this also, in these the dicrotic pulse rarely occurs, although many of the other inequalities happen, because on the whole some hardness of a vessel needs to be present for the dicrotic pulse. This is not likely to have been acquired from a phlegmatic flux. For **[182K]** it is the work of the more bilious flux to contract, dry, stretch and harden. Accordingly, since this is seldom disturbing to the lung, it is clear also that there will rarely be hardness. If there is however, there is also the dicrotic pulse. That on account of the magnitude of the damage, pulses will be either intermittent or intercident has also been shown in the second book. And anyone who doesn't remember those things is reading this in vain. On the other hand, to one who does remember, I think a more excessive explanation is not still required. For with respect to the things said subsequent to these in the work for beginners about the peripneumonias, they do not have causes that are obscure to anyone who remembers the first things said, so even here it is only necessary to subjoin the actual statement.

**Since all the fevers in the peripneumonias are acute and also have some depressions of consciousness/lethargic states (kōmatōdes), whichever of these should prevail,*

729 Added from *De pulsibus ad tirones.*

it is in relation to that particularly that the amount of the frequency is discovered. When the person with peripneumonia is more feverish, the pulse is very frequent; when the depression of consciousness is greater, it is less frequent. **[183K]**

13. *The pulse of those with lethargy is similar to that of those with peripneumonia in terms of magnitude, weakness, and softness; it is slower than this, more infrequent and less unequal, and is more intermittent or intercident. Sometimes it is also dicrotic. However, it is always undulant in the deep lethargic attacks (kataphoras), about which these things are said. For we shall go over the pulses of the diseases that have come to completion with all the signs of these, so that we are also able to recognise adequately the magnitude of those things that are lacking in what is not yet complete—as much as there already is, and as much as might possibly be added to them.*

That the lethargy arises from a phlegmatous humor has also been shown in the specific discussions about the diseases. Consequently, it is reasonable for the pulse to be very similar to that of those with peripneumonia, although slower to the extent that the brain is relatively far removed from the heart. For the lung, by its proximity to the heart, reasonably makes the pulses more rapid and more frequent, since it is hotter. And it is especially reasonable that **[184K]** they also become unequal due to the proximity—this is not obscure to those who remember the first [statements]. And because of this, the pulse of those with lethargy is intermittent more than intercident. For the disease of lethargy is much colder and, one might say, more corpse-like, than the disease of peripneumonia. The intermittent pulse is characteristic of a cold condition, and even more so of a capacity that has been cooled.

14.* *The pulse of those with phrenitis is small; sometimes very rarely it is seen to be large and to have moderate tonus. Also, it is hard and sinewy, very frequent and rapid. It is also somewhat undulant. Sometimes it will also seem to you to tremble somewhat, and sometimes to be cut off spasmodically (convulsively). The pulse of those with fevers clearly and typically acquires a specific symptom in the rapidity, particularly in relation to both the limits of dilatation, and more the external. And you discover the kind of inequality in relation to the position occurring sometimes strongly in these. But also, it will often seem to you that the whole artery, leaving its own position, is carried upward clonically, being thrown up more, more than dilating in a pulsatile manner. In the same way too,* **[185K]** *it withdraws back downward, being drawn downward more than drawn up (contracting). The great frequency of this lying in wait threatens syncope.*

And here there is nothing difficult in discovering the causes of the things described for someone who knows that phrenitis arises from a bilious humor, just as lethargy does from a phlegmatous humor, and knowing too that lethargy has its genesis more in the brain itself, whereas phrenitis has its origin in the leptomeninges and septum pellucidum.[730] Someone who remembers these facts has no need to learn that the puls-

730 There is again the problem of the meaning of διάφραγμα, whether it means the septum pellucidem (Galen, *De usu partium* IX.3 [II.719K]) or the actual diaphragm, which may well be intended here. On this matter, see Rocca 2003: 19n9, 108–109, 111.

es become small and hard as expected. And if the affection is hot, the pulses are small and doubtless also altogether frequent. It was shown in what has gone before that the deficiency of the first dilatation changes the pulses towards frequency. However, just as the peripneumonic pulse sometimes but rarely becomes dicrotic, because it partakes least of hardness, so too does the phrenetic pulse very rarely become undulant, because it partakes least of softness.

And it has been shown elsewhere that tremulousness is a sign of a weakened capacity. Therefore, whenever the capacity in the arteries attempts to distend them to a greater extent, for it requires this due to the amount **[186K]** of heat, and then is unable to do so due to the hardness of the arterial wall, somehow the pulse seems to tremble. And it was clear that under these circumstances particularly, such a pulse occurs, and when people are about to suffer from syncope, just as when they are going to convulse, the movement of the pulses is spasmodic. However, to have the movement cut off spasmodically (convulsively) and not to stop gradually exists continually in those with phrenitis, as also when the heat prevails and the arteries are hard. When the capacity is still reaching out to distend the arteries, while the hardness of the arterial walls acts in opposition,[731] even here the force and the kind of cutting off is spasmodic. In this way, hardness and the beginning of the dilatation in the arteries occurs suddenly and all at once, more like those starting up from sleep, than those dilating gradually. In this way too, in all those febrile from an inflammation, such an occurrence happens, appearing more in some and less in others, not only in the termination of the movement, but immediately at the beginning. The common cause of all these is the increase of the need of the pulses, together with the hardness of the arterial wall.

For whenever the need incites, and because of this **[187K]** the capacity bursts forth to a large and rapid dilatation, while the hardness of the arteries resists, inevitably the previously mentioned form befalls the movements. But also, that due to this same dryness, hardness and tension of the vessels, a pulse appears that is unequal in the position, is not a surprising concept to those who have remembered what was previously said, just as it is not that the threat of syncope lies in wait for excessive frequency. For it was shown that frequency is the offspring of a weakened capacity.

15. Next, the discussion is about an affection which seems to be the mixture of phrenitis and lethargy, because of which the pulses also have forms that are midway between each of the affections. Now what the nature of this particular affection is has been shown in other [writings].[732] On this account, it is reasonable for it also to have mixed pulses, as they would also have the same mixture in nature—there is nothing remarkable about this. As a result, for someone who knows this, no further discussion is required. It is enough for me to subjoin here too the statement from this introduction, which is as follows. **[188K]**

731 K: ἀντιβαίνει: F2, and Va1: ἀντιλέγει.

732 Τυφωμανίη: *Linguarum seu dictionum exoletarum Hippocratis explicatio*, sub verbo (XIX.147K); *In Hippocratis prorrheticum commentaria* I.1 (XVI.497K); *De pulsibus ad tirones* VIII.484K.

[733]*There is also another affection, which itself we must think of as either intermediate between lethargy and phrenitis, and not being the same as either one, or common to both, as a mixture of the forms of phrenitis and those of lethargy. We shall look at this specifically. We shall now speak about its pulses. And so that it is not brought forward as some riddle, I shall make it clear by those features that accompany it. In most instances [the sufferers] close their eyes and are sleepy and snore. Again, for the most part, they gaze in a strained way and continue unblinking like those with katochē.*[734] *And if someone should ask them a question or force them to converse, they have difficulty answering and are unresponsive. In many instances, they speak in a distracted way, don't answer correctly and talk nonsense heedlessly. Such is the affection I now wish to make clear by recognition of the accompanying features due to the lack of a specific name. The pulses of this disease are rapid and frequent, like in those with phrenitis, but less so. In the same way too, they have less strength than those. [In phrenitis] the pulses are broad and short and the part related to the outward movement is not suddenly cut off, but in another* **[189K]** *way as if striving to withdraw inward, speeding up the contraction and concealing the dilatation. These are not like those in relation to the dilatation of those with phrenitis—they don't have this kind of cutting off.*

16.* *The pulses of those with katochē—for the doctors of old called these katochos and katechomenos, while doctors of recent times term the affection "catalepsy"*[735]*—are like the others in the lethargies as to largeness, slowness and infrequency, just as the whole affection is not far removed in kind from the other affection. The pulse of those with katochē is not weak or soft; rather, in these features they differ greatly, just as also in the dissolution and swelling in the whole bodily state (hexis) in those with lethargy, there is binding and keeping together in those with katochē. In the same way too, they differ from each other in inequality and equality. Thus, the pulse of those with katochē is equal (even), while in those with lethargy it is unequal (uneven).*[736] *Archigenes says the place of the artery is found to be specifically hotter in these cases, just as in those suffering convulsions along with lethargic attacks (kataphora).* **[190K]**

Almost all these other things in those with *katochē* are as for those with lethargy, except when they are immobilised and as if dead in respect of the bodily state (*hexis*). From this also the pulses have alternating weakness and softness—the former because, since the *dyskrasia* has not yet got into the whole bodily state (*hexis*), the capacity is strong, while they don't have the softness itself, because the wall of the artery is not already soft. However, the place of the artery is not always distinct. To me, anyway,

733 There are some minor differences between the two versions, most notably in the verbs in the opening sentence.

734 The four Greek terms are, respectively: κατοχή, κάτοχος, κατεχομένος (from κατέχω) and κατάληψις. On these, see the appendix on medical terms.

735 On catalepsy, *Stedman's Medical Dictionary* 2000: 703 has: "the trance-like phase of catalepsy in which the patient is conscious but cannot move or speak".

736 This final clause is omitted in F1, F2, and Va1.

it usually seems hotter, as greater heat is manifestly found in most cases. The cause, both in this affection and in the others, is found to be the same as the symptom in these, in that the cooling of the bodily state has not attached itself to the arteries equally. For many such things happen in the non-uniform *dyskrasias*. Moreover, the place of the arteries seems hotter that those things surrounding; sometimes this is due to it itself being more heated and sometimes to those structures lying around having been cooled less.

17.* *The actual body of the artery of those convulsing seems to be contracted and constricted from all sides, but not like it has been compressed by something or confined. Nor is it altogether uneven like with a fever, and particularly in* **[191K]** *those who are symptomatic.*[737] *Nor is it very difficult to distend due to hardness, such as exists over a long period of time, and particularly with certain faults or bad states of the internal organs, but as if it is a hollow sinewy body, like the intestine or something similar, extending from both the ends. In the same way too, the movement is unequal when the artery changes position up and down like the string of a musical instrument. For there is no impression of either dilatation or contraction; it is more like clonus—like a leaping upward and in turn a drawing inward, and this is not an affection occurring separately, but in time often some part of it seems to be carried upward, as if shot out and carried in as though drawn by something, or one part is moved quickly and one part slowly.*

The pulse of those who are convulsing seems to be strong and large. It is neither weak nor small, nor as strong or large as it gives the impression of being. For the pulse-beat is deceptive due to the tension seeming to be strong and due to the clonus leaping out. Wherefore, there is sometimes the impression of greater height [and it brings a kind of rough sound to the touch.][738] **[192K]** *And this pulse is not something hidden from those who are practised. For it is like nothing else—not the tension in each direction, nor the spasmodic movement. When the movement*[739] *is mixed with that of the acute lethargic attack (kataphora), it is hard to detect. Only someone who has trained himself to recognise accurately each movement in it is also able to recognise the combination.*

This statement is the same as the one from the introduction. The cause of the things said is clear beforehand to someone remembering the principles[740] are affected in common with each other.

18.* *The pulse of paralysis is small, weak and slow. In some cases, it is infrequent and, in some cases, frequent, but is irregularly intermittent.*

The pulse of those who have become paralysed is small and slow because the affection is cold. It is weak because the capacity is also weaker. If the dissipation in it is

737 In Kühn's Greek text, οὐδ' ἐν ἐπισημασίας follows the first ἐν ἐπισημασίας but is absent in the manuscripts F1, F2, and Va1 and is not included in the translation.

738 The final clause, in parentheses, is not present in the manuscripts F1, F2 and Va1.

739 K: κίνησις; F1, and Va1: κρᾶσις.

740 For a summary of the principles, see Galen, *Ars medica* 5 (I.318–319K; Engl. transl. Johnston 2016: 177).

greater, [the pulse] is frequent and at the same time unequal and irregularly intermittent. I stated the cause previously, in the second of these books.[741] **[193K]**

19. *The pulses of those with epilepsy and those with apoplexy are similar.*[742] *Those things that will be said about those with epilepsy must be considered as said also about those with apoplexy, but extended further.*[743] *In the case where the disturbance is moderate and the affection is not yet much stronger than the nature, it is clear that there is no change to find in magnitude, strength, rapidity, frequency and hardness. Only in those convulsing is there a likelihood that the artery is extended in each direction. If, however, the affection is so strong as to weigh down the capacity, it takes on a certain inequality (unevenness), the tension is strong, and the pulse becomes smaller, weaker, infrequent and slow. When it is greatly compressed and the capacity is overthrown, it makes the pulses weak, small and frequent.*

This is the statement. There is no need of so much explanation for discovery of causes in these pulses because this is clearly evident from what was said in the first and second of these treatises.

20.* *The pulse in those with synanche (sore throat)*[744] **[194K]** *has a certain tension similar to that in the convulsion, but is large and undulant, as in those with peripneumonia. And which of the two prevails greatly in this is that from which you must expect the change. For if the peripneumonia is of the kind to predominate, the synanche will end in peripneumonia, whereas, if the spasmodic is, it will end in convulsion. In those who are choked strongly, the pulse becomes small and infrequent. In the late stages, it is already frequent and unequal.*

In every inflammation that leads the principle to *sympatheia*[745] a common symptom is the tense pulse. But those of the sinewy bodies and those that are near the principle show this more. Certainly, the pulse of those with *synanche* is not without tension, but partakes of this either more or less. For this occurs due to the inflowing humor, which is either more bilious or more phlegmatous, and on account of the difference

---

741 *De causis pulsuum* II (IX.55–104K).

742 There are two significant differences between the texts in the penultimate and final sentences respectively as follows:

- post μικρότερος: *De causis pulsuum:* καὶ ἀμυδρότερος γίνεται καὶ ἀραιὸς καὶ βραδὺς; *De pulsibus ad tirones:* γίνεται καὶ ἀμυδρότερος καὶ ἀραιότερος;
- post ἀμυδρούς: *De causis pulsuum:* καὶ πυκνοὺς καὶ ταχεῖς; *De pulsibus ad tirones:* καὶ μικροὺς καὶ πυκνοὺς.

743 Kühn's Greek text has ἐπιτεταμένα μάλλον following χρὴ δοκεῖν. This is not present in the manuscripts F1, F2 or Va1.

744 On the Greek term συνάγχη LSJ has simply, "a kind of sore throat"; the Latin is *angina* for which Lewis and Short (*A Latin Dictionary*, Oxford, Oxford University Press, 1993 impression, p. 118) has "the quinsy or suffocation."

745 The Greek term, συμπάθεια in a medical context is defined in LSJ as "a sympathetic affection of the body", opposite to ἰδιοπάθεια, with reference to Galen's *De locis affectis* I.3 (VIII.30K).

of the organs themselves in the neck. Those that are more sinewy and more under tension, strain the pulse more, whereas those that are fleshy and porous make it more undulant. **[195K]**

The other aspects of the statement are clear, apart from what was said at the end—that the pulse in those who are choking from these becomes small and infrequent, but at the end frequent and unequal. For this is so. In the beginnings of the choking, the dilatations of the arteries become smaller and more infrequent, while the capacity is already weakened, and the need diminished. Then next, when they are already near death, the capacity attempts to draw in more continuously, as the substance of the psychical *pneuma* is already deficient. Being weakened, however, it makes the dilatations small of necessity, and because they are small, they are also frequent. I have spoken about this before. Then suddenly constrained, as it often is, it is frequently on the point of stopping. One of these makes the pulse intermittent; the other makes it infrequent. Both are unequal in the *differentia* relating to this class.

21.* *The pulse in those with acute orthopnoea is unequal, irregular, and failing by degrees.*[746] *And that of the intermediate state in badness is frequent;* **[196K]** *when the patient is in extremis, it is forceful, slow and remittent; when death is already imminent, it is frequent and weak.*[747]

What people call orthopnoea are the acute paroxysms of the chronic disease, as of epilepsy. In these, then, and not least in all the other orthopnoeas arising suddenly in any way whatever, the pulses are altogether unequal, irregular and on the point of failing due to the causes mentioned a little earlier. Now, depending on the differences of the badness, they are either more frequent or more infrequent. They are never rapid due to the weakness of the capacity. In the intermediate states of badness, they are more frequent and at the same time are also clearly unequal and failing by degrees. At the beginning of the discussion this was described as inseparable. The reason frequency arises is either urgency of the need or weakness of the capacity, as stated previously. Furthermore, when the capacity is already moribund, the pulse is slow and at the same time deficient. Both these exist in a capacity that is not still able to act. With the approach of death itself, the **[197K]** craving for *pneuma* has already become very forceful in nature, while the capacity is obviously weak. When compelled by need, it comes to movement, but is not able to raise the artery to the extent the need[748] requires. Of necessity, this makes the pulses frequent. And this, displaying for a short time all the power it has, is ultimately extinguished once and for all.

22.* *The pulse of hysterical choking*[749] *is lengthened spasmodically and infrequent. In those who are terminal, it is frequent, irregular and failing by degrees.*

---

746 For ὑπεκλείπω LSJ has “to fail or flag by degrees”, citing Theophrastus, *De igne* 55, and this passage in *De pulsibus ad tirones* 1 (VIII.493K).

747 Καὶ ἀμυδρός, present in K, is absent from the manuscripts F2 and Va1.

748 K: χρείας; F2 and Va1: δυνάμεως.

749 The term used here is πνίξ. See Hippocrates, *Aphorismi* 4.34 (ed. and Engl. transl. Jones 1931: 143).

The cause in this is not problematical for one who recalls what has gone before. For due to the affected organ (artery) being sinewy, it is to be expected that the pulse is extended spasmodically. Because the affection is cold, infrequency is present with it. Since the capacity is already dissipated to the extreme, it is to be expected that it becomes frequent and unequal, and at the same time failing by degrees. I spoke about these things previously, saying that, whenever the need urges but the capacity is in a bad state, there is inevitably such a pulse.

23.[750] *When the stomachus is afflicted—for let the mouth of the stomach be termed thus by us in the present circumstances, following the* **[198K]** *customary usage of the majority—this doesn't change the pulse in one way. Rather, if it is only inflamed, it brings about such a change of the kind we said*[751] *occurs in an inflamed sinewy body. If, however, there is compression, biting, dissipation (lysis), hiccough, vomiting, nausea, anorexia or pain it will accord with the kind of the symptom. The bitings, vomitings, nauseas, hiccups and anguishes strongly condense the pulse and make it with this small and weak, and in some cases moderately more rapid. Compression alone, apart from any of these things, makes the pulse infrequent, slow, small and weak. Such a compression arises from burdensome foods which have no strong potency, but are disturbing due to the actual amount alone, and some non-biting fluids that flow into it. If it is cooled from these, then at that time there will also particularly be such a pulse. And that of those with bulimia*[752] *is of this sort.*

All the conditions that change the pulse towards frequency, if they are chronic or stronger, make the pulse vermicular. Those which change the pulse towards infrequency, in addition to extending the previously mentioned differences, **[199K]** *generate with themselves such a kind of pulse with inequality in a single pulse, so that to a great extent the body of the artery seems to have been perforated, such that it doesn't seem to be continuous; rather, in the dilatation, the perception to palpation is like sand falling.*

This is the statement from the introduction.[753] On the causes of the things mentioned, there are those that are very clear to people who recall what was said before; these don't need to be stated. However, those that are more obscure will be spoken of. When the capacity is weak in and of itself, clearly from a *dyskrasia*, or is weighed down or disturbed *kakoethically*[754] by something external, in the weighings down, the pulse becomes small, slow and infrequent—small as the capacity has been adversely affected and slow because of this and because the need has been diminished. For those things that burden the capacity are cold conditions, and because of this, the pulse is also infrequent. Such a pulse was shown to occur when the need is diminished. In the con-

---

750 There are predominantly minor differences between the two texts, reflected in the respective translations. Most notable is the difference in the list of symptoms in sentence 3.
751 εἴπομεν, present in K, is absent from F2 and Va1.
752 This is taken in its primary sense of 'ravenous appetite.'
753 There is some variation in the two statements, particularly involving the list of disturbed functions.
754 On κακοήθεια, we have avoided the rendering 'malignant' in view of its current medical connotations. It might be taken as a Galenic technical term meaning broadly, 'disturbed by a bad state.'

ditions that only irritate and disturb the capacity, the pulse is small and weak due to vitiation of the capacity. It is frequent because it is weak, and sometimes also a little faster.

For just as when we ourselves also **[200K]** often become weaker and thin, and are then compelled by some need, and are unable to progress with the legs to the greatest extent, we force ourselves as much as we can to step forward faster and more frequently, so too does the capacity move the arteries, whenever it is stirred by the need, to the movements, but being weak is not able to make the dilatations large. However, the dilatations do sometimes become a little faster and frequent to the extent that the need is more deficient. It stands to reason, then, that in the *stomachical* conditions changing the pulse towards frequency, when they are extended, the so-called vermicular pulse supervenes, for this was characteristic of a dissipated capacity. On the other hand, those that change it towards infrequency, as if broken down, are a product of a non-uniform *dyskrasia* of the arteries (organs) themselves.

24.[755] *Pulses of those with the fluid collections (dropsies) [are as follows]: of the ascites, small, more frequent [long, frequent]*[756] *and somewhat hard with a certain tension; of the tympanitic ascites, longer, not weak, more rapid, frequent, somewhat hard and with a certain tension; of anasarca,*[757] *undulant, broader and soft.* **[201K]**

The ascitic dropsy, which is no small excess of watery fluid collecting in the spaces of the abdomen, brings the large arteries to a sympathetic affection, so they are tense and at the same time also weighed down[758] and cooled. The tension goes along with the other things as follows: weighing down due to heaviness; cooling due to the superfluous fluid. Due to the tension, it is reasonable that it makes the pulse somewhat hard with a certain tension; the badness of the capacity and the cooling makes the pulse smaller. To the extent that it is smaller than it needs to be, to the same extent it is also more frequent, and more so when there is an accompanying fever.

The tympanitic dropsy (hydrops) weighs down the large arteries less and on the whole, the capacity, but creates no less tension. Because of this, then, the pulse becomes somewhat hard with a certain tension, just as in those with ascites. It is, however, more rapid than those and less weak, but frequent for the same causes to that, and longer, due to the lightness of the bodies lying on the arteries. For it was shown in the second book that the pulse becomes short when the arteries are compressed by the weight and amount of the bodies lying on and around them,[759] whereas it becomes of long duration from the **[202K]** opposite causes, the chief of which is the light-

755 There are two differences between the text, as indicated in the following two notes. The term *anasarca*, still in use, is preferred in the translation above.

756 In parentheses are the terms in *De pulsibus ad tirones*.

757 Galen has here κατὰ σάρκα instead of the more usual ἀνὰ σάρκα, as in *De pulsibus ad tirones* and in Kühn's Latin translation here. The meaning is presumably the same, as below.

758 K: βαρύνειν; F1, F2, and Va1: θλίβειν.

759 αὐταῖς, present in K, is absent from F1, F2, and Va1.

ness of the bodies lying on them, as occurs in the tympanitic dropsies, for the whole body is progressively filled with *pneuma.*

What remains is that dropsy termed *katasarca* and *anasarca*, when the whole bodily state (*hexis*) is soaked with copious fluid such as makes the walls of the arteries soft and in this way also renders the pulses soft, undulant and broader.

25. *The pulse of those with elephantiasis is small, weak,*[760] *slow and frequent.*

The pulse of those with elephantiasis is made small, weak, slow and frequent; this doesn't occur immediately at the start, but when [the affection] comes to a considerable degree of severity, so the capacity is weakened. For these were shown to be characteristic of a capacity that has been badly affected together with cooling.

26. *The pulse of those with jaundice without fever is smaller, harder and more frequent,*[761] *but neither weak nor rapid.*

The humor of yellow bile is somewhat drying in nature, just as salt and seawater are, and because of this, it also makes the wall of the arteries harder,[762] and along **[203K]** with this, also renders the pulse harder and immediately smaller as well because the hard arteries are obviously unable to dilate completely. Consequently, it is reasonable that, to the degree it becomes appropriately smaller, so to the same degree it becomes more frequent, but not in fact weak, for the capacity is not weak. Nor is the pulse rapid, because there is no fever; in those with fever, it is also rapid due to the urgency of the need.

27. *The pulse of those who have taken hellebore, a short time before the vomiting, at a time when there is compression, is broad, infrequent, weaker and slower. When they are vomiting and retching the pulse is unequal and irregular. When they are already becoming better, it is regular but still unequal, although less than before. When they are coming near to an accord with nature, it is equal, greater than before and stronger. In those among them who are suffering syncope, convulsing or hiccoughing, the pulse is small, weak, irregular, more rapid and very frequent. In those who experience choking, it is small,* **[204K]** *weak, irregular and unequal; it is not frequent or rapid, but slows still more. Also some undulation is apparent and breadth, and at times some small degree of tension of the artery.*

It is not without reason, in those who have taken hellebore, for the pulses, when they are still compressed, to become slower, more infrequent and weaker. These features are common to every compression. What is specific for these is the breadth, which happens to occur from the *pneuma*, when it is all drawn inward along with the heat. When the capacity is activated to the extreme, the characteristic pulse is high. However, when the capacity sinks down and draws into itself, the pulse is low. It is this, then, that creates the impression of breadth. All the other things that befall them are quite clear to those who recall what was said previously.

---

760 Here, there is ἀμυδρὸς while in *De pulsibus ad tirones* there is ἄρρωστος.

761 There is a minor variation in the list between the two texts.

762 In the manuscripts F2 and Va1, τε καὶ ξηρότερον is added after σκληρότερον.

VI *On Prognosis from the Pulses* (*De praesagitione ex pulsibus, libri IV* IX.205–430K)

Book I comprises 9 sections as follows:

1–3. These first three sections consider variations in the dimensional changes (length, breadth and depth) which are related to alterations in capacity, change of need, and variation in the state of the arterial walls. As well as considering what these changes signify prognostically (understanding the term "prognostic" in the broad sense, as defined by Galen), consideration is also given to causes. In section 1, after some preliminary consideration of the issue of prognosis from the pulses, Galen focusses primarily on the nature and significance of largeness of the pulses. He provides two illustrative cases, one real and one hypothetical. In section 2 he considers particularly smallness of the pulses. Section 3 focusses on the three dimensions of the distension of the artery—length, breadth and depth/height. He goes through the 27 pulses from the Table in Book I of *On the Differentiae of the Pulses.*

4. This is a long section ostensibly about the magnitude of the dilatation and the relation of this to strength of capacity and level of need of the pulses. He considers the prognostic relevance of the pulses in many conditions.

5. This deals with the prognostic significance of changes involving rapidity and slowness.

6. A brief digression on variations in dilatation—the quality of the movement, the amount of the movement, the indications pertaining to capacity, and those pertaining to the arterial wall.

7. This section considers the causes, combinations and significance of changes in strength/weakness and hardness/softness.

8. This brief section considers causes affecting the strength of capacity.

9. This final section, also brief, considers causes affecting the hardness/softness of the arterial walls and adds some concluding remarks.

Book II comprises 13 sections as follows:

1. This deals with the causes and significance of variations in the contraction.

2. This provides a rather complicated analysis of the causes and significance of various forms of frequency and infrequency.

3. A brief discussion of Herophilus' errors on the subject of pulse rhythms.

4–7. These four sections consider unequal pulses, including intermittent and intercident pulses. In section 4 there is a basic division into systematic inequality (i.e., that involving a series of pulses) and inequality in a single pulse. The systematic form is considered in detail, focussing on causes, associations and prognostic significance. In section 5 two oppositions are considered: frequency/infrequency and intermittency/intercidence after preliminary criticism of Archigenes on the interpretation of these pulse variations. Galen provides a comparison of the intermittent and intercident inequalities and then considers the extremes of the basic parameters. Sections 6 and 7 deal with inequality in a single pulse, section 6 being on intermittent and intercident inequalities and section 7 on a third form of inequality related to a non-uniform *dyskrasia* in the heart.

8–12. These five sections consider specific unequal pulses: bounding and dicrotic (8); undulant and vermicular (9); tapering (10 and 11); tremulous (12).

13. The final section gives brief consideration to the prognostic significance of regularity and irregularity.

Book III comprises 8 sections as follows:

1. A summary account of two of Galen's basic general divisions:

– things contrary to nature are divided into conditions harming function, causes of these conditions, and the resultant symptoms;
– *dyskrasias* are divided into 4 mono-*dyskrasias* and 4 compound *dyskrasias.*

https://doi.org/10.1515/9783110612677-012

2–3. On hot *dyskrasias*, which are basically associated with large, rapid and frequent pulses. In 3 there is particular consideration of plague.
4. This section begins with consideration of cold *dyskrasias*, which are basically associated with a small, slow and infrequent pulse. It goes on to consider non-uniform *dyskrasias*, particularly of the heart, the causes of clonic and dicrotic pulses, and hectic fevers in particular.
5. More on the pulses in fevers. The variations in the pulses are related to the discrepancy between the heart itself and structures contiguous with it, in terms of hot and cold. It ends with a brief critical statement on Archigenes and Erasistratus.
6. A brief digression on Archigenes' misuse of metaphorical names for pulses and the use of metaphorical terms generally—the pitfalls in using such terms.
7. This returns to the topic of pulses in fevers, with division of the latter into ephemeral, hectic and those due to putrefaction of humors. There is further criticism of Archigenes.
8. A short section on pulse signs associated with moist and dry *dyskrasias* of the heart.

Book IV comprises 12 sections as follows:
1. An outline of the plan of the book—to deal with affections of all parts of the body other than the heart and arteries, their effects on the pulses, and the prognostic significance of these.
2–10. The parts dealt with sequentially are as follows:
- (2) lungs and upper airways;
- (3) chest wall;
- (4) liver and thorax;
- (5) diaphragm and diaphragmatic pleura;
- (6) stomach, differentiating cardiac orifice, body of stomach and pylorus;
- (7) remaining organs of nutrition, urinary bladder and uterus;
- (8) axillary and inguinal lymph nodes, arms and legs, head and brain;
- (9) meninges;
- (10) testes.

11. A summarising section. Galen reiterates his view that prognosis includes recognition of what previously existed, what is presently existing and what will exist in the future. He summarises the key prognostic features relating to death and survival, and the times when these outcomes will occur. Here the pulses are not specifically considered.
12. Consideration is given to pulse variations in relation to the strength/weakness of the physical (vital) capacity and the excretion/elimination of superfluities, and what variations in these factors signify prognostically.

Book I

1. [205K] It is enough that long ago the followers of Hippocrates have shown prognosis to be not only about things to come but also about things present and things past. I don't now come to state what prognosis can do, and that it is not a secondary matter. But I shall point out what I know about the prognosis by means of the pulses to those who know what it can do. Since the teaching about the pulses has two components, just as that about remedies also has—in the latter case, one about the specific powers of these and the other about them in relation to the affection—I think it is better **[206K]** to be practised in both, so that whatever affection we are dealing with, we are able most readily to go through the pulses of this. In turn, I shall call to mind quickly, for each pulse spoken about, all the specific affections in which it customarily occurs. For if someone sufficiently practised in both were to train himself to recognise each of the pulses accurately, he would, I think, be completely learned in these. Consequently also, if someone is to be familiar with these treatises to the necessary extent, he must have first read that treatise I wrote about the recognition of the pulses, and before that the one I wrote about the *differentiae* of these, where I went over how many of these there are in all, of what kind each one is, and by what name we call it. There is also another third treatise about the causes in the pulses, all of which are needed for our present purposes.[763]

Without the first, it is not possible for someone, who doesn't know beforehand each[764] of the names in the case of the matter being discussed, to understand what is being said. When there is ignorance of the second, everything useful from the theory of the matter we now have in hand is lost with it. For if I should write the rapid, or the large, **[207K]** or the strong, or one of the other pulses to indicate what has previously happened, or what is present, or what will happen, the reader will not know how to recognise these from their actions, nor will he be able to judge the truth of what is being said, nor at any time will he be able to prognosticate anything from them. The third treatise, teaching the causes that change the pulses, is essential for demonstration of the matters which will now be spoken of, as well as for memory and recollection. And we shall know this more clearly, if we take up any one pulse whatsoever—let it be, for example, the large pulse. In this, we shall first distinguish the homonymy,[765] then the twofold sense in which 'a large pulse' is stated, then next, in relation to each one of the significations, we shall consider what it naturally indicates in relation to present, past or future.

So then, there are two such significations. A pulse is said to be large in relation to what is moderate and in accord with nature or in relation to anything else that might be met with. Of the things prognosticated from each, some are common to both pulses, while some are specific to one. Let us next go over everything in relation to each, beginning from the common. **[208K]** The discovery of these is very easy, if someone is mindful of those things we covered in the treatise about the causes in the pulses, when we showed there is not one class of the causes that change them, but the one essential, as one might say, for their genesis (the *synektic*), which is truly also in accord with nature, and another opposite kind of the causes contrary to nature, and in addition to this another, third kind, which is non-natural. It is these that relate to the true existence and nature of these causes, but there is another fourth that relates to the appearance, which it is now particularly necessary to explain precisely, as it is very specific for the treatise on prognosis. The beginning of prognosis is for us from such and such an appearance of each of the pulses. However, such an appearance is not always

763 These are the three preceding treatises in the present work: *De pulsuum differentiis* I–IV (VIII.493 – 765K), *De dignoscendis pulsibus* (VIII.766 – 961K), and *De causis pulsuum* (IX.1 – 204K).

764 The manuscripts F1, F2, and Va1 omit ἕκαστον.

765 Galen uses 'homonymy' here in the standard sense to refer to the different significations of a single term/name.

what it actually is. And I have already spoken about this also in the second of the books, *On the Causes of the Pulses* at the end of the book, and even briefly in the third book, where I also discourse about 'weak'[766] pulses; I shall speak [about this] a little later, after I first go over the causes that change them.

It was also shown in the first of the books of **[209K]** *On the Causes of the Pulses* that it is not possible for any pulse to be changed apart from there being some change and alteration in respect of the capacity creating it, or the wall of the artery, or the need through which it arises. In this way, then, in the change now before us, which is that to largeness of the pulses, we shall examine what it is at any time, of the things mentioned, that has undergone the change; that is, whether the capacity has become stronger, for we have learned how much by itself this makes the pulse larger, or when the capacity remaining the same, the wall of the artery has been softened, for we also learned that the softer wall raises the pulse to largeness, or such a change has occurred through increase of the need. If, therefore, the capacity is strengthened, the pulse would be altogether stronger. However, it is in fact assumed it has only become larger. On the other hand, if the wall of the artery alone is softened, and we have also learned this—that the softer wall of the artery raises the pulse to largeness—it would of necessity have become softer. If neither of these is apparent, it is clear that neither the capacity nor the wall of the artery has undergone change. What remains, then, **[210K]** is the necessity that the need of the genesis of the pulses has been changed in some respect and undergone an alteration.

The need (use) of these was shown to be twofold: first and greatest is preservation of the innate heat; second is generation of psychical *pneuma.*[767] I now refer in this way to what the psyche makes use of in regard to the voluntary functions. Therefore, when the pulse has been changed to largeness, one of these is altogether altered compared to before. But also, when the change occurring from an accord with nature due to the need is slight, only the largeness of the pulses will be apparent; the rapidity will not yet be apparent. If, however, it is great, the largeness will be increased still more and rapidity is clearly apparent. If there is further increase, both the largeness and the rapidity will be increased, and in addition also frequency, as is clear from the treatise, *On the Causes of the Pulses.* Moreover, if the pulse seems changed in largeness, for this is now assumed, of necessity the need is changed slightly. On the other hand, if the pulse has not become larger but seems larger, the body of the person himself has been rendered thinner than it was through any cause whatsoever.

Thus, knowing these things beforehand, let someone come to the diagnoses: **[211K]** first to those of absolutely healthy bodies; then next, to those associated with some newly acquired movement; then also to those of frankly diseased bodies; and finally, to those associated with some external change. For in the bodies that are absolutely

766 Galen uses the term ἰσχνός infrequently. See *De pulsuum differentiis* I.4 (VIII.506K). Alternative translations are "thin", "meagre" and "feeble".

767 The subject is covered in detail in Galen's short treatise *De usu pulsuum* (V.149–180K).

healthy, the large pulse will indicate the *krasis* of the person and the state (*schesis*)[768] of the body. In those with a newly acquired condition, it will indicate that movement which prevails of which the large pulse is characteristic. I spoke about all such movements in the treatise, *On the Causes of the Pulses.*[769]

In the same way too, in those who are frankly diseased, the large pulse will indicate the hot condition of the disease, while in those with some recent external change, it will indicate that prevailing movement, of which the large pulse is found to be characteristic. For example, of those who are healthy, some have a large pulse, obviously as compared to the moderate pulse. It must be remembered first with regard to this, what moderation is in this class and in relation to what nature of a body. It has been shown, in the second book of *On the Diagnosis of the Pulses*,[770] **[212K]** that the moderate pulse exists in the best conditions of the body but not in these continually; rather, it is when the body is freed of every acquired quality and movement arising from some exercise, nutriment, drink, baths, medications, or psychical affection. So then, when you have called to mind the moderate pulse, and compared the pulse you now find large to that, it is appropriate to consider from what cause it has increased to large; whether it is due to the thinness of the body, or to the heat of the *krasis*, or to the acquired movement. But [the changes due to] the acquired movements settle down immediately, whereas those due to the thinness and the *krasis* still remain. Thus, in this itself and in those I shall speak about a little later, it is necessary to make a distinction, having stated beforehand a little of what is essential for the whole subsequent teaching.

What it is that I wish to say first is this: Our most accurate diagnosis of all the changes occurring in the body is in those bodies of which we [already] know the nature. For the extent to which it departs from nature, following those causes we call contrary to nature, and in addition, those we call non-natural,[771] is impossible **[213K]** to discover accurately otherwise. We do not in fact always know this, therefore we shall not have an accurate diagnosis in all cases. Rather, the diagnosis will always fall short of accuracy to the extent that our knowledge of the natural state does. Now let it be assumed first in the discussion before us that there is the recognition of the nature of the person, and let this alone be unusual in him—that the pulse is larger than the moderate apart from any recently acquired change. It must be that such a person is either hotter or thinner than someone having the best constitution, if we call to mind my two previously written treatises—that is, the one about the diagnosis of the pulses and the one about the causes in these[772]—let us remember also that in all such discussions, we use the disjunctive conjunction, "or". It is possible for all these individually to come together at the same time, whether there appear to be two things bring-

768 The distinction made between *hexis* and *schesis* is that between a stable and enduring bodily state in the first instance and an unstable and temporary state in the second case, both here and elsewhere.
769 *De causis pulsuum* I–IV (IX.1–204K).
770 *De dignoscendis pulsibus* II (VIII.823–877K).
771 See Galen's *Ars medica* and Bylebyl 1979.
772 See Johnston 2006: 48–49 and 49n73 for references.

ing about the change, as now there are heat and thinness, or more. So then, the large pulse named absolutely, as was shown, when compared to the moderate pulse named absolutely **[214K]** will indicate either the *krasis* or the present state of the body (*schesis*), while what is large as relates to each accord with nature in that body will itself also indicate either a certain heat occurring in addition to that in accord with nature or thinness. Distinguish the changes of the pulses from some external cause as follows: the change occurring from baths, running, massages, and other movements is also rapid in this way and quickly returns to the original state, so it is possible for you, having palpated the pulses, to leave a short interval and then palpate them again, to recognise the change very clearly, for the second touch obviously differs from the first, even if only a very short time intervenes. However, the largeness of the pulses from wine and food persists for a longer time, and is distinguished by the associated strength, for the pulse becomes strong no less than it becomes large in these cases, but now is assumed only to have become larger. It is not, therefore, changed from these.

And in fact the change from anger also occurs along with strength, and would not otherwise escape someone of diligence **[215K]** paying attention to the eyes and the whole face. Of course, the pulse possessed by someone wishing to hide his anger becomes unequal and is more unequal in those contending and ashamed. The increase to largeness from semi-digested humors in the body, concocting and nourishing the nature similarly to those who have taken nourishment, jointly also increases the strength. In the same way, if someone is warmed in the sun or beside a fire, in addition to the change not lasting a long time, the skin also will be dry in these people and not moist as from baths. They are, then, to some degree oppositely affected to those who have bathed to an extent indicated by softness and moistness of the skin. Midway between both in the change relating to moistness and dryness of the skin are those who have been massaged and have exercised. Furthermore also, the heat itself is more around the chest in those who have exercised and those who are angered, whereas it is less so in those who have bathed, extending more equally to the extremities along with moistness and softness at the same time. Similarly, also, if the person was anointed in between times with some heating medication, a biting sensation from the heat **[216K]** will arise for you when you palpate, unless perhaps fat adherent to the skin of the person should conceal the effect of the medication.

So then, these are the systematic diagnoses of the causes changing the pulses to largeness in both those who are healthy and those who are diseased. The causes that are external are [discovered] in part from enquiry, as when you make trial of those who are among enemies, or those caught in an ambush, or those attempting malicious acts, and in part from the doctor himself not leaving the patient's side, as we say the best of the diagnoses are. However, there are those cases which occur when the treating doctor is not continuously present with those who are sick, or who do not wish to tell the truth. These are more difficult and require a very sagacious doctor. At all events, it occurred to me, of all the other aphorisms I have previously stated in which the search for diagnoses has failed, either the patient is providing for himself a heating medication in between times, when we are not present, or the pulse has taken to itself largeness by reason of the accession. Certainly, in such times, it is necessary to ascertain and conjecture about the truth from all quarters. The first objective in these cases is the time of exacerbation. Next, **[217K]** there are the customs and usual habitation of the patient. For if at some other time, and not at the time of the exacerbation, the pulse has acquired additional largeness, it is more plausible that the change will have occurred from the heating medication and not by reason of the exacerbation. On the other hand, if it is at the time of the exacerbation, it is more plausible that it has occurred by virtue of the accession. This will be distinguished accurately if, after a little time has elapsed, we again palpate [the pulse]. For what happens is that the heat from the medication, due to which the pulse increased, is always dissipated more with the progression of time, whereas the heat from the exacerbation is increased.

Therefore, the diagnosis itself is from the specifics of the art, whereas that from the customs and habitations of the patients comes as an addition. For some are unaccustomed and fearful as regards drinks of medications, whereas others are accustomed and cannot bear to live without medications. And moreover, the prognosis in those who have taken some heating medication in the case of those who are suspicious of them is altogether weak and brief, whereas it is great in the case of those who are accustomed to them. And the actual habitation of the patient contributes in no small degree to such a prognosis. For some are deceitful towards, and abusive of, those around them, criticising **[218K]** everyone as taking action without sound knowledge. Others, however, are simple and honest, and do everything openly, neither dissembling nor concealing. So then, in the case of the latter, one must not suspect them of having done anything covertly, whereas with those who are deceitful, it is appropriate to suspect everything, and to observe carefully and distinguish both those things related to the specific basis of the art contributing to the diagnoses and those appearing externally alongside these.

For example, I was compelled to act in the case of a rich man who loved medications and took them continuously. And his habit was to take particular delight in the criticising of those around him, as knowing nothing accurately. Discovering, then, that his pulses had increased in magnitude, and then eliminating all the other causes mentioned, there was a remaining distinction still left behind—whether such a thing had occurred due to a heating medication, or by reason of a pyrexial exacerbation, for this was somehow also the actual time of the attack of the fever, I directed him to show me his tongue, and seeing it to have been coloured by a medication, I again palpated his pulses, but did not display any suspicion. **[219K]** At this point, I also recalled that, on the previous day, I forbade him to drink the medication he wished to drink. I then immediately said to the man that he had taken a medication. He, realizing that I adduced this from the colour of his tongue, again on another occasion made small pills of the medicine to swallow, so his tongue was not coloured. But I also discovered his deceit on this occasion. For when I looked at the tongue, then again palpated the pulse and made the other distinctions which I mentioned a little earlier, I was led in my opinion in some way more to the medication, and again leaving an interval, palpated the pulse once more, for I was aware of the man's contentiousness. Having guessed what had happened, I ventured to say to him: “Nevertheless, you seem to me to have taken a medication.”

When he very strongly denied this and called the remaining house slaves as witnesses, I took his hand and palpated the pulses. “Well then,” I said, “would you be able to swear for me any oath which I bade you swear?” He did not readily reply to this and maintained an inequality in the movement of the arteries, as if in distress. And then **[220K]** I said: “Exactly! You took the medication, so don’t swear.” Certainly, towards those who are foolish in this way, one must be ingenious oneself. For the most part, however, laymen not only do not act in opposition to their doctors in the distinction from those things that are external, but they also co-operate spontaneously explaining everything.

I have sufficiently described the primary aspects concerning the change of the pulses to largeness. It is now time to pass on to the continuation of the discussion. Since the pulse is said to be large not in one way, but on the one hand, in regard to the moderate, and on the other, in regard to whatever happens, and since what happens is twofold—either lesser in relation to the moderate, or greater, the next thing to go over would be the progression of each of these pulses to largeness—I speak of what is less than the moderate and what is greater. Common, then, to every progression towards largeness is that only the heat is increased at this time, obviously; neither the capacity nor the arteries have been affected, while the sick person is immobile. Characteristic to the deviation from the accord with nature is that the largeness alone is increased, while of the other two, **[221K]** also the rapidity and sometimes also the frequency. Well then, if it should so happen that we were to find the man whom we left a little earlier having had an increase of the pulses to largeness, to have had a further increase of these, by this also the heat would be necessarily increased. For this was common to all the increases to largeness, but specific in addition to the common is that the rapidity is also changed, and even more in fact, if the increase to largeness of the pulses occurs, not only the rapidity but also now the frequency. It has been shown in the first book of the work *On the Causes of the Pulses*[773] when the heat is increased slightly, the pulse obviously increases towards largeness, but not obviously towards rapidity, whereas when the heat is increased still more, the pulse takes on still more largeness, but also clearly increases towards rapidity and gains some frequency as well, just as if the pulse increases to the greatest extent and the need of the genesis of the pulses exceeds what is the largest dilatation, the pulse does not increase in largeness, because it has increased to the limit. When the pulse has added as much as possible to the rapidity and at the same time to the frequency, and when the rapidity cannot be further increased, it hastens on to frequency.

Now let us assume another sick person, **[222K]** who has not been damaged at all, either in respect of either the capacity or the arteries, but has a pulse that is less than accords with nature. Then also let the pulse in such a person become larger apart from

773 *De causis pulsuum* I (IX.1–54K).

the external movement. It would be true to say that the heat in this person has also become greater than before, and in fact in relation to this, there will be a common assertion to what has already been said twice before. The specific feature of this will be that not only from the pulse being smaller has the increase to largeness occurred, but this is impossible without it also becoming slower and more infrequent, sometimes the pulse being made smaller than one that accords with nature, but it is not possible to become larger alone, apart from becoming more rapid and more frequent. For that there is such a condition in which the pulse that has become smaller than accords with nature co-exists with infrequency and slowness is quite clear to someone who recalls those things stated in the first book of *On the Causes of the Pulses.*

2. Passing on then, let us consider the change of the pulses to smallness. Now in all there are three *differentiae* of this also. What happens is that the *differentiae* to smallness arise either from what accords with nature and so-called moderation, or from what is larger than accords with nature, or from what is smaller. **[223K]** What is greater than the moderate is called simply (absolutely) large, while what is less is called small. Therefore, let us begin again here from those things that are common to these, then next pass on to those which are specific for each. For straightway in the consideration of the common, the path to discovery of the specifics lies hidden.

So then, let us assume some pulse in a particular person is smaller than it was before, but has no change in any other class, and let us enquire what the previous condition was in this person, and what it is now. The path of the enquiry is as follows: a pulse becomes smaller than it was, either when the capacity is rendered weaker, or the arteries are made harder, or the need is dissipated. But the capacity has not now become weaker, or at all events weakness would have come together with the smallness; nor has the body of the artery become harder, for in this way not only would the pulse be smaller; it would also be harder. What remains, then, is that only the need has changed, and if the person has become immobile in the meantime, it will have been in relation to the amount of the [cooling] (change).[774] Furthermore, it is also quite clear **[224K]** to those who recall the things that have been demonstrated in the first book of those on the causes in the pulses, that such a change has occurred from a pulse that is larger than, or in accord with, nature. For the change from an accord with nature to smallness due to cooling also immediately makes the pulse more infrequent and slower, while the reduction from a pulse which is large and increased above the moderate to the moderate is also immediately accompanied by rapidity and frequency, due to the fact that the pulse which is increased beyond an accord with nature changes in the three classes. Accordingly, the pulse is reduced in largeness only, while in other respects it is preserved unchanged, and is one that is slightly larger than accords with nature. For this consists of a slight increase of heat only.

774 Kühn's Latin text has *refrigerationis* instead of μεταβολῆς. The manuscripts F1 and Va 1 have ψύξεως.

However, of the other changes to largeness, none comes about without bringing about also some rapidity or frequency. As a result, nothing else will diminish, apart from the largeness alone, for only what is increased in this by heat will be diminished by cold, whereas in fact that which is changed to such a degree by heat as not to involve largeness alone, but also to change the rapidity and frequency, will have its return path through these same changes according to class. As a result, the pulse will not only be smaller, but also slower and more infrequent **[225K]** than before. The same applies also to the change from the moderate to the smaller, and the change from this in turn to the still smaller, which immediately makes the pulse more infrequent than it was and slower.

For a pulse which has acquired a certain moderation of dilatation has an accord with nature which is of necessity destroyed, if the body becomes hotter or colder than it was. But the manner of the loss is not in fact the same. For in the heating, the largeness is clearly added to it first, then the rapidity, then the frequency. Conversely, when there is cooling, the infrequency is clearly added first, then slowness second, and smallness last. I have said enough about these things in the first book of *On the Causes of the Pulses*. It is not, therefore, still necessary to speak at length about these matters, but to pass on to the continuation of the discussion, distinguishing next those things said with respect to the causes bringing about cooling which are, in terms of class, as many as those that are heating. For cold air contacting externally, as also cold water, or a bath,[775] or a medication anointing the body cools; also prolonged quietude or **[226K]** hibernation, as it were, make the body colder. And of the things effecting change internally, there are inspired air that is cold, nutriment that is phlegmatous, a medication that is cold in nature, and immoderate use of cold water; among the affections of the psyche, there is long-standing fear. Of the humors contained in the body, there are those that are cold in terms of *krasis*, which were first quiescent and then were moved, or have somehow become altogether thinner than the hot ones. In the case of the heating causes, we distinguish these individually analogously to those mentioned. It is not necessary to speak twice about similar things.

Now certainly the change of the pulses in terms of largeness and smallness is brought about by such kinds and numbers of causes, pertaining at all events to the nature of the matter itself. They are still needed for our diagnosis, as often the artery is dilated very greatly in itself, but the magnitude of the dilatation is concealed or hindered due to the amount, or the thickness,[776] or the hardness of the surrounding or overlying bodies. For the actual essence of the large and small pulses exists in the circumference of the artery, **[227K]** while the length is kept equal throughout, although it does not in fact seem equal, because some part of it is hidden. I have already spoken

775 F1, F2 omiserunt. ἢ λουτρὸν present in K.

776 Kühn's Greek text has τὸ ταχός here, whereas Kühn's Latin translation has *crassitie*. The latter is obviously preferable on grounds of sense, although the manuscripts F1, F2, and Va1 are all as Kühn's Greek.

about this in the second book of *On the Causes of the Pulses.*[777] I shall now say as much as is useful for present purposes for the sake of recall.

3. The height of the artery always dilates in proportion to the breadth. However, often either the whole movement, or some part of it, is concealed or prevented by the surrounding and overlying bodies, and this prevention appears either more in relation to the depth, or to the breadth, or in relation to both similarly. Therefore, whenever the movement is unrecognisable to our touch, and this occurs in the whole animal, so that no arterial movement appears anywhere, the affection is called *asphyxia* (pulselessness),[778] which in truth never occurs in a living body, for the heart continually moves along with all the arteries, as of course is often discovered by our touch applied externally. For lying over **[228K]** the bodies of the arteries everywhere are certain membranes and skin, and in many places also fat and flesh, by which not only is the pulse concealed, but also happens to be hindered. Thus, when the movement is completely concealed, if this should happen in all the parts of all the arteries, [doctors] call the affection *asphyxia*, whereas if movements should appear in some parts but not in others, the pulses have a difference in this in relation to the length. Furthermore, they apply the term 'long' to those pulses in which the length of the perceptible movement[779] exceeds that which accords with nature, and 'short' to those in which it is less than accords with nature, while those are called 'equal' which are moderate and in accord with nature. For always as much of the movement of the arteries appears as the bodies set before them allow. And because of this, the pulses are short in those who are fat, while they are long in those who are thin, just also as they appear moderate in those who are well-fleshed and moderate.

Nevertheless, the apparent movement itself is not always similarly hindered in breadth as compared to depth, but when the places of the arteries that are horizontal **[229K]** are filled with fat, flesh, vapour or moisture, while the skin lying above does not burden them at all, only the movement in breadth is obscured, and particularly to the extent that the pulse is strong and large of itself. When the places situated laterally are clear, while the skin overlying the arteries is wrinkled, the movement in depth in such a body is more obscured, and particularly if the pulse is neither strong nor large to such a degree in itself. Therefore, whenever you discover a pulse that is longer than accords with nature, if its breadth and depth should also be increased together, consider whether the person has become thinner than he was, or hotter, the added movement notwithstanding. For if he has become thinner, and if the pulse is larger such that it has changed this alone, or less, then not only is this a cause, but also heat is, whereas, if the state (*schesis*) of the body remains the same in him, only the heat is increased. If, however, he has also become fatter, there is a doubling of the estimation of the increase in heat. That it was increased by some cause is determined from the things said previous-

777 *De causis pulsuum* II (55 – 104K).

778 On the term ἀσφυξία see the Appendix on pulse terminology.

779 K: τῆς αἰσθητῆς κινήσεως μῆκος; F2, and Va1: τῆς αἴσθησεως μῆκος.

ly. If, however, the pulse has become longer and **[230K]** broader, although not in the same proportion to the increase of height, but remains moderate, then following the previously mentioned distinctions, this is also still necessarily exceptional in that the skin overlying the artery above has become more wrinkled. And if you should turn the limb up the other way, so that the part of it that was above at first is now below, it appears in this way that the depth has not been cut short.

These two pulses are written first of all in the table of the 27 pulses in the first book of *On the Differentiae of the Pulses*[780]—[a table] I made about the things existing in relation to the amount of the dilatation in the three distensions (dimensions) at the same time. Following these, I have listed a third, where the length and breadth exceed what accords with nature, while the distension in depth is low, which is rarely found. For the artery becomes greatly burdened by the underlying bodies, which would not happen at any time without the capacity also becoming weaker and the wall of the artery extremely soft. These things being weaker than accords with nature makes the pulse very soft. As a result, accompanying such things, there would at some time be **[231K]** length as well as breadth and lowness, which would not otherwise occur.

Furthermore also, when the artery seems like this, it does not at that time depart greatly from an accord with nature, either in being lower or in being broader. For it is not possible for the artery to be weighed down to an extreme degree in relation to the depth, nor wholly in relation to the breadth, just as it is not possible for it to be raised up to the maximum extent in height, nor altogether narrowed. For height increases more in most instances, and particularly when the pulse is strong, but not in fact to the same extent as it itself is very high and at the same time also very narrow. Of course, if the limb is turned up the other way in the case of this pulse, so that the parts previously overlying now become underlying, there will also be the increased distension of these in depth and the pulse will not still appear low. But just as it is long and broad, so too it is also high, which is large. For the increase in the three distensions (dimensions) is called large. Therefore, one must distinguish in this, in addition to those things being said, also those things said first of all. One must also consider further whether the limb being turned up the other way detracts somewhat from the breadth. For this also occurs **[232K]** sometimes and shows how much the vessel was broadened[781] above the proportion of the previous depth.

In addition to these, the fourth [pulse in the table] is long and at the same time high[782] and moderate in relation to breadth, and is broad as much as it is in itself, just as it is also high, but because the artery is constricted laterally, the movement in this direction has been curtailed. For generally speaking, when either the breadth is disproportional to the depth, or the depth is disproportional to the breadth, the other of these is hindered, either through the amount, or thickness, or hardness of

---

780 *De pulsuum differentiis* I.4 (VIII.504–507K).

781 The remainder of the sentence after 'broadened' is omitted in the manuscripts F1, F2, and Va1.

782 The section of this sentence from 'high' to 'but' is absent in manuscripts F1, F2, and Va1.

the bodies lying in front, which are between the artery and the applied palpation. For truly, the wrinkling makes the overlying bodies more, as the skin is folded in two by the wrinkles. Consequently also, such a pulse, although truly large, is hindered in relation to breadth, or could appear large in the case of a thinned body, although it doesn't appear so, because it is narrowed in one of the distensions (dimensions).

The fifth pulse listed in the table is longer than accords with nature, but **[233K]** moderate in the other two distensions (dimensions), which indicates that the person has become thinner than he was, while the pulse to the extent that in itself it is smaller than accords with nature. Since the long pulse appears to be so, either because of its own largeness or because of thinness, it has not now in fact become long because of largeness, for in this way, it would also be high and broad at the same time; if one or other of these is altogether lacking, it has become long due to the thinness. But since the pulses of those who are quite thin seem larger than accords with nature, it is clear that the pulse is narrower and lower than accords with nature to the same extent. If these things obtain, it is also to such an extent smaller in itself.

Next in the order listed in the table, the sixth pulse is long, low and moderate in breadth, being less than accords with nature due to the previously mentioned causes, to the extent that in itself it seems long due to thinness. But in it this is in common with the fifth: what is specific and remarkable is that the depth has been curtailed due to the cause I stated a little earlier. The seventh pulse in the table is long and at the same time also narrow and high. This would have appeared large, similar to the first of all, if it were not **[234K]** narrowed in breadth. However, the other aspects also pertaining to this similarity need to be distinguished from the first of all. Furthermore also, those things said in relation to the third pulse should also be added here: such a pulse neither frequently appears, nor when it does appear, does it exceed the moderate by very much in terms of height, while it falls short in terms of breadth. Next to this is listed the eighth pulse, which is long and at the same time also narrow and moderate in terms of depth. That it is small in relation to its specific essence is clear from the things previously said in the case of the fifth pulse. However, since the body is thin, it appears longer. The specific and notable feature in it compared to the fifth is the narrowness of the breadth, which shows the artery has been narrowed in this dimension. The ninth pulse is long, narrow and low; this is shown more to the degree that it is small to this extent in itself, while appearing long due to attenuation of the body, so associated with it for the most part are weakness and hardness. If, however, it is neither hard nor very weak, it is made like this from cooling. In this way, then, all the nine pulses mentioned now have a length that is longer than **[235K]** accords with nature, but vary in the remaining two distensions (dimensions).

The next nine pulses, being moderate in length, change in depth and breadth. The tenth pulse, then, which appears high and at the same time also broad in one person cannot exist when compared to the moderate, unless at sometime, albeit rarely, the whole position of the artery is changed, although if it is compared to the absolutely moderate, it is possible for it to occur. And a cause of its genesis relates to a specific feature of the conformation. For it is a symptom of an artery covered on each side by deep flesh, but protruding slightly in the middle, where it is exposed. The eleventh pulse is altered from those previously listed in one aspect only, having a depth equal to what accords with nature; it is by reason of its being low that it is out of the ordinary. The other features are the same as the tenth. The twelfth pulse is broad and low with moderate length, and occurs when the limb is turned over. For it truly appears in every respect moderate or large. If, then, it appears in every respect moderate,[783] it will indicate that the height has been previously compressed slightly, while the breadth is extended due to softness of the artery. **[236K]** If, however, it is large, the height has been compressed to the maximum. The thirteenth pulse is high and at the same time moderate in the remaining two distensions (dimensions), which makes it like the tenth in physical conformation, and in addition to this, there is a change of the position that accords with nature. What is added to it, which stands out compared to that (i. e. the tenth), is the narrowness of the breadth. The fourteenth pulse is moderate in the three distensions (dimensions), so clearly it itself appears to be in accord with nature. If, however, it is truly equal to an accord with nature, it is differentiated by the state (*schesis*) of the body. Thus, if the person is thinner than accords with nature, the pulse, to the extent that it is smaller in itself, appears equal to the moderate. If, however, the body is fatter than accords with nature, the pulse, to the extent that it is large in itself, appears moderate, being curtailed by the large amount of flesh. If, however, the body is changed neither to thinness nor to fatness, the pulse in itself not only of necessity appears moderate but is moderate. And in fact I have shown, in the second of the books of *On the Diagnosis of the Pulses*,[784] that all the other pulses when compared to this are said to be large or small, long or short, broad **[237K]** or narrow, and high or low.

The fifteenth pulse is low in respect of depth but moderate in the remaining two distensions (dimensions); it is the same as the previously mentioned fourteenth in every respect, except for those instances where its depth is compressed. Consequently, those things previously said in the case of that (the fourteenth), we must regard as said in this case. It is, however, appropriate to recall, concerning all such pulses, that when the limb is rotated, they no longer seem low. The sixteenth pulse is moderate in length but narrow and high; it becomes like this naturally, either in relation to the specificity of the conformation, or in relation to the breadth following compression of the arteries at the sides. The seventeenth pulse is moderate in length and depth, but narrow in the remaining distension (dimension), because from the moderation of the depth, it is clearly constricted from the sides, whereas in all other respects in this case, there is a similarity to those in the fourteenth and fifteenth pulses. The eighteenth pulse is moderate in length, but narrow and low; in respect of its own essence, it is small but appears moderate in length due to thinness. For to speak briefly, whenever

---

783 K: ἤτοι γὰρ φανεῖται πάντη σύμμετρος ἢ μέγας. εἰ μὲν οὖν πάντη σύμμετρος φαίνοιτο, δηλώσει ; F1, and Va1: ἤτοι γὰρ πάντη σύμμετρος φαίνοιτο, δηλώσει

784 *De dignoscendis pulsibus* II (VIII.823–877K).

[238K] the length is extended disproportionately to depth and breadth, thinness of the body is indicated. When it is reduced compared to the proportionality of the depth, it does not signify fleshiness, but either some physical conformation or position of the artery which is unequal.

The nineteenth pulse is short, broad and high, and either appears naturally like this due to the specificity of the conformation, or in relation to the affection in the position when the artery is distorted. And the twentieth pulse appears the same as the nineteenth in the other two distensions (dimensions), whereas it is moderate in depth. What is notable, as was said before in that case, is that the height has been curtailed. The twenty-first pulse is short, broad and low for the same reasons, and will further take on those things which make the pulse broad and at the same time low. Similarly, the twenty-second pulse is short and at the same time also high and moderate in terms of breadth, and for the same reasons will take on those things making a pulse moderate in breadth and high, just as in fact the twenty-third pulse is also short in terms of length, but moderate in the remaining two distensions (dimensions), and will be reduced for the same reasons. What is notable about the twenty-fourth pulse **[239K]** in addition is that, for the same reasons, it will take on the cause reducing the height. In the same way too, the two pulses next in order will be reduced in the same manner from the previously mentioned causes, since we remember that in those pulses where the length falls short of the proportionality in relation to depth and breadth, it is appropriate to attribute the cause in these either to physical conformation or a distorted position of the artery. It was also stated that, when the length exceeds the proportionality to the breadth or the depth, either one of these alone has been curtailed by constriction, or also the other has been increased, and that the curtailment often happens, whereas the increase seldom does, in the case of the breadth. When the height is weighed down due to the soft wall of the artery, straightway also such a pulse is not strong, whereas in the case of the depth, whenever the breadth is constrained, the pulse is to such an extent strong and at the same time large in itself.

Remaining following all these, the twenty-seventh pulse [is short, narrow and low and also small,][785] either appearing so or being so in relation to its own nature. For those same things distinguished in the case in the first [pulse] of all as large must be also distinguished in this case, **[240K]** for when the state (*schesis*) of the body is in accord with nature, such a pulse not only seems small but actually is small. When there is greater fatness, if the person has become to such a degree fatter than he was, to the same degree also the pulse is smaller, and would in fact be in this way moderate to such a degree in itself, due to the fatness of the body appearing less than accords with nature. If, however, the fatness is much more than the pulse is smaller, such a pulse would be larger than what accords with nature to the same extent in itself. If, on the other hand, the greater fatness is less than the pulse is smaller,

785 Absent in manuscripts F1, F2, and Va1.

such a pulse would be less than what accords with nature to the same extent in itself. Nevertheless, if the body is thinner than the pulse seems small, this pulse will also be small in nature but will seem less small than it really is. For the extent the thinness of the body departs from what accords with nature is the extent to which the largeness gives besides the impression to it.

The things differentiated in the case of the small and large pulses must also be differentiated in the case of all the others, since we know that each of those is either small, or large, or moderate to such an extent in itself. And these three pulses, in relation to the amount of the dilatation, are as accord with the **[241K]** nature of the matter itself. There seem to be twenty-seven pulses due to thickness or thinness of the body, or more wrinkling of the skin, or unequal constriction of the regions around the arteries, or the specific character of conformation. Accordingly, it is necessary for each person who does not simply seek an empty discussion, but prefers one sufficient for the actions themselves, to call to mind all the descriptions which I have just now have finished, when he considers one of the pulses constituted by the amount of dilatation, to consider first which of the twenty-seven it is; then next, separating it from the changes occurring due to the bodies lying adjacent, to discover whether it is moderate, large or small; then next in order, to discover the conditions from which such a pulse has arisen. For these same conditions teach both those things that have previously occurred and those that are going to occur. An immediate example is the case of the long, narrow and low pulse. Taking this as a paradigm, I shall teach the argument in this case. It is quite clear that it is small to such a degree in itself from the diminution of its circumferential distension, while the length is increased. For if, by two causes, the circumference of the dilatation appears less due to corpulence, and because **[242K]** the pulse has really become smaller, but now there is not corpulence, for at all events this alone would have added nothing to the length, but is also diminished, it is clear that the pulse has become small in relation to its own specific essence. Therefore, when the appearance of length which had occurred in it from the state of the body (*schesis*) is removed, it being from the small pulses in relation to its specific essence, we next consider through what cause it has become small. What contributes to such a very large extent is being able to recognise the amount of the smallness and of the length. For it has either become smaller than accords with nature to a small degree, or to a large degree, or, as one might say, to a moderate degree. Furthermore, in relation to length it exceeds what accords with nature either a little, or moderately, or a lot.

So then, let another of the pulses be extended in length to the greatest extent, and to have come to smallness to the greater extent; that is to say, it is narrow and low. Such a pulse is in nature much smaller than it appears, but such an appearance exists in it due to excessive thinness, while it is smaller in nature due to the causes making it smaller—cooling, weakness and thinness. Of these, then, what is or what are making the pulse now appear long, narrow, and **[243K]** low? It has already been said before that the hard pulse is an inseparable sign of the hardness of the arteries, just as the weak pulse is of the weak capacity. This is termed both *atonic* and weak, just as in fact the opposite to it is termed strong, *eutonic* and vigorous.[786] But in the case of the pulse under consideration, one must know that it is weak of necessity. For a pulse never becomes extremely small when the capacity is strong. However the hardness is not present of necessity.

It is not possible to speak absolutely about the cooling; it is necessary to be aware first, that it is one thing to know what was cooled is the heart along with the arteries, but another thing to know what is being cooled, and what was cooled are what have impaired the capacity, while what is being cooled changed the need. It is necessary to pay very careful attention to what is being said, knowing well that not only to the prognoses through the pulses, but also to almost all other matters, nothing is more useful than this. This whole argument is stated from knowing the essence of the capacity, which we omitted to speak about in the treatise *On the Causes of the Pulses*,[787] because there was no necessity. **[244K]** Here, however, this same knowledge contributes to understanding accurately those things that will be said. It seems to me to be time to go over this, although not here with specific demonstrations. These will be stated at greater length elsewhere, but as much is said in summary as is now useful to present matters.

4. The essence of the capacity of each of the parts is located in the *eukrasia* in each of them. For each part discharges its specific action strongly at the time when it is most *eukratic*. And of course, it also carries this out badly, when it becomes *dyskratic*, and badly to the extent that it is *dyskratic*. In this way, then, the actual body of the heart, when being cooled, heated, moistened or dried immoderately, has become to such a degree *dyskratic*, also makes the pulses weak to the same degree. And this is for the capacity to have been harmed by it. However, to have contact, either with the blood contained in the chambers of the heart itself and *pneuma* much hotter than accords with nature, or with that contained in the pericardial membrane, or in the lung, or also in the pericardium itself and the lung, is to have changed the need alone without **[245K]** having harmed the capacity. Similarly, in the cooling and chilling in the actual body of the heart, when the cold element is predominant in the *krasis*, there is damage of the capacity, while in relation to the pericardial membrane and the lungs, and the blood and *pneuma* in the chambers of the heart, there is dissolution of the need. The moistnesses and drynesses in relation to the actual body of the heart harm the capacity, but those in one of the associated structures don't change the need. For the *eukrasias* [and *dyskrasias*][788] occur in relation to the four elements, whereas the intensifications and abatements of the need (use) occur in the hot and cold [elements] alone. Therefore, when the need comes to the extreme of dissolution, and the structures in association with the heart have been strongly cooled, it is inevitable

786 This is simply a matter of using different terms to signify more or less the same thing. The Greek terms are, in the case of the first pair, ἄτονος and ἄρρωστος, and in the case of the second group of three, σφοδρός, εὔτονος and εὔρωστος.

787 *De causis pulsuum* I–IV (IX.1 – 204K).

788 Added on the basis of the manuscripts F1, F2, and Va1.

that the actual body of the heart is also *dyskratic* in every respect, which is nothing other than the capacity becoming weaker.

And this is why I said it is impossible to speak absolutely about the cooling. For it is not just from collapse of capacity alone, or hardness of arteries alone, that the pulse sometimes becomes smaller, but in this way also due to cooling alone, as **[246K]** has now been said, since the cooling is not understood as being due to the dissolution of the need. For it is not possible at any time for the heart to have come strongly to cooling, if it has not also been cooled itself. You would understand very clearly what I am saying, if you were to have examined at any time those dying from severe cold on journeys. All of them, immediately at the time of breathing out the cold air, also have a smaller pulse, although none of them have a very small pulse prior to coming to the extreme of cooling. Then not only has the need of the genesis of the pulses, which existed in them from the beginning, changed, but also the capacity is weak, since it no longer preserves the natural *eukrasia* in the heart. And for the most part this customarily befalls them all at once, and not gradually, as in other conditions, one of which is now before us. In this, the pulse appears small, narrow and low at the same time in chronic diseases. For in this way, those in such a state have danger lying in wait for them no less than those on journeys do. The pulse in them does not in fact suddenly become small; this occurs gradually in those who are wasting away. For **[247K]** this is one and the same kind of marasmus which was called by Philippus 'old age from disease'.[789] In fact, as the wasting marasmus is not due to cooling, but only to a weak capacity, it makes the pulses weak and small. It is distinguished from the prior marasmus by the rapidity of the movement, the frequency and the hardness, and by the fieriness of the heat. But it is not part of the present discussion to go over these now. I shall return to the matter before us. The small, narrow and low pulse, which deviates still more from an accord with nature in the three distentions (dimensions), when it appears as such without hardness, occurs due to kinds and numbers of causes we described sufficiently.

When, however, it occurs with hardness, there also exists immediately in it either the weak pulse or the pulse intermediate between weak and strong, which neither overthrows the touch nor is overthrown by this, although we must bear in mind that in one way there is no median of weak and strong. But it is shown that the condition prevailing[790] in the body makes the artery hard. This is either dryness, congelation by cooling, or tension. It is dryness in the most burning fevers, when they are

---

789 This is presumably Philippus of Rome (1st century A.D.), a Pneumaticist to whom Galen refers on a number of occasions. On the matter of "old age from disease" see *De methodo medendi* X.10 (X.706–707K), *De sanitate tuenda* V.9 (VI.357K), and *De marcore* 5 (VII.685K).

790 There is some issue about the verb here. Kühn has ἐπικροτέω (to rattle or chatter), as do the manucripts F1, F2, and Va1; Kühn's Latin text simply has *esse*. We have read the verb as ἐπικρατέω.

chronically *kakoethical*,[791] **[248K]** and particularly in the wasting marasmus, certain kinds of melancholias and quartan fevers. Congelation from cooling is from untimely or immoderate cold drinks, or similar baths, or from untimely or immoderate cooling fruits, or in general those things that are naturally sufficient to generate the cold phlegm, which Praxagoras was wont to call hyaloid (glassy).[792] There is tension in fact in the spasmodic (convulsive) conditions and the severe inflammations.

However, in the scirrhosities (indurations) also, congelation and tension of the arteries arise. Such a thing happens in the liver particularly and spleen. For from these internal organs the pulses appear very clearly and continuously hard, particularly in the scirrhosities, but now also even in the severe inflammations. But such affections are recognized both through the touch and through the urine and faeces, and through the colour of both the whole body and the tongue, and now also through the specific kind of fever. There are, however, those other conditions which make the pulse hard: those that have their genesis suddenly from recent faults, and all those others that are gradual. The faults **[249K]** that bring about such conditions are a drink of wine that is untimely and cold, often also cooling fruits, and a cold bath. You will recognize the harm from wine, whenever the pulse is not only suddenly hard, but also becomes larger, more rapid and stronger. In general, you discern all the other conditions from wine by neither magnitude nor strength nor rapidity being associated with them; specifically in each this is sometimes more evident, but at other times more obscure and conjectural. It is more evident when some symptom among those that naturally accompany fruits, follows—I speak of gastric inflation, or biting of those things that are destructive, or a certain compression relating to the oesophagus, or delirium, as with the winelike fruits, or vomiting, or faeces that is more moist and copious. These in fact are not by signs, but indicate directly the actual substance of the things eaten. However, when neither any one of these is present, nor one of the signs from wine, cold water is what brings about the harm, either when drunk or in those bathing in it inappropriately. Decisive in terms of greater certainty in all these cases, the pulses of those who are afraid and those who are contending is the greatest sign. **[250K]** And if to the things stated previously, you always bring this in later—that the disease was irregular[793] —it makes the prognosis more reliable and more remarkable, so that sometimes those listening are incredulous, and think you have not said any of these things from the pulses, but that someone among those observing reported what was done wrongly by the sick person.

It is necessary, having previously palpated the pulses accurately, to then give an account of what is shown by the signs, and to palpate the pulses again, considering whether the sick person has somehow been disturbed. If he is wholly undisturbed, con-

---

791 The basic Greek term is *kakoēthôs* for which LSJ has the primary meaning, "bad disposition, malignity" and the medical meaning, "malignant character" (Latin *maligne*). "Malignant" is avoided given the modern connotation—see appendix on terminology.

792 On Praxagoras' classification of the humors, see Steckerl 1958: 58 – 60.

793 The manuscripts F1, F2, and Va1 omit ὅτι ἠτάκτησεν ἡ νόσος.

sider whether or not you have either overlooked one of the previously mentioned signs or not evaluated it correctly. If he is disturbed, and seems to you distressed or fearful, don't immediately believe the sign, for often those who are sick are not distressed by seeing themselves to have done such a thing, but somehow by nature are easily afraid. In them, the pulse settles a little later, whereas in those who really have done something, to the extent it is revealed in this, the pulse is characteristic of the fear.

All the faults of those who are sick have now been stated: the inflammations, **[251K]** the scirrhosities (indurations), the drynesses or the congelations due to cooling of the internal organs, when they are increased, create the kind of change I have spoken about. In relation to the principle of the nerves, if one of these things is brought about, it creates a convulsive condition. I call those conditions convulsive in which inevitably some convulsive symptom occurs prior to death. And all such people that are about to die are still hot, contrary to those who have collapsed. For the latter are already very cold while still alive. However, those in whom the convulsive condition is in the principle of the nerves, even while lying completely dead, remain to the greatest extent warm.

To someone who has practised palpation, the pulses of all such conditions are immediately, in the first impact, indicative of the tension and reveal the kind of convulsive movement. If, however, in some of those having the previously mentioned pulse, the bowel movements become numerous, and this especially happens in the scirrhosities of the liver, the pulse in them is not still long, but either similar in length to that which accords with nature, or even shorter. And further also, this is good to know about all the pulses that are hard and at the same time also small—they **[252K]** become clonic very quickly, and those which are not clonic, either due to weakness of the vital capacity, or to dissipation of the need, or to both coming together, do not take on beforehand this same movement. So whenever the need requires a large dilatation and the vital (physical) *tonus* is able to produce it, while only the body of the artery counteracts this, being unable due to hardness to rise to a large dilatation, clonus occurs in such a condition. Otherwise, it is impossible for clonic movement of an artery to occur.

Consequently, the pulse under consideration which appears without clonus demonstrates either extreme weakness of the capacity or a very great dissipation of the need. However, if clonus does occur, this immediately indicates the capacity has acquired some *tonus*, and immediately also that the need of the genesis of the pulses has not dissipated to an extreme degree. Furthermore also, smallness does not come to a maximum extent, as the pulse under discussion was said to come to. In the same way, if the pulse is also longer than accords with nature apart from clonus, but narrower and lower, and in relation to each of these not departing to the maximum degree from an accord with nature but only to a small degree, it is appropriate to consider whether, **[253K]** although being hard, it is not still clonic or altogether does not partake of hardness. If being hard, it is without clonus, it is to be differentiated by those things that follow, whether cooling, weakness of capacity, or both together are indicated. Such a pulse in someone following cooling alone is immediately co-existent with slowness in it and infrequency. In someone following weakness of capacity, the weakness is inseparable. But if those things pertaining to need are preserved unchanged, the pulse will be to such a degree more rapid than accords with nature and more frequent to the extent the smallness deviates from the moderate. If in someone the capacity is weaker, the dissipation of the need will be commensurate in amount, and the pulse appears similar to that in accord with nature in relation to both *differentiae*, and is neither more rapid nor slower, neither more frequent nor more infrequent. If, however, one or other is excessive, it will be in proportion to the excess, as I said before. If there is hardness and clonus at the same time, it indicates the same conditions we spoke about before which will produce the hard pulse, and the *atonic*, as in the case of those in whom the capacity and not the need has been dissipated.

In the case of the capacity, the extent to which each of these has been damaged, the **[254K]** difference in strength and weakness will show together with the amount of clonus. In the case of the need, the difference in rapidity and frequency and here also the amount of clonus, and other matters of those previously mentioned, are evidence. Similarly, if a pulse departs from an accord with nature, neither by a little nor by a lot in the three *differentiae*, but moderately and in a middling way, it is distinguished from those things mentioned. However, often in the case of such pulses—I speak of those that have had a small change from an accord with nature towards length, narrowness and lowness—not only is there no hardness apparent, but also manifest softness is present in them. And what occurs in this is due to the same causes in terms of class, although it changes in amount in each. For either the pulse is rendered less than moderate due to the weakness of the capacity, the damage of the capacity being shown to be greater, or due to the dissipation of the need, this comes to a greater extent of badness. For to the extent the softness of the arterial wall is more favourable to largeness of the dilatation, to the same extent it will show the cause making the pulse small to have increased.

Therefore, the amount of weakness in the capacity **[255K]** is adequately shown through the amount of weakness [of the pulse]. Also, what is now said is something added. The dissipation of the need comes to such an extent of badness as is discovered from the stated signs alone. Something will be said about these in what follows, when we combine the classes of the pulses with each other. This is not what lies before us now but, having undertaken to speak about the amount of the dilatation, we were compelled to touch on the other matters. Let us, then, return again to what was set before us at the beginning, passing on to the remaining classes. For I do not think there is still any need to return similarly to the other pulses from the table, but in relation to the same method, which we have used now in the case of the long, narrow and low, to distinguish between each of the others, having first separated those that change in length due to the state (*schesis*) of the body, or breadth, or depth as to appear not only large or small. Second, we consider the causes through which they become either larger than accords with nature or smaller. **[256K]** Enough has been said about the present matters concerning the class of the pulses relating to the amount of the dilatation.

5. The next thing we must speak about is the change of the pulses towards rapidity and slowness, which is a very specific class of the actual movement of the arteries. And here again we must begin by distinguishing the homonymy. So then, a pulse is called rapid in one case as it relates to the moderate pulse, but in another case as it relates to any pulse whatsoever that happens to occur. Furthermore also, the moderate itself is either absolutely so, or in relation to what is obviously the best nature, or in relation to each person who is faultlessly healthy. This is determined by comparison to the absolutely moderate and teaches what kind the *krasis* is. Thus, the more rapid pulse indicates a hotter *krasis*, while a slower pulse indicates a colder *krasis*. In fact, a change only to rapidity or slowness is not found in bodies that are in accord with nature, and not in those that are not in accord with nature. If someone who is hotter than the absolutely moderate *krasis* seems to have a pulse that is more rapid, of necessity it is larger, **[257K]** and one who, at a certain time, has a pulse that is more rapid than accords with its own nature, also has a pulse that is altogether larger. In neither of these cases is it in fact necessarily more frequent. By increasing the heat still more, frequency is added. Moreover, sometimes the largeness is associated with the rapidity, whereas sometimes also the frequency is, just as conversely there is sometimes frequency alone, when we leave aside someone having in that time the largest allowable dilatation; after this, on returning, we find the largeness equal and the rapidity increased. For of necessity, in such a pulse the frequency is also increased at the same time. If, however, we were to leave someone who has had dissipation of the need of the pulses and then return, we would find the pulse to have increased towards rapidity. It would not at that time have increased towards rapidity alone, but also altogether towards frequency and possibly also largeness, since the dissipation of the need changes the pulse more clearly to infrequency, next to slowness, and third to smallness. In this way too, the changes of the pulses to slowness change them immediately from an accord with nature to infrequency, and often also to smallness, when the need has diminished to a greater extent. **[258K]** The changes from the states not in accord with nature after an increased need are often associated with the change to smallness, but sometimes with that to infrequency, and sometimes with that to both. On the other hand, the changes of the pulses after a diminished need are to slowness along with infrequency and smallness.

These are the primary and most simple changes of the pulses to rapidity and slowness, and the *proegoumenic* and *prokatarktic* causes[794] of these are discovered in the same manner to those mentioned in the case of largeness and smallness. The mixed and compound changes exist with strength and weakness or hardness[795] and softness. It is appropriate to consider and distinguish these by the same method we went over in the discussion about the long, narrow and low, for the sake of knowing for what reason

---

794 These may be understood as internal antecedent and external antecedent respectively. On the types of causes generally, see *De causis pulsuum* I.1 (IX.1–4K) and Johnston 2006: 81–128.
795 Hardness (*duritie*) is added from Kühn's Latin text. All three manuscripts (F1, F2, Va1) have σκληρότητος.

the pulse made the change to rapidity or slowness. It is possible, as I said, for the pulse to have increased to rapidity due to increased need, while it is not impossible also due to the softness of the artery and strength of the capacity. But when **[259K]** it is unchanged in terms of softness and hardness, and strength and weakness, the change would be due to the need alone. However, when it is also along with one of these, it is appropriate to consider and distinguish the prevailing cause, as I showed before. For example, if it immediately appears to be softer along with being more rapid, it is appropriate to consider whether it is to the same extent more rapid as it is softer, or more so. If it is to the same extent softer as it is also more rapid, it has the change from the softness alone, whereas if the increased rapidity is greater than the increased softness, it is not from the softness alone, but also due to the increased need. If, on the other hand, the increase in softness[796] is greater than the increased rapidity, the amount by which the softness is increased is taken away by the dissipation of need alone. In the same way, if along with becoming more rapid, it also becomes stronger, one must consider whether it is more rapid to the same degree as it is stronger. In this way, the strength of the capacity alone would be a cause of the rapidity. If the greater strength is less than the greater rapidity, the amount by which the rapidity is greater came from the need being increased. If, however, the increased strength is greater than the increased rapidity, it would not be the strength of the capacity alone, but also some cooling which changed the **[260K]** pulses.

If the pulse seems more rapid, and at the same time also softer and stronger, it requires a person sufficiently practised to recognise whether the amount by which rapidity is increased beyond an accord with nature has occurred to such an extent from both the causes—that is, the softness and the *eutonia*—or is not of such a degree. If it is of such a degree, it has the change due to those alone; if it is greater, it is also due to the increase of the need; if, however, it is less, it is due to the dissipation of the need. In this way, in the case of the change of the pulses to slowness, one must consider whether the change has occurred due to the need being dissipated alone, or due to weakness of the capacity, or due to hardness of the arteries. It is reasonable, then, for the pulse to be changed due to the need alone, when neither greater hardness nor greater *atonia* is found; it is due to the weakness of the capacity when the greater weakness and greater slowness are commensurate. In this way too, it is due to hardness of the arteries, when the greater hardness and greater slowness, are commensurate. If, however, it seems to be slower more than the moderate, but weaker less than the moderate, such a pulse has occurred from *atonia* and cooling.

And you will in fact discover which of the causes prevail in this from the **[261K]** comparison with what accords with nature and is moderate. If you reckon the weakness departs from an accord with nature by five units, as may happen, and the slowness by fifteen units, cold prevails as a cause in the generation of such a pulse; you must not suppose the *atonia* of the capacity to be a cause. On the other hand, if you

796 In the manuscripts F1, F2, and Va1 μαλακώτερος after πλέονι, present in Kühn, is absent.

reckon the slowness appears to be fifteen units, while you reckon the weakness to be ten, the *atonia* is more potent than the coldness in the genesis of the pulse. If, however, the greater slowness is less, while the greater weakness is more, such a pulse has arisen due to the *atonia* of the capacity, and in addition, to heat. And you will discover the magnitude of each of the causes from the excess in terms of the proportion, when we recall that all these things were said in the definition of the added movement.

Similarly also, when some hardness is discovered bringing about the change to slowness, one must consider whether the change has occurred due to the hardness alone, or due to both. For if the pulse seems harder to the same degree that it has become slower, the change is due to the hardness alone, whereas if the increased slowness is greater **[262K]** than the increased hardness, it is not due to the hardness alone, but also to the cooling. And even here you will discover the stronger cause, as I said a little earlier. If the greater hardness is more, while the greater slowness is less, such a pulse has not arisen due to the hardness alone, but also due to heat. And here again, you will know the magnitude of each of the causes from the excess in terms of the proportion. If, however, it is slower and at the same time also harder and weaker, there is need here of someone practised to recognise whether it is slower than accords with nature to a degree commensurate with the pulse that has arisen from both the causes—that is, the *atonia* and the hardness—or is not of such a degree. If it is of such a degree, it has the change due to those alone. If it is greater, it is also due to the dissipation of the need; if it is less, it is due to the increase of the need.

6. Since enough has been said about these things for our present purposes, let us put a brief stop to the discussion here, before going on to some other class, and let us say something **[263K]** useful about what was previously said and what is going to be said in the future. Now there are four generic differences in the dilatation of the artery: the first and most characteristic difference in the pulses is that relating to the quality of the movement; the second is that relating to the amount of the dilatation. We did not make mention either now or in the first book of *On the Differentiae of the Pulses*,[797] of the more important *differentia*, because it makes the diagnosis more difficult. But we did make mention of it being diagnosed more clearly in the work for beginners,[798] and because of this also will indicate something more useful. There is also in it, however, this specific remarkable feature, which is not in any of the others—there appear to be twenty-seven *differentiae* of pulses in this, but in truth there are three. In all the other classes, there are as many *differentiae* in nature as there appear to be.

There are two other remaining classes of pulses relating to the dilatation: the one is indicative of the capacity, while the other is indicative of [the state of] the wall of the artery. Ultimately, as someone might say, they are neither characteristic nor cognate, but are an added class of pulses. It is always the case in these two classes that the cause from which they arise **[264K]** is clearly indicated, whereas in the other two pre-

797 *De pulsuum differentiis* I (VIII.493–565K), particularly section 4 (VIII.502–508K).
798 *De pulsibus ad tirones* (VIII.453–492K).

viously mentioned classes, this is not always the case. For what actually happens in the two latter classes is that each is created by one cause, whereas in the first two [they are created] by all the causes, which we have often shown to be three in number. And this is why the discussion is more difficult in the case of the first two classes, and why there is need of more distinctions for the discovery of the cause. In the remaining two classes, however, the class of the cause is one in each case and a diagnosis is determined by the amount.

7. The strong pulse always arises through a strong capacity, and the amount by which it is stronger shows also the amount by which the strength of the capacity is increased. On the other hand, a weak pulse is due to the weakness of the capacity, and the amount by which it is weaker indicates the amount by which the capacity is weaker. In the same way too, the hard pulse is produced by a hard wall of the artery, while the soft pulse is produced by a soft wall. So then, the amount of hardness and softness in the pulse is indicative of the amount of the hardness and softness in the artery. **[265K]** But the median of hard and soft, which is also termed moderate, is the first of these in nature and alone in accord with nature. On this reasoning, nothing is the median of weak and strong. For the weak pulse is a product of the weak capacity and is contrary to nature, whereas the strong pulse is so disposed from a strong capacity and is in accord with nature. However, in another way, as we said in what has gone before, there is some median of weak and strong. The manner which relates to the diagnosis is that of the application of touch. When this is overturned by the pulse, we term this strong, whereas when it overcomes the pulse, we call it weak; when it is neither, we say the pulse is intermediate between the weak and the strong, and is itself contrary to nature, just as the weak also is.

However, the changes of the pulses to weakness and strength also often occur alone and from a single cause. For whenever the wall of the artery, together with the need, are in accord with nature, but somehow briefly there is a change of the strength (*tonus*) of the capacity, **[266K]** the pulse becomes in this weaker or stronger than before, whereas none of the other classes changes along with this. In this way, they only arise from a single cause, while in another way again they arise from one cause, but not only one cause. Thus, when the capacity is dissipated even more, or increased, even if it is not changed by any of the other of the *synektic* causes, the change of the pulses will happen to be from one cause, although not one class alone is changed. The capacity that is weaker to a greater extent not only makes the pulses weak, but also small, and because it is small, it is also frequent due to this. When the capacity is strengthened to the greatest extent, with this it makes the pulses very strong immediately and larger, and if the need doesn't change at all, also more infrequent.

The change of the pulses to hardness and softness, if it occurs from one cause, at all events also changes one of the other classes with this. If, however, it exists alone it is produced when several causes come together. At the same time, it in no way arises alone and from a single cause, unless **[267K]** it deviates to the very least extent from an accord with nature, so that for a very brief time the pulses become harder, but are nonetheless equal in magnitude. For in this way, neither the rapidity nor the frequency will be changed. If, however, the hardness becomes clearer and greater, of necessity the pulses also become smaller, and is more frequent to the same extent as it is smaller. On the other hand, if at the same time the wall of the artery were to become harder and the heat increased, the pulse would be harder compared to an accord with nature to the extent to which the wall of the artery is harder. Nevertheless, they would be equal in magnitude to what accords with nature, if the heat becomes as much compared to the moderate as the amount the wall of the artery becomes harder than accords with nature. Since I have also spoken about these things in due measure in regard to the present hypothesis, the next task would be to go over the two previously mentioned classes—that is, the *proegoumenic* and *prokatarktic* causes.

8. The *proegoumenic* (internal antecedent) causes increase the strength of the capacity; these are in the body itself and include certain semi-concocted and unconcocted humors which are now being concocted, the impulse of the nature being towards a turning point (crisis), unless **[268K]** this is already saying the same thing, and in addition to these, anger. The external (*prokatarktic*) causes are wine, nutriments and moderate exercises, and besides these those things that correct the *dyskrasia* of the body itself and of the heart and arteries, whether they be foods, drinks, or medications. Fastings dissipate the capacity, as do insomnias, anxieties, griefs, immoderate evacuations through the stomach, or urine, or vomiting, or uterus, or some hemorrhage, and also strong pains, or occurring with something specific, like the so-called stomachical (gastric) condition[799] in which there is sudden loss of strength (collapse, syncope).[800] Further, in addition to these, are all the things that create a *dyskrasia* in the heart or arteries, either contained in the body, some in fluids and some in solid structures, or also external factors preceding these. And all these dissipate the capacity by immoderately moistening, drying, heating, cooling and dispersing. Such *proegoumenic* and *prokatarktic* causes make the pulse harder, as was said before in **[269K]** the discussion about the long, and at the same time also the narrow and low pulses. Consequently, there is no need to go over these further now; once we add the causes of the soft pulses only, we may then finish the discussion there.

9. Thus, the moist foods and frequent baths, much sleep, a luxurious way of life, and rejoicing make the arteries soft. These are [in the category of] accord with nature. Those [in the category of] contrary to nature are deep sleeps (hypersomnia),[801] lethargies, dropsies, and those other things that soak the walls of the arteries with much moisture and particularly that which is phlegmatous. Still more will be said about these things in what follows when the discussion is about the affections. Now is already a proper time to put an end to the present discussion, having gathered together into

---

799 This term, στομαχικός appears to refer generally to disorders of the stomach, although given the range of meaning of the root term, στόμαχος, there must be some uncertainty here.

800 The Greek has συγκόπτονται.

801 The Greek term is κῶμα which is taken to indicate pathologically deep sleep. Perhaps "hypersomnia" would be appropriate. *Stedman's Medical Dictionary* 2000 has the following definition: "a condition in which sleep periods are excessively long, but the person responds normally in the intervals."

one summary as much from this as is useful. This is as follows: The pulses indicate the primary and present condition of each of the *synektic* causes. When this is recognised, the *proegesamena*[802] and *prokatarktic* causes of the first of these are taken, and in this is the prognosis of the things that have previously happened. On the other hand, the prognosis of the things that are going to happen is taken from the sequelae of these conditions. **[270K]** In the discussions following these, I shall distinguish in what manner particularly each of the things mentioned might occur, first going over, in the second of these books, as much as now remains for me to have added.

Book II

1. [271K] Concerning the primary and, as one might say, most elemental pulses, the recognition of which is in the dilatation of the artery, there is a definition in the book prior to this one.[803] Next in order, we must speak about the remaining pulses, beginning from the contraction. There are in this also four classes of pulses, as there are in the case of the dilatation. For the movement in the contraction becomes more rapid or slower, larger or smaller, and stronger or weaker in relation to the actual nature of the matter, just as also the body **[272K]** of the artery is rendered either hard or soft. It is not in fact possible to recognise any of the classes mentioned, apart from that of slowness and rapidity alone, and not even this in all cases, but [only] in those pulses in which there is perceptible recognition of the contraction, as has been shown in the first book of *On the Diagnosis of the Pulses.*[804] And certainly, in the work *On the Use of the Pulses,*[805] it was shown what the function of the arteries is, and of the contraction no less than of the dilatation. For both preserve the balance of the heat[806] in accord with nature. On the one hand, the dilatation is like inspiration, cooling the excess of the fiery heat and rekindling, by the use of fanning, what has been extinguished, whereas the contraction pours out what is smoky, like expiration. Therefore, the rapidity of the contraction will show a notable abundance of the sooty superfluities being gathered together and at the same time expelled, during that time in which they are found. The slowness, on the contrary, will show that not much of this superfluity is being gathered together, and because of this it doesn't need evacuation.

A large amount is gathered together in the putrefactions of the humors, from which **[273K]** fevers are inevitably kindled. However, the amount is altogether small,

802 This is a different form of the verb προηγέομαι, which is usually *proegoumenic* to indicate an internal antecedent cause.
803 *De praesagitione ex pulsibus* I (IX.205–270K).
804 *De dignoscendis pulsibus* I (VIII.766–822K).
805 *De usu pulsuum* (V.149–180K).
806 K: θερμότητος; F2 and Va: κράσεως; Kühn's Latin text: *caloris.*

when the body is *euchymous.*[807] In those other conditions which are in between these, the amount will be proportional to each of the conditions—more in those who are eating and sleeping but less in those who are awake or fasting. It is the same in relation to the actual nature of the foods—more in those who are *kakochymous* and less in those who are *euchymous.* It is analogous also in the fevers—more in the accessions and less at other times. And it will be more with the thick and viscid humors, as the phlegmatic humors which they term crude are, but less with the picrocholic and melancholic humors, and in between these with blood. For surely this is also putrefied in the inflammations, in which the patients are febrile, and the others of which the burning inflammation reaches such an excess as to burn the skin and create an affection which is called anthrax.[808] Of the same nature in some way also are those that accompany eroding herpes, which Hippocrates termed "devouring".[809] In all such things, the humors are putrefying.

In turn, the humors putrefy in another way in the rough arteries (upper airways),[810] [for they distribute **[274K]** their badness to the lung, and in time this organ putrefies too. They collect them in the rough arteries][811] due either to the downward flow from the head, or from pleurisy, or peripneumonia, or pulmonary affections, or *synanche*, or by being transferred from an empyema. They take the origin of the suppuration from not being properly expectorated, for they need to be evacuated quickly due to their own bad condition and the heat in the region. Often also, rupture of a vessel occurring in the lung pours forth some of the blood into the rough arteries, which, putrefying in time, brings about *phthoē.*[812] It comes more quickly to this, when it is poured out in relation to erosion of the vessels.

In all these conditions which I have listed, and those others which have some putrefaction of humors, the inward movement of the arteries is increased, and in relation to the amount, they contract to a greater extent, but you would not diagnose this accurately by perception. Anyway, the progression to rapidity is clear and easily diagnosed by someone practised in the action, although it is not possible for him to begin practice before **[275K]** learning the aspects of diagnosis we have written about, which none of the doctors before us had taught. But some go so far as to present a credible theory, writing of certain systematic Herophilean treatments about the

---

807 Here and elsewhere εὐχυμία/εὔχυμος is retained in a transliterated form as part of a pair with κακοχυμία/ κακόχυμος. The LSJ definition for this usage is, "productive of healthy humors" with reference to "Hippocrates, *Aff.* 55, Galen 17.2.876".

808 The medical sense of ἄνθραξ (anthrax) given in LSJ is "carbuncle, malignant pustule" with reference to Hippocrates, *Epidemiae III* 7 (ed. and Engl. transl. Jones 1923: 230–233), and Galen, *De tumoribus praeter naturam* IV (VII.719K).

809 Hippocrates, *Aphorismi* V.23 (Engl. transl. Jones 1931: 141), *De humidorum usu* 6 (ed. and Engl. transl. Jones 1931: 53), and *De ulceribus* 3 (ed. and Engl. transl. Potter 1995: 347).

810 *De usu partium* VII.8 (Engl. transl. May 1968: 336–339).

811 The section in parentheses is not present in the manuscripts F1 or Va1.

812 *Phthoē* (Latin *tabes*) is broadly consumption as a wasting with pulmonary signs and symptoms, possibly covering a number of conditions including tuberculosis.

rhythms in the pulses, although others dub such a theory idle chatter and completely distance themselves from it. Neither, however, saw the truth, which is that one must look at the rapidity and slowness of the dilatation and contraction, as each is able to signify many things—something we have already said in summary, and will speak about more fully in what follows in the discussion about the diseases.

2. At the present time, we must pass on to the discussion about frequency and infrequency. Each of these is twofold, because the pauses of the artery also happen to be two; the one after the dilatation and before the contraction, and the other after the contraction and before the dilatation.[813] One thing is common to both, while one is specific to each. What is common is that they are shortened due to a twofold cause—either the antecedent movement is delayed or the following movement begins earlier **[276K]**. Conversely, they are prolonged, if either the antecedent movement is shortened or the following movement comes late; what this naturally signifies is specific for each. Clearly, they signify that which relates to the other movement from its having been prolonged or are indicative of the other movement which begins too soon, or arising in turn from the slowing of the other movement, slows the remainder.[814] Since the purity of heat and minimal superfluity make the dilatation larger and at the same time more rapid, while the contraction becomes turbid with the putrefying superfluities, when the dilatation is earlier or large, the pulse becomes more frequent, which will indicate that the large amount of pure heat has been increased in the body, whereas, when it is the contraction that begins earlier, this will indicate an abundance of impure and excrementitious heat.

In like manner, the pulse becomes more infrequent in relation to the external pause, if it is due to shortness of the dilatation, and will show the heat in the animal to have been cooled, whereas, if due to the contraction being slowed, it will show the smoky superfluities to have become less. Contrariwise, in relation to the internal pause, if due to the contraction being prolonged, **[277K]** the pulse is made frequent, it will show an abundance of smoky superfluities growing up, whereas, if it is due to the dilatation beginning too soon, it will show the heat to be increased. Conversely, the infrequent pulse relating to this same pause, if due to the contraction having been diminished, will show the superfluities have been reduced, whereas, if due to the dilatation being slow, it will show the heat to have been diminished. If both come together at the same time—that is, the first movement being prolonged and the second beginning too soon—this happening either in relation to the external pause or to the internal pause, it will indicate a combined condition from both of those, which each of them customarily indicates separately.

Similarly, if the pulse becomes infrequent in relation to either one of the pauses coming together with both the same time, the first movement being brought to an

813 Galen elsewhere terms the former "external" and the latter "internal."

814 A complex Galenic sentence! The meaning seems to be that whether there is delay in one movement or prematurity in the other, the overall effect is to slow the pulse.

end and the other beginning much later, the condition will also be in this way twofold and compound from each of the factors already indicated, when there is diminution of one or other of the movements, while when delayed, the other movement also begins much later. But the diagnosis in fact does not come to an equal action in relation to each of the pauses. For the external pause is clearly perceptible, while the internal pause is not clear in the same way, although **[278K]** it is possible to recognise it by perception and not obscurely. Someone might, however, discover this by guesswork alone, conjecturing from the impulse in relation to each movement, and particularly in those pulses in which the contraction is clearly recognised by sense perception. I have said enough for the present about this class of the pulses.[815]

3. The next thing in the order of the discussion is to go over rhythms, about which still more was said by Herophilus, who gave an account of observation and experience more than he taught logical method. He wrote for the most part of the apparent rhythms of the pulses in relation to each stage of life (age), at first not saying anything to us of what he observed of the pulses in certain natures. Then, from those things which he taught, it is clear that he was confused and disorganised about the recognition of the contraction and the pauses. If he thinks that sometimes it is possible for a contraction in those who have grown old to become extended up to ten of the primary times, it is clear that he really never perceived the contraction. For sometimes this is of shorter duration than the dilatation, sometimes of an equal time, and sometimes, as **[279K]** the man himself writes, of longer duration, although not as he thinks five times as long but greater by a little. Those other things said by Herophilus about rhythms in regard to prognoses are wrong, disorganised and impossible, as can be learned from what is said in the third of the books about the diagnosis of the pulses,[816] and also from what I wrote specifically about Herophilus' *On the Art of the Pulses.*[817] Now, however, as I am not creating a work of history or refutation, but one which teaches what I have come to know, I shall already pass on to the discussion about the unequal pulses.

4. The class of these is twofold. Some have the inequality in a single beat, while others have it in a collection of beats. And the custom among more recent doctors is to call such an inequality "systematic", because, I think, they also call the collection a system. One must, then, know this about the unequal pulses—something which is common to them all and which has been shown in the second book of the work *On the Causes of the Pulses.*[818] This is that they exist due to obstructions, or compressions of arteries, or an excess in regard to the supposed capacity, or the unequal (non-uni-

---

815 "Concerning the prognosis from the *ekrhyhmic* pulse according to Herophilus" is added in the manuscript F2.

816 None of Herophilus' writings on the pulse and related matters survive—indeed, none of his works survive at all. The fragments on the pulse in von Staden 1989 run from T144 to T188.

817 There is an addition here in manuscript F2 which has some indecipherable words and has not been included.

818 *De causis pulsuum* II (IX.55–105K).

form) *dyskrasia* of the heart, **[280K]** while there is greater difficulty in relation to the inequality in one pulse than with that in a collection of pulses. And certainly also, the complete absence of movement in the artery, which is called “to intermit”,[819] and which occurs in both a single pulse and in a collection (system) of pulses, is a problem like no other of all the kinds of the unequal pulses. However, it is necessary to know separately about all the kinds in relation to each class, beginning from the inequality in a collection (system) of pulses.

[Let us put the discussion into practical terms, and first in relation to the collective form,][820] since this happens to be clearer than that in the single pulse. Now the capacity creates the intermittent pulses due either to being weighed down by the excess, preventing by thickness the initiation of the movement at the appropriate time, or an excess of humors obstructing either the openings of the arteries adjacent to the heart through which it draws in and in turn sends out the materials, or even the whole lumen of the vessels up to some point from the mouths. Often, however, such an excess is like a binding poured around them externally somehow weighing them down like a burden, and sometimes also **[281K]** occupying the spaces into which they dilate. Sometimes, however, some inflammation, or induration or some other such mass (swelling) exists in the bodies surrounding the arteries and stands as an obstacle to their dilatations. If such a thing also exists in the walls of the arteries themselves, the danger is greater, while also, if it is in the body of the heart itself, this is extremely bad, for such people collapse suddenly. The *dyskrasia* of the heart is not similarly deadly, but is much milder than inflammation. However, weakness of capacity follows this. And in fact the weak pulse is nothing other than a sign of such a condition. For the heart which has a *krasis* in due proportion creates the strong pulse, while when the *krasis* is abnormal it creates the weak pulse. There is less and more in each of the two due to the increase and abatement of the *eukrasia.* For it was shown to have breadth and due to that of the *dyskrasia*, there is nothing less in the pulses. But the *dyskrasia* alone makes the pulses small and weak, although not yet also unequal, unless the capacity throws in its lot with the humors, both those in relation to the heart itself **[282K]** and those in the arteries and veins near to it, these seemingly being less in terms of strength, which is nothing other than the excess concerning the capacity. Some inequality will necessarily coexist with small and weak pulses in such conditions. Where only the capacity of the heart is weak, and it is not weighed down by the humors, the pulses are altogether small and weak, due to the weakness (*atonia*) of the capacity, whereas no inequality coexists with these, unless at some time also the actual body of the heart should exhibit an unequal *dyskrasia,* which I shall speak about a little later.

---

819 There is an important distinction which Galen makes, signalled by the verbs διαλείπω and παρεμπίπτω, the first meaning, ‘to intermit/be intermittent/discontinue/interrupt’ and the second, ‘to intercide/be intercident/cut through/fall between’. See the appendix on pulse terminology.

820 This part of the statement is absent from the manuscripts F1 and F2.

Now, however, I shall return to the intermittent pulses, which lie side by side with those that are more infrequent, but can be distinguished from these by the whole rhythm of the pulses having been destroyed in relation to the intermissions. On the other hand, if they pause not only for the time of one pulse, but also of many, such a pulse would appear more clearly intermittent, and will be more dangerous to the degree that it pauses more than the one prior to it. Consequently, I do not know if it is possible for someone to be saved, when the artery has paused for the time of two pulses. However, we have often seen, **[283K]** when the movement of the arteries intermits for the time of one pulse and a little bit more, the person restored to strength, and particularly when he comes to old age, although perhaps such a thing would seem to be contrary to reason. For if the old man is otherwise weak and the capacities in him are afflicted even before he is diseased, what then must we expect will eventuate, if the suffering due to age is added to by disease? But as it would seem, there is no difficulty in this, but it is more reasonable than that, because what happens is that the weak capacities are often threatened by small causes, whereas the strong capacities are overcome only by the largest causes. And this much is certain—those who are strong in their bodily states (*hexis*) and have been disease-free for a very long time are endangered when they are diseased, whereas those who are weak and continually harmed by every cause are very easily preserved, because clearly none of them await a large cause, being threatened beforehand by every cause, even the smallest, whereas the strength of those who are strong fights off every small badness, so that it is overcome by the large ones only. Therefore, because of this, old people have intermittent pulses that are less destructive than those in their prime.

Now children in turn, because they are very strong in the digestive capacity, **[284K]** are very soft in the body and perspire easily, therefore concoct and disperse the unequal (non-uniform) *dyskrasias* and the excess and thickness of the humors. So, what also happens is that they are less endangered by the intermittent pulses than those in their prime for two reasons:

1. Because they have a weaker pulsatile capacity than that of those in their prime, they are, as a consequence, more readily overcome by smaller causes, just as old people are.
2. Because they have a stronger digestive capacity, those things that are distressing are more easily corrected.

And in addition, the ease of transpiration is of no small benefit towards correction of diseases, as Hippocrates also said.[821] If, therefore, someone is satisfied by the aforementioned considerations, it is possible to use them, whereas if he is not satisfied, let him at least know that from experience the appearance is such as was stated.

Therefore, those in their prime are particularly endangered in the intermittent pulses, while children are endangered less than they are; least endangered are those

821 See Hippocrates, *Prognosticon* 3 (ed. and Engl. transl. Jones 1923: 10–15).

who are old. It is not surprising that among the inequalities the intermittent is the most life-threatening. For the other movements are to some extent deficient, whereas the deficiency in relation to **[285K]** intermittency arises in deprivation of movement. During the time the pulse intermits, the artery is not moved at all. And health, death and disease hardly differ as much from each other as do normal movement, absence of movement and abnormal movement. For movement in accord with nature is like health, while abnormal movement is like disease, and absence of movement is like death. Moreover, to those to whom it seems that health and disease occur in relation to the actions themselves, and not in relation to the conditions of the body from which we act, the normal movement of the arteries is not like health, but will be said to be health outright. In the same way too, the deficient movement is not like disease, but is in fact disease itself, while absence of movement is a sort of death, if either it should become more complete or if one pulse seems to pause for the duration of two, like torpor, or some apoplexy, or paralysis of the voluntary actions. For just as these involve the other principle which is in the brain, so too the pause of the pulses involves the principle in the heart. And just as **[286K]** it is rightly said in that case to resolve apoplexy, that is impossible if it is severe and not easy if it is weak, so too will it be correctly said here that to resolve a prolonged immobility of the pulses is impossible, while to resolve an immobility of short duration is not easy. Prolonged, then, is that which exceeds the time of two pulses, while of short duration is that which is within the time of two pulses. The pause involving the times of one pulse is the most moderate of the intermittencies and there are many who have been saved from this, both old people and children, although none of those in their prime. But some, due to ignorance, think the long infrequency is an intermission, and believe some in their prime are saved from such a *kakosphyxia.* This is, then, also serious, and particularly in those in their prime, but is not altogether fatal; of course, in children and old people it is more mild. Death occurs suddenly with the intermittent pulses analogous to apoplexy. In both, the heat in the heart is quenched, being deprived of respiration.

Infrequency is distinguished from the intermittent pulse by the length of the time, while sometimes it will also be defined besides by the infrequency **[287K]** coexisting with equality of all the beats, although the intermittent pulse of necessity becomes unequal; it is sometimes found intermitting three beats, sometimes four, and sometimes five or more and pausing one time or more in the movement, while also one of the other classes immediately changes with it, turning to inequality. But also, this is not inseparable from the infrequencies extending for a longer time, so the movement is made slower than is appropriate, sometimes to the same extent as the infrequency, but sometimes still more or less, as the condition is one that creates such a *kakosphyxia.*[822]

Severe cooling is the cause of the genesis of such pulses, either in the body of the heart, or in the *pneuma* and blood in the chambers of the heart, or in one of the asso-

822 This is taken as a general term for an abnormality of the pulse.

ciated parts. Thus, the major cooling in the actual body of the heart makes the pulses weak, slow and small to the extent to which it is dominant. That which is stirred up from one of the other factors **[288K]** makes the infrequency more than the other classes mentioned which depart from an accord with nature. As a result, some of those who have pulses that are not much less than accords with nature and are slower and weaker, nevertheless die quickly. Death is more inevitable, however, when the body of the heart itself takes on the cooling, and occurs suddenly in relation to the other variety. The former particularly die in their sleep, so they never get up again, while the latter are also like this, although some of them die while conversing, so as to seem to laymen not to have any problem, but then are suddenly dead, as if swooning; some sweat, but others are without this. This symptom and the condition itself also seize some of those who are diseased, when they become unexpectedly afebrile, and some of those who are healthy, particularly in old age. Therefore, infrequency of pulses is to this extent fatal, and certainly, the intermittent pulses are even more so, if in fact their genesis is an extension of the infrequency, and the resolution is a return to the original moderation **[289K]** through infrequency, for none of those who are saved from the intermittent pulses are saved otherwise. In all cases, it is in the manner we described: first the infrequency extends sufficiently, then the intermittent pulse comes next, after which the pause always becomes more of shorter duration up to the point of recovering the moderation of accord with nature.

5. There are two other oppositions in the pulses mentioned: frequent to infrequent and intercident to intermittent, which Archigenes seems to think are more serious than those previously mentioned, being deceived by the frequent, because they both precede and are accompanied by collapses (syncopes), while in the case of the intercident, it is because they sometimes come on suddenly with the peripneumonias that are fatal and sometimes in fevers in which there is obstruction or compression of major arteries. To someone looking at the genesis of such pulses, the opinion of Archigenes seems not to be true. For the intercidence arises from unequal frequency, often due to the heart having carried out its use too deficiently. This itself occurs either when the heart is weighed down by excess, or when the arteries **[290K]** have been obstructed or narrowed, which I think are also causes of the intermittent pulses. These, then, differ from each other by the intercident pulses having the capacity to fight against and contend with the troublesome causes, whereas the intermittent pulses resile from doing this. Of course, the intercident pulses often produce a crisis. It is as if nature were in ignorance preparing for a contest. However, if some people die from such crises, this doesn't mean the condition of the intercident pulses is more serious than that of the intermittent pulses. For many more people die from the intermittent pulses, and in the same way from those that are significantly infrequent compared to those that are frequent. Anyway, while all the burning fevers have the frequent pulse, some of these are without danger. On the other hand, the frequent pulses are dangerous in the collapses (syncopes), due to the weakness which is inseparable from all collapses. And it is necessary, I think, to regard the very weak pulses as dangerous, if any of the others are, as they are the pulses of those who are suffering collapses. In general, it is good to consider the extremes in all the classes of the pulses. **[291K]**

That they are fatal was also agreed by Archigenes. However, neither that man nor anyone else considered accurately what kind the most fatal of all is. And indeed, it is clear what extremity of pulses seems the most deadly compared to the others; also, the class in relation to that will probably be considered the most deadly compared to the other classes. In like manner too, if we compare one class to another, and find the peak in one of these is more deadly, it would be reasonable for us to say the whole class is the more deadly of the classes.[823] Thus, extreme frequency exists no less in those who are collapsing than in those with burning fevers, who are not altogether fatally disposed, whereas infrequency is not sometimes without danger, just as neither weakness nor smallness are not. For these three *differentiae* of the pulses, when they come to an extreme, are continuously dangerous. This does not apply to largeness, just as it does not to strength, but this [latter] alone of all the extremes is not only not dangerous, but is greatly beneficial, should someone know how to recognise this. However, the extreme largeness, whenever it coexists with morbid softness, is dangerous, **[292K]** although when it exists with a healthy consistency of the arterial wall, it is not unnatural. For countless times in the fevers without danger, which are hot and burning, the pulse becomes very large, whenever the capacity is strong. Therefore, someone would not show, from the comparison of the extremes, that the class of the frequent pulses is more deadly than that of the infrequent pulses, nor that the class of the intercident pulses is more deadly than that of the intermittent.

Since, however, the discussion was striving towards the demonstration of these things alone, while secondarily it was also giving consideration to certain other classes, it would not be a subsidiary task to undertake to put aside in the memory those things learned about the large and small, and the strong and weak pulses. It would be better, however, if I also added the remaining *differentia* to these, which is that relating to slowness and rapidity. For it seems here also that the class of the rapid pulses is less dangerous than that of the slow pulses, and the same applies certainly to extreme rapidity compared to extreme slowness. Thus, extreme slowness, just as also extreme infrequency, come about when the innate heat is quenched, while an increase to rapidity occurs due to the need urging things on and altogether by **[293K]** the capacity being strong. And even Archigenes errs in this, taking the formicant pulse to be rapid. For such a pulse on the whole is neither rapid nor comes to a peak of frequency. It is also possible for you to learn here that the extremes relating to frequency, largeness and rapidity are mild.

Think about the healthy young man whom you order to run very swiftly and you will discover immediately the three extremes occurring, without there being any present or future harm to the person. The very hard pulse, for it is reasonable also for something to be said about this *differentia* so nothing is omitted from the discussion,

823 Kühn's Greek text does not seem to make sense here; Kühn's Latin text is followed.

is similar in danger to the very soft. For the former is due either to induration or a major inflammation of an internal organ, or congelation from cooling, or dryness occurring from a very burning fever, or a strong spasm (convulsion), while the latter is due to deep sleep (hypersomnia), the *leukophlegmatic* dropsies, and to speak collectively, such things as bring about immoderate moisture of the bodily state (*hexis*). Therefore, after taking up again those things said under the chief points, let us now put an end to the discussion.

All the pulses deviating from the moderation of an accord with nature to a great extent are not good, but the weakest, **[294K]** slowest and most infrequent are worst of all. Next in order to these are the smallest, softest and hardest, or the most frequent, but not in fact the most rapid, nor the largest, although these are moderately dangerous. Of all the extremes, only the strongest is good. Archigenes spoke otherwise about these, but did not add any demonstration to the argument, for he had none, nor did he have confirmatory experience, but drew false inferences from them, as I said a little earlier. Therefore, someone who wishes the actions of the art to be diligently followed should examine closely the things said by us now, based on the experience of those who are sick. But I know those who are zealous about the truth will do these things. And since enough has been said about these matters, it is now an appropriate time to say something also about the inequality in one pulse.

6. And even here the intermittent pulse is the most grievous of all. This occurs when a pause cuts across the dilatation. There are many kinds of this, all of them grievous; they differ from each other in terms of more and less. When the second movement becomes stronger than the first, **[295K]** it is more mild, whereas when it is weaker, it is more grievous. In this way also, the slower is a more grievous sign, while the more rapid is more mild. For such pulses occur when the nature is contending with and at the same time is also hindered by the morbid causes. Therefore, whenever the nature is able to shake these off somehow, and as it were be freed from a certain bond, it will make the movement along with the pause better than before, and it is more mild. However, when the movement is made worse, it is more grievous. The strong and rapid movement is better than the weak and slow. However, the *differentia* related to largeness and smallness of the first movement has no notable superiority regarding the second. It is also similarly possible, when the condition is equal, that sometimes the part of the movement before the pause is greater and sometimes the part after the pause is, while sometimes both become similar. Nevertheless, the *differentia* in relation to hardness and softness of the two impacts of the artery would sometimes not occur due to the greater time always required for such a change, when in fact even in the systematic inequalities, it is rare in one examination to encounter an unequal pulse in relation to hardness and softness.

Moreover, the second **[296K]** impact often appears more clonic than the first in the inequality in one pulse and that in the collection. And because of this, it would perhaps deceive those who are unpractised as being harder. Clonus is a symptom of an artery that is hard and is compelled to dilate forcibly to a greater extent. However, it does not in truth become harder under these circumstances, but being compelled to the greatest extent to dilate forcibly, it suffers such a symptom. And the clonic impact is worse than the even (equal) impact, whenever the hardening of the wall of the artery is not due to some recent cause, while it is worst when the time of the pause is longer. Therefore, the intermittent inequality in one pulse is the most grievous of all inequalities. For it occurs from the same causes according to class from which the systematic inequality occurs, but is much worse than this. That, if nothing else, at least allows the simple movements to be completed free from interference, whereas this does not allow these movements to occur in accord with nature, but stands as a hindrance to, and deprives the arteries of movement, not through three, four, or an even greater number of beats, but immediately in the very first pulse itself. Therefore, all the things said about that must be considered as also said about this, but **[297K]** increased and changed in this alone—they carry the threat of imminent death. The bounding (gazelle-like) pulse is also from this class of the pulses. We have explained the form of this, just as we have of all the others, in the first book of *On the Differentiae of the Pulses.*[824]

7. There is another class of unequal pulses in relation to one dilatation, having in common a double beat; what is not common is the cause. But as was also said in the second book of *On the Causes of the Pulses,*[825] there are three primary causes of the genesis of these. Now certainly something was said about two of these in that work, and will be also said now for the purpose of making mention of the third cause. This much has already been said about this cause: it is an unequal (non-uniform) *dyskrasia* of the actual body of the heart, although we postponed the complete discussion about this to make it here. And accordingly, the appropriate time has come for this, so let it be spoken about. Due to the commonality of the discussion with all the other unequal pulses—that is, those that follow the unequal (non-uniform) *dyskrasias* of the actual body of the heart—I shall go over them in this discussion, beginning **[298K]** here again from the more obvious inequalities, which they term "systematic". The body of the heart has in itself a certain dilating and contracting capacity, which as it happens is neither given by Nature nor functions in vain, but as has been shown in other writings is for the preservation of the innate heat. Thus, in the contractions, the smoky superfluity is poured out, while in the dilatations what cools this heat is drawn in. Therefore, when there is equality in respect to heat, or cold, or purity, or turbidity, all the parts of this are similarly stirred towards the function. On the other hand, whenever it happens that some parts of it are hotter while some are colder and that some parts are perfectly pure while others are made turbid by the abundance of the superfluities, they do not all have the same impulse towards the function; rather, some of them precede, while some of them delay. And even in this, those that are fewer and weaker follow the majority and are urged on towards stronger action. For it is not possible for them to be scattered and separated from each other, as they are parts of one organ.

824 *De pulsuum differentiis* I (VIII.493 – 565K).

825 *De causis pulsuum* II (IX.55 – 104K).

However, just as the greater number **[299K]** and the stronger stir up action, urging on the whole heart, so too conversely, when there are fewer things stirring up the movement while more remain as they are, the whole heart is constrained in a longer pause. And in particular, the quality of the movement in the whole organ arises from the dominating parts in it. For if the hotter lead the way for the movement, the dilatation occurs earlier, and at the same time is quicker and larger, whereas when the colder lead the way, it is later, slower and smaller. Often then, what leads the movement follows the majority of the parts, whereas often it follows the magnitude of the need. For those things that sufficiently need something are stirred to action more readily and more strongly, even if they are equal in number to those that do not need anything. Therefore, when some of these are much more than the others, the internal organ in relation to those is moved and functions continually. However, when some are slightly more than others, or precisely equal, the former are not able to continually prevail, but are variously overcome in part; this creates unequal movements of the whole internal organ.

Let it be the case then, should it so happen, that the cold **[300K]** parts are more than the hot parts, but not by much; of necessity, the first action becomes less and slower, and furthermore the pause becomes longer in those when the cold parts prevail than when the hot parts prevail. And accordingly also, let the second action of the heart occur when the cold parts prevail, and after this again pause. And let the third and fourth actions and pauses in addition to these be accomplished when the cold parts are prevailing. Would it not be quite clear that, when the hotter parts prevail, there will be need of a larger and, at the same time, more rapid and frequent movement to provide the benefits for what is lacking of the need of these themselves? As a result, in this they will become hotter and will impart something of the warmth to those parts that are adjacent. Therefore, the original condition of the heart will still be preserved, even if the warmer parts become more or the heat more than before.

Moreover, if this always happens, there will be a time in which these parts will prevail over the opposites, so they will also initiate the action earlier, due to the urgency of the need, making the movement itself more rapid and larger than all the previous movements, which the cold parts brought on. What will occur in this, therefore, is that the prevailing parts gain the benefit of **[301K]** cooling moderately, whereas those that were colder will now be cooled more, and in this way in turn, over time the condition of the whole heart is changed, such that the colder parts will prevail. For when the hotter parts have sufficiently enjoyed the benefit of cooling, while the colder parts themselves are cooled more than they need to be and cool with them the parts adjacent, it is inevitable in this way that the colder parts become more numerous. In turn, in this time, it is inevitable that the movement of the whole heart will be adjusted to the need of such things, and its dilatations will become infrequent, slow and small, such as they were assumed to be at the beginning.

Therefore, let these things at least be said now which have been said a little earlier about the change of these. Similarly also, let those things now be recalled which have been said previously about the return to the original state. For in this way, I think you will be persuaded about the continuous change of the pulses, as long as the aforementioned *dyskrasia* should remain in the heart. By the same token, when, in the excess of the smoky superfluities, the unequal (non-uniform) *dyskrasia* in the internal organ and the **[302K]** movement of the pulses will be led to the systematic inequality. Of course, it is no surprise that the inequality in one pulse occurs following the *dyskrasias* of the heart. For the parts of the arteries will straightway become hotter from the parts of the heart that are hotter than the others and will distend more and be moved more rapidly. On the other hand, these will be colder from the parts of the heart that are colder, will be moved slower, and will distend less. However, when the whole vessel arises directly from the hotter parts of the heart, it will be moved quicker and distend more than those parts which extend directly from the colder parts of the heart. And apart from the heart, certain characteristic *dyskrasias* occurring in the arteries produce the aforementioned inequalities; that is, those relating to large and small, rapid and slow, and infrequent and frequent, and if they should deviate greatly from an accord with nature, also those relating to strength and weakness. Furthermore, in relation to the heart itself, and besides this, to all the arteries, from the unequal (non-uniform) *dyskrasia* in the internal organ many differences of the movements will occur in one dilatation, there being some that stir it to rapidity, and some **[303K]** that hinder it and compel it to slow. As a consequence, it also sometimes stops midway on the rise, as happens to those on journeys when otherwise in any way whatsoever they drag along others who are slow. For they are drawn back in the opposite direction by them, irregularly stopping and interrupting the onward movement, so as not to make a similar movement continuously, often being held back. And if, when they are drawing something, they are counteracted and drawn in the opposite direction, not only are they hindered in the movement, so as to walk slower or sometimes not at all, but they are also sometimes dragged backwards. All these things occur, then, in the heart and the arteries in the unequal (non-uniform) *dyskrasias.*

8. And, in fact, the bounding pulse is from the class of those that are interrupted in a pause, sometimes supervening in such conditions of the heart, just as the dicrotic pulse also is from the class of those that are drawn in the opposite direction inward, and itself presents together sometimes in the unequal (non-uniform) *dyskrasias* of the heart, and particularly in fact when the abundance of the sooty superfluities in this is noteworthy in many of its parts. **[304K]** For if it should happen at the same time that some parts are absolutely very hot, while others are choked with smoky soot, movements will start off towards the opposites, as some attempt to dilate while others attempt to contract. And in this way, in truth, either when the artery dilates, if parts are actively contracting, the contraction sometimes supersedes the dilatation, and apart from the external pause, if it is contracting when parts are actively dilatating, the dilatation supersedes the contraction without the internal pause. However, each of these will exist in a number of ways, for either the second movement will completely overcome the first, or it will draw back a little, while the first, having overcome, will complete the prescribed measure in itself, so that it creates three *differentiae* of the pulses in all in the dilatation and the contraction. The first is when, in the dilatation before it completes the measure, the contraction supersedes; the second is when it awaits exactly, completely destroying the external pause; the third is when the dilatation is still occurring, the opposite movement draws back, and becoming weaker will make way for the remaining filling. In the same way too, the dilatation sometimes **[305K]** awaits the contraction when it has already occurred, not allowing it to stop, and sometimes, being cut off and overcome, it supersedes before there is filling, and sometimes again it draws back a little and is overcome. Such movements of arteries sometimes occur apart from the heart having been affected, either due to an unequal (non-uniform) *dyskrasia* existing in themselves, or due to a weakness of the capacity, or due to hardness of the arterial wall. We have already said something about these matters in the second of the books *On the Causes of the Pulses*,[826] and will speak now on the major points, after we have first indicated that the discussion about the unequal (non-uniform) *dyskasia* in the heart, which we were going over just now, we put this matter off because we were still enquiring into and examining it closely through long experience, just as also all the other things that were said and will be said, and because, in relation to that matter, we have not yet said anything about the essence of the capacity. For we have not dared to make any statement on this doctrine before finding agreement on all the parts individually.

Now, however, since we have been persuaded that the essence of the capacities is nothing other than the kind of *krasis*, **[306K]** we find that all the things relating to the diagnoses, prognoses, and treatments are bought together harmoniously in this doctrine, so it is, I think, reasonable also that here we go over the pulses which follow the unequal (non-uniform) *dyskrasias* of the heart. There are, as I said, some of these which also exist apart from that. Thus the dicrotic pulse is a pulse that is clonic, which at first I was accustomed to think is only a double beat, being misled by those things Archigenes said, likening this to the double blows of the hammer against the anvil. For I assumed, in relation to such blows, that something hard rebounded from another hard object, from which I erred concerning the concept of the matter; from these things, I was also mistaken about the diagnosis. However, since in the time devoted to the recognition of the contraction, always sitting and watching diligently, I clearly perceived the artery at some times soft and without clonus (*aclonic*),[827] sinking down gradually, and then filling again in the dilatation, and from this I closely watched the conditions in which such pulses occur. After a long time and with much observation, I was enabled to firmly state this, which I am now going over.

It seems to me that Archigenes also **[307K]** has carefully observed the twice beating pulses, although he has not interpreted them correctly, nor has he recognised the conditions which they follow. If he had, he would not simply liken them in this way to the double beats of the hammer against the anvil, nor would he have omitted to state what the manner of their genesis is. But in fact this too was discovered by us after a

---

826 *De causis pulsuum* II (IX.55–104K).

827 LSJ gives the meaning of this term ἄκλονος as “steady, regular” citing this passage.

long time, just as many other aspects of the theory also were. And you would find the matter to be like this, if only you were willing[828] to bring a certain labour of love and zeal to the task, such as we brought. Apart from this, you will not be able to know the truth in these things, nor will you yourself be benefitted in any way.

Therefore, up to this point, we have discovered three conditions and three causes productive of the double-beating (dicrotic) pulses. One is that of the clonic pulses which separates itself from these in this alone—that when the pulse becomes short in length relative to the parts on either side of the artery, the clonus is obscure, and because of this it is dicrotic, and does not appear clonic. The second is that in which the pulse is *aclonic*, while the artery is hard and the capacity weak to such a degree as to be weighed down by the overlying bodies. Nevertheless, in relation to some part in the **[308K]** skin, when it is stretched out, it rises up unhindered. Since this alone comes to the perception of those palpating, the pulse will be at that time dicrotic. Starting out like this due to the lightness of the overlying bodies, it in turn holds back, being constrained and not following together the parts on either side, these being weighed down by the overlying bodies, and because of this also making the movement wholly imperceptible, which is not the same as in the apparent part of the artery, but hesitating and delaying. What happens in this is that what starts them off beforehand draws them back towards those things, because the artery is hard. Otherwise, it is not possible for the retraction to occur in such conditions.

These two pulses, then, are dicrotic and hard, whereas the third, occurring due to a *dyskrasia*, is not dicrotic but only twice-beating, and is not necessarily hard. If, however, someone should also wish to term this dicrotic, it makes no difference. However, let it be remembered only that this is not like hard bodies rebounding from hard bodies, but as was said a little earlier, it is what occurs when two capacities (forces) are contending with each other—that is, the dilatation and the contraction in respect of the **[309K]** heart. There is no urgency to say now in what diseases and in relation to what kinds of causes and conditions the heart comes to an unequal (non-uniform) *dyskrasia*; these will be stated in the things mentioned subsequently. It will be enough to say this much for the present—that in the unequal (non-uniform) *dyskrasias* of the heart and those existing in the actual body of the internal organ, the double beating pulse occurs, and what also occurs is that the external pause is lost, and in addition to these things, it also cuts short the dilatation. It was shown that those interruptions of the movement by a pause also all occur in the unequal (non-uniform) *dyskrasias* of the heart. But also, such inequalities in relation to slowness and rapidity do not appear apart from interruption of the dilatation, and it is clear that these, in addition to the other causes by which they were shown to occur, will have the same cause. It is about this that discussion has arisen—that is, the unequal (non-uniform) *dyskrasia* of the heart when the hot parts are to such a degree in contention with other cold

828 K: καὶ σὺ οὕτως ἔχειν τὸ πρᾶγμα μὲν εὑρήθεις, εἰ μόνον ἐθέλησάς τινα …; F1 and Va1: καὶ παρ’ ὑμῖν γε δυνατόν ἐστι εὑρεθήναι τὰ τοιαῦτα, εἰ μόνα ἐθέλησάς τινα … K has been followed.

parts that sometimes the hot parts are compelled to slow, being restrained by the cold parts, while sometimes the cold parts move more rapidly, being drawn on by the hot parts.

9. There is another class **[310K]** of inequality in relation to one pulse in addition to those mentioned, which appears not infrequently, nor in a small number of affections, but on the contrary often and in many affections. In this, the part of the artery nearer its origin is found to be moved first. And in fact the undulant (wave-like) and vermicular pulses are from this class, existing with a soft artery and a capacity that is not strong. The vermicular pulse shows not only that the capacity is not strong, but also that it has been dissipated, whereas the undulant pulse is sometimes present when the capacity is only not strong but has not yet already been dissipated, while it sometimes also partakes of a moderate *tonus*, at which time particularly it is a prodrome of a critical sweat. This is found in the moist and phlegmatic diseases,[829] when the solid bodies themselves become soft, being moistened by the abundance of the fluid. This happens particularly in the lethargic affections and the quotidian fevers, and besides these, the so-called marshy and delirious fevers.[830] Furthermore, the pulse takes on slowness, infrequency and largeness in all such things as these. If it also sometimes gains height or strength besides, it will indicate a critical sweat, **[311K]** and this will be particularly expected, if it appears more high and strong than the opposites. If, however, such pulses appear equal, they indicate the sweat will be useful and at the same time fleeting, and particularly if there are also the signs of concoction of the disease. It is quite clear, I presume, that neither some inflammation, nor induration, nor congelation from cold, nor immoderate dryness, nor spasmodic (convulsive) condition exists in the body of the person when the pulse is undulant, because the artery is neither able to harden, nor to be stretched tight in such a pulse. It is necessary, as I said just now, for the artery itself to be absolutely soft, so that when the raising and dilating *tonus* sinks down somewhat, in relation to that and to itself falling down, it makes a configuration like a wave. However, the hard and strained artery, even if it should abate somewhat because the capacity dilating it is weakened, this part itself is raised together with those on either side. Wherefore, if such a pulse should not appear at sometime in relation to the hard artery, so that the first part of it always dilates first, it will not at that time become undulant or vermicular, but rather will have an inequality **[312K]** occurring in relation to the first and last parts of the wall of the artery being moved. The undulant and vermicular pulses also have this, but they take on the configuration

829 Kühn's Greek text has σφυγματικοῖς while Kühn's Latin text has *pituitosis*. The manuscripts F1, F2, and Va1 all have φλεγματικοῖς. The latter are followed.

830 The Greek word rendered "delirious" above is τυφώδης. See Hippocrates, *Epidemics* 4.2 (ed. and Engl. transl. Smith 1994: 93); Galen, *De morborum differentiis* VI.2 (VI.850K; Engl. transl. Johnston 2006: 141–142). It seems probable that typhoid fever was not considered as a clinical entity in classical antiquity. Instead, it was confused with phrenitis and other types of fevers. The Hippocratic doctors mention a continuous fever accompanied by diarrhea, watery stools, red rashes, nose-bleeding, delirium, abdominal pain and emaciation, which is said to resemble typhoid fever.

like a wave over and above this. These are found in the moist affections, as was said, when there is neither induration nor inflammation involving any internal organ, since such a pulse would appear continually in the dropsies and particularly the so-called *anasarka* and *leukophlegmasias*,[831] unless the dropsies are due to some internal organ becoming indurated, which happens predominantly.

We have often seen genuine undulant and vermicular pulses following from retention in haemorrhoids or suppression of menstruation, since often in such dropsies none of the internal organs are indurated. Therefore, we must take up again the discussion under the chief points: the undulant pulse is born of moist and soft arteries together with a moderately distressed capacity, whereas the vermicular pulse is born of soft arteries but a significantly weakened capacity. Anyway, because of this, and apart from fevers, it appears in most instances to occur in **[313K]** profuse emptyings—haemorrhages, diarrhoeas, choleras, strong evacuations, severe vomiting, profuse urination and disproportionate menstrual discharges. In this way too, such a pulse supervenes in the emptyings of haemorrhoids and in the haemorrhages from the uterus, as well as in the so-called *flux muliebri* (physiological menstrual flow) and in abnormal sweats, and no less in the putrefactions involving the lungs with much moisture, at which time their capacity would already be damaged. Nevertheless, if some inflammation has supervened in such a condition, having taken on hardness, the vermicular pulses are no longer apparent. At all events, because of this, the vermicular pulses do not occur in the so-called hectic fevers, which include all the marasmic fevers and those due to the consumptions, although they would in general occur to the extent that the capacity is hard pressed, but the dryness existing in the affections makes the arteries hard. However, the inequality of this class of pulses always befalls them, when they are sufficiently hard pressed in the capacity, in which case we said the first movement appears in the first part of the artery. This is what **[314K]** Erasistratus always understood to occur, although it does not appear to happen in this way, but rather as has now been stated.

10. Next, after the pulses mentioned, there are the so called tapering (mouse-tailed) pulses.[832] The class of these is twofold, the one being in a single dilatation of the artery, while the other is in a collection [of dilatations]. The one occurring in a single dilatation indicates weakness of capacity along with thinness of the body. However, it is necessary for the one applying his fingers to palpate the artery, whether the pulse in this appears larger, or stronger, or both together in the first parts of the artery, but in the more distal parts is smaller, or weaker, or both, to raise the first of the applied fingers, and to consider whether the remaining fingers perceive the movement similar to that before the finger was lifted, or stronger and larger, for it is impossible for this to be

---

831 For the classification of the dropsies (*kata/anasarka, tympanitic, ascitic* and *leukophlegmatic*) see Appendix 3 on diseases and symptoms.

832 The Greek term is μύουρος (Kühn's Latin: *decurtatos*), for which, in relation to the pulse, LSJ gives "dying away gradually" citing Rufus, *Synopsis de pulsibus* 8.1, and several passages from the Galenic treatises.

perceived as weaker or smaller, when the first finger is raised up. If the fingers still perceive a similar movement after the first finger has been raised up, the artery was dilating in a tapering fashion by reason of itself, and **[315K]** this would be a sign of a weak capacity, which we showed to flow to the arteries from the heart. For when there is accord with nature, the *tonus* dilating the arteries comes to the end of all the arteries similarly, whereas in the tapering pulses it is gradually dissipated, becoming weaker than it was the more distant it is from the origin.

If, on the other hand, after you lift the first finger, you perceive the movement stronger and larger than before, it now of course also indicates weakness of the capacity in the artery, but less than previously mentioned. In that case, the capacity was being dissipated as it became more distant from the origin, whereas now it remains equal by reason of itself, but the artery, burdened by the application of the fingers, is moved more weakly. This, however, will be less dangerous. For it indicates the condition is not so much of the heart itself as it is of the artery we are palpating. The other however indicates more that the heart itself is weak and is not able to extend the action from itself up to the extreme parts of the animal, as it did before. Therefore, it is necessary in such pulses to apply the fingers in various ways for the purpose **[316K]** of learning precisely the whole condition of the arteries. For the application of the fingers compressing the artery will give the impression of pulselessness (*asphyxia*) in such conditions, whereas a superficial touching will give the impression of the tapering pulse, whether it falls upon two fingers only, or three, or even four. This will be dependent on the thinness of the organism and the weakness of the capacity. But the thinner the organism is, the more the artery shows, whereas the less thin it is, the less it shows. And the same also applies to the capacity—to the extent it is stronger, so it is more, while to the extent it is weaker, so it is less.

If in such an application, when the first finger is lifted, the pulse no longer remains similar in relation to the second, but appears larger and stronger, the *tonus* from the heart is stronger, for it flows equally to the end, while when the artery is weighed down by the application of the first finger, it dilates worse after this, and because of this, when what is weighing it down is lifted, it is better. If, however, the pulse appears similar to how it appeared in relation to the second finger, before the **[317K]** the first finger was raised, there is tapering of the movement in the arteries themselves, and not in the recognition by the touch. And this occurs when the capacity flowing to the arteries from the heart is gradually dissipated. On the other hand, if the movement of the artery seems better when the first finger is lifted away, but is nevertheless not precisely equal to that in relation to the first finger, it indicates both conditions exist at the same time. And for the most part, in fact, this is what customarily happens in all such inequalities. Therefore, after this, we must look at many pulses in order, and compare the movements accurately with one another.

Sometimes all such movements occur as they did in relation to the first application. Often, however, they come to be smaller and weaker. In these cases, the capacity is obviously also weaker. And if they completely fail, they indicate still more the weakness of the capacity. In this case, it is necessary to lift all the fingers, then after leaving a short interval, to apply them again. In the majority of instances, the arteries will be found to be pulsating on the second application. So if in fact at any time absolute pulselessness (*asphyxia*) should appear on this application, one must know that at **[318K]** the very time in which the first touch was made, in which it happened to the arteries to come to pulselessness, that a little of the movement was still preserved. But this, as I said, is rare, as for the most part they fail, and then, when we palpate again, we find them pulsating. From this it is clear that, when they do not bear the application of the fingers, they have come to extreme weakness of movement. And this was the reason why the pulse escaped our notice, as the pulselessness itself occurred in a twofold manner—either that the arteries did not pulsate at all, which would be extremely bad, or that they did pulsate but were weak and small, so as to escape the perception. So then, there is such a condition, and if it occurs apart from some recent swooning (*lipothymia*),[833] it is very serious, while much worse still is the other condition in which they truly do not pulsate, although much more, one must say that is not much worse but absolutely fatal and the worst, as the arteries would already have died. Of course, if after a longer time, when the fingers are reapplied without force so as merely to touch, the movement should return again, such a condition is less fatal. Sometimes, however, it **[319K]** either comes to extreme pulselessness (*asphyxia*), or sometimes to *mikrosphyxia* and weakness. Moreover, both are also called recurrent,[834] the difference being that in one case they are tapering recurrent, and in the other failing recurrent.

It is better in such observations to arrange the part which we happen to be palpating in such a way that our touch is underneath the artery, for in this way the compression of the fingers and of the overlying skin[835] itself is least. However, it is possible for each of those who wishes to learn by practice, how much such an application of the fingers contributes to recognition of a movement that is unseen. Anyway, often when we have applied the fingers to the arteries from above, we have not perceived a single movement, whereas if the limb is turned up the other way, we have perceived it immediately.

Worst of all are those which have come to complete absence of movement which doesn't return again when the limb is rotated. Less bad than these are those in which there appears to be movement when the limb is rotated, and less bad still than these are those that have become tapering but have not yet come to complete lack of movement. Less bad than these again are the returning tapering, and more when they fall on several fingers at the same time. I think it particularly worthwhile to pay attention to this **[320K]** and to consider precisely the amount of thinness of the body. For if the thinness of the body is extreme, the movement of the arteries falls on a number of fin-

833 For λιποθυμία LSJ has "swoon", citing Hippocrates, *Aphorismi* 1.23 (Engl. transl. Jones 1931: 109). For the verb, λιποθυμέω, LSJ has: "to fall into a swoon, faint" citing Hippocrates, *De articulis* 68 (ed. and Engl. transl. Withington 1928: 361).

834 The verb is παλινδρομέω; this is listed in LSJ as meaning "recurrent, recovering". See also *Synopsis de pulsibus* 23 (IX.510K; Engl. transl. in present volume).

835 K: δέρματος; Kühn's Latin translation: *cute:* F2 and Va1: σώματος.

gers; it is not due to the strength of the capacity, but to the attenuation of the body that it happens in this way. However, when the thinness is moderate, it is no small good for the movement of the artery to fall on three or four fingers, as in such cases, this indicates that some of the *tonus* is preserved in relation to the action. But the number of fingers on which the movement is going to fall is specific to each of the arteries. For the movement of all [arteries] is not equal in being prominent and being situated adjacent to the skin. And is better in this to have experience of the nature of the sick person himself, for in some the pulse is naturally smaller and in others shorter. In each case, the measure of the amount is in relation to the nature of the person himself. Therefore, when we are ignorant of the specific nature of the actual person who is sick, we conjecture from the commoner variations, which are those in relation to the *krasis* of the body and the bodily state (*hexis*). In the *krasis*, we particularly consider the *differentia* in relation to hot and cold, whereas in the bodily state (*hexis*), we consider that in relation to thinness and fatness. **[321K]**

Sometimes, then, the two inequalities of the pulses occur together, either in one movement, or in a collection. Often, however, it is not in a single movement, but is clearly found in a collection. And the reason why it doesn't appear at that time in a single beat is twofold: the minor nature of the condition or the shortness of the pulses themselves. For if the condition producing the tapering pulses in a collection is not very severe, there is equality in a single dilatation, while the tapering inequality is picked up in many. But if the condition is major, the pulse is short in length so as to fall on one finger alone, and particularly even if it really is tapering, it will not have the perceived inequality. It is necessary for the inequality that is tapering to fall on many fingers in one pulse.

Therefore, from the things said, it is quite clear that both classes of the tapering pulses indicate weakness of capacity, but that in one dilatation jointly indicates thinness of the affected body, whereas that in a collection can sometimes also **[322K]** occur apart from thinness. It has also been shown that someone might discover the amount of the harm in each class. These things are then enough about the tapering pulses.

11. Some have written that sometimes they find a kind of pulse that is the reverse of those mentioned. Anyway, in the case of the artery at the wrist, which they are most accustomed to palpate, they say a pulse has appeared to them broader in the end-point below, this obviously also being tapering, but the thinner part of it not to have been changed downward towards the fingers, but upward and towards the region of the elbow. I think the people in whom they say they have found such pulses don't have some temporary cause, but something in the physical conformation of the parts in the wrist. For a kind of conformation is discovered sometimes, albeit infrequently, in which the part of the artery towards the wrist is placed under the skin itself, whereas the whole area above it, which is higher, has been covered by abundant flesh, while **[323K]** below and in relation to the wrist itself, it sinks all at once to the depths. Therefore, the pulse appears tapering in such people, and not only in those who are diseased but also in those who are healthy, having the narrower part above as towards the forearm, while in the parts below and towards the wrist, it is to the same extent broader.

It is quite clear that those who write about such pulses as being morbid have not accurately applied their intelligence to the matter, judging from the other things they write, and from their not writing in addition to their description what they are indicating. For it is obvious they found nothing exceptional signified by these in the case of such things occurring due to physical conformation and not a disease condition. And on this account also, I have often said, regarding the diagnoses of the pulses, that there is nothing as useful as knowing what kind the pulse is in each nature. For in truth, there are certain other arteries disposed oppositely to those mentioned, that regarding what is above are placed under the bare skin alone, whereas lower, in coming towards the wrist, they sink down into the depths. And because of this also the pulse in these cases appears tapering. In other cases, however, it is nodding (compressible)[836] **[324K]** and tilted on each side more than accords with the general nature. It is not produced by disease or in general any other cause contrary to nature, but by physical conformation.

Unless we know these things are natural, we shall make very great mistakes. For as I said, the tapering pulse occurs in certain causes contrary to nature. Nevertheless, in some it is also discovered in relation to the specific nature of the conformation, as the so-called tilting and nodding pulses are—pulses in which the artery is arranged in such a way that only the middle of it is placed under the bare skin, whereas those parts on either side go down into the depths in a more concentrated fashion. It also occurs due to weakness of capacity being unable to lift up those things weighing it down. And whenever in fact such an affection becomes too strong for it, the pulse appears to have some bend in relation to the height, and not the circumference of a circle. Archigenes terms this bend not badly "angular".[837] For the height of the artery is no longer like the circumference of a circle but like the apex of a triangle which moves the touch. This happens to it still more with shivering and freezing due to cooling, **[325K]** or it is hardened in any other way whatsoever, and because of this it is very often found in the advent of strong symptoms associated with cooling, although not in all these in fact, or only these, as some think. For either the hardness of the arteries makes it difficult for them to bend, or the capacity being weak is weighed down on either side, so the circumference of a circle is, as it were, destroyed, and it becomes like the apex of a triangle, since the artery is unable due to hardness to be gradually bent and to incline towards the depths on either side, but vibrates more than it bends, due to the hardness. Therefore, these affections often occur in it in the strongly cooling symptoms, but

---

836 See Appendix 1 on pulse terms regarding περινευκῶς.

837 Kühn's Greek has γωνίωσιν, while Kühn's Latin translation has *angulotatem*. According to the *Dictionnaire Universel de Médecine* 1746: 4.127, *goniosis* was the name given by Archigenes to one type of pulse whose beating, according to Galen, can only be felt in one angle of the artery, as if the artery was not forming a circle but a triangle. The cause of this pulse was the inability of the vital spirit to go through the whole length of the artery.

sometimes occur also apart from an attack of paroxysms, just as conversely such a pulse does not occur at all in the symptoms that are free from chilling. It is perhaps better not to spend further time on this point, since we are going to go over these matters to a greater extent in what follows.

12. The trembling pulse is also from this class of those spoken of—something not very clearly apparent. Since, however, just as has been shown in the second book of the work **[326K]** *On the Causes of Symptoms* that tremor follows weakness of capacity,[838] it is inevitable because of this that the trembling pulse is also weak, and at the same time small, while the tremulous movements are concealed in such a pulse, as they require a noteworthy dilatation to be discerned. Sometimes the trembling pulse appears as clonic to those who are unpractised, and because of this they think they have often seen such a pulse. But there is no need to go over these at greater length in the present discussion. For it was stated in the second book of the work *On the Causes of Symptoms* to what extent tremor and clonus differ from each other, and someone will know from which things the clonic pulse begins, and what kind it is precisely, and how infrequently the trembling pulse is seen in relation to those times only in which the capacity is not extremely weak, the wall of the artery soft, or the state of the body (*hexis*) thin. What will happen in this way is that the pulses are not very small, but are near to those which accord with nature; nor are they extremely weak, but such that they tolerate without distress application of the fingers, which is a necessary requirement for recognition. And if ever a trembling pulse appears, it shows the capacity to be weak, the wall of the **[327K]** artery to be soft, and the state of the body (*hexis*) to be thin. These things are, then, enough about the inequality and the causes producing this. And if something has been omitted, it will be very easily discovered by someone who has learnt these things now stated, and not as a secondary matter, and those things stated in the second of the books of *On the Causes of the Pulses.*[839]

13. What remains is to go over regularity and irregularity. It seems, certainly, that every regularity is better than irregularities of the same class. "Of the same class" is added to the statement because sometimes it can be that the irregularity from one class is preferable to the regularity from another class. For example, as may happen, the irregularity in largeness or rapidity is less serious than intermittency in any regularity. In this way too, the infrequency that is prolonged, if it also preserves a certain regularity, is worse than the inequality relating to strength, largeness and rapidity. But in fact the regularity of the same class is more mild than the irregularity seems to be, because the regularity is more characteristic of our nature than anything else. So the statement should not be made simply in this way, but along with such a distinction. **[328K]**

Regularity seems somehow to show the cause to be fixed, whereas irregularity shows it to be fluctuating and not to remain uniform in size, kind or place. However,

838 *De symptomatum causis* II.2 (VII.154–160K).
839 *De causis pulsuum* II (IX.55–104K).

such a cause is not more fatal, but must be considered to be more dangerous. For it is possible for this to fluctuate, sometimes coming to a less important and sometimes to a more important region, so that either death is sudden or safety is attained. Moreover, the cause of the regularity is not unexpected but has a defined deviation. The regularity in the case of a serious inequality, as if it has acquired intermittency in this way, is certainly fatal, whereas that which is moderate, as in the case of inequality relating to largeness and smallness, is certainly mild. However, the irregularity is also able to make the mild serious, if the cause creating the irregularity changes to a more important region. On the other hand, it can also make the serious mild, if the cause is now moved from the more important regions to the more unimportant. Nothing, then, is safe or certain in the case of the irregularity, which is not like that of the regularity, which is neither good nor bad. Nevertheless, should some irregularity occur along with a strong pulse, whether **[329K]** the strength is equal or unequal, if the signs of concoction are also present, it must be hoped there will be a crisis, and particularly if the other signs of a crisis should also occur. Sometimes also, in relation to this, the irregularities differ from the regularities to the extent that, if all other things are the same, Nature urges more readily towards a crisis in the irregularities.[840]

Book III

1. [330K] What the capacity of each pulse is was stated previously. In this book, we shall speak about the characteristics in the affections. For present purposes, I term an affection anything contrary to nature. In each case, I call characteristic pulses those which are continually present in the affection and those which are often present. Since, as has been shown in other works, of all the things contrary to nature, some are conditions by which function is harmed primarily, some are causes of these, and some are symptoms, I must speak about the primary conditions.[841] Also, since some of these were shown **[331K]** to be specific for *homoiomerous* bodies, while others were specific for organic bodies, I must start from those existing in the *homoiomere*.

They are eight in number and all occur in relation to a *dyskrasia*, but differ in that some are simple, while others are compound.[842] The simple *dyskrasias* are those of heat, dryness, moistness and coldness; those that are compound are of coldness together with dryness, heat together with dryness, coldness together with moistness, and heat together with moistness.[843] We have set out the *dyskrasias* in the treatises *On the Kraseis*,[844] and distinguished how we say the hot and moist [*krasis*] is sometimes healthy,

840 Kühn's Greek text: κατὰ τὰς ἀρτηρίας: Kühn's Latin translation: *in ordinis perturbationis.*
841 See, for example, Galen's *De methodo medendi* I (X.1–77K).
842 See Galen's *De morborum differentiis* (VI.836–880K; Engl. transl. Johnston 2006: 180–202).
843 Only the first and the fourth are given in Kühn's Greek text as is the case in the manuscripts F1, F2, and Va1. Kühn's Latin translation has all four and is followed here.
844 *De temperamentis* (I.509–693K).

and how disease harms function through immoderate moisture and heat. In addition to these things, we must bear in mind that the pulse is not at anytime changed through a *dyskrasia* apart from some affection involving the arteries or heart, and that it is possible sometimes for the arteries to have been affected, while the heart is without affection. If, however, the heart is affected, it is not possible for the arteries to remain unaffected, but of necessity they "enjoy" some of the *dyskrasia* themselves, either less or more. Therefore, the **[332K]** *dyskrasia* of the heart in relation to coldness and heat is readily distributed to all the arteries, whereas that in relation to moistness and dryness is more difficult [to distribute].

2. We must begin, then, from the *dyskrasia* relating to heat and we must distinguish in this first that it is possible during the time this occurs for the arteries to be in accord with nature or to have been affected somewhat, and that this applies to all of them or to some. Certainly, those that are in accord with nature will take up the *dyskrasia* in the heart similarly, whereas the condition of those that are in a state contrary to nature will be mixed, either in part preserving the characteristic state or in part being involved along with the heart. First, let it be assumed the arteries are unaffected at the beginning, while the heart becomes hotter. Then next, we consider which pulses exist of necessity in such a condition and which do so for the most part. In turn then, bearing in mind those things we determined in the book before this one and in the treatise *On the Causes of the Pulses*,[845] let us go over the pulses characteristic of heat that is predominant, so we may consider which of these are inseparable from the condition in question and which occur for the most part. **[333K]**

Thus, the characteristic pulses due to an excess of heat are large, rapid and frequent, but the largeness prevails not with the heat alone, but also requires arteries that are soft and a capacity that is strong. As a consequence, this would not be inseparable in the *dyskrasia* related to heat. In this way too, the rapid pulse requires a strong capacity, and arteries that are readily dilated, so this is not inseparable. The frequent pulse, however, if the need is sufficiently fulfilled by the largeness and rapidity, would not exist at the beginning, for this has also been shown, whereas it would if the need were not fullfilled. As a consequence, this is not inseparable, nor would any of the characteristic pulses be inseparable. Of those that are characteristic but not inseparable, sometimes the larger will prevail, sometimes the more rapid and sometimes the more frequent. For if the heat alone is increased, while the capacity and arteries are in accord with nature, when a brief change occurs, the pulses will increase to largeness alone, but when [the change is] of longer [duration], also to rapidity, and if it is particularly marked, also to frequency.

It is necessary in this to direct your attention precisely to the matter of more and less in each class of pulses. For if **[334K]** they change to largeness more than to rapidity and frequency, consider the heat in the heart not to be very great. If, however, they change similarly in the three parameters, and these immediately are obviously greatly

845 *De causis pulsuum* (IX.1–204K).

changed, they immediately also indicate great heat. If, on the other hand, the rapidity is increased to a greater extent than the largeness, the heat is also still more. And if the frequency is increased, then the heat is still more. If, however, the pulse itself is at the same time very large, very rapid and very frequent, it will show the heat to be increased to a peak. On the other hand, if it is very rapid and very frequent, but not in fact also very large, this could not have happened otherwise apart from the capacity being weaker, or the artery being harder. It is easy to distinguish each in the inseparable pulses, or the actual indication here of the heat in the heart. Furthermore, when the frequency reaches a maximum, but the rapidity and largeness are not increased together equally to this, of necessity at that time the heat is great, but there is either weakness of capacity, or hardness of arteries. With weakness [of the capacity], the pulse will be weak, whereas with hardness [of the arteries], the pulse will be hard and at the same time clonic. If, however, it is hard without clonus **[335K]** and extremely frequent, it will indicate a weak capacity along with abundance of heat. Altogether then, such a pulse will be weak and will bring danger of collapse (*synkope*). It must be remembered here also that it is possible in a certain way for there to be very great heat in the heart, while the capacity is strong. For if the strength follows the *eukrasias*, while weakness follows the *dyskrasias*, a *dyskrasia* will take hold of the heart along with the excess heat. As a result, the capacity also will be weak of necessity.

How then, when the pulse is at the same time very large, very rapid, frequent and strong, do we say that not only is there very great heat in the heart, but also strength of capacity? We often declared in other places and in the diagnoses of the fevers that being heated differs from from having been heated. For when the body of the heart is heated by the actual substance contained in the chambers (cavities), it is possible for the capacity to be strong, and because of this itself such people breathe in maximally, very rapidly and very frequently, while they pulsate very greatly, very rapidly and very frequently, since the heart is still fighting against the heating substance, and **[336K]** the other cooling substance which counteracts the heat is drawn after externally. However, when at times it happens to be overcome and is heated together with the substance in the chambers (cavities), then, under these circumstances, it has already been heated, and is not only still being heated, while weakness of capacity follows the *dyskrasia* of this. And if it is heated still more, and the feverish heat is received throughout the whole of it, it brings with it the so-called hectic fever. Sometimes, however, the converse of what was just now stated happens, when the substance contained in the chambers (cavities) of the heart is colder than accords with nature, while the body of the heart has been heated still more. This occurs in those who have collected crude, cold and phlegmatous humors, or have been made very angry, or have lost sleep to an immoderate extent along with extreme anxiety, or [in whom] the actual body of the heart is affected by some pestilential inhalation, or because in general some specific deleterious quality has come to the heart in any way whatsoever. I have mentioned these things, then, so no one wonders how it is possible for the *tonus* of the heart to reach a peak, when heat prevails. Therefore, let us return once more to those things we were saying when we digressed. **[337K]**

For certainly, heat in the heart changes the pulses, as was said. But this is twofold in kind: in one instance, it is a product of putrefaction of humors, the quality of which is acrid and smoky, whereas in the other instance, it is with sweetly pleasant moisture, deviating from an accord with nature in its actual amount only. The first spoken of is stable and difficult to resolve, whereas the second is of short duration and is an accompaniment of baths, exercises, heating food, drinking wine, and the affections of the psyche with shame and anger accompanying. Whether from this class of heating, there are certain fevers as in the case of those that follow fatigues, heatstrokes, coolings, buboes or anger is a matter of dispute. More has been said about these in the work on fevers.[846] Let this much also be said now—the pulse associated with heterogeneous heat compared to that in accord with nature has a short duration in respect of the external pause, and a contraction which is rapid to a greater extent than would be proportional to the dilatation, whereas that differing in amount alone has a dilatation that is larger and at the same time also more rapid than the contraction. Such heat by itself requires cooling, **[338K]** which we showed occurring in the dilatation, while the smokiness still more than this, but to the greatest extent the pouring out of the superfluity, which was the task of the contraction. As a result, such heat will also make the dilatation more rapid to a greater extent and the contraction to the greatest extent. However, such heat will not change the contraction at all, whereas it will change the dilatation to the extent that it has overstepped the measure of the *eukrasia* according to nature.[847] But we shall say more about these things subsequently in the considerations of the fevers.

3. For the present let us go over the discussion about the arteries, when they are opposite to the heart—I speak of when the latter is hotter than accords with nature while the arteries are colder, either when certain cold humors that have not yet putrefied are contained in them, or when some other cooling prevails. It is necessary under these circumstances for the arteries to have been cooled in relation to the specific condition they are in, but to have acquired in addition heating from the heart. But since they also have the capacity for movement flowing to them from that organ, and the *tonus* dilating and contracting the arteries together with the heart is one, it is necessary that the same *tonus* exists in the **[339K]** movement of the heart and in that of all the arteries individually, although there is some difference [between them]. Certainly, the first beginnings of the movements will be at the same time, whenever the arteries receive the capacity from the heart faultlessly. But neither the largeness nor the rapidity nor the strength will be similar in both. Rather, under these circumstances, the heart will have a pulse that is larger, more rapid and stronger than all the arteries individually. For the beginning of the movement in them is from the heart, while the manner

---

846 *De febrium differentiis* (VII.273–405K).

847 This and the previous sentence, as translated from Kühn's Greek, seem contradictory. Kühn's Latin translation has: "And so this will also restore the dilatation very much to greater rapidity and will change the contraction very little; the dilatation to the extent it has overstepped the measure of a moderated temperament." Sense obviously favours the latter.

of the characteristic movement in each happens as jointly determined with the heart. For it is jointly determined,[848] when being similar in *krasis* the heart functions in the same way, or oppositely to the degree it happens to receive the change from the heart. It is certainly sometimes possible for the cold artery, when heated by the heart, not to make the pulse small, like that of the cold artery is, nor large like that of the hot artery, but intermediate between both, which is in fact moderate and in accord with nature. In this way too, the pulse will be moderate in rapidity in such conditions and moderate in strength. You will clearly understand what I am saying, if at the same time **[340K]** you were to touch the heart with one hand and one of the arteries with the other. The beginnings of the movements will appear to you to occur at the same time. However, they will differ in magnitude, rapidity and strength under these circumstances. On the other hand, what sometimes happens is that all the arteries do not receive the capacity which flows to them from the heart at the same time, but those nearer to the internal organ receive it first, while those further away receive it later.

Such a thing particularly occurs, when at one time the arteries come to be colder and the heart *atonic*. If, however, the coldness of the arteries themselves is greater in the parts further from their origin, as happens in certain accessions and in the malignant intermittent fevers,[849] not only will the movement be sooner, but also the parts nearer the heart will make the movement itself more rapid and larger. Furthermore, the *tonus* in such circumstances appears weak in all the arteries, but more in those further from the heart, so that clearly the pulse in these is weaker than accords with nature compared to those more proximal. Sometimes it also happens that there is such an opposition of the *kraseis* in the heart. For sometimes the actual body of the **[341K]** organ is colder than accords with nature, whereas the substance contained in the chambers (cavities) is hotter; conversely, sometimes the substance is colder, while the heart is hotter. And at that time also, the pulses of the heart are like those in accord with nature, for somehow the state in this is made intermediate between the two opposites. Such conditions deceive even the best doctors, as recently happened in the Great Plague.[850] Some from the beginning and a very long time thereafter, and some throughout the whole disease were *eusphyktic*[851] having pulses that were little changed from an accord with nature. But these [people] died more than the others and we recognised them immediately in the first days—that they were suffering the

848 The Greek verb here is συνδιατίθημι which LSJ lists as equivalent to συμπάσχω in the medical sense, that is, a sympathetic affection. The use here seems more general. Kühn's Latin translation has *consentio* (determine together, feel together, be in accord with).

849 On the term here, λιποπυρίαις πυρετοῖς, LSJ gives the reading used above, citing Hippocrates, *On Crises* 11. See also Galen, *In Hippocratis aphorismi* 76 (XVIIB.768K), and *In Hippocratis prognosticum* 4 (XVIIIB.121K).

850 The Great Plague is taken as referring to what was probably an epidemic of smallpox that hit Rome in A.D. 166, while Galen was there.

851 The three terms *eusphyxia, kakosphyxia* and *asphyxia* constitute a transliterated technical triad used by Galen to denote respectively a normal pulse, an abnormal (pathological) pulse and an imperceptible pulse.

plague, and that they were in the worst state due to the acridity of the heat and the actual kind of the heat, and that their pulses were only slightly changed from an accord with nature, and always remained so. For pulses of this kind exist particularly in the so-called hectic fevers. And in those who were sick, but said they did not have a fever, the fever in them was precisely in their state (*hexis*), **[342K]** having taken hold of the actual body of the heart. For certainly these two signs are characteristic of the hectic fevers: [those affected] are continuously like themselves, having no attack of any paroxysm, and no progression, peak or abatement, while the one suffering is not aware that he is febrile. Such pulses as occur are not of necessity larger than those which accord with nature, while sometimes they are not more frequent, but of course always more rapid. And what is correctly known by most good doctors is that the pulses of those with fever have an inseparable rapidity. But more about these things in due course.

What we were saying when we digressed is that, when the inspired air is defiled by a putrefying quality, and what is putrefied touches the body of the heart, such pulses occur. And it is entirely unremarkable that those who have the plague in this way have pulses like those we have spoken of, that the breathing is pestilential, and that in all cases they will die. For what remedy will be found for the putrefaction in the heart? Many of those in whom this putrefactive heat was deflected to the humors in the chambers (cavities) of the heart more than **[343K]** to the actual body of the organ were saved.[852] The diagnosis of those who have the plague is made, if first they are perceived to be febrile, and then if the fever is not continuously the same, and in addition to these things, if they have a pulse that is not vigorous. For in a *dyskrasia* of the actual body of the heart all are *atonic*. Certainly, enough has been said about what the hot *dyskrasias* of the heart make the pulses like.

4. The cold *dyskrasias*, of course, always produce the opposite pulses—those that are small, slow and infrequent, which change either in proportion to the three *differentiae*, or to some of these, or to one, more or less. Some, however, are at one time less and another time more. But it is necessary here in the same way to know the deficient and the excessive, as was also said previously. First, it will change the frequency; second, the rapidity; and next, the largeness, at least if the capacity is not affected. If in fact this is also somewhat affected, it will be recognised by the specific and inseparable quality of the pulses, which we call weakness, or towards smallness and slowness, more than **[344K]** it makes the change of the pulses to infrequency. So when the capacity is suffering to the greatest degree, and if chilling is present with this, the pulses become extremely small but not extremely slow; sometimes they are like those that accord with nature. However, not only do they put aside the former frequency, but also they increase it. Moreover, when the aspects of the capacity become slightly weak, while those of the chilling prevail to the greatest extent, the change of the pulses

852 The translation of this sentence is based particularly on the manuscript F2.

to infrequency becomes very prolonged, while that to smallness is of very short duration, and that to slowness, intermediate between both.

Clearly, however, all the arteries will be moved with the same movement, if they are normal in *krasis*. But if the heart has become colder, while the arteries are hotter than accords with nature, they will change the pulse of the heart to the extent that they have changed from an accord with nature. Consequently, sometimes the pulse in the arteries will be precisely similar to that which accords with nature, as much as pertains to largeness and rapidity, when that in the heart becomes less and slower. It is quite clear, however, that such conditions without fevers exist only when the heart itself is cooled by cold humors that are at the same time both viscous and thick. Therefore, such **[345K]** a condition is rare, but does occur sometimes, and particularly when such humors are contained in the lung. What happens to the pulses in such a condition is that for the most part they are immediately also unequal due to the obstructions which the previously mentioned humors make in the openings of the vessels arising from the heart.

If, however, it should happen on some occasion that the arteries also become colder than accords with nature, similarly to the heart, they will increase the *atonia* of the movements of the heart, remaining in the same class, so that when the heart pulsates slower, to a smaller degree and more infrequently than accords with nature, the arteries themselves pulsate still more slowly than that, and to a smaller degree, and of course immediately also more weakly. But sometimes it happens that the heart is colder than accords with nature and at the same time is full of biting[853] superfluities, at which time what happens is that the dilatation becomes smaller and at the same time slower, while the contraction becomes more rapid. The artery would seem to move inward to a greater extent, if the movement were to be perceptible at all, although this is now unrecognisable to perception. That the time of the internal pause becomes longer than accords with nature is clear also **[346K]** to those palpating. However, it was shown that this whole time is a compound from three times: the first is of the last parts of the contraction, the second of the internal pause, and the third of the first parts of the dilatation. This time will be significantly extended in relation to the condition mentioned, while the external pause will be either shorter than accords with nature or nearly equal. Since the capacity creating the pulses is always striving to separate the superfluities, and is hastening towards that action through which this exists for it to do, what happens is that the pause before the action becomes of shorter duration, just as that before the other action (I speak of the dilatation) becomes longer, because the heart being colder comes to the action more freely. This is because there is only the one need, that of the fanning, since the other and greater need, which is that of the cooling, has already been done away with.

853 K: δακνωδῶν; Kühn's Latin translation: *fumosorum*; F1, F2, and Va1: καπνωδῶν.

The remaining or third in those solitary conditions stirs the heart towards action, whenever either some significant movement moves the animal or a large evacuation occurs, since the arteries are opened up to a greater extent towards the exterior, as in **[347K]** baths, and if certainly some others due to any cause whatsoever pour out much sweat. If, however, these are internal, they will be in the purifications (catharses), and of the affections, in choleras, dysenteries and strong bitings of the intestines. Of course, when it should also happen that the actual body of the heart has an unequal (non-uniform) *dyskrasia* in itself, it will be inevitable also that the pulses of the heart itself and of all the arteries in the animal are turned aside variously to inequality. Furthermore, the so-called systematic (collective) inequality and that relating to one pulse will be of all the classes, when the change becomes continuous to largeness and smallness, rapidity and slowness, [strength][854] and weakness, and frequency and infrequency. However, change to hardness and softness would not at any time become continuous. This has been frequently shown.

Sometimes, in such conditions, aclonic and clonic [pulses] succeed one another continuously, when either hardness or tension of the arteries comes together with the *dyskrasia* of the heart. For it was shown, if I recall, in relation to such conditions, that whenever the heart urges the arteries to dilate still more, clonus befalls the movement. **[348K]** When this happens, it is easy for the dicrotic pulse to arise. And it becomes difficult in this distension to discover accurately the condition, as the inequality mentioned would exist not only in a *dyskrasia* of the heart but also in badness of the arteries and in abundance. For since such inequalities occur in obstructions of the openings in the heart, while they also follow the non-uniform *dyskrasias* of this, it is not easy to diagnose the condition. Thus, the double-beating pulse is a characteristic symptom of the *dyskrasia* involving the body of the heart, as has been shown in the book prior to this one, and it differs from the dicrotic pulse in softness. However, when the previously occurring hardness is lost, the diagnosis is lost along with this, for it changes to the dicrotic pulse, which is common to many conditions. Of course, the clonic dicrotic was shown to have its genesis when the simply clonic pulse becomes short in length, while the aclonic dicrotic, pulse similarly to the clonic, is hard, but due to the slowness of this and the weight of the overlying bodies is rendered of such a kind. As a consequence, each of these will follow many causes, **[349K]** while they differ from each other in the magnitude of the powers and causes[855] creating them. For the clonic dicrotic pulse follows very many, whereas the aclonic follows few.

It must be remembered first how many causes the absolutely clonic [pulse] follows; all have been stated previously. Next, there is the kind of *synektic* cause of this. For it occurs when the hard or strained body of an artery is compelled to dilate still more. Third, in addition to these, the clonic dicrotic pulse was shown to occur because a weak capacity comes together with this or physical conformation. And in addition to

854 Added from Kühn's Latin translation.
855 "and causes" is omitted in F2 and Va1.

these, the dicrotic aclonic pulse occurs due to hardness of an artery, slowness[856] of the pulse, or weight of the bodies overlying it above and on either side of the perceived movement. In such states, when what is still rising in the middle is held back by the parts on either side, and then again comes together with them, the dicrotic pulse occurs. These, then, are their *synektic* causes, as it were, and often[857] precede each. It is necessary for anyone who is going to discover these antecedents to have those things ready to hand in the memory **[350K]**.

Therefore, the double-beating pulse follows non-uniform *dyskrasias* of the actual body of the heart, while it requires a practised touch to recognise accurately the contraction, so if someone is not practised in such recognition, a pulse of this kind will seem to him not to differ at all from that which stands still during the rising and pauses a little. It will, however, be recognised, if it partakes a little of *eutonia.* That which is altogether *atonic* has the contraction unrecognisable. However, there is for the most part *tonus* in all the double-beating pulses. For they would not rise to such a height, if they did not partake of some *eutonia.* Otherwise, of necessity, they would become much softer than accords with nature, so they would appear weak and at the same time small. These pulses also arise in the lethargic conditions, but it is not possible under these circumstances to distinguish the slight pause of the double beat, so that it is not possible to distinguish the conditions in relation to this. It is, however, possible through another sign. This sign is specific for the body of the heart that has been mixed badly (i. e. has a *dyskrasia*). And it is no bad thing that there is a certain fever and pulse [called] hectic; in this way **[351K]** we also name an inequality hectic for the purpose of concise teaching. Let anyone who is not pleased with the name, call this 'stable'. Stable is when equality is sequentially preserved. And certainly, should all the other signs indicate the fever is hectic, you will I think recognise with much greater certainty that there is an unequal (non-uniform) *dyskrasia* in the body of the internal organ (heart). These matters were spoken about in the work, *On the Differentiae of the Fevers.*[858]

And of course you must look at all the other things—those that are present, those that are *proegoumenic* (internally antecedent), and those that are *prokatarktic* (externally antecedent). Thus, if peripneumonia, empyema, pleuritis, or *phthoe* (consumption), or in general one of the other abscesses has occurred on the spot, or asthma, a catarrhal disease, haemoptysis or some pulmonary disease, or to speak in summary, if there is excess, thickness or viscidity of humors contained in the lung, you must give careful attention to whether or not the unequal pulse arises due to obstruction of arteries. On the other hand, if none of these is present, give attention to the internal organ itself, and much more so when, during the life the sick person previously led, neither thick humors were collected in the body, nor viscous humors were apparent,

856 K: βραδύτητα πολλαὶ πολλάκις; Kühn's Latin translation: *tarditatem*; F2, and Va1: βραχύτητα.
857 K: πολλαί; Kühn's Latin translation: *multae*; F2 and Va1: πολλάκις.
858 *De febrium differentiis* (VII.273 – 405K).

by which, when they either spread around externally **[352K]** or block up the mouths (openings) of the arteries, the inequality exists. For it has been shown that, if one of these, either inflammation or simply a swelling (tumor) contrary to nature, exists on the spot, or an excess weighing down of the capacity, inequalities befall the pulses. Moreover, if neither through these, such affections are recognised as being present, nor a hectic and stable inequality persists, one must suppose there is a non-uniform *dyskrasia* in the body of the heart. And this itself has two aspects, one being in the characteristic fluids of the heart and the other in the solid parts. The one arising from the fluids occurs with palpitation of the whole heart, whereas that from the solid parts is without palpitation.

If someone wishes to learn thoroughly and scientifically those things now being said, let him carefully read the discussion about hectic fevers from the work on fevers and from the work on marasmus.[859] Certainly, the unequal (non-uniform) *dyskrasia* of the heart is diagnosed and distinguished from the associated conditions by the signs I have spoken of. However, every other *dyskrasia* involving the organ, when the body itself has one kind of *krasis* and the blood and *pneuma* in the chambers (cavities) of the heart have another, **[353K]** never creates the unequal pulses, but, as was said previously, from the mixture of the conditions, the pulses come close to some median state. Sometimes the mixtures of the conditions are precisely the opposites to each other, whereas sometimes they are not precisely so. I speak of conditions as precisely opposites of each other in relation to the *krasis,* when one happens to be as much hotter than accords with nature as the other happens to be colder. Not precisely, on the other hand, is when one is changed more than the other. It is sometimes possible, however, when both are of the same class, for the difference to be in the quantity, either the heart itself being hotter to a greater extent than the materials in the chambers (cavities), or these materials deviating more from an accord with nature, while the material of the heart deviates less. In this way too, it is possible for either the body of the heart or the materials to become colder. Diagnosis of the stated combinations is not similarly clear in all cases; sometimes in those that are not clear, it is made inferentially from other things.

Thus, the first mixture stated, in which the body of the heart is hotter than accords with nature, while the materials are colder, has three *differentiae.* Either the body is hotter by as much as **[354K]** those materials are colder, or it is hotter to a greater extent than they are colder, or it is hotter to a lesser extent than they are colder. In the first *differentia*, the pulse appears precisely similar to that which accords with nature, in largeness, rapidity and frequency. However, it changes towards *atonia* to the degree that the organ departs from the *eukrasia* that is in accord with nature. But it often seemed to me to have clearly come somewhat to rapidity in such conditions, although this does not always obviously appear, and it seems to me also that someone might find

859 The two Galenic works are *De febrium differentiis* referred to in the previous note and *De marcore* (VII.666–704K; Engl. transl. Theoharides 1971).

available another greater sign from this. For it is possible to form a judgement from the rapidity on the extent to which the actual body of the heart has departed from the *eukrasia* that accords with nature. When other things are equal, the hotter body is always more ready for movement. Thus, when the actual body of the heart has an accord with nature but is heated, either by the materials[860] in the chambers (cavities), or by the association of inflammation of the lung, or from one of the other of the adjacent parts being heated, it will change the pulses towards largeness, rapidity and frequency for the purpose of cooling. Under these circumstances, the first change is to largeness, the second is to **[355K]** rapidity, and next in turn to frequency. However, when it becomes hotter than the structures in contact with it which are in accord with nature, the movement advances first to rapidity, then second to largeness. As a consequence, whenever we discover that the pulse has clearly advanced to rapidity, but has had no change to largeness and frequency, we conclude that in such a body a great change towards heat has occurred in the heart and towards coldness of those structures contiguous to it. Immediately, I presume, the weakness of the pulses is also clear and appears significant in such bodies. If, however, neither the rapidity nor the weakness is clear, assume that a little heat has been added to the body of the organ. Such fevers deceive many doctors inasmuch as the patients themselves say they are without a fever, for they do not perceive the fever because of its smallness and the heat of the organ, and because the fever has taken hold of the body of the organ.[861] I have spoken about this often. In general, then, none of the hectic fevers is a large fever. This is still more so, when the heat in the organ is slight.

Therefore, to the extent **[356K]** that the heat from the heart flows to the whole body, this only carries a perception of its distasteful (nauseating)[862] inequalities, due to which people desire baths, just as with the other distasteful (nauseating) inequalities due to fatigue, heatstroke, lack of bathing, or lack of exercise. And in fact the majority delight in the bathing and say they become better. Inasmuch as they dispel the inequality from the whole body and nothing noteworthy harms them in the immediate term, they think it is because of the bath. Perhaps some of them are not harmed at all. But now is not an appropriate time to consider such matters. We distinguish all these things in the treatments of the fevers, where it seemed good to write about them separately, either in that or in the writings on the therapeutic method.[863] Regarding the present subject, let the recognition alone be dealt with. And furthermore, it will perhaps seem to some that doubt and uncertainty exist, when no *differentia* of pulses changes from an accord with nature, and the patient himself says he is unaware of the fever, and when the inequality which alone he blamed is cured by baths. However,

860 K: ὑλῶν; F2: ὑγρῶν

861 K: διά τε τὴν μικρότητα καὶ τὴν τοῦ σπλάγχνου θερμασίαν, καί ὅτι τοῦ σπλάγχνου τὸ σῶμα κατείληφεν; F1, F2, and Va1: διά τε τὴν μικρότητα καὶ διότι τοῦ σπλάγχνου τὸ σῶμα κατείλ η φεν

862 It is not entirely clear what Galen means by ἀσώδης here. Kühn's Latin translation has *fastidiosus.* Perhaps the term specifically means "nauseating". See the references in the following note.

863 *De febrium differentiis* (VII.273–405) and *De methodo medendi* VIII–XII (X.530–873K).

so that I shall make the discussion still more difficult, it is not always possible to take a precise diagnosis from the urine in such conditions, but this is ambiguous also **[357K]** in the majority of these cases, and similar to those in accord with nature in both colour and consistency. I saw countless such cases in the Great Plague, and among them I saw doctors similar to laymen, and some were much worse than an intelligent laymen. And if nothing else, it was possible to see many of the laymen considering at least the exhalation of those who were suffering, and if this seemed foul-smelling, to expect the worst. Therefore, such a sign was found by them from experience in the pestilential fevers. However, let this be apparent to you not only from experience but also from reason.

Certainly, the parts around the mouth are also examined, and if some of the pestilential skin colour should appear in them even slightly, this makes known that those affected in this way are suffering the plague. But in fact I think these things are also more apparent to the doctor than the laymen, for the colour appears like those with erysipelas in some cases and in others like those with eroding herpes, being scattered in many parts from the beginning, as I think. And if the experienced doctor should place his hand on the chest in the diagnosis of febrile heat, he would not allow the condition to escape him. Not least the urine must also be examined. **[358K]** In some of these cases the urine appears thickened and in others more watery and very much thinner[864] than accords with nature. These findings are, then, indisputable, just as also those in cases where the urine seems exactly in accord with nature in terms of both colour and thickness are, these being also indisputable with regard to the second expectation. And if the urine should make good suspended matter, this is still more indisputable, when the sediment is white, smooth and even. However, one must pay particular attention to those things contained (suspended matter). For some of these that are somewhat black in colour escape notice, just as others seem like broad webs, as when pieces of wool have been thrown together with each other.

If you should find some such thing, or nothing at all, being carried as suspended matter in the urine, place all these cases in the worse group. If, however, the opposites appear, of the kind I went over just now, it is not possible for the person to be in a bad condition. From these things, then, distinguish whether there is some inequality in him, or an attack of a fatal pestilential disease having a hectic fever, and in addition to these, whether after the bath, [the people] are hungry and thirsty more than is fitting, and desirous of cold. **[359K]** It is also appropriate to observe their eyes accurately as well. For in bathing, they become particularly clear, being made hot and bloodshot, while some cases beginning here remain so, and these we clearly recognise to be suffering the plague. But the kind of plague that occurred among us and is still present would never be established in such people. Moreover, it must also be known that these same fevers occur otherwise without plague, as has been written about them

864 "and very much thinner' is omitted in manuscript F2.

by all noteworthy doctors, and they call them pestilential. More has been said about every kind of these in the work *On the Differentiae of the Fevers.*[865]

The discussion now is not how someone might best diagnose fevers, but how they might best do so from the pulse. And perhaps more than is needed was said about the fevers kindled from the actual body of the heart. Let us, then, bring the discussion back to the pulses once more, expounding on what there is remaining of the hectic fevers. As was said, not a few of the pestilential fevers are immediately hectic from the beginning, differing from all the other fevers which occur in severe putrefaction. Sometimes, however, **[360K]** some of the fevers arising due to grief and anger are hectic from the beginning. Otherwise, I have not seen any hectic fever occur immediately from the beginning, although with the progression of time, they do supervene in many of the others. And the stable equality clearly distinguishes these, if they are also combined with one of the others, as I showed, and those other things stated on the matter in the work *On the Differentiae of the Fevers.*[866] As I said a little earlier, the diagnosis from the pulses is hard to recognise in the beginning, and in the first days, when they still have the form common to the ephemeral fevers. After the third day, however, they already become clearer for diagnosis through the pulses. For they clearly have pulses that are more rapid than those that accord with nature, even if they do not differ in other respects. And the heat in the chest is also clearly feverish in them, and the pulses are weaker than those which accord with nature to the degree that the heat is increased.

These things then are sufficient about the fevers in which the heart has become hotter than accords with nature to the degree that the materials in it have become colder. For the sake of certainty, let all those bodies contiguous to the heart be added to the materials. **[361K]** These are, of the internal organs, the lungs, while of the membranes, that which is termed pericardial and that which divides off the chest near the artery[867] of the heart (i. e. the diaphragm). Certainly, the surrounding air is in communication with this organ via the pericardial membrane, as also are the arteries and veins of which the largest orifices appear in the chambers of the heart. All these same things change the pulse of the heart analogously to the materials, and let all the things enumerated be called thus, for the sake of brevity of teaching.

5. Next, therefore, let us speak about those fevers in which the heart is hotter by a greater measure while the structures contiguous to it are colder by a lesser measure. The pulses in such conditions will become more rapid, and at the same time also larger, more frequent and weaker than those which accord with nature, but they depart from an accord with nature more towards weakness and rapidity and less towards frequency and largeness. However, the genesis of these is stable, as is the genesis of the other hectic fevers also. But if the heart should change less to heat than accords with nature, while those things contiguous with it change more to coldness, **[362K]** the pulses in

865 *De differentiis febrium* (VII.273 – 405K).
866 *De differentiis febrium* (VII.273 – 405K).
867 “the artery” is added following the manuscripts F1 and Va1.

such conditions will become smaller and more infrequent, while in rapidity they will either be similar to those which accord with nature or slower; they are similar when the contiguous structures change not much more to colder while they are slower when these deviate much more from an accord with nature towards the cold or the heart towards heat. And here too the weakness relates to the amount of the heat. If, however, the actual body of the heart is colder than accords with nature, whereas the materials [in it] and the other things contiguous to it are hotter, then here somehow all the three *differentiae* will occur. The pulses, however, will be changed in proportion to those things previously mentioned.

In the first place, let the heart be cooled to the degree that the contiguous structures have been heated; such a pulse will be slower than accords with nature whenever the change is noteworthy. Analogous to this also is that of the weakness. In magnitude, it will not be changed from an accord with nature, just as there clearly will not be anything worthy of note in frequency.

However, the distinction[868] is harder to detect in the slight deviations from an accord with nature in either direction, as was said previously, and there is particularly difficulty of recognition **[363K]** in those bodies of which we have no prior knowledge of what is normal. On the other hand, it is possible to detect a slight change of this when someone is practised in the recognition of pulses. Just as a little earlier, when the heart is hotter, while those bodies contiguous to it are colder, we considered the pulses and said the fever to be hectic, so now, if the heart is altogether colder than the whole measure according to nature, while the contiguous structures are hotter, the person will be febrile, but we don't term the fever hectic; rather we term it either ephemeral or that due to putrefaction of humors. Each has specific signs.[869] If the ephemeral fever is combined with a colder body of the heart, when each departs equally from an accord with nature, the pulse doesn't change at all in magnitude from an accord with nature, just as it clearly doesn't in frequency or rapidity before there is a notable deviation of each. It will, however, be weaker to the degree that the heart is colder.

In the fever due to putrefaction of humors, the pulse is always equal to that which accords with nature in magnitude, whereas it is not equal to those that accord with nature of the other classes, but when it does deviate a little from an accord with nature, it is nothing **[364K]** notable. However, when the deviation is clearly more, the pulse becomes more rapid in the contraction, just as it was before in the dilatation. And it is certainly also clearly more frequent in relation to the external pause, while it appears weaker to the degree that the heart is cooled. Now, conversely, let the heart be cooled still more, while the contiguous structures have been heated, and let the fever be ephemeral; the pulses of these will clearly become still more weak and

868 F2: διάγνωσις; K: διάκρισις; Kühn's Latin translation: *discrimen*.

869 Kühn's Latin translation punctuation in this and the following sentence is taken as correct, rather than the Greek, on the grounds of sense.

smaller, although not clearly similarly slow, but to the degree that the body of the heart was cooled still more than those structures contiguous to it. Furthermore, the pulses will also be infrequent and small in proportion. Contrariwise, let the fever be due to putrefaction of humors; the pulse of these will be weak, similar to that previously mentioned, but still smaller than those in relation to the dilatation and slower. However, the contraction of this will be more rapid than accords with nature. Still more will this be so in those times of the paroxysm, in which there is both a beginning and a progression, but after that, less at the peaks and still less in the abatements.

Suppose now there is the third kind of change, which is **[365K]** the heart being less changed to coldness than those structures contiguous to it are to heat, and the fever to be ephemeral. The pulses in such a body will be weaker than accords with nature by as much as the body of the heart is more *dyskratic* and will be larger and more rapid than accords with nature, being larger by as much as they are more rapid. They will also be more frequent. All these things are said of the pulses that have manifestly not changed to hardness or softness, because if in fact they are changed in this respect, their change to softness will increase altogether the magnitude and the infrequency, while sometimes it will also add slowness. If, however, the change is to hardness, there will be the opposites, smallness and frequency, and indeed also rapidity will also sometimes be added.

I certainly seem to have already stated all the *differentiae* of the fevers from pulses, if in fact we correctly attribute the existence of these to the increase of the heat in the heart. And with the fevers, I have also mentioned quite a few other conditions opposite to the fevers, in which the condition of the heart is a cooling. But I also spoke about the mixed conditions in which the mixture **[366K]** occurs from the combination of opposites with each other. And it is now time for me to pass on to some other discussion. However, since I don't know how those writing about the fevers have fallen into so great a disagreement, I have resolved because of them to take up, under the chief points, the things previously said in this book, in which [I have written] about the diagnosis of the fevers from pulses. And if the truth must be told, I have resolved to do this particularly because of Archigenes. For although he seems to be the most careful in the matter concerning the fevers, he is in danger of teaching us nothing, just like Erasistratus.

The manner of those teaching nothing is twofold: some take refuge in an unspeakable peculiarity, while others though stating names, neither preserve what they signify nor explain what they wish them to signify. Thus, the hard pulse, which Archigenes claims to be inseparable from the fevers, is only a name, not signifying any matter which can be understood scientifically and securely. I have spoken at greater length about such a pulse in the work *On The Differentiae of the Pulses*, and further also in that on the **[367K]** diagnosis,[870] but not least through these, I also explained and at the same time evaluated those things which Archigenes wrote in his book on the puls-

870 *De pulsuum differentiis* (VIII.493 – 765K) and *De dignoscendis pulsibus* (VIII.766 – 961K).

es. It is, however, necessary, it would seem, to also say something now about these under the chief points.

6. Just as among all the Greeks each of the other names indicates a specific matter, so too people gave certain names to those things falling upon the senses; such things indicate separately the different things heard, seen, smelt, tasted or touched. They do not, then, ascribe to another sense what is white, black, yellow and red, or pale yellow, grey and dark blue, nor are there other names of colours apart from those that are seen, nor in turn others [such as] sour, sweet, bitter, harsh, and astringent, or salty and salt-like that are not of things tasted. Nor are hot and cold, dry and moist, and soft and hard, of some other things; in all cases they are of things touched. If, however, someone transfers any one whatsoever of those things said to something else, the manner of such usage is called **[368K]** a metaphor. However, this certainly occurs similarly somehow in other metaphors in relation to some likeness and analogy—for example, in the case of likeness, when feet, heads and flanks of mountains are spoken of. For as in the case of animals, the feet are the lowest, while the head is the highest of all, so likewise in the case of mountains, we allow the poets to call the highest parts "heads" and the lowest "feet". Therefore, if someone, making the statement in the case of an animal, says either feet or flanks, he will expect them to neither understand the feet as all understand them, nor the flanks as is customary, but in this way understand the saying of flanks as Callimachus has said in the case of the mountain of Brilessus:[871]

The flanks of Brilesseus are pastureland which they acquired.

This would be absolutely laughable, using the metaphor in a double and reciprocal sense from those things that are not important to those that are. And in the same way too, someone, when speaking of customs or wine as hard metaphorically, not making the discussion about bodies that are touched as is customary to all, in this way expects them to understand the term, but as wine is said to be hard, **[369K]** or a person, or the custom is hard. Similar to that is this: when someone is discoursing about visible matters, then says white, he expects it to be understood in the way that is customary to all, but when Archigenes himself says certain fevers are white, it is necessary also to understand the colour white in this way. Such a person would not seem to be in his right mind, if speaking properly the name of a colour, to transfer from this in a twofold metaphor from those things they do not properly understand, and in this way, I think, whoever is making the discussion about a body that is touched will direct us to understand from the metaphor[872] the term hard. For what happens in all such discussions is that those listening learn nothing, when they are excluded from what is said properly, and are unable to understand anything else. For what other colour would someone under-

871 Callimachus of Cyrene was a prolific writer and poet who flourished in the third century B.C. Brilessus was a mountain in Attica, north-east of Athens. It is mentioned in Thucydides II.23 (ed. and Engl. transl. Smith 1956: 302).

872 K: μεταφορᾶς; F2 and Va1: διαφορᾶς here and below.

stand on hearing the name white, when excluded from understanding what was customary,[873] and was directed to understand it from the metaphor of the white fevers?

Therefore, in this way too, the hard pulse, when introduced by Archigenes in the case of a palpable matter, if it does not signify the proper matter, which is customarily said by all, but as that man wished from the metaphor of wines, **[370K]** it becomes altogether obscure and unintelligible and of such a kind as if he said *skindapsos.*[874] In addition to this too, he would be able to apply the term to anything else whatsoever in relation to the pulse characteristic of the fevers, and seem to say something, so that to men unpractised in distinctions of significations, but reasonable and trusting, because an old man having so great a reputation, who had contributed much of benefit to the art, would not be willing at anytime to bring such obscurity to the discussion as to put forward names signifying nothing. Well then, if someone were to say the pulse of a person with a fever is 'bitter' (*pikros*), thinking it right in this way to understand the term 'bitter' as we sometimes say it in the case of taste, would we grasp the concept clearly? No more than if we were to hear that the pulse is harsh, or rough, or sharp. For in all these, something unpleasant and contrary to nature is indicated. What this is, is not clear, and yet in such metaphors the transfer occurs from those things that are proper to something else that is not proper, a matter agreed upon and customary to everyone. However, in relation to the signification of the hard pulse, which Archigenes framed for himself, **[371K]** this is neither agreed upon [nor customary,][875] nor in general has occurred, or is able to occur. It is necessary to understand the signification, reverting from those things that are metaphorical to that which is proper. On which account, it is possible to hear the name but not understand the matter. But we must delay no longer on this, since we have discoursed on it at greater length elsewhere. We must take up those things previously said, going over all the fevers.[876]

7. It has been shown through other [writings] that there are in all three classes of these [fevers]. At this point, we must state what kind of diagnosis there is in each from the pulses. All the ephemeral fevers change the pulses to largeness, rapidity and frequency to a degree dependent on themselves. If some other condition is mixed, the change produced is inevitably mixed, and from that and the existence of the fever according to reason. Similarly too, the hectic fevers change the pulses to the same *differ-*

---

873 K: συνήθως; F2 and Va1: κυρίως

874 The word σκινδαψός can mean a four-stringed musical instrument, but it is also used as a nonsense word. See Galen, *De febrium differentiis*, particularly II.6 (VII.348K).

875 F1, F2, and Va1 omit "nor customary".

876 Galen's argument here is presented in a rather convoluted fashion and there is some doubt about the accuracy of the translation. What he seems to be saying, stated in simple fashion, is that when a term naming an attribute whose perception is restricted to one sense modality is applied metaphorically to a putative attribute where it is not perceived by this sense modality, confusion is likely to arise. His particular target here is Archigenes, and such terms as "white pulse." It is not clear which other work or works Galen is referring to—whether those about fevers or one of his works on language, most of which are no longer extant.

entiae to a degree dependent on themselves and in fact in these the rapidity is changed more. The fevers due to putrefaction of humors have both movements rapid in the peaks; that of the **[372K]** dilatation and that of the contraction. In the same way too, they change both the frequencies equally; that occurring in relation to the external pause and that occurring in relation to the internal pause. But in the attacks, they do not change things similarly; rather, they increase the rapidity of the contraction more. In the periods of increase, they work together on this and the external pause to the greatest extent, so that the pulse appears very frequent in relation to this. These alone are characteristic pulses of fevers by reason of themselves. All the others that befall those with fever follow some other condition rather than that of the fever. There are two classes of those conditions also; either an affliction of the capacity or of the arteries. For it was shown that it is necessary to look to these three things to discover the causes changing the pulses; organ (artery), use (need) and capacity, and it has been shown how many in each class of these, deviating from an accord with nature, are necessary to befall the pulses together.

It is not, therefore, surprising that wise doctors differ from each other regarding the characteristic pulses of the fever, and that they are all mistaken inasmuch as they neither differentiate the classes of the fevers, **[373K]** nor consider whether the changes of the pulses are due to the capacity, or the defect of the organs (arteries), or are specific to the fevers themselves. Signs are established: for example, when the paroxysmal attacks occur, in some cases the pulses are frequent and small, and in addition to these also weak, while in some cases they are also unequal,[877] and some are also slow, just as others are rapid. And yet in fact some of the fevers are "not oppressed",[878] as they call them, and make the attacks from which the primary change of the pulses occurs to largeness and rapidity, and then a little later frequency is also added to the increases, while of the other pulses mentioned, none appears in them when the paroxysm is either beginning, or increasing, or reaching a peak. But in these, neither rigors, nor shivering, nor inequality are clear, there is not cooling of extremities or of the external surface, nor is there biting, heaviness or compression of the oesophagus[879] occurring together.

In the majority, however, either one or more of the things mentioned do come together. And in some of these, vomiting is clearly apparent a little later, as if from the flow of deleterious humors to the stomach, some of them having been stung, while **[374K]** some were compressed, and in some cases nausea occurred. It is very clear in the majority of onsets the blood moves away from the whole external surface and flows to the internal organs. From this, compressions, obstructions, stoppages and dilatations of the important arteries arise, in which the unequal pulses are asso-

877 F1, F2, and Va1 omit καὶ τοὺς ἀμυδροὺς, ἔνιοι δὲ.

878 The Greek term is ἄθλιπτος (ἀθλιβής, ἄθλιβος) for which LSJ has "not oppressed" citing the present passage. Kühn's Latin translation has *incompressas,* not listed in the main Latin-English dictionaries.

879 Here again there is the issue of what στόμαχος is referring to—stomach or oesophagus. It is taken to be the latter in this instance.

ciated with the attacks of such paroxysms, just as in turn the weak pulses due to the affliction of the oesophagus and of the capacity are. In this way too, the small pulses arise due to the weakness of the capacity and due to cooling, as in fact do the slow pulses following the severe coolings. But none of these customarily make the contraction slow in like manner to the dilatation; rather, the contraction of the artery becoming more rapid is inseparable from and very characteristic of the attacks of fever due to putrefaction of humors. And anyone practised in the recognition of a contraction will have this as an inseparable sign of the onset of the paroxysm, whenever, as I often said, the pulse has a certain *tonus*. For the contraction was shown to be unrecognisable in the weak pulses. Therefore, someone who wishes to recognise accurately the onset of a paroxysm must be particularly practised in these. How someone **[375K]** might begin the practice has been stated to the fullest extent in the first book of *On the Diagnosis of the Pulses.*[880]

This sign is absolutely infallible, and one must have confidence in it, particularly when it appears clearly in relation to the diagnoses. Therefore, I myself often, in many cases in whom I had not previously palpated the pulses either when [the person was] healthy or diseased, never made a mistake when relying on this sign alone. For in all the others, whether there is cooling of the whole body, or of the cardiac orifice of the stomach, which we term catachrestically *stomachos,* or the downward movements of the blood to the depths, or occurring due to grief, fear, or any other cause, there is a similar slowness of the pulse in dilatation and contraction of the artery, while immediately the infrequency is joined with this also in relation to both the pauses. However, in the febrile attacks alone, even the dilatation sometimes becomes slower than before, but in fact the contraction gains rapidity. Also, something of the external pause is held back, so here the pulse becomes more frequent, but this does not happen in relation to the internal pause, which is either similar to what it was before or appears to a small degree more infrequent during this time. **[376K]**

It is obviously necessary to compare the previous pause, as regards the touch, with that now appearing, having called to mind that it is compounded from the end of the contraction, the beginning of the dilatation and the pause in between these, while calculating how much was associated with this to either side from the non-apparent part of the movements, and calculating this in turn by measuring it against the *tonus.* For if the pulse has a similarity of *tonus* now and previously, to the extent that the time of the perceived pause becomes longer, conclude also that the real pause has been lengthened to the same extent. If, however, the pulse were to become weaker than before, we should conclude that also something of the movements has been concealed because of this, which must be taken away from the perceived pause for the discovery of the real pause. If the pulse has become a little more *atonic* than before, but much more infrequent, we must conclude that the real pause has been extended. If, however, it is much more *atonic*, but only slightly more infrequent as regards the perceived recog-

880 *De dignoscendis pulsibus* I (VIII.766 – 822K).

nition, it must no longer be concluded that the real internal pause has been extended. In the same way, if, when the pulse takes on a slight *atonia*, **[377K]** the extension of the frequency would appear to someone to be noteworthy, one must deduce that the real pause is reduced.

From these things, then, one must deduce, as was said, that the pulse has become weaker than at the time of the diagnosis. However, in the case of those that have a similarity of *tonus*, the clearest diagnosis to those considering the matter is from these and from the internal pause, as has been said. Whenever, at some time, the pulse which is unchanged in other respects, becomes more rapid during the contraction, put this down as an inseparable sign of a beginning paroxysm. Such a pulse immediately becomes more frequent in relation to the external pause also, being stirred on by Nature to wipe away the smoky superfluity, on which account it contracted more rapidly. Therefore, in relation to the first beginning of the genesis of this, the pulse seems to change only in the rapidity of the contraction compared to before, while a little later it is also more frequent in relation to the external pause, when the artery begins to contract earlier. In the ephemeral fevers, however, this is not so, because they occur apart from putrefaction of humors. Nor does it exist in the hectic fevers, because these never begin, unless by chance some other condition comes together with them, so that in the case of the fevers in which the beginning of the paroxysms is particularly sought, **[378K]** we need to have a reliable sign. If, on the first day, when some person has just undergone a change from health to disease, we palpate one of the pulses, this sign will also immediately indicate to us the class of the fever, for when it attacks with the paroxysm, they say the fever is due to some putrefaction of the humors. However, among these *differentiae* of the fevers which are from this class, those following the inflammations were also shown by this sign. But these must be distinguished from those due to putrefaction of humors by the hardness alone, for the impact of those due to inflammations is hard. If, then, you are able to distinguish the hard impact of the artery due to tension from that due to cooling and hardening, you will have from this three prognoses of antecedent causes—a cold drink, wine that is more unmixed, and inflammation. We shall say this is not only the prior cause of the paroxysm, but also a continuation. When drinks of cold water and of more unmixed wine preceded this, the condition is also maintained.

Therefore, the walls of the arteries are hardened by the damaging cold drinks, while they are dried by **[379K]** drinking too much wine. And if the wine should affect the principle of the nerves, and these become similarly tense, they foreshadow the convulsion. However, in the inflammations, they become tense without the characteristic hardness of the wall, and at that time, the pulse is called hard, due to the forceful nature of the beat, the artery not being hardened by reason of itself, as it is when it is either dried in some way, or is strongly cooled, or acquires some kind of scirrhous condition. Let me distinguish these in this way. Those fevers which, on the first day, do not have the stated sign at all, must be separated from those due to putrefaction of humors, and must be considered as either ephemeral or hectic. Very rarely, then, does a hectic fever exist right from the beginning, although such a genesis of this fever is not impossible; anyway, such fevers were also seen by us. All the other signs of these were stated in another treatise, in which we considered the *differentiae* of the fevers.[881]

The signs from the pulses will be stated now: Neither the large, nor strong, nor hard pulse[882] ever occurs together with the hectic fevers, but **[380K]** if these features should be present in them, they are not hectic. If, however, they are all present at the same time, the fevers are clearly ephemeral, since their contraction, as was said, is not made rapid, rapidity being specific to that [fever] due to putrefaction of humors. I went over all the *differentiae* of these individually in the previously mentioned treatise. In summary, these are all those in which the paroxysms are periodic and all those which are continuous, whenever they are extended beyond the third day. Of the same class as these are also those supervening in the inflammations. Thus, in the abatements, the rapidity of the contraction is dissipated, and particularly when patients come to be afebrile, or near to this, although it is engendered in relation to the beginnings of the paroxysms, when the humors become smoky in some way, and is increased during the periods of increase.[883] However, when the fever is already near its peak, and the humors are fixed and seething, the rapidity of the dilatation takes on a sudden increase, while that of the contraction remains equal to what it was before. When the fever finally reaches its height, the dilatation comes to a peak of rapidity in relation to that paroxysm, while the rapidity of the contraction is diminished.

Therefore, the **[381K]** majority of doctors, and particularly those who say the contraction is unrecognisable by perception, do not come near to the stated sign of all the fevers due to putrefaction of humors, and the onset in each paroxysm of these. Some, but very few, seem to have seen a faint likeness of the matter. Anyway, they write vaguely and obscurely to such a degree that no one forms a picture of what they wish to make clear. One of these is Archigenes; it is possible for someone who wishes to know this to read his book about the pulses and that concerning the observed symptomatology of the fevers, or the summary of these. I have also reviewed at the same time the things that are obscure in the explanations in eight commentaries on his work on the pulses.[884] Some seem to me to have spoken excellently, if we add to the discussion itself the following distinction: the change in each of the pulses is a sign of an incipient paroxysm. The distinction which I say it is necessary to add is this: "if the change should occur apart from another cause." Some causes are external, **[382K]** and familiar not only to doctors but also to laymen, [such as] administration of foods and drinks, movement of any kind whatsoever, and a psychic state provoked

881 *De febrium differentiis* I–II (VII.273–405K).

882 "Hard" is included in parentheses in Kühn's Greek text but is not present in Kühn's Latin translation.

883 The Greek term ἀνάβασις with the primary meaning of "going up" or "away up" is listed in LSJ as having a medical meaning as above, citing Galen, *De crisibus* I.3 (IX.556K).

884 Archigenes (fl. 95–115 A.D.), a Pneumaticist, wrote a number of works apart from his oft-referred to work on the pulses, none of which is extant. Galen's commentary is listed in *De libris propriis* 5 (XIX.33K), but is also no longer extant.

in those who are angry, afraid, contentious or otherwise stirred up in any way whatsoever. Some [causes], however, start from the body itself. I spoke about these a little earlier, when I mentioned shivering, rigors and chilling and also those things flowing into the stomach. If nothing of this sort befalls the person who is sick, a sudden change of any sort heralds a paroxysm, when the pulse becomes more rapid and larger, and heat grows up in the principle of the arteries, or slower and smaller, when cold does. In this way too, frequency will be consistent with the abundance of the heat or infrequency with the cold.

And weakness [of the pulse], although being a symptom due to something else, is characteristic in the attacks of the paroxysms, when abnormal fluids flow together into the stomach under these circumstances, and these through injury of the [cardiac] orifice afflict the capacity. The inequality of the pulses, which makes inroads with many of the paroxysms, is a sign of sudden compression, or recent obstruction of arteries, or excess [of humors][885] in regard to the capacity, and those **[383K]** capacities occurring by reason of the paroxysm. Therefore, those changes of pulses brought about, either by the increase of heat in the heart, or by the inward flow of the humors, clearly seem to indicate an incipient paroxysm, whenever, as I said, other causes are absent. Those of cold, however, do not still of necessity signify an attack of fever. But even here, if we exclude all the other causes, we shall also have the cooling giving a warning of how much the fever has not yet been kindled. For it is stifled in relation to the cold of the material flowing to the depths, when this is cold enough to carry a danger of the principle being quenched. This is the kind of thing that often happens externally, when all of a sudden we throw much material that is difficult to burn on a small fire, for there is a danger of it being immediately quenched, or if it is not quenched, but holds out, the material begins to smoulder a little later, and generates a certain heat, which at first is smoky, but as time advances is also fiery, until it overcomes the material completely and creates a pure flame. What flows away from such a fire is not smoke, but **[384K]** soot and thick smoke, for both are names for the burnt remnants of the material that is carried upward, just as the remnants carried downward are called ashes and embers.

Now consider for me as analogous in the case of the fevers, that when an abundance of cold material comes to the heart in the beginning of a paroxysm, the primary danger being death, the change of the pulses to infrequency, slowness and smallness will clearly presage this danger to the greatest extent. For if significant weakness were to supervene in these [people], they would die immediately. However, if the weakness has not occurred, Nature in contending to overcome this, urges on the movement of the arteries to largeness, greater rapidity, and greater frequency so that both the movements become more rapid, but that of the contraction more obviously so, and this perceptibly increases very quickly until the paroxysm is increased, and at the time when the heat has overcome the material, a kind of flame is lit. And this is the

885 Added following Kühn's Latin translation.

peak of the paroxysms, when Nature pours out the sooty superfluity through expiration and the contraction of the arteries, while what is like ashes cannot be expelled similarly due to thickness, but is drawn off through certain perceptible **[385K]** channels. In the inflamed parts this is pus, whereas in the vessels themselves it is what settles in the urine. Much more is said about these in the treatise on the fevers.[886] In respect of the pulses however, as we said, it is inevitable that there are the changes mentioned in each time of the whole paroxysm. Furthermore, in the beginning, as we said, it is not bad to state these for every one of their changes in each direction.

With a view to alertly being on the lookout for the previously mentioned diagnoses, it is necessary to be practised in separating all the changes of the pulses from the other causes, and in looking out for them. Archigenes often lights upon [these other causes], so the teaching is prolix and at the same time confused. He often writes equally about the specific and inseparable signs of each of the times and those things that accompany them, and yet are taken to lie in the smallest enquiry. For the movement of the dilatation indicates the pure heat, while that of the contraction is as it were smoky. The other changes of the pulses occur due either to the change of the capacity or that of the arteries. **[386K]** However, each of these follow certain other *proegoumenic* causes, the majority of which follow in turn other *prokatarktic* causes. For the sake of clear teaching, let those causes arising from the body which change either the capacity or the arteries be termed *proegoumenic*. Conversely, there are those causes that exist outside the body, which were causes of the change; let these be called *prokatarchonta*. The whole method concerning these has been stated by me in other books, which I wrote about the causes of the pulses.[887] It would not be surprising, then, if someone who makes all these available to his memory, which are all the practices for readily diagnosing the pulses scientifically, will sometimes discover those that are antecedent among the causes in the case of those who are ill. Therefore, let this discussion be brought to a close here.

8. Next I shall go over the moist and dry *dyskrasias* of the heart itself and of the arteries as well, each individually and all in general. I have said something about these **[387K]** previously, but the order of the discussion demands a specific teaching of these, which I shall carry out as if nothing has been said. Sometimes, when the heart becomes drier than accords with nature, it will change all the arteries with itself, whereas sometimes, either in some one part of the animal or also in more, this will happen to the arteries. Sometimes, on the other hand, the arteries will have in themselves the opposite condition to the heart, and will also receive the condition from the latter, so there exists in them some median condition between moist and dry, just as I said earlier in the case of the *dyskrasia* relating to hot and cold. However, the *proegoumenic* causes which render the arteries drier, such as sulphurous and astringent waters, and to speak simply, those things that dry by nature, not only waters, but also medications

886 Presumably the books of *De febrium differentiis* (VII.273 – 405K).
887 The four books of *De causis pulsuum* (IX.1 – 204K).

when associated with some part of the body over a long period, in which such a *dyskrasia* occurs in relation to the arteries, will make the pulse itself harder.

Not least, in the majority of the atrophies, the arteries of the atrophic parts become drier and **[388K]** render the pulse harder. And in the heart itself it creates the *dyskrasia* related to dryness.[888] And the regimen, if followed over a long time, leads to greater dryness, and of course also the same applies to medications that are drying, and foods and drinks, among which are also unmixed wine and lack of nutriment, or in relation to some other cause, or due to widespread famine and sometimes also drinking, as in long sea voyages, and further immoderate labours, old age, anxiety, grief and insomnia over long periods, and the hectic fevers, particularly those that are marasmic. All these things mentioned, occurring either individually or as two, three or more coming together, create a drier heart, and make the pulse harder than accords with nature to the degree that the dryness is increased.

Causes of moist *dyskrasias* are the chronic oedemas supervening in the parts and the *leukophlegmatic* dropsies which also change the heart, and also of course the oedemas of the important internal organs. All these make the arteries moist, just as they also do the heart, while they make the pulses soft. However, there are causes of softness of pulses **[389K]** that are moderate and not very obvious. Such causes are meals of many moistening foods, much idleness, prolonged sleeps and baths after food. It was shown before that smallness follows the hard pulses, just as largeness follows the soft pulses. Rapidity and slowness, and in addition frequency and infrequency follow these themselves in turn. The conjunctions of the other causes were also shown. I speak now of the primary, and as it were *synektic*, causes of the genesis of the pulses—the need (use) and the capacity.

As a consequence nothing still remains of those things happening to the heart following a *dyskrasia.* Certain swellings (tumors) occur in these but do not change the pulses to hardness, just as the inflammations and scirrhosities (indurations) of other organs do not. For the animal dies beforehand, collapsing following the genesis of such swellings (tumors). Therefore, everything that relates to the heart which changes the pulses has been stated. Consequently, the time has come for me now to put an end to the present book on this, and to move on to the fourth book explaining **[390K]** the conditions of all the other parts of the body, both those occurring in relation to some *dyskrasia* and those due to swellings (tumors) contrary to nature, and what kind of pulse follows each.

---

888 The translation here follows the punctation in Kühn's Latin translation as opposed to that in the Greek text.

Book IV

1. This, our fourth book, is on the prognosis by pulses; in the first two books, we considered the generating conditions in relation to each pulse put forward, while in the third book, we sought out the characteristic pulses in each condition. We began from those things happening to the heart, among which are all fevers, as has been shown in the books about these. What, then, still remains are the conditions in all the other parts of the animal, in which the heart, being affected sympathetically in common, changes the pulses. Let us go over all these **[392K]** in order, bringing the other things into the enquiry, just as previously.

It is certainly necessary to know this one thing alone,[889] which is common to all these—that is it is never possible for the pulses in the whole animal to be brought to a common change before the heart is affected sympathetically with the first part that has been affected, although in certain parts of the body, some movements of arteries can be changed even without the harm reaching the heart. After mentioning this, let us go over first conditions that arise in the respiratory organs; then next, those in the so called trophic (nutritive) organs; then in turn, those that involve the head; and then finally, those that involve the limbs and external parts of the animal, and after that those that involve the genitalia.

2. The respiratory parts of the animal are two only; lungs and chest (thorax). The nutritive organs, on the other hand, are far more than two, being liver and stomach, intestines and spleen, the kidneys, and the two bladders.[890] The lungs are the quickest and largest of all to bring harm to the heart. For surely in position, they are nearest of all to it and to the vessels which are most similar in kind, and flow together through the largest openings **[393K]** into both chambers [of the heart]. Anyway, if at any time the lung becomes hotter than it normally is, it immediately also heats the heart, and even if it is colder, this is not delayed. In the moist and dry *dyskrasias*, unlike those mentioned, these [structures] send quicker than all the other organs to the heart. And the phlethoric diseases in the lung, particularly when it is erysipelas or inflammation that takes hold of the organ, immediately change the actions of the heart. The same applies also to those affections due to the existence of a phlegmatic flux in it, which is drawn on to sympathetically affect the heart.

But one must start from the *dyskrasias* which arise without a flux, and the first of these we must deal with is that relating to heat. Therefore, if the heart is jointly warmed pathologically with the lung that has been warmed, this produces the pulses that occur due to their need being increased. If this happens over a long time, there is a danger in this of taking on the hectic fever, the signs of which were stated previously. Certainly, the same change of the pulses will occur, as when the heart itself alone and of itself is *dyskratic*, following **[394K]** the joint affection involving the lung. And of

---

889 K: μόνον; F2 and Va1: πρῶτον.

890 The gall bladder and the urinary bladder.

course also, the kind of respiration will have the same condition, whether the heart alone is affected with a febrile *dyskrasia*, or is jointly (sympathetically) affected with the lung. However, if the one affected through the lung were to perceive greater heat in the whole cavity of the thorax, he would exhale the heat to a large extent, as also he would have a greater thirst than someone else, and would have both the pharynx and the tongue hotter, as they would for the most part also be drier. If the affliction comes to be still greater, the tongue will be tinged with a black colour and will be extremely rough.

When the lung is cooled, the heart, being jointly affected from the beginning, will bring the pulses to that change which we showed before exists when the need is dissipated, while with the progression of time, having taken on the cooling hectically will create the pulses said to occur due to this prevailing. Therefore, the other things common to those cooled in respect of the heart alone are also common to those in whom the lung is affected alone, for of necessity the one is jointly affected along with the other. What stands out as exceptional in those in whom the lung was cooled first is a perception of cold respiration. If, however, the lung also becomes *dyskratic* in relation to dryness, at the outset **[395K]** the pulses will not clearly change, whereas over time it will sometimes make them harder, when the heart is jointly affected with it. In these cases, the pharynx is readily dried, and yet people are not thirsty. The *dyskrasia* in relation to moistness makes the pharynx damp and without thirst, whereas in time it changes the pulses to softness. Such are the simple *dyskrasias* of the lung. The compound *dyskrasias* are to be recognised from these.

It is time now for me to go to the other class which I said is a product of fluxes. Common to all the pulses in this is an inequality. Blockage, obstruction and compression of the most important arteries readily occur following such fluxes. Not less than these, however, there is also the excess regarding the capacity, as the heart lies near to the lungs, and due to there being the confluence through large openings. Thus, the inequality is common to all the fluxes that rush down to the lung, and there is more or less in this, pertaining to the composition and amount of the humor occurring. For they are less with those that are thin and small in amount but more noteworthy with those that are thick and viscid and create many fluxes, **[396K]** for these obstruct, compress and weigh down more. I said before that the changes to the other *differentia*e of the pulses exist following the hotter fluxes, like those due to an abundance of heat, or following the colder fluxes, like those due to a deficiency. The pulses become harder than those which accord with nature in the dry *dyskrasia*s and the inflammatory swellings (tumors) or indurations (scirrhosities) of the lung, just as they also become softer due to the oedematous conditions. But the kind of respiration is not the same in the *dyskrasias* as it is in the fluxes. The cold *dyskrasias* make the respirations small, slow and infrequent, just as they do the pulses. Certainly, the fluxes that are cold change the pulses similarly, whereas they do not make the breathing small, slow or infrequent. For due to the confined space only a little air (*pneuma*) falls on the lung, and because of this the nature receives this deficiently, so what happens is that the second action is immediately made rapid, and at the same time also large, and in these alone of the dyspnoeas, as was also shown in the treatise about these.[891] What has happened is that the amount of the inspiration is opposite to the dilatation of the thorax, **[397K]** for the thorax is dilated very greatly, while the inspiration is not very great, just as when the lung is not obstructed.

Therefore, the form of the respiration is alike in all the fluxes. The suffocation, however, is not similar in all, but follows all the hot fluxes more, since under these circumstances Nature needs a greater inspiration. Only due to the *dyskrasias* does no danger of suffocation follow. And in the pulses themselves another *differentia* supervenes alongside those previously mentioned; in this, the flux sometimes falls on the rough arteries (trachea and upper airways) themselves alone, but sometimes on the smooth arteries and the veins, or also takes over all the spaces between these. For when the rough arteries alone are filled, no change in hardness occurs to the pulses. With the smooth, however, the arteries in the whole animal are stretched and make the pulse harder. On the other hand, the oedemas are lacking this same hardness to such a degree that they also create a greater softness, moistening over time the walls of the arteries, while creating no tension, as is the case with the inflammatory and scirrhous swellings (tumors). Accordingly, do not still seek to hear names **[398K]** of peripneumonias, asthmas or pulmonary affections, having learned the conditions of the lung. For it has been shown often in many treatises that the teaching through the names has first the dispute about the signification and second the concept of the number of the diseases.

3. But it is now time to go on to diseases of the chest (thorax) according to the same method. Certainly, the *dyskrasias* bring about the same changes as were stated a little earlier in the case of the lungs. There are swellings (tumors) in relation to amount and kind of each of the humors creating them. They differ, however, according to whether the heart is jointly affected quicker and more with the lung, or less and slower with the thorax. They also differ in that sometimes the affected part is now of the thorax, and sometimes the membrane underlying the ribs (the pleura). The tension is greater after the hard and dense parts are raised to a swelling (tumor), and because of this also, those with pleurisy always have a harder pulse than those with peripneumonia. In general, the pulses are equal in the swellings (tumors) involving the ribs, but not **[399K]** in those involving the lungs when they are unequal in all kinds of ways. For there is no commonality through great vessels or proximity of the heart to the thorax in these. Concerning the form in the dyspnea, I don't need to say anything in this particular discussion, as there was need to distinguish in the case of the lungs. Sometimes, the heart is affected first, and sometimes the lungs. The diagnosis is easy with the affected thorax.

4. If the liver is diseased, pulses change in relation to the *dyskrasias*, as previously stated, whereas in the swellings (tumors) contrary to nature, [the change] is somehow proportional in these, but the pulses are less hard when the inflammations involve the

---

891 *De difficultate respirationis* (VII.753 – 960K).

chest to the extent that they are more hard when they involve the lungs. The pulse never becomes soft with a swelling (tumor) in the case of the thorax or of the liver, because phlegmatic oedema does not ever supervene in these organs. For the thorax is too dense to receive such a flux, whereas the liver is not only too dense, but also with its innate capacity easily changes the phlegm, while the heart is more readily jointly affected with the liver, **[400K]** and indeed is situated further away than the thorax[892] through the commonality of the hollow vein (inferior vena cava).

5. There is also a sympathetic (joint) affection with the diaphragm more than all the other parts of the chest because the veins in it are from the inferior vena cava itself. No other part produces tension in the arteries in this way, and especially when the sinewy part of it is itself affected. If, then, it is stretched by the superfluities in it, apart from fevers, pain and dyspnoea supervene from the outset. As time progresses, people become febrile when the superfluities putrefy. If, however, it is from a flux attacking it, this must at all events be either yellow bile or blood that is hot and thin, for it would not at any time receive thick blood and viscid phlegm, due to the thickness. Of necessity, these people will be delirious and will have the pulse markedly hard, and because of this also, small. Because there are both these things and the considerable heat, the pulse is also very frequent. With a diaphragm that is adversely affected without a fever, the pulse is hard and small, although not to a greater extent frequent, but **[401K]** comes to such a change either altogether slightly, or not at all. With the major pleuritides everything follows analogously to those in the sinewy part of the diaphragm, because the membrane underlying the ribs (the pleura) has a share of very many nerves and is closely bound to the diaphragm, and to the membranes dividing the thorax, due to which it continually comes near to the heart. And if the pulse becomes sufficiently hard, this is the greatest sign of a fatal pleuritis, and because of this, the pulse is also small, and due to both things and the heat, very frequent. No one with such a pulse and pleuritis has been saved, although they become delirous less than those with inflammation of the diaphragm. For the outgrowth of the diaphragm is nearer to the principle of the nerves. But having set out to speak about pulses, we must not speak at length on the other diagnoses. I have distinguished such things at greater length in the treatise *On Affected Places.*[893]

6. The stomach, when adversely affected by a *dyskrasia*, will change pulses analogously to those things previously stated, but when it is due to some swelling (tumor) occurring in relation to the fundus, it will be analogous to other things. If, however, it should occur in the opening (cardiac orifice), **[402K]** the pulse will be harder due to the abundance of nerves, but not in fact hard in the way it is when the diaphragm is involved. It falls far short of this, although the part doesn't fall short in the magnitude of the nerves. But due to these nerves, it has a superfluity of sensation. For the

892 K: θώρακος; F2 and Va1: ἥπατος.

893 *De locis affectis* (VIII.1–452K; Engl, transl. Siegel 1975).

nerves are soft and large, and have their origin from the principle itself.[894] Due to the fleshy nature of the substance of the body, it is lacking in hardness compared to the pleural membrane and the diaphragm. For those parts, being stretched tight, are hard and dense, and accordingly are stretched more when raised to a swelling (tumor). And their nerves (sinews), similar to the organs themselves, are hard and stretched tight, whereas those of the stomach are soft, as I said.

Certainly, then, the pulses are very hard with involvement of the diaphragm and pleural membrane, and because of this also become small. And if some heat is joined with them, as in inflammations and fevers, they also immediately become very frequent. With involvement of the opening of the stomach,[895] the pulse is never hard in this way because of the nerves themselves or the part is strained. Nevertheless, they are often small and frequent due to the excess of sensation, because **[403K]** the capacity very quickly wears out and breaks down. For we learned what kinds of pulses pain produces, and particularly that which affects the capacity. However, just as with the pains in relation to the opening of the stomach, the capacity very quickly suffers, by the same token, it doesn't tolerate excess, even for a short time, but is weighed down and suffers, and makes the pulses unequal. It has been said in what has gone before that these kinds of pulses occur when the capacity is weighed down by abundant material, just as they also are when it is weak. Therefore, with the opening of the stomach having both these harms of the capacity associated, it is to be expected that it sometimes makes the pulses weak, small and frequent, and sometimes unequal, in the three previously mentioned classes and besides these, in that relating to rapidity and slowness.

7. And accordingly, there is no further difficulty in discovering, in the case of the remaining organs of nutrition, the changes of the pulses due to all those things previously mentioned, looked on as elements, so none of them are omitted unobserved. Those structures nearer the heart and connected through the great vessels **[404K]** will lead it more quickly and more to a a joint affection, whereas those further away and connected through smaller vessels will do so less and more slowly. In the same way too, those structures that are hard and under tension and are provided with many hard nerves readily make the pulses hard, while those that are the opposite make them soft. However, those that are sensitive will harm the capacity either way. As a result, in respect of some particular thing, there is no difficulty in discovering, in relation to each of the parts, what kind of pulse it will produce in each affection. For example, bladder and uterus do not differ from each other in position, but the bladder is harder and more sinewy. Consequently, when inflamed it also makes the pulses harder than the uterus does, whereas the latter will change them more rapidly and more to largeness and smallness and the other *differentiae*. For it has larger arteries and veins which come off the large vessels in relation to the spine much earlier than

894 Presumably the vagus or tenth cranial nerve.

895 It is not entirely clear whether the cardiac orifice or the pylorus is being referred to here.

those which come to the bladder. On the matter of pain in both organs, inasmuch as the nerve supply is equal, they are similar, but inasmuch as the body of the bladder is denser and harder, it is more [affected] by pain. For more dense bodies, **[405K]** when they are blocked up and obstructed with fluids, are more painful than less dense and soft bodies, as they make the tension stronger and the magnitude of the pain follows this. In this way, then, it is necessary to seek by scientific enquiry all the *differentiae* of the parts.

8. But for purposes of practice, I now come to the *differentiae* of the other external glands of the groins and the axillae, since in these also there is something different from all those things previously mentioned, and just as they differ from each other individually, there is, on the other hand, also something common between them and those things previously mentioned. What is common is that both are a long way from the heart, and more so with those below. However, they differ from each other in their substances and in their functions, just as something common to one another and to all the previously mentioned structures has been acquired in relation to all the aforementioned *differentiae*. This, certainly, is that of these themselves, some are harder and more dense, while others are less dense and softer; also there are those that partake of vessels and nerves that are more and larger, and those that partake of fewer and smaller vessels, while some **[406K]** happen to be more sensitive and others more insensitive.[896] No one is ignorant of the difference in the functions of these, and particularly of those in the axillae above.

Legs and arms exist for the purpose of voluntary actions and are similar to each other in class. However, eyes, ears, tongue, meninges, and the brain itself, and those parts in relation to the neck, certainly differ greatly in their functions from those parts in relation to the limbs. Accordingly, it is probable that following each there is a characteristic form of symptoms, which does not alter the pulses similarly; for example, although the brain and meninges lie in the head in the same place with all these things, following inflammations of these, people are of necessity febrile and delirious, whereas with inflammations of the other parts, this is not so of necessity. Anyway, when inflammations involve the muscles of the temples, they often bring on fevers, deliriums, convulsions and torpors, when they are made major, but not in all cases, nor do they bring such symptoms when they happen to be small.

However, the brain itself, when it falls into a morbid *dyskrasia*, of necessity brings the symptom characteristic of the *dyskrasia*. In relation to the immoderate heatings, **[407K]** if at any time these should exist alone and of themselves, this symptom is delirium. If, however, it is with dryness, there will at the same time be the insomnias. This symptom is characteristic of dryness, just as hypersomnia or coma is of moistness. Of cooling itself, the inseparable symptom is the so-called *morosis*. If, however, it occurs

896 F1, Va1 omit δ' ἀναισθητικώτερον present in K.

with moistness, deep *kataphora* follows in these people.[897] On the other hand, if it is made much more moist and hotter than accords with nature, it brings *kataphora* along with insomnia and delirium. The whole form of this has been stated in another book, which is that about coma according to Hippocrates.[898] However, strong cooling of the brain, occurring along with dryness brings immobility (*akinesia*) of the whole body; these cases are not comatose but look with open, unblinking eyes.

The pulses, on account of which these things were said, are, when the brain is very hot without dryness, moderate in the consistency of the artery, similar to those in accord with nature, but exceeding this in magnitude, rapidity and frequency to the extent that the morbid heat exceeds the natural heat. If **[408K]** it is not only hotter, but also morbidly drier, the membrane encompassing the brain, that is, the thin membrane (arachnoid/pia mater), will of course immediately also be similarly disposed. With it also all the nerves, and the heart and arteries have the pulse harder, because they are stretched in respect of the wall and dried. The sympathetic (joint) affection of the principles has been shown elsewhere. If, however, the brain is brought to a hot and dry condition from an excess of biliary humor, then the tension supervenes even more in the arteries. In this way too, from blood that is hot and as it were seething, the dryness is less, but the tension is greater in the meninges themselves and in the nerves and arteries. The bilious humor, or at least the yellow variety, which we mentioned just now, being fine-particled does not similarly cause obstruction in the soft bodies, so that it is not able to raise these to a swelling (tumor). However, blood that is hot blocks up and obstructs more, so that it also makes the swelling (tumor) longer and the tension stronger. Accordingly, we must pay careful attention to the consistency of the artery, and whether, due to dryness or tension, the wall seems hard. **[409K]** If it seems to be due to dryness, one of two things is considered to exist in relation to the brain—either a hot and dry *dyskrasia* or an excess of bile; if due to tension, an inflammatory condition. If, however, it is due to both, we must consider the condition itself to be a compound from bloody and picrocholic humor.

All such patients are delirious with a continuous delirium, but differ from one another in that some suffer from the *dyskrasia* alone, and suffer this moderately, becoming weak in all the voluntary actions, whereas others with biliary humor are not weak and are out of their minds during the strong deliriums. Those who suffer such a thing due to seething black bile from putrefaction are easily mentally deranged, while the pulses in these people are hard due to both the causes, dryness and tension. They are also, of course, immediately small due to the hardness, more rapid than those in accord with nature because the patients are febrile, and frequent because the need is not fulfilled.

---

897 The two symptoms referred to here are: μώρωσις and καταφορά. They have a variety of meanings —see the appendix on medical terms; torpor or sluggishness might suffice for the former and lethargic attack for the latter.

898 Galen, *De comate secundum Hippocratem liber* (VII.643 – 665K).

The moist *dyskrasias* in the brain render the pulses soft over time and the person lethargic. If feverish heat is joined with these as well, the disease is termed lethargy, while the pulse not only becomes soft but also large. If, however, **[410K]** such a condition occurs in the brain due to putrefying phlegm, the disease would also in this way be lethargy, while the pulse will have an appearance of tension of the artery and yet it will be soft. In all such diseases, the pulses are large to the extent that the heat is increased. The rapidity on the other hand is either altogether least or not at all increased in these, and frequency similarly. All these things increase together with the amount of the heat. Certainly, when the condition arises from a mixture of bile and phlegm, the patients are neither sleepless altogether nor in a deep sleep. When lying down, such people are delirious and unable to open their eyelids. To speak in general, such are the lethargies and insomnias. And accordingly also, the pulses in these are intermediate in form between the conditions interwoven with each other.

If morbid cooling occurs in the brain, when the heart comes to a sympathetic (joint) affection, the pulses will become smaller, slower and more infrequent than accords with nature. And if it is without material, with only the *dyskrasia* of the brain, it will be without any tension. If, however, there is material, and if this is black bile (melancholic), **[411K]** what will happen is that the arteries are under tension, and because of this, the pulse appears harder. On the other hand, if the material is phlegmatic, and not simply so, but if the phlegm is thick and viscid, they will be made tense, for such a humor produces scirrhous swellings (tumors). However, if the phlegm is moist and airy, from which particularly the so-called oedema arises, the pulse will be soft apart from tension. But if the phlegm is mixed, partaking of both the previously mentioned kinds, the pulse will be at the same time soft and with a certain tension. It is quite clear that when the pulse becomes soft, it will also be rendered larger than before. What kinds of other symptoms will occur in these, it is possible for someone to calculate for himself, starting from those things previously mentioned.

It is not necessary to delay still further here, having brought all such things to completion to the greatest extent in the treatise *On Affected Places.*[899] For now we did not set out to consider every diagnosis but only that coming from the pulse. Consequently, something was said of the other symptoms, because it was advantageous to do so in a secondary manner, not incidentally. Therefore, having now put an end to the current discussion, I shall move onto something else of use for our present purposes. **[412K]** This is something that has received a longer consideration in the work *On Affected Places*, and which it is now necessary to explain also.

9. The thin membrane (arachnoid/pia mater) which invests the brain does not seem to me to be able to suffer alone any of the previously mentioned diseases at anytime separately from the brain, just as the brain does not without the membrane. But if a *dyskrasia*, either without a flux or with a flux were to fall upon either one of these, immediately and inevitably, I think, there would be transmission of the affection to the

---

899 *De locis affectis* (VIII.1–452K).

remaining one. However, the hard membrane (dura mater), which is for most part separate from the brain is affected alone. And if the condition from this is substantial and should draw the heart and arteries into a sympathetic (joint) affection, the pulse will be changed in proportion to the conditions involving the brain, since the hot conditions in it increase the need, while the cold conditions dissipate and at the same time make this and the arteries slightly harder, and particularly those that are above the heart; the dry conditions make the pulse tense and hard, while the moist conditions make it soft, although not soft and hard to the degree that the brain is, but **[413K]** still more hard but less soft.

For those bodies that are hard in nature, if they have incurred hardening conditions, come to very great hardness, just as also in fact those bodies that are soft in nature, when they incur moistening and liquefying conditions, become completely dissipated and soft. However, when those bodies that are hard have incurred the moistening conditions, or those that are soft, the hardening conditions, they still do not come to the peak of the condition that is opposite to them. In this way, then, in oedema of the meninges, the pulse would never become as soft as it does in the case of [oedema of] the brain and lungs, nor in inflammations of brain and lungs would it become hard in this way as with the meninx, but in those that are hard due to the meninx, the pulse is similar to that due to the diaphragm and the pleural membrane, while in those due to [oedema of] the brain and lungs, none is similarly soft, but nearer to the others which are in the liver and spleen, in relation to the oedematous conditions, as when these internal organs are indurated, it makes the pulses very hard, although less in the inflammations compared to the indurations (scirrhosities). Brain and lungs would never equally incur **[414K]** a scirrhous condition, for such a condition is characteristic of hard and dense bodies and not those that are loose-textured and soft. This is known to be common to all those stated, even if does not happen to have been said in each case, and that in respect of the cold conditions, when they are made excessively so, they make the pulses hard, although not otherwise, and change those arteries that are near more to softness and hardness, whereas those that are either further from the affected places or [are affected] through the mediation of the heart less.

Someone who knows this will not need other instructions about the other structures such as those in the face, or the neck, or the limbs when they become diseased. For if the affected part is dense and hard, it will make the pulses hard, whereas if it is inflamed, indurated or excessively cooled, it will bring the arteries to a sympathetic (joint) affection. If, however, it is fleshy, and is either inflamed or indurated, it will make the pulses hard but less than in those previously mentioned conditions. If, on the other hand, it is oedematous, it will make them soft. All the arteries that are nearby and the heart will be more quickly involved in a sympathetic affection with artery-like structures, while those that are vein-like will follow these, and least those that are nerve-like. For in time these also **[415K]** are greatly affected and change the movement of the arteries.

It is not at all surprising then, that since fevers readily follow the inflammation of muscles in the epigastrium and thorax, and because of this the pulses also become large, rapid and frequent, the pulse never becomes significantly hard in these cases, or spasmodic. However, when the muscles in the forearm and shank are inflamed, and particularly in that part of them where the tendons take their origin, and are readily seized by spasm, the pulses are very easily strained and hardened. For the tissues around the thorax and abdomen are fleshy, whereas those in the hands and feet are significantly sinewy. Still more, if a sinew of those relating to the extremities of the hands or feet is inflamed, the pulse in these readily becomes hard but readily goes into spasm. However, pulses become significantly hard also due to induration of liver or spleen. But in those hard in this way, there is no fear of them being made spasmodic. Thus, in a major induration of the spleen, we often saw no dropsy supervening, whereas it follows of necessity in the liver, and the pulse under these circumstances has to a remarkable extent the signs of both the conditions, the actual wall of the arteries **[416K]** being more moist and softer, indicating clearly the tension from the induration (scirrhosity), and because of this, it is also harder. However, dropsy in and of itself, without inflammation of the internal organs, never creates tension involving the pulse. Such dropsies exist with immoderate evacuation through haemorrhoids, or unwonted stoppage. In the same manner, the stoppages of menstruation and immoderate evacuations bring dropsies. How those things that women suffer in relation to the uterus change the pulse was stated previously, when we compared that organ to the bladder.[900]

10. It is fitting then to call to mind what still remains about those things involving the testicles. For these parts, although they are external to the groins, immediately bring fevers when inflamed, due to the proximity of the vessels. From above, a vein and artery are carried down from the regions of the kidneys through the openings of the peritoneum to the testicles; these vessels are of significant size. And the *tunica vaginalis testiculi* springs from the peritoneum itself, so it is reasonable, when this is inflamed, for the pulse to become hard, and because of this also smaller, although it is not similarly hard and large **[417K]** due to fever involving the testicles. For those parts that are artery-like and vein-like stir up fevers more. On the other hand, those parts that are sinewy do so not at all or slightly, whereas they generate tetanic and spasmodic movements. Therefore, the whole matter put before us now has an end, for reviewing all the conditions in each part, we have shown what kind of pulse each change creates.

11. What is most necessary is that we attempt to bring the synopsis of these [matters] under a few headings. For surely the diagnoses of those things that presently exist is nothing other than the discovery of conditions. The diagnoses of those things that previously existed is, in turn, nothing other than that of the causes creating the conditions. Accordingly, if all the conditions are differentiated in this work, while all the ef-

900 See the present book IV.7 (IX.404K).

fective causes are stated in other books—those in the treatise preceding this on the causes in the pulses—nothing would still remain of those things necessary for prognosis, either of what previously existed or of what is presently existing. And because the prognosis of those things that are going to occur arises from those things that presently exist, as we have already shown often through what was previously said, it is now necessary to go over the whole discussion under the chief points, **[418K]** making of these this beginning which will be a starting point of the recognition of what is going to happen.

One thing and first is to what end will the disease come—whether to death or salvation. Another and second to this is in what time particularly will this be. And third in addition to these, what will the manner of death or salvation be. The prognosis of all these things is from the conditions. For from these, as we showed, the form of the disease and the weakness and strength of the capacity will be recognised. When we compare these with each other, we shall know beforehand those things that will come about. For the knowledge of the disease is from the affected place and from the condition in it; knowledge of the capacity on the other hand is from the *eukrasia* and *dyskrasia* of the solid bodies. For if some weaknesses (*asthenias*) occur due to the psychical *pneuma*, these are of short duration and incidental.[901] When we compare these with each other, if they are of the kind where the capacity is able to bear the peak of the disease, of necessity the person will be saved, as long as no mistakes are made externally. If, however, they are not of this kind, the person will at all events die.

The prognosis of the time in which each of the things mentioned **[419K]** will come to be, will be recognised through the following things. Those who are going to be saved will be saved in that time in which Nature prevails completely over the morbid condition. On the other hand, someone will die in that time in which the disease will be to such an extent stronger than the capacity that one of the functions essential to life is destroyed. Furthermore also, the manner of death and that of salvation will be known beforehand from these same things. For when the functions essential to life are destroyed, some bring a choking, some a collapse of the physical (vital) capacity, and some what is analogous to a collapse of the physical capacity and is like a kind of choking. Those that bring on a choking of the physical capacity are conditions arising from those of the thorax and lungs, and further, from the rough arteries, larynx and pharynx. Some also exist immediately in relation to the heart itself, while some that bring on collapse are from the so-called stomachical conditions,[902] from which patients collapse, and besides this, from chronic conditions involving the liver from which patients suffer atrophy, and from chronic or severe pains, and further, from all the hectic fe-

901 There is some issue about the meaning of συμπτωματικός here. Kühn's Latin translation has *symptomatum* which is not listed as such in standard Latin-English dictionaries. LSJ gives meanings of "accidental" citing Theophrastus, *De igne* 7.15.1 and "casual" citing the present passage. In the present context, "incidental" seems appropriate.

902 See Appendix 3 on medical terms on the range of meanings of this term; here these conditions should be considered to include at least those of the oesophagus and stomach.

vers. To speak collectively, **[420K]** the major *dyskrasias* of the body of the heart, which in some manner are either of themselves immediately from the onset, or are effected through other internal organs, lead to collapse of the physical capacity. [In the same way too, the weaknesses of the psychical capacity][903] follow the major *dyskrasias* of the brain, the chokings as it were follow the fillings of the cavities (ventricles) of the brain, and the obstructions of the channels coming to them.

Consequently, whoever accurately diagnoses the measure of the capacity, the affected place, and the condition in this, is the best prognosticator of those things that will come about. How one must diagnose the affected parts has been shown in other treatises.[904] How also one must diagnose the conditions of these has been jointly shown in that treatise and in those on crises, and indeed, not least in this treatise.[905] I have also spoken about the capacities often in many places and how the diagnosis is from the functions of these and from the *dyskrasia* of those things carrying out the functions.[906] I have defined the capacity of the other [functions] in another work, while regarding that which creates the pulses, which we in fact also call physical (vital), I have spoken about in these works to no small extent, and will also speak about them now. **[421K]**

12. When all the pulses happen to be weak by reason of their own capacity being weak, and if this occurs in relation to the whole body, it is that the body of the heart is *dyskratic*; when it is in one part, it is of that part alone. When it is in relation to the arteries themselves, sometimes the pulse is weak while sometimes it is strong, and not by reason of its own capacity being weak, but by its being weighed down by an excess. I have, however, never found, either in the whole body or in some one part, such an inequality alone, but the other three classes are also changed along with it and become unequal. The three classes I speak of are that in relation to large and small, that in relation to rapid and slow, and that in relation to frequent and infrequent. For large, small and intermediate, rapid, slow and intermediate, and frequent, infrequent and intermediate, just as in fact also, weak, strong and intermediate occur due to the capacities being weighed down by an excess of humors. Therefore, if the pulses are strong and large more than their opposites, the capacity has suffered less due to the excess, whereas if they are small and weak, it has suffered more. In this way too, of the **[422K]** inequalities in relation to a single beat, some were said to occur when the movement is cut through by a pause and some when it remains continuous. All those that make the ends of the movements more *eutonic*[907] and more rapid indicate

903 K: Οὕτω δὲ καὶ τίς ψυχικῆς δυνάμεως; omitted in F1, F2, and Va1.

904 Presumably *De locis affectis* in part.

905 *De crisibus* (IX.550–760K).

906 See particularly *De naturalibus facultatibus* (II.1–204K), and *De temperamentis* (I.509–694K). In addition, the four treatises on the *differentiae* and causes of diseases and symptoms (Johnston 2006) are relevant here.

907 The various meanings of the term εὔτονος include the following: vigorous, well-strung, distended, active, energetic and yielding. In the present context it probably indicates better arterial *tonus*.

the capacity is stronger compared to all those that make it weaker and slower. Enough was said about the *differentiae* in the tapering pulses.

Therefore, when at some time the pulse appears high in such a condition of pulses, it is clear that such a pulse is sufficiently large, as was said previously, and will show Nature already prevailing over the excess. A weak pulse never becomes high, although [it may become] more or less strong to the extent that it is made higher and stronger at the same time; this indicates the strength of the capacity, and that the movement is unhindered. However, immediately in such a change, it is inevitable that the previous inequality changes with it, and first that the large and high pulses become much more than their opposites, then subsequently also all equal, and of necessity, these will be determined through the excretions. **[423K]** This follows not only the high and strong pulses, but also the large together with the strong pulses. It is more certain and more trustworthy, whenever high pulses are produced. Under these circumstances, the movement of the contraction is to be considered, for you will always find it setting aside the previous rapidity, suffering this less or more. The distinction here for you will be of the customary nature of the crisis. Thus, it is better for the crisis to customarily follow, when the rapidity in the contraction is dissipated to a greater extent. If, however, the other signs of concoction also exist, (I have written all these in the work, *On Crises*)[908] such a condition presages the best crisis. If, on the other hand, the rapidity of the contraction is dissipated only a little, or also the disease appears to have not yet been concocted, the crisis will occur along with excretion and will not be altogether good; rather, there will be one or two bad crises, or also all these together at the same time. Nevertheless, when the large and strong pulse occurs, apart from the aforementioned inequality preceding it, it doesn't always signify the excretion; the high pulse always does however, but without the inequality occurring prior. If it does exist, it signifies that there will be moderate excretions. They will, however, be copious, when they are produced following many **[424K]** inequalities, indicating the excess with regard to the capacity.

Perhaps, then, it will seem remarkable that the pulse which is at the same time high and moderate in rapidity in relation to the contraction precedes the best excretions, if in fact the excretions do not follow in the dilatations of the arteries, when they draw to themselves, but in the contractions, when they expel. And it seems to me, this particularly persuaded the followers of Archigenes that the arteries draw to themselves in the contractions and expel in the dilatations. But that this is impossible has been shown in my work, *On the Use of the Pulses.*[909] Whatever the cause is, due to which the high and at the same time rapid pulse in the contraction is a harbinger of critical excretions, will be stated as follows. The actual capacity is not the so-called physical (vital), and the nutritive, and the natural. For its action is in this alone—

908 *De crisibus* (IX.550 – 760K).

909 *De usu pulsuum* (V.149 – 180K; English translation by Furley and Wilkie 1984, and in the present volume).

the genesis of the pulse. The natural capacity is to draw what is suitable for the nourishment and eliminate what is not, and with this obviously to retain and change what is suitable. Bodies need all these for nutrition, just as in fact they also need **[425K]** the arteries and heart to dilate and contract to preserve the moderation of the innate heat. Such a capacity is not found either in plants or in those animals that are cold, like oysters and those things that grow on rocks, or on some other solid bodies. However, those animals that are hot, and particularly those that are sanguineous, have such a capacity that is very obvious and very strong. This, then, in the contractions of the pulses, pours off the sooty superfluity alone, as much as would stand as a hindrance to the moderation of the innate heat, just as also during the dilatations, it draws in what cools this [heat]. The other capacity, the nutritive, draws what is suitable for nutrition and expels what is useless.

The principles of the capacities are, then, different; this was also shown in the treatise on the opinions of Hippocrates and Plato.[910] In the individual organs, however, they are intermixed with each other. Anyway, the arteries, inasmuch as they nourish, are governed by the nutritive capacity, and an action of this is excretion of the superfluities **[426K]** of the nutriment, just as that of the other, physical (vital) capacity is the excretion of the smoky superfluities. Whenever these are in excess, the arteries make the contraction very large and at the same time also rapid, and if we were to apply the finger to them from within to come to the diagnosis of the pulses, we would accurately perceive that they are carried inward to the greatest extent. However, since this is impossible, one must direct one's attention to the rapidity first and foremost, and next to the amount of the contraction, if one is sufficiently practised in recognising this. On the other hand, when these superfluities no longer exist, and the previously putrefying humors have been concocted, the superfluities of those things concocted are excreted by another capacity which is the natural.

This capacity is distributed through every part of the animal and is extended through the flesh, sinews, ligaments, and bones, and not only through the arteries and veins. Moreover, this is the same capacity that also produces the crises, and insofar as the arteries partake of this, they get rid of the superfluity. However, it is through the other capacity, the pulsatile, of which they alone partake, that they eliminate **[427K]** the smoky superfluity in the contractions, just as in the dilatations they draw what cools themselves. The need of each of these is not continuously equal, but sometimes that of the more powerful prevails. And moreover also, it may form the pulse[911] by itself, if the aspects of the capacity should be similar. Nevertheless, when the smoky superfluity prevails, it contracts to the greatest extent, whereas the heart does not dilate to the greatest extent along with the arteries, but only as much as is sufficient for cooling from the last contraction in these. But if the smoky superfluity is slightly hot and to the greatest extent free of superfluity and pure, and fiery, as it were, the arteries are

910 *De placitis Hippocratis et Platonis* (V.181–804K; Engl. transl. de Lacy 1978).
911 “the pulse” omitted in F2 and Va1.

raised maximally when dilating but contract only as much as is sufficient to fulfill the other need. Consequently, the dilatation is always equal in magnitude to the contraction, except when, by some chance, we apply our fingers suddenly, the change should occur. It differs, however, in that sometimes the movement outward prevails and sometimes the movement inward, and signifies, as was said, a certain smoky **[428K]** heat to be in the arteries, when the inward movement prevails, and a fiery and pure heat, when the outward movement does. For this reason, certainly, the high pulse is indicative of excretion, occurring after the humors producing the fever have already been concocted. However, every excretion brought about by reason of Nature always follows concoctions and in relation to the stomach, Nature appears to eliminate the superfluity at that time when it concocts. But in respect of all the other excretions, there are those that occur before the nutriment has been concocted, sometimes due to the biting by those things contained in it, and sometimes by being weighed down, it is compelled to do this before the proper time of the concoction.

On the same basis too, the capacity in the whole body, when it sufficiently concocts the characteristic humor of each part, then eliminates the superfluity, whenever in fact, as was said, nothing irritates it before the appropriate time. However, the abundance of the excretion is prognosticated from the plethoric signs occurring beforehand and from the fiery heat present. Due to this same heat, the pulses appear large and high.[912] So, for example, the worst of the excretions **[429K]** are those in which the pulse is not large and appears rapid in the contraction. If, however, the smoky superfluity should draw back even more inward, this particularly signifies that the superfluity prevails, and because of this, the disease is unconcocted. If in this the pulse is also *atonic*, there is danger of collapse. If, however, the disease is more phlegmatic, when the change is brought about by the unequal pulses to greater strength and size, the crisis is due to abscess formation, and particularly when it is already chronic, and more so still in winter, at a cold stage of life, and nature, and place. In what part of the body particularly an abscess will start cannot be known beforehand from the pulses, but as has been said in the work *On Crises*, it is necessary to consider the decisive influences of Nature, which, in the reckonings on crises, we said indicate the manner of the movement. For when the pulse is completely eliminative and there are present in it the so-called haemorrhagic signs, patients will bleed from the nostrils. **[430K]** However, when these are absent, look for the elimination from some other place. But enough has been said about all such things in the treatise, *On Crises*.[913]

912 "and high" omitted in F2 and Va1.

913 *De crisibus* (IX.550 – 760K).

Part Three: **Appendices**

I Lexica

1 Pulse Terms

1.1 Special Terms

The following 3 works are the major sources of reference:

1. Adams, Francis: *The Seven Books of Paulus Aegineta.* London: The Sydenham Society, 1844, vol. 1.
2. Karl-Gottlob Kühn (ed.), *Claudii Galeni Opera Omnia*, 20 volumes in 22 tomes (Medicorum graecorum opera quae exstant 1–20). Lipsiae: C. Cnobloch, 1821–1833. (works referred to by title followed by chapter number, then Kühn vol. and page number).
3. *Stedman's Medical Dictionary* 27th edition. Baltimore, MD: Lippincott, Williams and Wilkins, 2000.

1. Angular pulse (γωνίωσιν ἔχειν ὁ σφυγμὸς): LSJ has, "a name for a pulse coined by Archigenes apud Galen IX.324K." It is not a term Galen himself uses.

2. Bounding (gazelle-like) pulse (δορκαδίζων σφυγμός): the term is derived from the verb δορκαδίζω which means to bound like a gazelle or antelope. It has been described as the most moderate of the pulses in which the dilatation of the artery is cut in two by a pause, and also, when associated with concoction of a disease, as a harbinger of a good crisis. Adams writes: "… in the Latin translations of Avicenna and Haly Abbas, it is rendered *gazellans.* It is said that when this animal leaps upwards, it at first takes a short spring, then seems to make a sudden stop, and afterwards takes a much larger and swifter bound. This character was applied to the pulse when an imperfect dilatation of the artery is succeeded by a fuller and a stronger one" (Adams 1849: 216).

3. Clonic pulse (κλονώδης σφυγμός): the Greek term is defined as "shaking, agitation" in LSJ. Adams calls it "vibratory", and refers to Le Clerc's term, "le pouls dardant". The term "clonus" is defined in Stedman as, "a form of movement marked by contraction and relaxation [of a muscle] occurring in rapid succession. Also termed 'thrilling'." (*Stedman's Medical Dictionary* 2000: 364) Adams states: "According to Actuarius, it is hard, quick and dense, and indicates inflammation of nervous parts" (Adams 1849: 212–213). See also Galen, *Synopsis de pulsibus* 22 (IX.502–508K).

4. Dicrotic pulse (δίκροτος σφυγμός): "A dicrotic pulse is one in which the apparent retreat of the artery doesn't become complete, but there is a delay in the amount of withdrawal, or of the *pneuma* being pushed back, then a doing away with completion of the contraction that ought to happen." (*Definitiones medicae* 226 [XIX.410K] = 249

https://doi.org/10.1515/9783110612677-013

CMG [94]). In Galen, *De pulsibus ad tirones*, Galen speaks of two instances of inequality in a single pulse—bounding and dicrotic. The former starts slower and ends faster, while the latter, the dicrotic, starts faster and ends slower. Adams expands on this, writing: "An inequality may take place in respect to one pulsation, for the dilatation of the artery may be interrupted, and then completed, when it is called *dorcadissens, caprizans* or *goat-leap*; or the stroke may be suddenly repeated, when it is called *dicrotos*, which may be translated the *double, reverberating* or *rebounding* pulse." (Adams 1849: 216). In *Synopsis de pulsibus*, Galen writes: "The dicrotic and clonic pulses are linked: Nevertheless, some inequality occurs due to the wall being hardened in different parts of the artery, whenever the need of the genesis of the pulses urges and the capacity is strong. For if this itself compels dilatation, while the wall of the artery is hard, a clonic pulse is produced. Sometimes the pulse itself becomes dicrotic, when the apparent part of the artery, in so far as it is not weighed down by the bodies lying on it, is raised to the previous height, whereas by those either side of it, it is slowed in its journey and is drawn back inward again due to the hardness of the wall, and then in turn raised up. On this account also, it appears such a pulse particularly arises in those cases where the length is contracted. When it itself is long, the dicrotic pulse is never seen, just as the clonic pulse is not when it is short." (*Synopsis de pulsibus* 22 [IX.502–508K]).

5. **Formicant pulse (μυρμηκίζων σφυγμός):** the modern definition of formication is: "a form of paraesthesia or tactile hallucination; a sensation as if small insects are creeping under the skin." (*Stedman's Medical Dictionary* 2000: 704). Adams writes: "According to Galen, it is a difficult pulse to detect. He says: 'Anyway, the so called formicant pulse requires precise perception for recognition, as it falls to extreme weakness and smallness, so that it is at the beginning altogether hidden from those unpractised in palpation. Consequently, they often say people are pulseless in whom such a pulse is discovered. Accordingly, if I should contradict those who say it is rapid, and indicate the source of their error, who would be my judge? For how would I teach those who are not able to recognise at the beginning, whether this pulse exists in the sick person, what kind of pulse it is?' Haly Abbas states that it is the termination of the vermicular." (Adams 1849: 218, citing Galen, *De dignoscendis pulsibus* II.1 [VIII.823–841K]).

6. **Hectic pulse (ἑκτικός σφυγμός):** Galen's definition in *De pulsibus ad tirones* is as follows: "Hectic, like a fever, is how a pulse is named which makes no great change but remains always similar; it stays together and never resolves. The whole bodily state (*hexis*) is turned towards disease in such fevers and pulses" (*De pulsibus ad tirones* 8 [VIII.460–462K]). Adams has: "The *hectic* pulse, it is well known, is small and dense (frequent). Alexander Trallian says, small and feeble." (Adams 1849: 218).

7. **Inclining/Nodding pulse:** Galen uses two compound forms of the verb νεύω which, according to LSJ, means "to incline" in compounds. The forms are ἐπινενευκώς σφυγμός (verb ἐπινεύω) for which LSJ has "a name coined by Archigenes (apud Galen

VIII.479K)", and περινενευκώς σφυγμός (verb περινεύω), for which LSJ has "perhaps compressible pulse, Archigenes apud Galen, VIII.479K and IX.86K." The Latin terms are *innuo* and *circonnuo* respectively. In *Synopsis de pulsibus* Galen has: "These things, then, are enough about infrequency and frequency. Let us return to the congeners of the pulses we spoke about a little earlier. There is certainly something which is higher than the median, as considered in terms of length, while those on either side of this are lower. We call this pulse 'nodding' and 'inclining'. It seems this pulse arises from weak capacities, whenever the constitution of the body is such in respect of nature, that some skin is small [in amount] and light when pressed on in the middle of the length of the artery, while that on either side of this in the length is weighed down by the parts above and below. Because of this same thing certainly, the pulse of those who are obese seems short in length, even if it is strong, while the pulse of those who are thin appears long, even if it is weaker" (*Synopsis de pulsibus* 28 [IX.532–533K]). Adams has: "(from Haly Abbas) *Inclinus* autem pulses est qui sub duobus mediis movetur digitis grossus et sub extremis hinc et inde subtilis et tenuis; aut medium eius elatum est, ut extremitas demisse, videturque tangenti arteriae extrema ad inferiorem declinari." Sprengel appears to think that the *pulsus inclinus* of Haly Abbas is one first described by him, but there can be no doubt of its being identical with the *innuens et circumnuens* of Galen's system. It is likewise the same as the *pulsus inclinatus* of Rhazes, who says it occurs in hectic fevers and consumption. Galen attempts to explain this pulse by referring it to the weakness of the vessel and an unequal wasting of the parts around, which convey the impression, when several fingers are applied together, as if the vessel swells out in the middle." (Adams 1849: 217).

8. Intercident pulse (παρεμπίπτων σφυγμός): the definition in *Definitiones medicae* 229K is: "A pulse is intercident when, between two beats, it moves in relation to a specific order, from which some middle beat is intercident." (*Definitiones medicae* 229 [XIX.411K] = 252 CMG [96]) It is also called "intercurrent"—the *Oxford English Dictionary* 1933, sub verbo, has: "Of the pulse: having an extra beat" citing Floyer's 1707 work on the pulse watch, and "intercadent" for which the *Oxford English Dictionary* 1933, sub verbo, gives the general meaning of "irregular or uneven in rhythm" (*Oxford English Dictionary* 1933: 5.387), while the noun "intercadence" has the following definition applying specifically to the pulse: "An irregular rhythm of the pulse, so that there seems to be now and then an additional interposed beat." (*Oxford English Dictionary* 1933: 5.384).

9. Intermittent pulse (διαλείπων σφυγμός): "A pulse is intermittent whenever, in the accomplished contractions and dilatations of one pulse, the artery leaves out (intermits) an interval of time. Or in this way: A pulse is intermittent in the case when not only in two but in three or still more beats, the artery leaves out one dilatation, or two, or more." (*Definitiones medicae* 228K [XIX.411K] = 251 CMG [94]). In *Synopsis de pulsibus* Galen writes: "Thus, infrequency of the pulses is not good, whether it is equal or unequal. However, it is much worse when it is intermittent. For what occurs, as was shown, is that the physical capacity is burdened by the abundance of humors, or some arteries are obstructed by thickness and viscidity of humors. And this occurs due to a major inflammation, when the lumina in these arteries are narrowed, or there is some external compression, or they are weighed down, or are confused in any way whatsoever." (*Synopsis de pulsibus* 17 [IX.476–482K]). Adams states: "The intermittent and intercurrent (i. e., intercident) pulses are opposed to one another, there being a complete failure of pulsation in the one, and an accessory or superfluous one in the other." (Adams 1849: 218). In modern terms, the former equates with "dropped beats" and various forms of heart block, while the latter equates with extrasystoles.

10. Nodding pulse: see 7 above (Inclining/Nodding pulse).

11. Non-straight (crooked, oblique) pulse (ἐσκολιωμένος σφυγμός): a non-straight (crooked, oblique pulse). The Greek term ἐσκολιωμένος (Latin obliquatus), presumably comes from the verb σκολιάζω. This pulse is one which is not brought to completion in a straight artery, on account of the position.

12. Recurrent pulse (παλινδρομῶν σφυγμός): the Greek verb means go back, return or recur. It is applied by Galen to the tapering and remittent pulses. In *Synopsis de pulsibus* he writes: "Sometimes, in fact, in such pulses, when the slightly reduced dilatation is destroyed and returns again, we call them 'tapering recurring'. Therefore, with regard to extreme brevity of movement coming again to increase itself, we now have this same stated name—'tapering' ('mouse-tailed'). If, however, they come to an end at complete immobility and stay there, and will not yet have made a perceptible movement, we term such pulses 'remittent'. If, on the other hand, after the immobility, they start to move again, we call them 'remittent recurring'. It is quite clear that all such pulses occur due to weakness of capacity—sometimes when there is a specific weakness of this alone and sometimes due to some excess, or compression, or obstruction of the arteries near the heart, such as bring about a mixed condition." (*Synopsis de pulsibus* 23 [IX.508–513K]).

13. Remittent pulse (ἐκλείπων σφυγμός): "A pulse is called remittent in which, not in two only, but in three or more beats, the artery remains immobile. Or like this: remittent is when, from a large and strong beginning, the pulse always drops down more, until it has the sort of size that is called dying away (tapering); when it drops down and is smaller, as when it inclines downward to become smaller, it is called remittent." (*Definitiones medicae* 230K [XIX.411–412K] = 253 CMG [96]).

14. Serrated pulse (ἐμπρίων σφυγμός): in *De pulsibus ad tirones* Galen writes: "The pulse common to every inflammation is saw-like (serrated), as one part of the artery seems to be dilated but another does not, and this latter seems clearly to be harder. This pulse has some clonus and is rapid and frequent, although not always large. The specific characteristic of each is that, at the beginning, the pulse is larger than accords with nature, and stronger, quicker and more frequent." (*De pulsibus ad tirones* 12 [VIII.474–475K]). Adams states: "The serrated is hard, quick and unequal in its feel. Rhazes says that it is principally observed in inflammations of the pleura and diaphragm. According to Actuarius, the pulse characteristic of inflammation is serrated and hard, more especially if seated in a tendinous part. Le Clerc renders it *le pouls en manière de scie.* (*Hist. de la Méd.* p. iii, 1, iii, c. 3). It appears to correspond with the pulse called *harsh* in the works of English writers on medicine." (Adams 1849: 218–219).

15. Tapering (mouse-tailed) pulse (μύουρος σφυγμός): in relation to the pulse, LSJ gives the meaning of the verb μυουρίζω as "to die away gradually." The adjective means "tapering" (lit. "mouse-tailed"). *Definitiones medicae* has the following definition: "A tapering pulse (mouse-tailed, dying away gradually pulse) is one in which, when comparing the position, some parts of the artery are brought upward and some conversely downward. Here the distension reaches a maximum in the parts raised upward. The majority call a pulse tapering when, from the larger size of the beat that is brought about, those following in relation to the reduction are always made increasingly smaller. They are also called tapering (mouse-tailed) when, from the frequent pulse, more infrequent pulses are brought about. [A pulse] is also called tapering when, from being rapid, it becomes very much slower." (*Definitiones medicae* 225K [XIX.410K] = 248 CMG [92–94]). The most complete description is given in *De pulsuum differentiis* as follows: "Thus, suppose the second pulse is a little smaller than the first, while the third is smaller again than the second by the same amount, and also the fourth is equally smaller than the third. And suppose this occurs for several subsequent pulses; people call such pulses "tapering" and "dying away gradually",[914] transferring the name from figures that come to an end in a point. Those of them that diminish continuously, and there is not yet cessation of this affection so as to end in complete absence of movement, we call "deficient tapering". Those that do stop have two *differentiae*; some of them end first in diminution and smallness and maintain this continuously, whereas others increase again—people call these "recurring tapering. Of these same pulses, some return to a magnitude equal to that at the beginning, and some to less. And some are diminished in terms of their previous excesses, while others again are increased in these; some use lesser or greater additions. Such are some of the equal *differentiae* of the inequalities." (*De pulsuum differentiis* I.10 [VIII.523–524K]). Adams states: "When there is a succession of pulses which diminish in magnitude regularly, such a system of pulses is called *myuri*, that is to say, *decurtate* or *sharp-tailed* from their resemblance to the tapering tail of a mouse. Those who are familiar with the Scholiast on Hephaeston, and the other ancient writers on prosody,

---

914 The two terms—μειουρίζοντας (μυουρίζω) and μειούρους (μυούρος), Latin *mutilos, decurtatos*—are essentially the same, tapering like a mouse-tail or dying away gradually in relation to this passage and elsewhere in these treatises.

will readily recognise a term with which they are familiar. Galen inclines to the opinion that this state of the artery is connected with an intemperament of the portion where the expansion is limited. Of these pulses, some end in complete asphyxia, and hence are called *swooning, failing* or *fainting* myuri … and others again gradually recover their former magnitude, and are called *recurrent* myuri." (Adams 1849: 216 – 217).

16. Tremulous pulse (τρομώδης σφυγμός): a tremulous pulse is one which yields a beat that is neither equal nor unchanging but seems to change.

17. Undulant pulse (κυματώδης σφυγμός): In *Synopsis de pulsibus* Galen writes: "What is taken from the clonic pulse in relation to the inequality in different parts is very similar; the opposite relates to the quality of the wall. For the clonic is hard, while that is not hard, but is still itself softer than accords with nature. We call this "undulant" (wave-like) because it has an inequality like waves. Sometimes it seems moderately e*utonic* and sometimes between *eutonic* and *asthenic.* It was said that we call the *eutonic* pulse, "strong", following our predecessors, just as we also call the *asthenic,* "weak".[915] This wave-like pulse in general proclaims sweats, and more so to the extent that it is softer, but not *atonic.*" (*Synopsis de pulsibus* 22 [IX.502 – 508]). In *De pulsuum differentiis* I.25 – 26 (VIII.549 – 554K), Galen considers the series, undulant, vermicular, formicant. Adams states: "The *undulatory* [pulse] derives its name from a supposed resemblance to the rolling of the waves, and can only take place when the artery is particularly soft, being accompanied with an unequal elevation of it in the diastole." In his translation of Paul's text he has: "The undulatory pulse most commonly attends the more humid affections, such in particular as anasarcous swellings, lethargy and peripneumonia. When the powers of life are beginning to fail, but are still making a faint struggle, the undulatory pulse sinks into the vermicular, but when they decline to their lowest ebb, this last passes into the ant-like or formicant." (Adams 1849: 212).

18. Vermicular pulse (σκωληκίζων σφυγμός): this, in which the movement of the pulse is likened to that of a wriggling or crawling worm, is the middle term in the triad, undulant, vermicular, formicant (see undulant above). The basic characteristics of the triad are: softness, smallness, weakness and frequency. Galen, in *De pulsibus ad tirones* 8, writes: "The vermicular is when there is an impression of a worm crawling along the artery, which is stirred up like a wave, and of the whole artery not dilating at one time. If this occurs with a small dilatation, it is called "vermicular"; if it is with a large dilatation, it is called simply wave-like. It is clear that the vermicular is weak and frequent." (*De pulsibus ad tirones* 8 [VIII.460 – 462K]). In *Synopsis de pulsibus* 22, he has: "Whenever this wave-like pulse is weak, it becomes vermicular (worm-like),

915 In this passage, εὔτονος is equated with σφοδρός and ἀσθενής is equated with ἀμυδρός.

and has taken this name from the likeness it has to the movement of worms.[916] There are three of these pulses which those before us have given names corresponding to animals—bounding (like a gazelle), vermicular (worm-like), and formicant (ant-like), applying the names due to the similarity of the movement in this way. But concerning the bounding pulse, this has been shown at the end of the discussion; I have spoken just now about the vermicular, since it is smaller than the wave-like, always weak, and because of this, easily undergoes a change to the formicant, indicating a very complete weakness of capacity, just as in fact also, in the first part of the artery, the first movement seems like a creeping creature." (*Synopsis de pulsibus* 22 [IX.502–508K]). Adams quotes Rhazes as follows: "Vermiculosi forma est ut forma undosi, in elevatione quae secundum diastolem eadem invenitur, non tamen est latus, neque plenus, cujus inundatio existit debilis, qui vermibus per foramen arteriae discurrentibus assimilatur: formicans vero pulsus adeo invenitur in ultimo parvitatis, et spissitudinis consistere, ut pulsus pueri recens nati assimiletur." (Adams 1849: 218).

19. Web-like (arachnoidal) pulse (ἀραχνοειδής σφυγμός): a cobweb-like (arachnoidal) pulse is one which is small and not settled but is moved in the manner of a spider's web, oscillating due to a slight breeze.

Other abnormal pulses have been described by other writers, most notably Archigenes; Galen is scathingly critical of these descriptions and their nomenclature.

1.2 Archigenes' Terms:[917]

apokekrēmnismenos (mixed), *truzōn* (murmuring), *hygrophanēs* (moist appearing*), *karōdes* (drowsy), *bombōn* (rumbling), *ektethambēmenos* (alarming), *analuthēis* (dissipated), *atenēs* (strained), *adranēs* (feeble), *apopepēgōs* (congealed), *diapephusēmenos* (inflated), *diēgkōnismenos* (disrupted), *egkaluptomenos* (concealed*), *exōstēs* (expulsive*), *embrithēs* (weighty), *egkrekōn* (falling away*), *notulos* (moist*), *katanenagmenos* (stamping*), *dianenagmenos* (caulking*), *lagarizōn* (nudging*), *hypokleptomenos* (concealed), *hrakōdes* (ragged), *sesobēmenos* (agitated), *skuthrōpōs apoxōrōn* (sadly withdrawing) and that which befalls a drawn-out artery, and a very great number of other such names which signify something specifically and of themselves, but there is nothing common between them and the pulses, just as saying some pulse does not thrust more moistly.

916 Here there is a hiatus in the Kühn Greek text which is filled in the Latin text by *gressus*. Presumably the missing Greek word is κίνησις.

917 What follows is the list Galen gives in *De pulsuum differentiis* III.4. (VIII.662–663K). He lists 24 terms, for which a possible English translation is given, and which presumably Archigenes applied to pulses. An asterisk indicates those for which the meaning is given in LSJ.

2 General Terms

2.1 Transliterated Terms

The eight terms or groups of terms considered here are terms which have been transliterated in the translations. The main reason for doing this is the absence of satisfactory terminological equivalents in English and that attempts to devise such are likely to be misleading. These terms are essentially technical terms in Galen's concepts of anatomy, physiology and pathology and need to be understood in that context. Transliteration helps to identify them as such. In some instances, the same term remains in use, albeit with a somewhat different meaning.

Asphyxia, kakosphyxia **(ἀσφυξία, κακοσφυξία):** in LSJ, *asphyxia* is defined as "stoppage of the pulse." The term would probably be better defined as, "absence or apparent lack of pulsation" meaning that the pulse is impalpable for a certain time for whatever reason. In recent times, the term has taken on a rather different meaning. Thus, Stedman (27th edition) has, "impaired or absent exchange of oxygen and carbon dioxide on a ventilatory basis; combined hypercapnia and hypoxia or anoxia." *Kakosphyxia* is listed in LSJ as "a bad pulse." In my translation of Galen's *Hygiene* (Johnston 2018), I have used the transliterated term with "pulse abnormalities" in parentheses, with an accompanying note referring to Galen's, *On the Differentiae of Symptoms* (VII.63K), where both terms are found together—*asphyxias* and *kakosphyxias* are symptoms involving the functions of the pulses. Now I think *asphyxia* should be understood as an apparent cessation of arterial pulsation for an unspecified time, while *kakosphyxia* should be understood as a general term for abnormalites of the pulses adversely affecting their functioning.

Euchymia, Kakochymia, Plethora **(εὐχυμία, κακοχυμία, πληθώρα):** these terms apply to the amounts and states of the four humors in the body. The terms and what they represent are relevant generally to the maintenance of health and specifically to the aetiology of fatigues. Thus, a person may be described as *euchymous* if the four humors (blood, yellow bile, black bile, phlegm) are normal in quality and quantity. If there is excess of one or more, the term *kakochymia* applies; if all four are in excess, the term *plethora* applies. The terms *euchymous* and *kakochymous* are also applied to foods;[918] in this context, they can be understood as relating to the power of a particular food to produce *kakochymia* in the body. Both *kakochymia* and *plethora* can produce a disturbance of health which falls short of being an actual disease, and then come within the province of hygiene. An example is the non-spontaneous fatigues. The following quote is from Galen's *Method of Medicine:*

918 The Greek title of Galen's short work on this subject is *On Euchymous and Kakochymous Nutriments* (VI.749–815K). The Latin title is given as *De probis pravisque alimentorum succis.*

> How you must take care of the whole body when it is in a pathological state is something I have spoken about at length, both throughout what has gone before and also in the work *On Plethora.*[919] Now I shall speak of the chief points of the discussion. When the humors are increased to an equal degree to each other, [doctors] call this "abundance" or *plethora.* On the other hand, when the body is already full of yellow or black bile, or phlegm, or the serous humors, they call such a condition *kakochymia* and not *plethora. Plethora* is treated by the letting of blood, and by numerous baths, exercises and massages, as well as by dispersing medications, and, in addition, by all fastings, which I covered fully in the treatise on hygiene. *Kakochymia,* however, is treated by the specific evacuation appropriate for each of the humors in excess. There was also discussion about this in the section on prophylaxis in my work on hygiene. (*De methodo medendi* 13 [X.891–892K]).

***Homoiomeres*, Organ, Organic Part (ὁμοιομέρεια, ὄργανον):** there are several places where Galen clearly defines what he means by *homoiomeres.* The list varies somewhat but in essence it includes arteries, veins, sinews, ligaments, cartilage, bones, membranes and flesh as *homoiomerē* in humans. The defining feature is uniformity of structure. Bodies in one outline (*prosgraphē*) are often called *homoiomerē* because all their parts are similar to each other and to the whole, and they are also often called simple or primary, but may be components of organic bodies and are themselves formed from the primary elements. Finally, in *De methodo medendi* XII.6 (X.848K) he writes: "A part is *homoiomerous,* as the name itself clearly shows, which is divisible into similar parts throughout, like the vitreous and the crystalloid and the specific substance of the membranes of the eye." Galen wrote a book on the subject, *De partium homoiomerium differentia* which only survives in Arabic.[920]

The basic meaning of "organ" as instrument is retained in Galen's use of the term, which he defines as follows:

> I term an organ a part of an animal which carries out a complete function, like the eye with respect to vision, the tongue with respect to speech, and the legs with respect to walking. In this way too, artery, vein and sinew are organs and also parts of the animal. And according to this usage of terms, at least as defined not only by us but also by the Greeks of old, the eye will be termed a 'constituent part', a 'part', and an 'organ' (*De methodo medendi* I.6 [X.47K]).

As can be seen from the quotes above, there is some overlap of these terms, *homoiomerē* and organs, and this applies to arteries, for which Galen variably uses them in the pulse treatises.

***Kakoētheia* (κακοήθεια):** this term is retained in the sense of "chronic", "intractable" or "incurable". LSJ's translation for medical purposes is malignant, referring to several texts. We have avoided this rendering, given the modern connotation of the epithet when applied to an ulcer. The adjective *kakoethical* has been formed from the noun for use in the translations.

---

919 *De plenitudine* VII.513–83K.
920 Strohmaier 1970.

***Krasis, Eukrasia, Dyskrasia* (κρᾶσις, εὐκρασία, δυσκρασία):** *krasis* has a basic meaning of mixing, combining or blending. In certain contexts, it is translated as "temperament" and in fact Galen's treatise on *kraseis* is titled in Latin *De temperamentis.* In his system of medicine, *krasis* refers specifically to the mixing or blending of the four elemental qualities (hot, cold, wet and dry). If these are properly blended (this is a conceptual issue rather than an observational one, although inferences can be drawn from signs), the body or parts thereof are in a state of *eukrasia.* If there is a preponderance of one quality, or one of the four possible combinations of two qualities, this constitutes a *dyskrasia.* There are, then, four possible mono-*dyskrasias* and four compound *dyskrasias* in addition to *eukrasia.* In any given structure (whole body or body part), there can be a single mono-*dyskrasia* or a compound *dyskrasia*, or more than one type of *dyskrasia* can be present, creating an irregular (non-uniform, anomalous, unequal) *dyskrasia* in that structure. The other point of particular relevance in his *Hygiene* is Galen's identification of what he calls a "healthy *dyskrasia*"—that is, a *dyskrasia* that is compatible with normal function. Once in *Hygiene* Galen uses the term *akrasia*, but the context suggests that this is used here for ἀκράτεια, meaning debility or lack of self-control.

***Pneuma* (pneuma):** in ancient physiology (e. g., Hippocrates), this term had a number of meanings. Particularly with Erasistratus it was used to describe an air/spirit-like material, derived from the inspired air, which travelled in the arteries either alone (Erasistratus) or with blood (Galen) and was distributed throughout the body. In Galen's system, there were three forms of *pneuma*, one associated with each of the three "systems," which he describes in several places. *Pneuma* became particularly important in the physiological and patho-physiological theories of the Pneumaticist school.

***Proegoumenic*, *Prokatarktic*, *Synektic* (προηγούμενον, προκαταρκτικόν, συνεκτικόν):** The term προγούμενος-η-ον is transliterated in the text as "proegoumenic", a term created by the translator so as to avoid declension of the adjective and retain the consistency of one term throughout the text. The opening sentences of Galen's *On the Causes of the Pulses* summarise the use of these three terms in relation to the pulses: "Of the causes changing the pulses, some are causes of their genesis, while some are only causes of their change. Those of genesis are the use (need) for which they arise, the capacity by which they arise, and the organs (arteries) through which they extend. All the rest are causes of change—both those called *proegoumenic* (internal antecedent) and those called *prokatarktic* (external antecedent).[921] For the class of causes is threefold, and not in pulses alone but also in all other things. One class, which is primary and most significant, and which people call *synektic* (containing),[922] is named

921 On this Stoic-based classification of causes which Galen employs in certain instances, see Johnston 2006: 31–37, 110–112.

922 On this term, see von Arnim 1903–1905: 2.121, in addition to the references above.

from its holding together of the essence of the pulses which, as was said before, is a cause of genesis. The other two classes are not causative of the coming into being of the pulses but causes of change in those that have come into being." The following two brief definitions of the first two terms are taken from the commentary on Galen's four books on the classification and causation of diseases and symptoms prepared by J. du Bois and published in Paris in 1539.[923]

- προηγούμενος—that is, preceding or internal is an affection or movement outside nature occurring in the animal itself.
- προκαταρκτικός—that is, evident and external which, approaching externally, forcefully alters and changes bodies.

***Tonus, Eutonus, Atonia* (τόνος, εὔτονος, ἀτονία):** τόνος and its cognate verb τείνω have a range of meanings, the most basic of which relates to stretching, straining or of being under tension. The definition of *tonus* in Stedman differentiates *myogenic tonus* (contraction of a muscle caused by the intrinsic properties of the muscle or by its intrinsic innervation) from *neurogenic tonus* (contraction of a muscle caused by the influence of its extrinsic nerve supply). The *Oxford English Dictionary* 1933 has this definition pertaining particularly to arteries: "The condition or state of muscle tone; the proper elasticity of the organs," referring to Wagner's *General Pathology* of 1876 as follows: "In a reflex manner the arterial *tonus* is reduced or increased." In more modern terms, Galen's *tonus* would seem to relate to *vasomotor tonus. Eutonus* is then a good state/normal *tonus*, with *atonia/atonus* referring to lack/absence of tone. All three terms may of course have a more wide-ranging application. In one instance at least, in the pulse treatises, Galen equates εὔτονος with σφοδρός and ἀτονία/ἄτονος with ἀσθένεια.

2.2 Translated Terms

Accord with nature, non-accord with nature, contrary to nature (κατὰ φύσιν, οὐ κατὰ φύσιν, παρὰ φύσιν): the first and third of these terms are of particular importance in Galen's basic definitions. The slightly cumbersome translations are used to identify and emphasise the technical nature of these terms and are preferred over such possible alternatives as "normal/abnormal" and "natural/unnatural" although in certain non-definitional contexts one or other of these pairs may be used. The middle term above—non-accord with nature—is not used in either the *Method of Medicine* or the four treatises on diseases and symptoms. It is, however, prominent in Galen's *Art of Medicine* and is used in the pulse treatises in the same sense. The following passage from the pseudo-Galenic *Definitiones medicae* 131 (XIX.384–385K) = 132 CMG (54–56)

923 du Bois 1539: 4. See also Johnston 2006: 33–35.

clarifies the meanings of these terms and the relationship between them, and also further clarifies Galen's concepts of health and disease.

> Health is that which is in accord with nature. Disease is that which is contrary to nature. What is "natural" (φύσει) but neither "in accord with nature", nor already "contrary to nature", is like someone being very thin, or dry, or thick-set, or fat, or sharp-nosed, or grey, or snub-nosed, or grey-eyed. Those who are thus are not in a condition "in accord with nature" for they have gone beyond "balance" but neither are they "contrary to nature" for they are not hindered in function. Something that is "non-natural" (οὐ κατὰ φύσιν) is neither "contrary to nature", nor "in accord with nature", nor "natural". Examples are those who have *leukē*, leprous warts, warts and the like. Such people are not "in accord with nature" as they are outside what accords with nature but they are not "contrary to nature" as they are not impeded in the functions that accord with nature. These things are not, however, "natural" in that they are not present *ab initio*, i. e., from the initial genesis. They remain, therefore, "non-natural". What is "non-natural" by definition is close to what accords with nature and what is contrary to nature.

Action (ἔργον): we have taken *ergon* to be the action carried out by the performance of a function. On function expressed in action, Galen writes in Book 1 of *De methodo medendi:* "For it is agreed, then, in this case, by all men and not only doctors but also by those they meet, that it is [the eye's] action (*ergon*) to see. And whether I say 'action' (*ergon*) or 'function' (*energeia*) certainly makes no difference now in this case." The interchangeability of these two terms is found in the pulse treatises.

On the inter-relation of the three terms—capacity, function and action—Galen writes in *De facultatibus naturalibus* I.2: "… I shall call the action (*ergon*) what has already been brought about and 'filled up' by the function (*energeia*) of these, for example the blood, flesh and nerve. I term the active movement the function and the capacity (*dunamis*) the cause of this." Brock 1916: xxx-xxxi, in the introduction to his translation of Galen's *On the Natural Faculties* (*Physical Capacities*), summarises the use of the three terms as follows:[924] "Any of the operations of the living part may be looked on in three ways, either (a) as a *dunamis*, faculty, potentiality; (b) as an *energeia*, which is the *dunamis* in operation; or (c) as an *ergon*, the product or effect of the *energeia*."

Affection (πάθος, πάθημα): in Urmson's modern definition, relating particularly to Aristotle's *Metaphysics* V.21 (1022b15–21), an affection is "… what happens to anything that undergoes, suffers or experiences anything."[925] In this sense, an affection is not necessarily pathological. Indeed, the distinction between health, disease and affection is not altogether clear, despite Galen's statement to the contrary (*De symptomatum differentiis* I.3 (VII.44K). Nor is the distinction between affection (*pathos*) and condition (*diathesis*—vide infra) clear, although it might be said that the former is a change in progress whilst the latter is an established change.

924 Brock 1916: xxx-xxxi.
925 Urmson 1990: 126.

Capacity (δύναμις): this is an important term for Galen in general and in the pulse treatises in particular. Not only is it used frequently, but also it is found in three distinct contexts. First, in the strictly "physiological" sense, as in Galen's important and fundamental work, *On the Natural Faculties.* Second, in relation to a patient's ability to tolerate a particular treatment, and third, in relation to the strength, power or potency of medications. The term is given detailed consideration by Aristotle in *Metaphysics* V.12, where he writes:[926] "Capacity, then, is the source, in general, of a change or movement in another thing or in the same thing *qua* other, and also the source of a thing's being moved by another thing or by itself *qua* other." Phillips says of Galen's use of *dunamis* in his *On the Natural Faculties:*[927] "The notion of *dunamis* in this book is very pervasive and mostly verbal, being a development in medicine, not of *dunamis* as known in *Ancient Medicine*, but of the Aristotelian *dunamis* as potentiality contrasted with *energeia*, activity or actuality, also Aristotelian." Galen himself recognises the difficulty, writing in his work, *The Faculties of the Soul follow the Mixtures of the Body:*[928]

> Many of the wise are openly in confusion on this matter, having an incorrect understanding of 'capacity'. They seem to me to wrongly conceive of 'capacity' as something which dwells in substances, as we do in houses, not being aware that the effective cause of each thing that comes about is conceived of in relation to something else, and that there is some name for this cause as of such a thing that is separate and *per se.* But in it, in relation to what is brought about from it, the 'capacity' is of what is brought about, and because of this we say that substance has as many capacities as it has functions (*energeiai*).

Condition (διάθεσις): a modern English dictionary definition of the term "diathesis", very definitely in current medical use, is as follows: "A particular condition or habit of the body, esp. one predisposing to certain diseases."[929] This would serve for Galen's use and might favour simply transliterating the term which, for him, is a key element of several of the other definitions. We have, however, opted for "condition" as the translation.

Function (ἐνέργεια): This term is taken to represent the functional expression of the capacity (*dunamis*). LSJ lists one meaning as "physiological function" with reference to a passage in Galen's *Hygiene.* This is taken to be the sense intended in Galen's series of definitions foundational to his method in the *Method of Medicine.* The relevant passage reads as follows:[930]

> Certainly one must not determine those who are healthy and those who are diseased simply by strength or weakness of functions, but one must attribute "in accord with nature" (*kata phusin*) to

926 *Metaphysics*, 1019a18–21 (ed. and Engl. transl. Ross 1984: 2.1609).
927 Phillips 1987: 176.
928 *Quod animi mores* II (IV.769K).
929 *The Chambers Dictionary* 2008: 200.
930 *De sanitate tuenda* I, VI.21K.

> those who are healthy in contrast to "contrary to nature" (*para phusin*) to those who are diseased; that is, for the former to be a healthy condition (*diathesis*) in accord with nature effecting functions, and for the latter to be a diseased condition (*diathesis*) contrary to nature harming function.

Debru has argued that "the concept of function (from the Latin *functio*) does not occur in antiquity in the sense with which it has been used since the seventeenth century."[931] It is, however, clearly the modern concept of the term "function" that applies to what Galen is describing. Linacre, for example, in his Latin translation of the *Method of Medicine* and of the four treatises on the classification and causation of diseases and symptoms uses *actio* and *functio* interchangeably for *energeia*.

Innate Heat (ἔμφυτον θερμόν): this is a feature of all animal bodies whilst they are living. In sanguineous animals its source is in the heart from which it is distributed to the other parts of the body. Galen's views on innate heat were derived from Hippocrates and Aristotle. For those men, it was formed at the time of the organism's initial development but was maintained and cooled as required by respiration. Others (notably Erasistratus and Asclepiades) considered that the body heat was not innate but acquired from outside sources.

Putrefaction: the process of decay or fermentation. In Galen, the term is generally found in one of two contexts: in reference to abnormal change in a humor and in reference to presumed inflammatory change in a wound or ulcer. "Putrefaction" is used in both contexts although "sepsis" would serve in the second case and even "digestion" perhaps in the first.

State (ἕξις) and State (σχέσις): the distinction between "state" and "condition" is clearly expressed by Aristotle as follows:[932] "A state differs from a condition by the latter being easily changed whereas the former is longer lasting and more difficult to change. States are also conditions but conditions are not necessarily states. For those who are in states are also in a certain condition in some way on account of these, but those who are in a certain condition are not always also in a state." On σχέσις, LSJ has: "state, condition of the body, much like διάθεσις, which is alterable, opp. ἕξις (state), constitution or temperament, which is permanent," with reference to Hippocrates. So, in broad terms, state (*hexis*) and constitution are relatively stable states of the body and its components, while state (*schesis*) and condition (*diathesis*) are relatively unstable states, readily susceptible to change.

Superfluity (περίττωμα): what is over and above—particularly after the digestion of food. We have almost invariably rendered περίττωμα as "superfluity"; other possibilities include "residue", "surplus" and even "excretion" or "excrement".

---

931 Debru 2008: 265.
932 Aristotle, *Categoriae* VIII, 9a8–13 (Engl. transl. Ackrill in Barnes 1984: 14).

Symptom (σύμπτωμα): Galen provides a detailed definition of "symptom" in Book 1 of his *Method of Medicine* which prompts Hankinson to write: "This (i. e., symptom) is a type of [condition/*diathesis*], that is of a body or part of a body constituted in a certain way, but one which has no causal bearing on any of the *energeiai* (i. e., functions). It will be apparent from this that a *sumptōma* differs from its contemporary homophone, either in its general or its medically exact use (although in many cases it reasonably approximates to the former, as Galen's examples suggest)."[933] However, in his *De symptomatum differentiis* I.8 (VII.51–52K), Galen states unequivocally that "the specific characteristic of a symptom is this; it is contrary to nature." This suggests a closer correspondence with modern usage than Hankinson claims and is, in fact, how the term is used by Galen in many texts, including the pulse treatises. In effect, a definition might be that a symptom is a condition contrary to nature that does not adversely affect function. Two other, related terms are *epigennema* (epiphenomenon) and *sundrome* (syndrome). As neither is used to any extent in pulse texts, they are not considered here.

Use (χρεία): this is a term frequently used in the pulse treatises, which is important for an understanding of Galen's anatomical and physiological views generally, forming as it does part of the title of one of his major works, *On the Use of the Parts*. May's clear definition, which includes the distinction between "use" (χρεία) and "function" (ἐνέργεια), is as follows: "The Greek χρεία of the title ... does not mean function as one might naturally suppose. Function is more nearly ἐνέργεια or "action" in Galen's terms. Χρεία means for him rather the suitability of a part for performing its action, the special characteristics of its structure that enable it to function as it does."[934] In our translations of the pulse treatises, where the term is used primarily in the physiological sense, we have variably rendered it "use" or "need". These are uses III and I in LSJ's extensive entry on the word.

3 Medical Terms: Diseases and Symptoms referred to in the pulse treatises

Abscess (ἀπόστημα): this includes the modern pyogenic abscess but is not confined to this. Galen classifies it as a disease of "excess".

Apoplexy (ἀποπληξία): sudden collapse with paralysis; presumably close to modern usage as a descriptive term. Ascribed to a cold *dyskrasia* by Galen.

933 See Hankinson 1991: 152.
934 See May 1968: 1.9.

Anasarca, Katasarka (ἀνὰ, κατὰ σάρκα): a form of dropsy—see Dropsy. *Stedman's Medical Dictionary* 2000 has: "a generalised infiltration of edema fluid into subcutaneous connective tissue."

Anthrax (ἄνθραξ): listed in LSJ as "carbuncle or malignant pustule"; the possibility of smallpox is raised. Other sources describe it as an infected wound. Whether there is any connection with anthrax in the present sense is unknown.

Apnoea, Dyspnoea, Orthopnoea (ἄπνοια, δύσπνοια, ὀρθόπνοια): failure of respiration, disordered or difficult respiration, and difficulty breathing when lying flat—all as in modern usage.

Ascites (ἀσκίτης): a form of dropsy as a collection of abnormal fluid, similar to modern usage—an accumulation of serous fluid in the abdominal cavity.

Atrophy (ἀτροφία): a non-specific loss of substance; similar to modern usage.

Buboes (βουβών): glandular swellings in the groin; the equivalent to modern inflammatory lymphadenopathy or lymphadenitis.

Bulimia (βουλιμία, βούλιμος): *Definitiones medicae* has the following: "a condition in which a desire for food comes at short intervals, the extremities are feeble, failing, discoloured and cooled, the stomach is compressed and the pulse is weak." (*Definitiones medicae* 256K [XIX.418 – 419K] = 285 CMG [108]). In current usage, the term has acquired additional meaning under the name *bulimia nervosa* with forced vomiting and purging.

Catalepsy (κατάληψις): a term with several meanings including collapse, rigidity and mental disturbance.

Catarrh (κατάρροος): a nasal discharge or flux from the head.

Cholera (χολέρα): a disease marked by simultaneous diarrhoea and vomiting.

Coma (κῶμα): a reversible loss of consciousness, or perhaps hypersomnia.

Consumption (φθίσις): wasting—usually due to pulmonary tuberculosis.

Convulsion (σπασμός): listed by Galen along with apoplexy, epilepsy and tremor as a primary *dyskrasia* of cold.

Cynanche (κυνάγχη): a non-specific term for an acute infection involving the structures of the throat—i.e., tonsils, uvula, pharynx, larynx and epiglottis (this term is included for comparison with synanche—it does not itself appear in the pulse treatises).

Dropsy (ὕδερος): a general term for accumulation of fluid in the body. Four types are referred to in the pulse treatises, as follows: (i) ana/kata sarca as defined above; (ii) ascites as defined above; (iii) tympanitic ascites in which there is said to be a mixture of fluid and *pneuma* in the peritoneal cavity; (iv) leukophlegmatic dropsy as defined below.

Dyspnoea (δύσπνοια): see Apnoea.

Dysentery (δυσεντερία): probably a non-specific term to indicate frequent watery stools, possibly containing blood and mucus (similar to present usage). For Galen's classification, see his *De symptomatum causis* III,VII. 246–47K.

Elephas (ἐλέφας): taken to be the same as elephantiasis.

Elephantiasis (ἐλεφαντίασις): this is not elephantiasis in the modern sense. It may refer to true leprosy.

Empyema (ἐμπύημα): an internal collection of pus, as in present usage. The term is particularly associated with collections in the pleural cavity.

Epilepsy (ἐπιληψία): similar to present usage.

Erysipelas (ἐρυσίπελας): a disease characterised by skin discolouration (redness, yellowness). The modern erysipelas is defined as: "A specific, acute, superficial cutaneous cellulitis caused by β-haemolytic streptococci and characterized by hot, red, oedematous, brawny and sharply defined eruptions usually accompanied by constitutional symptoms" (*Stedman's Medical Dictionary* 2000). Galen considers erysipelas in Book XIII of *De methodo medendi.* The term erysipelas in Greek medical parlance may have designated various diseases that 'redden the skin' and also diffuse, purulent inflammations of internal organs, but in its commonest sense, it designated a group of skin diseases with hot, painful reddish swelling, now thought to be streptococcal dermatitis.

Fatigue (κόπος): Galen distinguishes spontaneous fatigues from the non-spontaneous variety (for example, after exercise or hard physical work). He divides the former into three basic types: wound-like, tensive, and inflammation-like. These may be combined in various ways to give four compound fatigues so there are seven possible kinds of spontaneous fatigue. Galen devotes the major part of Book 4 of his *Hygiene* to this topic.

Fever (πυρετός): as in modern usage but without the actual measurement of body temperature.

Heatstroke (ἔγκαυσις): presumably the same meaning as now. Interpreted by Galen as a hot *dyskrasia.*

Herpes (ἕρπης): a skin affliction of uncertain nature.

Hydrops (ὕδερος): see Dropsy.

Hysterical choking, Suffocation (ὑστερική πνίξ): in his *Commentary on Hippocrates, Aphorisms* V, Galen writes: "In the *Aphorism* (35) the term hysterical is unclear. Some understand it as applying to all conditions involving the uterus, while some say it applies only to those with the affection of hysterical suffocation/choking, and this is not true suffocation/choking, but apnoea." (XVIIB.824K).

Induration (σκίρρωσις): ancient usage is probably not far removed from the current sense—a hardening of tissues typically related to inflammation.

Inflammation (φλεγμονή): inflammation is discussed by Galen in detail in Book XIII of the *Method of Medicine* and is characterized by firmness, tension, redness and a throbbing sensation as well as a tendency to go on to suppuration. All are key features of inflammation as understood today. What is different is the concept of mechanism which Galen identifies as an influx of blood primarily. See Phlegmone and Phlogosis.

Insomnia (ἀγρυπνία): linked with coma by Galen as a symptom of the sensory component of the soul. Both, he says, are "… due to involvement of the 'primary sense' itself which is … common to all sensations." (see Johnston 2006: 190).

Jaundice (ἴκτερος): in *Definitiones medicae, icterus* is defined as "an effusion of bile into the skin so that the visible surface of the body is changed in colour and made yellow; the skin is harmed and the excretions from the belly (faeces) become white in those affected." (*Definitiones medicae* 276K [XIX.423K] = 276 CMG [104]). This would suffice for a definition of obstructive jaundice in modern terms.

Kataphora (καταφορά): this term has the basic meaning of "bringing down, descent or falling." LSJ lists two meanings under "medical". One is καταφορὰ κοιλίης meaning diarrhoea (with reference to Hippocrates' *Aphorisms*), and one is a lethargic attack (with reference to Hippocrates' *Epidemics*). The term is defined in Galen's *De comate secundum Hippocratem,* where its relation to coma is considered (VII.643K). We have either transliterated the term or rendered it "lethargic attack" following LSJ.

Katochē, Katochos (κατοχή, κατόχος): the basic meaning of these terms is "holding down, being overcome, possession." It later came to mean catalepsy, a term presently defined as follows: "A condition characterised by waxy rigidity of the limbs, which may be placed in various positions that are maintained for a time, lack of response to stimuli, mutism and inactivity; occurs with some psychoses, especially catatonic schizophrenia" (*Stedman's Medical Dictionary* 2000).

Lethargy (λήθαργος): *Definitions medicae* has the following definition of lethargy: "Lethargy is a *kataphora* from which it is hard to be roused, occurring with pallor and flatulent oedema, and with weakness of the solid parts themselves and of the pulsatile *pneuma.*" (*Definitiones medicae* 235K [XIX.413K] = 258 CMG [98]).

Leukophlegmatic dropsy (λευκοφλεγμασία): a form of dropsy thought to have been caused by a distribution of phlegm through the whole body.

Marasmus (μαρασμός): wasting or withering. Also applied to a type of fever.

Oedema (οἴδημα): this appears to be a general term for a type of swelling defined by its characteristics on palpation and not limited to local or generalized fluid collections. Galen lists it as a disease of *homoiomerous* bodies due to inflowing material.

Orthopnoea (ὀρθόπνοια): see Apnoea.

Paralysis (παράλυσις): similar to the modern sense but includes sensory as well as motor functions.

Peripneumonia (περιπνευμονία): see Pneumonia.

Phlegmonē (φλεγμονή): the standard term for inflammation and inflammatory swellings. It is the major topic in Book XIII of the *Method of Medicine.* The term phlegmon remained in use until recently to denote a localized inflammation.

Phrenitis (φρενῖτις): inflammation of the brain. Probably phrenitis indicates what was essentially symptomatic febrile delirium but may have included meningitis and cerebritis.

Phthisis (φθίσις): see Consumption.

Phthoe (φθόη): taken as equivalent to phthisis.

Phyma (φῦμα): tumor or tubercle. Grmek 1991: 188 writes: "The word *phuma* corresponds primarily to our terms 'abscess' and 'tubercle', but also functions as the name for certain forms of cancer and hydatid cyst."

Plague (λοιμός): essentially the same meaning as currently: "any disease of wide prevalence and/or excessive mortality" (*Stedman's Medical Dictionary* 2000). The Great Plague of Athens (430 – 427 B.C.) described by Thucydides, may have been a mixed epidemic. More recent evidence suggests it was typhoid fever. The plague in Rome, referred to by Galen, may have been true bubonic plague (due to *Pasteurella pestis*).

Pleuritis (πλευρίτις): pleurisy, as understood today although it may also have applied to lobar pneumonia.

Pneumonia (περιπνευμονία): a term for inflammation of the lungs which may equate with the modern lobar pneumonia.

Scirrhus (σκιρρός): listed in LSJ as a "hardened swelling or tumor, induration." This is similar to modern usage, i. e., "obsolete term for any fibrous indurated area, especially an indurated carcinoma" (*Stedman's Medical Dictionary* 2000).

Sunstroke: see Heatstroke.

Swooning (λειποψυχία): probably identical in meaning to λειποθυμία and meaning fainting in the modern sense.

Synanche (συνάγχη): LSJ has simply, "a kind of sore throat" with reference to Aretaeus, *Acute Diseases* 1.7, where he differentiates synanche from cynanche.

Syncope (συγκοπή): a sudden generalised collapse. Much like the present use of the term but with broader connotations.

4 Individuals and Schools

4.1 Individuals

This list is limited to those mentioned in the translated pulse treatises. Those men whose fragments are collected in von Staden 1989 are indicated by an asterisk and the page references given in parentheses at the end of their entry.

Aegimius of Elis (325–300 B.C.): little is known about him. According to the *Anonymus Londinensis* (13.21–14.3), he thought diseases arose from superfluities (*perissōmata*) which were not adequately expelled by natural excretion. He may have been the first to write on the pulses.

Agathinus of Sparta (1st century A.D.): he was a pupil of Athenaeus of Attaleia and may have taught Archigenes. Considered a Pneumaticist, he practised medicine in Rome. He may also have embraced elements of Empiricism and Methodism. None of his writings, including those on the pulses, have survived. Galen seems to have been well disposed towards him, although he doesn't escape censure for prolixity, obscurity of explanation, and undue emphasis on names rather than matters.

* **Alexander Philalethes (20 B.C.– A.D. 25):** he was a follower, and possibly a pupil of Asclepiades of Bithynia. He became head of the Asian branch of the school of Herophileans and may have taught Aristoxenus and Demosthenes Philalethes. His definitions of the pulse are known through Galen. He was also credited with influential views on invisible channels (*poroi*) and the condition of lethargy (von Staden 1989: 532–539).

Apollonides: this is probably Apollonides of Cyprus (*fl.* ca. A.D. 150) and so a contemporary of Galen, by whom he is categorised as a Methodic (*De Methodo Medendi* I.7 [X.54K]).

* **Apollonius Mus (50 B.C.-A.D. 30):** an Herophilean from Alexandria, and a student of Chrysermus, along with Heracleides of Erythrea, he wrote extensively on Herophilean physiology and medicine and is credited with more than twenty works, none of which have survived apart from his commentary on Hippocrates' *On Joints*. However, his writings are extensively quoted in other early writings, not only by Galen but also by Celsus, Soranus, Oribasius and Aetius of Amida. On the pulses, he ascribed to the view that they arose through the agency of physical and psychical capacity, as did his teacher and fellow student (von Staden 1989: 540–554).

Archigenes of Apamea (*fl.* A.D. 95–115): he practised in Rome during the reign of Trajan. He may be categorised as an eclectic with Pneumaticist tendencies, or vice-versa. He is the doctor most referred to in Galen's treatises on the pulses, and is himself credited with a work on the pulses, on which Galen wrote a commentary in eight parts. Archigenes identified eight qualities of pulses. Only fragments of his writings have survived.

Arcesilaus of Pitane (316–242 B.C.): the head of the Academy from around 268 B.C., he is associated with the introduction of Scepticism to it.

Aristotle (384–322 B.C.): although not accorded the same respect as Plato in Galen's writings, he nevertheless influenced the latter profoundly. This can be seen particularly in Galen's dedication to taxonomy, his account of causation, his concepts of bodily structure (notably the idea of *homoiomeres*), and his acceptance of teleology. Aristotle's specifically biological works were in fact of considerable importance, as was his *De anima*, although Galen disagreed with his cardiocentric account of the soul (*psyche*). In Galen's *On My Own Books* eleven works on Aristotle are listed, none of which is extant.

* **Aristoxenus (ca. A.D. 25–50):** he was an Herophilean physician who wrote a polemical work on the school, criticising prominent members, including Bacchius, Zeno, Chrysermus, Apollonius Mus and Heracleides of Erythraea, all listed here.

Asclepiades of Bithynia (*fl.* 120–90 B.C.): he developed a theory of basic structure in which fragile corpuscles (*anarmoi ogkoi*) were thought to move through non-observable channels (*poroi*) in the body. Ill-health arose when this movement was reduced or increased. He is regarded as the forerunner of the Methodist school. Although he is not credited with any work on hygiene, his treatment methods for diseases arising from variations in this process consisted largely of methods associated with hygiene—massage, diet, wine, mild exercise and moderation in personal habits. Galen mentions Asclepiades several times in his pulse treatises.[935]

Athenaeus of Attaleia (A.D. 30–70): a Greek doctor who practised in Rome, he was perhaps the founder of the Pneumaticist school of medicine. He accepted the theory of the four elemental qualities (hot, cold, wet and dry), and gave particular prominence to *pneuma* as a fifth "elemental" component of the body. He defined health as equilibrium between *pneuma* and the four elemental qualities and disease as a disequilibrium or imbalance of the five factors (i. e., a *dyskrasia*), rather than of the four only. His theoretical formulations were based on Stoic concepts, and he was linked by Galen with the Stoic, Posidonius. Whether he was, in fact, a student of the latter is uncertain, even unlikely. It seems that Galen admired Athenaeus although there were points of disagreement, both theoretical and practical.

* **Bacchius of Tanagra (250–200 B.C.):** he was a notable early Herophilean who ascribed to the school's emphasis on the value of the pulses for diagnosis and prognosis. He is most noted for his Hippocratic lexicon, but is cited by both Galen and Marcellinus in their writings on the pulses (von Staden 1989: 484–500).

* **Chrysermus of Alexandria (*fl.* 70–30 B.C.):** he was a Herophilean physician and teacher of Heracleides of Erythrae and Apollonius Mus. His definition of the pulse as a dilatation and contraction of arteries through physical and psychical *pneuma* was rejected by Galen (and others). None of his writings survive. (von Staden 1989: 523–528)

Chrysippus (250–205 B.C.): Galen's attitude to the Stoics in general may be characterised as ambivalent. He has recorded his own early exposure to Stoic doctrines and, indeed, his first written work was said to have been a commentary on a work by Chrysippus: "When I was still a boy and my father first handed me over to someone who taught the logical theory of Chrysippus and the notable Stoics, I prepared, in relation

935 For a general account of Asclepiades' theory and therapeutic methods see Vallance 1990.

to this, notes of the syllogistic books of Chrysippus…" (Galen, *De libris propriis* 11 [XIX.43K]). In fact, the five works on the Stoics which Galen lists in *On My Own Books* all bear on logic and related matters.[936] Of the Stoics, it is predominantly Chrysippus who is mentioned in Galen's writings, although his two predecessors, Zeno of Citium and Cleanthes, as well as the later Stoic, Posidonius, are also mentioned.[937] The ambivalence referred to may be revealed by comparing Galen's observations on Chrysippus in the *Method of Medicine* with those in *On the Opinions of Hippocrates and Plato.* In the former, the Stoic is linked with Hippocrates, Plato and Aristotle in espousing what Galen himself accepts as the correct explanation of matter, whereas, in the latter he is roundly criticised in several passages.[938]

* **Demetrius of Apamea (ca. 200 – 100 B.C.):** an Herophilean physician, he is cited by a number of later authors apart from Galen, including Soranus and Caelius Aurelianus (von Staden 1989: 506 – 511).

Democritus of Abdera (440 – 380 B.C.): he was one of the founders of Atomism as concerns the basic structure of matter, including of course the human body. It is possible, however, that the man referred to by Galen is pseudo-Democritus (ca. 150 – 50B.C.). There are no extant writings.

* **Demosthenes Philalethes (50 B.C.-A.D. 25):** one of the last identifiable members of the Herophilean school, he is most noted for his work on ophthalmology. However, as von Staden writes, he was, "… not a narrow specialist. He also wrote a work in three volumes on pulse theory, aligning himself closely with the sphygmological views of his teacher, Alexander Philalethes but without merely repeating Alexander's definition of the pulse. Like Alexander, he offers two pulse definitions, one 'subjective' or from the point of view of the observer's experience, the other 'objective' or from the point of view of what 'actually happens' in the arteries and the heart" (von Staden 1989: 570 – 578).

Dion, also known as Dion (Medicus): essentially nothing is known about him. This is reflected in the entry where his dates are given as ca. 120 B.C.-A.D. 120. He is, however, also referred to by Soranus, Oribasius and Aetius of Amida.

---

936 *De libris propriis* 11 (XIX.47– 48K).
937 All three early Stoics are mentioned together in *De placitis Hippocratis et Platonis* (V.283K). Posidonius is quoted contra Chrysippus in the same work (V.397K).
938 The passage in the *De Methodo medendi* (X.16K) is as follows: "For Hippocrates first put forward the hot, cold, dry and moist, whilst Aristotle demonstrated [these] after him. And the followers of Chrysippus took these up as already given and did not dispute, but said that all things are mixed from these, and that these affect and act on each other and that Nature is systematic. They accept all the other doctrines of Hippocrates about Nature, apart from there being some small differences between them and Aristotle." On the criticisms, see de Lacy 1973: 33.

Erasistratus (*fl.* 260 – 240 B.C.): he provides quite a contrast to his fellow Alexandrian, Herophilus, in terms of Galen's later evaluation. Whilst he too made significant contributions to anatomy, mainly with regard to the cardiovascular system and peripheral nerves, in a number of his theoretical formulations, he was strongly criticised by Galen. First, in his physiology and pathology, he moved away from the concept of four elemental qualities, basing his physiology on a corpuscular theory following, it is said, Straton of Lampsacus. Longrigg 1993: 214 writes that Erasistratus, "...conceived of his particles as very small imperceptible corporeal entities partially surrounded by a vacuum in a finely divided or discontinuous condition." In his explanation of disease he gave particular importance to blood and *pneuma*, invoking the concepts of *plethora* and *paremptosis*. In short, what was involved in these presumed pathological processes was an increase in blood in the veins to an abnormal level causing a spill-over into the arteries with a resultant displacement of *pneuma*. Other aspects of Erasistratus' theorising to which Galen took exception were his concept of *horror vacui* (or "following toward what is emptied"), his theory of digestion, his departure from the idea of "complete" teleology, and some of his attitudes to therapy. Also, on the issue of causation, there were certainly substantial differences, although it may be said that Erasistratus did clearly accept the need for causal explanation.[939]

Favorinus (Φαβωρῖνος) (A.D. 85 – 155): he was a philosopher categorised as a Sophist. Born in Arelate, he learned Greek in Marseilles, and worked exclusively in that language throughout his career. He may have studied with Dio Chrysostom in Rome.

* **Hegetor (220 – 50 B.C.):** an Herophilean doctor cited by both Galen and Marcellinus (*On Pulses* 3) (von Staden 1989: 512 – 514).

* **Heracleides of Erythrae (ca. 30 B.C.-A.D. 30):** he was a Herophilean physician and pupil of Chrysermus (von Staden 1989: 555 – 558).

Heracleides of Tarus (95 – 55 B.C.): he was an Empiricist physician, praised by Galen. A considerable number of fragments of his writings survive.[940] He wrote on the pulses in opposition to the Herophileans, Baccheus and Zenon.

Herophilus of Chalcedon (*fl.* 280 – 260 B.C.): along with his close contemporary Erasistratus, he has been classified as a Dogmatic, although with Herophilus at least this is not entirely uncontroversial. Both also have been subjects of recent detailed studies and collections of *fragmenta*.[941] Although precise dates have not been established for either, it is probable that Herophilus was slightly earlier. His primary contribution

939 See Garofalo 1988 for the edition of the fragments.
940 Edition by Deichgräber 1930: 172 – 204 (fragments), 258 – 261, and more recently Guardasole 1997.
941 von Staden 1989: 115 – 124 for the edition of the fragments.

was to anatomy, especially of the nervous system, liver and heart. From the viewpoint of his importance for Galen, he subscribed to the same physiology based on the four elemental qualities with a significant role for *pneuma*, the latter being importantly implicated in neurological function by Herophilus. He also played a major part in establishing "pulse theory", a subject enthusiastically and extensively taken up by Galen. In terms of methodology, Herophilus argued for attention to be directed to phenomena but did, it is thought, accept the importance of causal explanation. It may be said that, if he was indeed a Dogmatic, he had significant Empiric tendencies and was arguably less committed to detailed causal explanation than Galen was subsequently.

Hippocrates of Cos (440–370 B.C.): it is now recognised that the *Hippocratic Corpus* was compiled over an extended period of time, most probably from the middle of the 5th century B.C. to the middle of the 3rd century B.C. from writings by various people including Hippocrates himself.[942] Regardless of the origins of the various works, Galen clearly identified Hippocrates as the major precursor of his own views, which he saw himself as systematising and extending where necessary. There is, however, little about the pulses in the Corpus, and the references to Hippocrates in Galen's writings on the pulses are about other matters related to his discussions of various issues pertaining to the pulses.

Lysias (ca. 100 B.C.-A.D. 35): he is considered to have been an Asclepiadian. Little is known about him apart from his medicinal preparations.

Magnus of Ephesus (A.D. 50–100): he has been classed as both a Methodist (Caelius Aurelianus) and a Pneumaticist (Galen). He is thought to have paid particular attention to the heart, vascular system and pulse. With regard to the last, he attached great importance to magnitude, strength, rapidity and fullness.

Moschion (ca. 90 B.C.-A.D. 80): mainly noted for his pharmacology, he is regarded as a follower of Asclepiades. He held that "pulsation ... arising from the heart, veins, arteries and the brain emerges as a single pulsation via the meninges."

Phillipus of Rome (A.D. 45–95): he is often mentioned in conjunction with Archigenes, both identified as Pneumatists. He is said to have written extensively on medications, although none of his writings survive.

Philonides the Sicilian (10 B.C.-A.D. 25): a physician from Catina, he is known from brief quotations in other works. He is said to have produced works in both Latin and Greek.

942 For an excellent summary of the *Hippocratic Corpus*, including issues of authorship, see Craik 2015.

Plato (390–348/7 B.C.): he was, on Galen's own say-so, the major influence on his philosophical views, both generally and as they related to medicine. For Galen, Plato stands together with Hippocrates on the highest pedestal. De Lacy writes: "Plato is repeatedly praised. He is first among philosophers, as Hippocrates is the best of all physicians. Like Hippocrates, he is 'divine.' He is a member of the 'chorus' that is closest to God, whose members are devoted to the pursuit of the highest arts and sciences and are honoured equally with the gods."[943] In the pulse treatises Plato is only referred to briefly and unimportantly.

Praxagoras of Cos (ca. 325–275 B.C.): from a notable medical family, relatively little is known about him. None of his works survive. He has been identified as a follower of Diocles of Carystus and perhaps a teacher of Herophilus,[944] giving him particular relevance in the early studies of anatomy. He espoused theories which attracted Galen's criticism. This was certainly justifiable in relation to the cardiocentric view of mental and emotional function, shared by other notables including Aristotle, and also to his belief that the arteries carried *pneuma*, a concept developed by Erasistratus in his theories of disease causation. On the positive side, however, he is credited with having made the structural distinction between arteries and veins. His particular similarity to Galen, who identified him as a Dogmatic, was his belief in the importance of seeking a causal explanation for disease. Indeed, he is said to have written a book on the subject.[945] In addition, there are his early studies of the pulse and its abnormalities, another subject substantially developed by Galen, and his incorporation of *pneuma* into theories of physiology and pathology.[946] Praxagoras also expanded on the concept of humors, subdividing the basic four on the grounds of colour, taste and other aspects to make ten. He particularly associated disease with alteration of the humors and specifically fever with putrefaction of humors.

Pyrrho of Elis (360–279 B.C.): he is known as the founder of Scepticism. As a matter of principle, he wrote nothing. Timon of Phlius (*vide infra*) is the main source of information on Pyrrho's views.

Themison of Laodicea (90–40 B.C.): he is, in one place, credited by Galen as being the founder of the Methodic sect and as formulating the concept of "communities" (of diseases). He may have studied under Asclepiades but may also have been linked with the Empiric, Theodas of Laodicea, whom Galen mentions once in the *Method of Medicine* II.7 (X.142K). Themison is credited with developing certain compound medications, particularly utilising beeswax. Four books attributed to him (but no longer extant) are *Acute Diseases*, *Chronic Diseases*, *Rules of Health* and *Letters*.

---

943 See de Lacy 1973, esp. 32–33.
944 See Steckerl 1958: 34 on this matter.
945 See Steckerl 1958: 5.
946 See Lewis 2017.

Timon of Phlius (ca. 290–240 B.C.): an exponent of the ideas of Pyrrho of Elis, he is said to have written prolifically but only fragments of his works survive. Striker 1999: 1529 writes: "In view of the connection between Pyrrhonism and Empiricist medicine, it is perhaps significant the Timon had his eldest son, Xanthus, trained as a doctor."

Zeno of Citium (335–263 B.C.): he is regarded as the founder of Stoicism and taught at the Stoa Poecile. Two aspects of Stoicism that had a particular relevance to medicine were its empiricism and the central place given to *pneuma* in physiology.

*** Zeno(n) the Herophilean (ca. 200–150 B.C.):** he is possibly Zeno of Laodicea, most noted as a pharmacologist, and cited in Celsus and Galen. (von Staden 1989: 501–505).

4.2 Sects or Schools

By Galen's own time, Roman medicine seems to have become divided into four main sects or schools—Empiricists, Rationalists, Methodists and Pneumaticists. Quite how important these recognised divisions were is hard to determine accurately, given that surviving evidence is limited to very few contemporary sources, the two notable ones being Celsus (1st century A.D.), and Galen himself. The latter certainly attached considerable significance to these sects, writing two works specifically on the subject —*On the Sects for Beginners* and *On the Best Sect*—and mentioning them frequently in other works. Moreover, in his late work, *On My Own Books*, he writes, regarding his *On the Sects for Beginners*, that this "...should be the first book to be read by students of the art of medicine."[947] Although Galen doesn't discuss the issues of the sects or schools in any detail in the pulse treatises, there is frequent reference to one or other of them. The essential features of each of these schools are as below.

Rationalism/Dogmatism

- A theory of the basic structure of matter which is applicable to the human body.
- A theory of the nature of health and disease in terms of concepts of structure and function founded on the theory of basic structure.
- A detailed knowledge of human anatomy.
- Acceptance of the idea of causation—that all events have a cause or causes which are at least potentially identifiable.
- Acceptance of the view that measures to maintain health and cure disease can be determined theoretically (at least in part) on the basis of this theoretical foundation.

947 The two books referred to are: *De sectis ad eos que introducuntur* (I.64–105K) and *De optima secta* (I.106–223K). The latter is now regarded as spurious. The reference to *De libris propriis* is XIX.12K (Engl. transl. Singer 1997: 4–5).

Empiricism

- There is no need of a foundational theory of structure and function or a detailed knowledge of anatomy.
- There is no need to search for causal explanations—indeed, this is not only unnecessary but fruitless.
- Recognition of diseases and their treatment is based on experience (*peira*), observation (*tērēsis*), history or inquiry (*historia*—or, in medicine, collected case histories), and inference from analogy (*metabasis*).

Methodism

- A theory of basic structure which involved identical atoms/corpuscles (*anarmoi ogkoi*) and channels/pores (*poroi*), inaccessible to observation, through which the corpuscles move.
- The view that disease was due to disturbance of this normal movement—either constriction of the channels obstructing flow, or dilatation of the channels allowing excessive flow, or as a third possibility, coexistence of constriction and dilatation giving a combined condition.
- The view that these states were readily recognisable and were the basis for treatment.
- The claim that all parts of the body were similarly affected by these processes and that there was no need for a detailed knowledge of anatomy.

Pneumaticism

- Acceptance of the four element/four quality theory of basic structure, as in Rationalism.
- Postulation of particularly important role for the *pneuma* in health and disease.

II Bibliography

1 Sources

1.1 Galen

Opera omnia

Kühn: Karl-Gottlob Kühn (ed.), *Claudii Galeni Opera Omnia*, 20 volumes in 22 tomes (Medicorum graecorum opera quae exstant 1–20). Lipsiae: C. Cnobloch, 1821–1833.

Single works

Ad Glauconem de methodo medendi: XI.1–146.
Engl. transl.: Johnston 2016: 321–560.
Ars medica: **I.305–412.**
Engl. transl.: Johnston 2016: 137–320.
De bono habitu: IV.750–756
Engl. transl.: Johnston 2020: 452–461.
De causis pulsuum: IX.1–204.
De comate secundum Hippocratem: VII.643–665.
De constitutione artis medicae: I.224–304.
Engl. transl.: Johnston 2016: 2–136.
De crisibus: IX. 550–768.
De differentiis febrium: VII.273–405.
De differentiis morborum: VI.836–880.
Engl. transl.: Johnston 2006: 180–202.
De difficultate respirationis: VII.753–960.
De dignoscendis pulsibus: VIII.766–961.
De inaequali intemperie: VII.733–752.
Engl. transl.: Johnston 2020: 282–325.
De libris propriis: XIX.8–48.
Engl. transl.: Peter N. Singer, *Galen: Selected Works* (World's Classics). Oxford: Oxford University Press, 1997, pp. 3–22.
De locis affectis: VIII.1–451.
Engl. transl.: Rudolph E. Siegel, *Galen on the Affected Parts. Translation from the Greek Text with Explanatory Footnotes.* Basel: S. Karger, 1976.
De marcore: VII.666–704.
Engl. transl.: Theoharis C. Theoharides, "Galen on marasmus", *Journal of the History of Medicine and Allied Sciences* 26 (1971), pp. 369–390.
De methodo medendi: X.1–1021.
Ed. and Engl. transl.: Ian Johnston, and Greg H. R. Horsley, *Galen, Method of Medicine*, 3 vols (Loeb Classical Library 516–518). Cambridge, MA: Harvard University Press, 2011.
De naturalibus facultatibus: II.1–214.
Engl. transl.: Arthur J. Brock, *Galen On the Natural Faculties* (Loeb Classical Library 071). London: William Heinemann, and Cambridge, MA: Harvard University Press, 1916.

https://doi.org/10.1515/9783110612677-014

De optima corporis nostri constitutione: IV.737 – 749.
Engl. transl.: Robert J. Penella and Thomas S. Hall, "Galen's *On the Best Constitution of Our Body*", *Bulletin of the History of Medicine* 47 (1973), pp. 282 – 295;
Engl. transl. Johnston 2020:434 – 451.
De placitis Hippocratis et Platonis: V.181 – 805.
Ed. and Engl. transl.: Phillip De Lacy, Galeni *De placitis Hippocratis et Platonis* (Corpus Medicorum Graecorum V 4,1,2). Berlin: Akademie Verlag, 1978.
De praesagitione ex pulsibus: IX.205 – 430.
De pulsibus ad Antoninum: XIX.629 – 642.
De pulsibus ad tirones: VIII.452 – 492.
Engl. transl.: Peter N. Singer, *Galen: Selected Works* (World's Classics). Oxford: Oxford University Press, 1997, pp. 325 – 344.
De pulsuum differentiis: VIII.769 – 941.
De sanitate tuenda: VI.1 – 452.
I–IV: Engl. transl.: Ian Johnston, *Galen. Hygiene, Books 1 – 4* (Loeb Classical Library 535). Cambridge, MA: Harvard University Press, 2018.
V–VI: Engl. transl.: Ian Johnston, *Galen. Hygiene, Books 5 – 6, Thrasybulus, On Exercise with a Small Ball* (Loeb Classical Library 536). Cambridge, MA: Harvard University Press, 2018.
De sectis ad eos qui introducuntur: I.64 – 105.
Engl. transl.: Richard Walzer and Michael Frede, *Three Treatises on the Nature of Science*, Indianapolis: Hackett Publishing Company, 1985, pp. 4 – 20.
De symptomatum causis: VII.85 – 272.
Engl. transl.: Johnston 2006: 203 – 304.
De symptomatum differentiis: VII.42 – 84.
Engl. transl.: Johnston 2006: 180 – 202.
De temperamentis: I.509 – 694.
Engl. transl.: Johnston 2020: 2 – 281.
De tremore, palpitatione, convulsione et rigore: VII.584 – 642.
Engl. transl.: David Sider and Michael McVaugh, "Galen, On tremor, palpitation, spasm and rigor. Galeni De tremore, palpitatione, convulsione, et rigore", *Transactions and Studies of the College of Physicians of Philadelphia* 5th series, 1 (1979), pp. 183 – 210.
De tumoribus praeter naturam: VII.705 – 732.
De usu partium: III.1 – 939, IV.1 – 366.
Engl. transl.: Margaret Tallmadge May, *Galen on the usefulness of the parts of the body.* Translated from the Greek with an Introduction and Commentary, 2 vols. Ithaca: Cornell University Press, 1968.
De usu respirationis: IV. 470 – 511.
De usu pulsuum: V.149 – 180.
Engl. transl.: David John Furley and James Sterling Wilkie, *Galen on Respiration and the Arteries.* Princeton, Princeton University Press, 1984.
De venae sectione adversus Erasistratum: XI.187 – 249.
Definitiones medicae: XIX.346 – 462.
Ed. and Germ. transl.: Jutta Kollesch, *[Galen], Medizinische Definitionen.* Herausgegeben und übersetzt (Corpus Medicorum Graecorum V 13,2). Berlin and Boston, MA: De Gruyter Akademie Forschung and De Gruyter, 2023.
In Hippocratem de acutorum morborum victu: XV.418 – 919.
In Hippocratis aphorismi: XVIIIB.345 – 887
In Hippocratis librum de officina medici commentarii III: XVIIIB.629 – 925.
In Hippocratis prorrheticum commentaria: XVI.489 – 840.
Linguarum seu dictionum exoletarum Hippocratis explicatio: XIX.62 – 157.
Synopsis librorum suorum sedecim de pulsibus: IX.431 – 549.

1.2 Other medical and scientific works

Aretaeus, *De causis et signis acutorum morborum:*
Greek text and Engl. transl.: Francis Adams, *The Extant Works of Aretaeus the Cappadocian.* London: The Sydenham Society, 1856.
Ed.: Karl Hude, *Aretaeus* (Corpus Medicorum Graecorum II). Berlin: Akademie Verlag, 1958.

Aristotle
Engl. transl.: Jonathan Barnes (ed.), *The Complete Works of Aristotle: The Revised Oxford Translation* (Bollingen Series 71), 2 vols. Princeton, NJ: Princeton University Press, 1984.

Aristotle, *Analytica Posteriora:*
Engl. transl.: Jonathan Barnes in Barnes 1984: 1.114–166.

Aristotle, *Analytica Priora:*
Engl. transl.: Arthur J. Jenkinson in Barnes 1984: 1.39–113.

Aristotle, *Categoriae:*
Engl. transl.: John Lloyd Ackrill in Barnes 1984: 1.1–24.

Aristotle, *De caelo:*
Engl. transl.: John L. Stocks in Barnes 1984: 1.447–511.

Aristotle, *De Juventute et Senectute:*
Engl. transl.: George R. T. Ross in Barnes 1984: 1.745–763.

Aristotle, *De Motu Animalium:*
Engl. transl.: Arthur S. L. Farquharson in Barnes 1984: 1.1097–1110.

Aristotle, *De Respiratione:*
Engl. transl.: William D. Ross in Barnes 1984: 1.764–773.

Aristotle, *Ethica Nicomachea:*
Engl. transl.: William D. Ross and James O. Urmson in Barnes 1984: 2.1729–1867.

Aristotle, *Historia Animalium:*
Ed. and Engl. transl.: Arthur Leslie Peck, *Aristotle, History of Animals* (Loeb Classical Library 438). Cambridge, MA: Harvard University Press, and London: William Heinemann, 1979.

Aristotle, *Metaphysica:*
Engl. transl. William D. Ross in Barnes 1984: 2.1552–1728.

Aristotle, *Problemata:*
Engl. transl.: Edward S. Forster in Barnes 1984: 2.1319–1527.

Aristotle, *Topica:*
Engl. transl.: Arthur W. Pickard, in Barnes 1984: 1.167–277.

Aristoxenus, *Rhythmica:*
Ed. and Germ. transl.: Rudolf Westphal, *Aristoxenus von Tarent: Melik und Rhythmik des classischen Hellenenthums.* Übersetzt und erläutert, 2 vols. Leipzig: Ambr. Abel, 1883–1893.

Caelius Aurelianus:
Engl. transl.: Israel E. Drabkin, *Caelius Aurelianus on Acute Diseases and Chronic Diseases.* Edited and translated. Chicago: University of Chicago Press, 1950.

Erasistratus:
Ed. and Ital. transl.: Ivan Garofalo (ed.), *Erasistrati Fragmenti* (Biblioteca di Studi Antichi, 62). Pisa: Giardini, 1988.

Harvey 1628/2005:
William Harvey, *Exercitatio Anatomica de Motu Cordis et Sanguinis in Animalibus.* Francofurti: Guilielmus Fitzerus, 1628.
Emerson Thomas McMullen, *William Harvey's De Motu Cordis.* A New Translation and Latin Edition. Herndon, VA: Academica Press, 2005.

Heracleides:
Ed. and Germ. transl.: Karl Deichgräber, *Die griechische Empirikerschule: Sammlung der Fragmente und*

Darstellung der Lehre. Berlin–Zürich: Weidmann, 1935, pp. 172 – 204 (fragments), 258 – 261.
Ed. and Ital. transl.: Alessia Guardasole, *Eraclide di Taranto. Frammenti. Testo critico, introduzione, traduzione e commentario* (Speculum 18). Napli: M. D'Auria, 1997.

Herophilus:
Heinrich von Staden, *Herophilus: The Art of Medicine in Early Alexandria.* Cambridge: Cambridge University Press, 1989.

Hippocrates:
Jones 1923: William Henry Stuart Jones, *Hippocrates*, Volume I (Loeb Classical Library 147). London: William Heinemann, and Cambridge: Harvard University Press, 1923.
Jones 1931: William Henry Stuart Jones, *Hippocrates*, Volume IV (Loeb Classical Library 150). London: William Heinemann, and Cambridge: Harvard University Press, 1931.
Withington 1928: Edward Theodore Withington, *Hippocrates*, vol. III (Loeb Classical Library 149). London: William Heinemann, and Cambridge, MA: Harvard University Press, 1928.

Hippocrates, *Aphorismi:*
Ed. and Engl. transl.: Jones 1931: 97 – 221.

Hippocrates, *De aeribus, aquis, locis:*
Ed. and Engl. transl.: Jones 1923: 65 – 138.

Hippocrates, *De alimento:*
Ed. and Engl. transl.: Jones 1923: 337 – 361.

Hippocrates, *De articulis:*
Ed. and Engl. transl.: Withington 1928: 201 – 398.

Hippocrates, *De humidorum usu:*
Ed. and Engl. transl.: Jones 1931: 61 – 96.

Hippocrates, *De ulceribus:*
Ed. and Engl. transl.: Paul Potter, *Hippocrates* Volume VIII (Loeb Classical Library 482). Cambridge MA: Harvard University Press, 1995, pp. 339 – 376.

Hippocrates, *Epidemiae I* et *III:*
Ed. and Engl. transl.: Jones 1923: 139 – 288.

Hippocrates, *Epidemiae IV–VII:*
Ed. and Engl. transl.: Wesley D. Smith, *Hippocrates*, vol. VII (Loeb Classical Library 477). Cambridge, MA: Harvard University Press, 1994, pp. 18 – 415.

Hippocrates, *Prognosticon:*
Ed. and Engl. transl.: William Henry Stuart Jones, *Hippocrates*, Volume II (Loeb Classical Library 148). London: William Heinemann, and Cambridge MA: Harvard University Press, 1923, pp. 1 – 56.

Praxagoras:
Ed.: Fritz Steckerl, *The Fragments of Praxagoras of Cos and his School.* Leiden: E.J. Brill, 1958.

Rufus of Ephesus, *Synopsis de pulsibus:*
Ed.: René Chartier, *Operum Hippocratis Coi, Et Galeni Pergameni, medicorum omonium principum*, Tomus VIII. Luetetiae Parisiorum, 1639, pp. 230 – 232, sub titulo *Compendium pulsuum Galeno adscriptum.*
Ed. and French transl.: Charles Daremberg, *Traité sur le pouls attribué à Rufus d'Ephèse.* Paris: imprimerie Dupont, 1848.
Ed. and French transl.: Charles Daremberg and Charles-Emile Ruelle, *Oeuvres de Rufus d'Ephèse.* Paris: Jean-Baptiste Baillière, 1879, pp. 219 – 232.

Stoicorum Veterum Fragmenta:
Ed.: Hans von Arnim, *Stoicorum Veterum Fragmenta* (Biblioteca Teubneriana), 4 vols. Leipzig: Teubner, 1903 – 1905.

1.3 Others

Cicero, *Epistulae ad Atticum:*
Ed. and Engl. transl.: Eric O. Winstedt, *Cicero, Letters to Atticus with an English translation, in three volumes*, vol. 1 (Loeb Classical Library 7). London: William Heninemann, and New York, NY: Macmillan, 1903

Euripides, *Phoenissae:*
Ed. and Engl. transl.: Arthur S. Way, *Euripides with an English translation in four volumes*, vol. 3 (Loeb Classical Library 11). London: William Heinemann, and New York, NY: Macmillan, 1912, pp. 339 – 491.

Homer, *Odysseia:*
Ed. and Engl. transl.: August T. Murray, *Homer, The Odyssey with an English translation in two volumes.* Revised by George E. Dimock, 2 vols (Loeb Classical Library 104 – 105). Cambridge, MA: Harvard University Press, 1995.

Pindar:
Ed. and Engl. transl.: John Sandys, *The Odes of Pindar including the principal fragments* (Loeb Classical Library 56). London: William Heinemann, and New York, NY; Macmillan, 1915.

Plato:
Engl. transl.: Edith Hamilton and Huntington Cairns (eds), *Plato: The Collected Dialogues.* Princeton, NJ: Princeton University Press, 1989.

Plato, *Hippias Major:*
Engl. transl.: Benjamin Jowett, in Hamilton and Cairns 1989: 1534 – 1559.

Plato, *Phaedrus:*
Engl. transl.: Reginald Hackforth, in Hamilton and Cairns 1989: 475 – 525.

Plato, *Respublica:*
Engl. transl.: Paul Shorey, in Hamilton and Cairns 1989: 575 – 844.

Plato, *Theaetetus:*
Engl. transl.: Francis Macdonald Cornford, in Hamilton and Cairns 1989: 845 – 919.

Plato, *Timaeus:*
Engl. trans.: Benjamin Jowett, in Hamilton and Cairns 1989: 1151 – 1211.

Plutarch, *Alcibiades:*
Ed. and Engl. transl.: Bernadotte Perrin, *Plutarch Lives*, Volume IV, (Loeb Classical Library 80) Cambridge MA: Harvard University Press, 1916, pp. 1 – 116.

Theocritus:
Ed. and Engl. transl.: John Maxwell Edmonds, *The Greek Bucolic Poets with an English translation* (Loeb Classical Library 28). London: William Heinemann, and New York, NY: G. P. Putnam's sons, 1912.

2 Literature

Adams 1849: Francis Adams, *The Genuine Works of Hippocrates.* London: The Sydenham Society, 1849.

Allbutt 1890: T. Clifford Allbutt, *Greek Medicine in Rome*, London: MacMillan and Company, 1921.

***Allgemeine Deutsche Biographie* 1876:** *Allgemeine Deutsche Biographie*, herausgegeben durch die historische Commission bi der königlichen Akademie der Wissenschaften, vol. 4. Leipzig: Duncker und Humblot, 1876.

Amati et al. 1800 – 1834: Girolamo Amati et al., *Inventarium codicum Vaticanorum Graecorum 993 – 2160.* Città del Vaticano, Biblioteca Apostolica Vaticana, Manuscript *Vaticanus graecus* 2664.

Amber and Babey-Brooke 1993: Reuben Amber and Anna M. Babey-Brooke, *Pulse Diagnosis.* Santa Fe, NM: Aurora Press, 1993.

Bacalexi 2014: Dina Bacalexi, "Le Traité de Galien *De pulsibus ad tirones:* Pratique Médicale et Représentation du Corps Humain", in Demetrios Michaelides (ed.), *Medicine and Healing in the Ancient Mediterranean World, including the Proceedings of the International Conference with the same title, organised in the framework of the Research Project 'INTERREG IIIA: Greece-Cyprus 2000–2006, Joint Educational and Research Programmes in the History and Archaeology of Medicine, Palaeopathology and Palaeoradiation' and the 1st Internaional CAPP Symposium 'New Approaches to Archaeological Humain remains in Cyprus'.* Oxford, and Philadlephia, PA: Oxbow Books, 2014, pp. 92–104.

Bandini 1764–1770: Angelo Maria Bandini, Catalogus codicum manuscriptorum Bibliothecae Mediceae Laurentianae ..., 3 volumes. Florentiae: Typis Caesareis (volume 1); Typis Regiis (volumes 2–3), 1764–1770 (reprint: Fridolf Kudlein [ed.], *Angelo Maria Bandini, Catalogus codicum manuscriptorum Bibliothecae Mediceae Laurentianae. Accedunt supplementa tria ab E. Rostagno et N. Festa congesta, necnon Additamentum ex inventariis Bibliothecae Laurentianae depromptum*, 2 volumes [Catalogi codicum graecorum lucis ope reimpressi]. Leipzig: Zentral-Antiquariat der Deutschen Demokratischen Republik, 1961).

Barnes 1984: Jonathan Barnes (ed.), *The Complete Works of Aristotle: The Revised Oxford Translation* (Bollingen Series 71), 2 vols. Princeton, NJ: Princeton University Press, 1984.

Baumstark 1922: Anton Baumstark, *Geschichte der syrischen Literatur mit Ausschluss der christlich-palänensischen Texte.* Bonn: A. Marcus und E. Webers Verlag Dr. jur. Albert Ahn, 1922.

Bergsträsser 1925: Gotthelf Bergsträsser, *Hunayn ibn Ishāq über die syrischen und arabischen Galen-Übersetzungen zum ersten Mal herausgegeben und übersetzt* (Abhandlungen für die Kunde des Morgenlandes, XVII. Band, No. 2). Leipzig: Brockhaus, 1925.

Berrey 2015: Marquis Berrey, "Three notes on Galen's pulse lore", *Classical Philology* 110(1) (January 2015), pp. 74–77.

Bhayro and Hawley 2014: Siam Bhayro, and Robert Hawley, "La littérature botanique et pharmaceutique en langue syriaque", in Villey 2014, pp. 285–318.

Boudon-Millot 2007: Véronique Boudon-Millot, *Galien*, Tome I. *Introduction Génerale, sur l'ordre de ses propres livres, sur ses propres livres, que l'excellent médecin est aussi philosophe* (Collection des Universités de France). Paris: Belles Lettres, 2007.

Boudon-Millot, Cobolet, and Jouanna 2012: Véronique Boudon-Millot, Guy Cobolet, Jacques Jouanna (eds), *René Chartier (1572–1654) éditeur et traducteur d'Hippocrate et Galien. Actes du colloque international de Paris (7 et 8 octobre 2019)* (Collection Medic@). Paris: Bibliothèque Interuniversitaire de Santé, Paris Descartes, CNRS, and Paris Sorbonne, 2012.

Boudon-Millot, Garzya, Jouanna, and Roselli 2010: Véronique Boudon-Millot, Antonio Garzya, Jacques Jouanna, and Amneris Roselli (eds.), *Storia della tradizione e edizione dei medici greci. Atti del VI Colloquio internazionale, Paris, 12–14 April 2008* (Collectanea 27). Napoli: D'Auria, 2010.

Bouras-Vallianatos and Zipser 2019: Petros Bouras-Vallinatos and Barbara Zipser (eds), *Brill's Companion to the Reception of Galen* (Brill's Companions to Classical Reception 17). Leiden, and Boston, MA: Brill, 2019.

Boutineau 1904: F. Emile Boutineau, "Notice sur la vie et les oeuvres de Martin Grégoire, médecin à Tours au XVIe siècle", *Bulletin de la Société française d'histoire de la médecine* 3 (1904), pp. 35–64.

Broadbent 1890: William Henry Broadbent, *The Pulse.* London: Cassell and Co, 1890.

Brock 1916: Arthur John Brock, *Galen: On the Natural Faculties*, (Loeb Classical Library 71). London: William Heinemann, and New York, NY: G. P. Putnam's Sons, 1916.

Browne 1921: Edward G. Browne, *Arabian Medicine.* Cambridge: Cambridge University Press, 1921.

Burnett 2005: Charles Burnett, "Marc of Toledo", in Glick, Livesey, and Wallis 2005, pp. 327–328.

Bylebyl 1979: Jerome J. Bylebyl, "Galen on the non-natural causes of variation in the pulse", *Bulletin of the History of Medicine* 53 (1979), pp. 482–485.

Cataldi Palau 1998: Annaclara Cataldi Palau, *Gian Francesco d'Asola e la tipografia Aldina. La vita, le edizioni, la biblioteca dell'Asolano.* Genova: sagep, 1998.

***Chambers Dictionary* 2008:** *The Chambers Dictionary*, 11th edition. London: 2008.

Conrad et al. 1995: Lawrence I. Conrad, Michael Neve, Vivian Nutton, Roy Porter, Andrew Wear, *The Western Medical Tradition, 800 BC to AD 1800.* Cambridge: Cambridge University Press, 1995.

Cortés Gabaudan and Martínez Manzano 2013: Francisco Cortés Gabaudan and Teresa Martínez Manzano, "Galeno según Galeno: *Synopsis de Pulsibus* IX, 531–549 Kühn", *Revue d'histoire des textes* n.s. 8 (2013), pp. 61–110.

Coxe 1853: Henry O. Coxe, *Catalogi Codicum Manuscriptorum Bibliothecae Bodleianae Pars Prima Recensionem Codicum Graecorum Continens.* Oxonii: E Typographo Academico, 1853 (reprint: Oxford: Bodleian Library, 1969).

Craik 2015: Elizabeth M. Craik, *The 'Hippocratic' Corpus: Content and Context.* London: Routledge, 2015.

Cronier 2021: Marie Cronier, "Les manuscrits grecs des 'Definitiones medicae' pseudo-galéniques", in Caroline Petit, Cimon Swain, Klaus-Dietrich Fischer (eds), *Pseudo-Galenica. The Formation of the Galenic Corpus from Antiquity to the Renaissance* (Warburg Institute Colloquia 34). London: Warburg Institute, 2021, pp. 47–67.

Cronier et al. 2015: Marie Cronier, Alessia Guardasole, Caroline Magdelaine, Antoine Pietrobelli, "Galien en procès à Byzance. L'Antirrhétique de Syméon Seth", *Galenos* 9 (2015), p. 71–121.

d'Alverny 1989/1994: Marie-Thérèse d'Alverny, "Marc de Tolède", in *Estudios sobre Alfonso VI y la reconquista de Toledo 3. Actas del II Congreso Internacional de Estudios Mozárabes, Toledo 20–26 Mayo 1985, Serie histórica 5.* Toledo: Instituto de estudios Visigótico-Mozárabes, 1989, pp. 25–59 (reprint: Marie-Thérèse d'Alverny, *La connaissance de l'Islam dans l'Occident médiéval* [Variorum Collected Studies Series CS 445]. Aldershot: Variorum, 1994, no. VII).

de Andrés 1965: Gregorio de Andrés, *Catálogo de los códices griegos de la Real Biblioteca de El Escorial*, t. 2. Madrid: [Real biblioteca de El Escorial], 1965.

de Lacy 1973: Philip de Lacy, "Galen's Platonism," *American Journal of Philology* 93 (1973), pp. 27–33.

de Lacy 1996: Phillip de Lacy, Galen, *On the Elements according to Hippocrates. Edition, Translation and Commentary* (Corpus Medicorum Graecorum V 1,2). Leipzig: Akademie Verlag, 1996.

de Meyïer 1955: Karel Adriaan de Meyïer, *Codices Vossiani Graeci et Miscellanei* (Bibliotheca Universitatis Leidensis, Codices manuscripti 6). Lugduni Batavorum: In Bibliotheca Universitatis, 1955.

Debru 2008: Armelle Debru, "Physiology", in Robert J. Hankinson (ed.), *The Cambridge Companion to Galen.* Cambridge: Cambridge University Press, 2008, pp. 263–82.

Degen 1981: Rainer Degen, "Galen im Syrischen: Eine Übersicht über die syrische Überlieferung der Werke Galens", in Vivian Nutton (ed.), *Galen: Problems and Prospects. A Collection of Papers submitted at the 1979 Cambridge Conference.* London: The Wellcome Institute for the History of Medicine, 1981, pp. 131–166.

Diels 1905: Hermann Diels, *Die Handschriften der antiken Ärzte. I. Teil. Hippokrates und Galenos. Im Auftrage der akademischen Kommission herausgegeben von –* (Abhandlungen der Königlich Preussischen Akademie der Wissenschaften vom Jahre 1905, Philosophisch-historische Klasse, Abhandlung III). Berlin: Königliche Akademie der Wissenschaften, 1905.

Diels 1906: Hermann Diels, *Die Handschriften der Antiken Arzte. II. Teil. Die übrigen griechischen Ärzte ausser Hippokrates und Galenos im Auftrage der Akademische Kommission herausgegeben von –* (Abhandlungen der Königlich Preussischen Akademie der Wissenschaften vom Jahre 1906, Philosophisch-historische Klasse, Abhandlung I). Berlin: Königliche Akademie der Wissenschaften, 1906.

Diels 1908: Hermann Diels, *Bericht über den Stand des interakademischen Corpus Medicorum Antiquorum und Erster Nachtrag zu den in den Abhandlungen 1905 und 1906 veröffentlichten Katalogen: Die Handschriften der antiken Ärzte. I. und II. Teil. Zusammengestellt im Namen der Kommission der Königlich Preussischen Akademie der Wissenschaften von –* (Abhandlungen der Königlich Preussischen Akademie der Wissenschaften vom Jahre 1907, Philosophisch-historische Klasse, Abhandlung II). Berlin: Königliche Akademie der Wissenschaften, 1908.

du Bois 1589: Jacques du Bois [Iacobus Sylvius], *Methodus sex librorum Galeni in differentiis et causis morborum et symptomatum in tabellas sex …* Paris: apud Andream Wechelum, 1589.

Durling 1961: Richard J. Durling, "Renaissance editions and translations of Galen", *Journal of the Warburg and Courtauld Institutes* 24 (1961), pp. 230 – 305.

Eloy 1778: Nicolas François Joseph Eloy, *Dictionnaire historique de la médecine ancienne et moderne. Mémoires disposés en ordre alphabétique pour servir à l'histoire de cette science et à celle des médecins, anatomistes, botanistes, chirurgiens et chimistes de toutes nations*, 4 vols. Mons: H. Hoyois, 1778.

Fortuna 1993: Stefania Fortuna, "Galen's *De constitutione artis medicae* in the Renaissance", *Classical Quarterly* 43 (1993), pp. 302 – 319.

Fortuna 1999: Stefania Fortuna, *Galeno, A Patrofilo sulla costituzione della medicina. Testo e traduzione* (Corpus Medicorum Graecorum V 1,3). Leipzig: Akademie Verlag, 1999.

Fortuna 2012a: Stefania Fortuna, "René Chartier e le edizioni latine di Galeno", in Boudon-Millot, Cobolet, and Jouanna 2012, pp. 303 – 324.

Fortuna 2012b: Stefania Fortuna, "The Latin Editions of Galen's Opera omnia (1490 – 1625) and their Prefaces", *Early Science and Medicine* 17 (2012), pp. 391 – 412.

Fortuna and Urso 2009: Stefania Fortuna and Anna Maria Urso, "Burgundio da Pisa traduttore di Galeno: nuovi contributi e prospettive", in Ivan Garofalo, Alessandro Lami, and Amneris Roselli (eds), *Sulla tradizione indiretta dei testi medici greci. Atti del II Seminario internazionale di Siena, Certosa di Pontignano, 19 – 20 Settembre 2008.* Pisa and Roma: Fabrizio Serra, 2009, pp. 139 – 175.

Furley and Wilkie 1984: David John Furley and James Sterling Wilkie, *Galen on Respiration and the Arteries.* Princeton, NJ: Princeton University Press, 1984.

García Ballester 1993: Luis García Ballester, "On the Origin of the Six Non-natural Things" in Jutta Kollesch and Diethard Nickel (eds), *Galen und das hellenistische Erbe. Verhandlungen des IV. Internationalen Galen-symposiums veranstaltet vom Institut für Geschichte der Medizin am Bereich Medizin (Charite) der Humboldt-Universitat zu Berlin 18.–20. September 1989* (Sudhoffs Archiv Beihefte 32). Stuttgart: Franz Steiner, 1993, pp. 105 – 115.

Gärtner 2015: Florian Gärtner, *Galen, Über das Erkennen erkrankter Körperteile I–II. Herausgegebn, übersetzt und erläutert.* Teil 1: Text und Übersetzung (Corpus Medicorum Graecorum V 6,1,1). Berlin and Boston, MA: De Gruyter, 2015.

Garofalo 1988: Ivan Garofalo, *Erasistrati Fragmenti* (Biblioteca di Studi Antichi 62). Pisa: Giardini, 1988.

Garofalo 1998: Ivan Garofalo, "Le commentaire ravennate au De pulsibus ad tirones de Galien: le commentaire et ses sources", in Carl Deroux (ed.), *Maladie et maladies dans les textes latins antiques et médiévaux. Actes du Ve Colloque International "Textes médicaux latins" (Bruxelles, 4 – 6 septembre 1995)* (Collection Latomus 242). Bruxelles: Latomus, Revue d'études latines, 1998, pp. 382 – 392.

Garofalo 2004: Ivan Garofalo, "Agostino Gadaldini (1515 – 1575) et le Galien latin", in Véronique Boudon-Millot, and Guy Cobolet (eds), *Lire les médecins grecs à la Renaissance. Aux origines de l'édition médical. Actes du Colloque international de Paris (19 – 20 septembre 2003)* (Collection Medic@). Paris: Bibliotèque Interuniversitaire de Médecine-Paris 5, CNRS-Paris 4, and De Boccard, 2004, pp. 283 – 321.

Garofalo 2010: Ivan Garofalo, "Prolegomena à l'édition du De pulsibus ad tirones de Galien", in Boudon-Millot, Garzya, Jouanna, and Roselli 2010, pp. 89 – 108.

Garofalo 2019: Ivan Garofalo, "Galen's Legacy in Alexandrian Texts Written in Greek, Latin, and Arabic", in Bouras-Vallianatos and Zipser 2019, pp. 62 – 85.

Ghasemzadeh and Maziar Zafari 2011: Nima Ghasemzadeh and Abfrmard Maziar Zafari, "A Brief Journey into the History of the Arterial Pulse", *Cardiology Research and Practice* 2011, article 164832.

Glick, Livesey, and Wallis 2005: Thomas Glick, Steven J. Livesey, and Faith Wallis (eds), *Medieval Science, Technology, and Medicine. An Encyclopedia.* New York, NY, and London: Routledge, 2005.

Gossen 1907: Iohannes Gossen, *De Galeni libro qui Σύνοψις περὶ σφυγμῶν inscribitur. Dissertatio inauguralis philologica quam consensu et auctoritate amplissimi philosophorum oridinis in alma litterarum universitate Friderica Guilelma Berolinensi ad summos in philosophia honores impetrandos scripsit.* Berlin: Friderica Guilelma Berolinensis Universitas, 1907.

Grmek 1991: Mirko D. Grmek, *Diseases in the Ancient Greek World.* Baltimore and London, Johns Hopkins University Press, 1991.

Hamilton and Cairns 1989: Edith Hamilton and Huntington Cairns (eds), *Plato: The Collected Dialogues.* Princeton, NJ: Princeton University Press, 1989.

Hankinson 1991: Robert J. Hankinson, *Galen on the Therapeutic Method. Books 1 and 2.* Oxford: Oxford University Press, 1991.

Hankinson 1998: Robert J. Hankinson, *Galen on Antecedent Causes.* Cambridge: Cambridge University Press, 1998.

Hanson and Green 1994: Ann Ellis Hanson and Monica H. Green, "Soranus of Ephesus: Methodicorum princeps", in Wolfgang Haase (ed.), *Aufstieg und Niedergang der römischen Welt (ANRW).* Teil II: Principat, vol. 37: Philosophie, Wissenschaften, Technik. 2. Teilband: Wissenschaften (Medizin und Biologie [Forts.]). Berlin: De Gruyter, 1994, pp. 968–1036.

Harris 1973: Charles Reginald Schiller Harris, *The Heart and Vascular System in Ancient Greek Medicine.* Oxford: Oxford University Press, 1973.

Haskins 1924: Charles Homer Haskins, *Studies in the History of Medieval Science.* Cambridge, MA: Harvard University Press, 1924.

Hirsch 1884–1888: August Hirsch (ed.), *Biographisches Lexikon der hervorragenden Aerzte alter Zeiten und Völker,* 6 vols. Wien and Leipzig: Urban & Schwarzenberger, 1884–1888.

Hunger and Kresten 1969: Herbert Hunger, and Otto Kresten, *Katalog der griechischen Handschriften der Österreichischen Nationalbibliothek,* Teil 2: Codices juridici, codices medici (Museion, Veröffentlichungen der Österreichischen Nationalbibliothek, Neue Folge, Vierte Reihe: Veröffentlichungen der Handschriftensammlung, Erster Band: Katalog der griechischen Handschriften der Österreichischen Nationalbibliothek, Teil 2). Wien: Georg Prachner, 1969.

***Hurst's The Heart* 1990:** Robert C. Schlant and R. Wayne Alexander (eds), *Hurst's The Heart.* New York, NY: McGraw-Hill, 1990.

Jones 1923: William Henry Samuel Jones, *Hippocrates* I (Loeb Classical Library 147). London: William Heinemann, and Cambridge, MA: Harvard University Press, 1923.

Johnston 2006: Ian Johnston, *Galen on Diseases and Symptoms.* Cambridge, MA: Cambridge University Press, 2006.

Johnston 2016: Ian Johnston, *Galen, On the constitution of the art of medicine; The art of medicine; A method of medicine to Glaucon* (Loeb Classical Library 523). Cambridge, MA: Harvard University Press, 2016.

Johnston 2020: Ian Johnston, *Galen, On Temperaments; On Non-uniform Distemperament; The Soul's Traits Depend on Bodily Temperament* (Loeb Classical Library 546). Cambridge, MA: Harvard University Press, 2020.

Johnston and Horsley 2011: Ian Johnston and Greg. H. R. Horsley, *Galen, Method of Medicine*, 3 vols. (Loeb Classical Library 516–518). Cambridge, MA: Harvard University Press, 2011.

Keyser and Irby-Massie 2008: Paul T. Keyser and Georgia Irby-Massie (eds), *Encyclopedia of Ancient Natural Scientists. The Greek tradition and its many heirs.* London and New York, NY: Routledge, 2008.

Köhler and Milchsack 1913: Franz Köhler and Gustav Milchsack, *Die Handschriften der Herzoglichen Bibliothek zu Wolfenbüttel.* Vierte Abteilung. Die Gudischen Handschriften. Die griechischen Handschriften bearbeitet von Franz Köhler. Die lateinischen Handschriften bearbeitet von Gustav Milchsack. Wolfenbüttel: Julius Zwissler, 1913.

Kollesch 1973: Jutta Kollesch, *Untersuchungen zu den pseudogalenischen* Definitiones medicae (Schriften zur Geschichte und Kultur der Antike 7). Berlin: Akademie Verlag, 1973.

Kollesch 2023: Jutta Kollesch, *[Galen], Medizinische Definitionen.* Herausgegeben und übersetzt (Corpus Medicorum Graecorum V 13,2). Berlin and Boston, MA: De Gruyter Akademie Forschung and De Gruyter, 2023.

Lamagna 2010: Mario Lamagna, "Gli Scoli a Magno di Stefano d'Atenea", in Boudon-Millot, Garzya, Jouanna, and Roselli 2010, pp. 241–257.

Lamoreaux 2016: John C. Lamoreaux, *Hunayn ibn Ishāq on His Galen Translations. A parallel English-Arabic text edited and translated* (Middle Eastern Texts Initiative, Eastern Christian Texts 5). Provo, UT: Brigham Young University Press, 2016.

Le Coz 2004: Raymond Le Coz, *Les médecins nestoriens au moyen âge. Les maîtres des Arabes* (Comprendre le Moyen-Orient). Paris: L'Harmattan, 2004.

Lewis 2015: Orly Lewis, "Marcellinus' *De pulsibus:* A neglected treatise on the ancient art of the pulse", *Scripta Classica Israelica* 34 (2015), pp. 195 – 214.

Lewis 2017: Orly Lewis, *Praxagoras of Cos on Arteries, Pulse and Pneuma. Fragments and Interpretation* (Studies in Ancient Medicine 48). Leiden: E.J. Brill, 2017.

Lin 2004: Zheng-Hong Lin. *Pocket Atlas of Pulse Diagnosis*, New York: Thieme, 2004.

Lloyd 1978: Geoffrey E. R. Lloyd, *Hippocratic Writings.* London: Penguin Books, 1978.

Longrigg 1993: James Longrigg, *Greek Rational Medicine. Philosophy and Medicine from Alcmaeon to the Alexandrians.* London: Routledge, 1993.

Maillard, Kecskeméti and Portalier 1995: Jean-François Maillard, Judit Kecskeméti, and Monique Portalier, *L'Europe des humanistes (XIVe-XVIIe siècles)* (Documents, Etudes et Répertoires publiés par l'Institut de Recherche et d'Histoire des Textes). Tunhout: Brepols, 1995.

Martínez Gázquez 2001: José Martínez Gázquez, "Observaciones a la traduccion latina del Coran (Qur'an) de Robert de Ketene", in Jacqueline Hamesse (ed.), *Les traducteurs au travail, leurs manuscrits et leurs méthodes. Actes du Colloque international organisé par le "Ettore Majorana Centre for Scientific Culture" (Erice, 30 Septembre—6 Octobre 1999)* (Fédération Internationale des Instituts d'Etudes Médiévales, Textes et études du Moyen Age 18). Turnhout: Brepols, 2001, pp. 115 – 142.

Martini and Bassi 1906: Aemidius Martini and Dominicus Bassi, *Catalogus codicum graecorum Bibliothecae Ambrosianae.* Digesserunt, 2 volumes. Mediolani: Impensis U. Hoepli, 1906 (reprint: Hildesheim and New York, NY: Georg Olms, 1978).

May 1968: Margaret Tallmadge May, *Galen on the usefulness of the parts of the body. Περὶ χρείας μορίων. De usu partium.* Translated from the Greek with an Introduction and Commentary, 2 vols. Ithaca, NY: Cornell University Press, 1968.

McKendrick 1999: Scot McKendrick, *The British Library. A Summary Catalogue of Greek Manuscripts*, volume 1. London: British Library, 1999.

Menétrier 1924: Pierre Eugène Menétrier, "A propos du traité du pouls attribué à Rufus, et de la sphygmologie ancienne", *Bulletin de la Société Française d'Histoire de la Médecine* 18 (1924), pp. 97 – 98.

Mercati and Franchi de' Cavalieri 1923: Giovanni Mercati and Pio Franchi de' Cavalieri, *Bibliothecae Apostolicae Vaticanae codices manu scripti recensiti, Codices Vaticani graeci*, Tomus I: Codices 1 – 329. Romae: Typis Polyglottis Vaticanis, 1923.

Mettler 1947: Cecilia C. Mettler, *History of Medicine.* Philadelphia, PA: The Blakiston Company, 1947.

Mioni 1981: Elpidio Mioni, *Bibliothecae Divi Marci Venetiarum, Codices graeci manuscripti*, volumen I, Thesaurus antiquus 1 – 299 (Indici e cataloghi, Nuova serie 6). Roma: Istituto Poligrafico e Zecca dello Stato, Libreria dello Stato, 1981.

Molhuysen 1910: Philipp Christiaan Molhuysen, *Codices Scaligerani (praeter Orientales)* (Bibliotheca Universitatis Leidensis, Codices manuscripti 2). Lugduni-Batavorum: E. J. Brill, 1910.

Moran 1990: John F. Moran, "The Pulse", in H. Kenneth Walker, W. Dallas Hall, and J. Willis Hurst (eds), *Clinical Methods. The History, Physical and Laboratory Examinations*, 3rd edition. Boston, MA: Butterworths, 1990, pp. 98 – 100.

Muraviev 2014: Alexey Muraviev, "La médecine thérapeutique en syriaque (IVe-VIIe siècles), in Villey 2014, pp. 253 – 284.

Ni 1995: Ni Maoshing, *The Yellow Emperor's Classic of Medicine (A New Translation of the Neijing Suwen with Commentary).* Boston, MA, and London: Shambhala, 1995.

Nutton 2002: Vivian Nutton, “In Defence of Kühn”, in Vivian Nutton (ed.), *The Unknown Galen* (Bulletin of the Institute of Classical Studies, Supplement 11). London: Institute of Classical Studies, School of Advanced Studies, University of London, 2002, pp. 1–7.

Olivieri and Festa 1895: Alessandro Olivieri, and Nicola Festa, “Indice dei codici greci delle biblioteche Universitaria e Comunale di Bologna”, *Studi italiani di Filologia classica* 3 (1895), pp. 385–495 (reproduced in Samberger 1965–1968, volume 1, pp. 1–113).

Omont 1886–1888: Henri Auguste Omont, *Inventaire sommaire des manuscrits grecs de la Bibliothèque nationale et des autres bibliothèques de Paris et des Départements*, 3 volumes. Paris: Ernest Leroux, 1886–1888.

O’Rourke, Kelly and Avolio 1993: Michael F. O’Rourke, Raymond P. Kelly, and Alberto Avolio, *The Arterial Pulse.* Philadelphia, PA: Lea and Febiger, 1993.

***Oxford Classical Dictionary* 1996:** Simon Hornblower and Antony Spawforth (eds), *Oxford Classical Dictionary*, 3rd edition. Oxford: Oxford University Press, 1996.

Palmieri 2005: Nicoletta Palmieri, *Agnellus de Ravenne, Lectures galéniques: le “De pulsibus ad tirones”. Introduction, texte latin, traduction et notes explicatives* (Centre Jean Palerne, Mémoires 28). Saint-Etienne: Publications de l’Université de Saint-Etienne, 2005.

Papavramidou 2004: Niki Papavramidou, “Galen on Obesity”, *World Journal of Surgery* 28(6) (2004), pp. 831–835.

Perilli 2017: Lorenzo Perilli, Galeno, *Interpretazione delle parole difficili di Ippocrate. Testo, traduzione e note di commento* (Corpus Medicorum Graecorum V 13,1). Berlin and Boston, MA: De Gruyter, 2017.

Petit 2007: Caroline Petit, “Gadaldini’s Library”, *Mnemosyne* 60 (2007), pp. 132–138.

Phillips 1987: Eustace D. Phillips, *Aspects of Greek Medicine.* Philadelphia: Charles Press, 1987.

Pietrobelli 2013: Antoine Pietrobelli, “Une nouvelle copie d’impression de l’aldine de Galien: le Guelferbytanus Gud. Gr. 11 (= 4198)”, *Galenos* 7 (2013), pp. 137–149.

Pithis 1983: John A. Pithis, *Die Schriften Περὶ σφυγμῶν des Philaretos. Text, Übersetzung, Kommentar* (Abhandlungen zur Geschichte der Medizin und der Naturwissenschaften 46). Husum: Matthiesen Verlag, 1983.

Puntoni 1896: Vittorio Puntoni, “Indice dei codici greci della Biblioteca Estense di Modena”, *Studi italiani di filologia classica* 4 (1896), pp. 379–536 (reproduced in Samberger 1965–1968, volume 1, pp. 293–452).

Rathbone and Nutton 1996: Dominic R. Rathbone and Vivian Nutton, “Archigenes”, in *Oxford Classical Dictionary* 1996: 145.

Rocca 2003: Julius Rocca, *Galen on the Brain. Anatomical Knowledge and Physiological Speculation in the Second Century AD* (Studies in Ancient Medicine 26). Leiden: Brill, 2003.

Rodriguez Suarez 2016: Alex Rodriguez Suarez, “From Greek into Latin: Western scholars and translators in Constantinople during the reign of John II”, in Alessandra Bucossi and Alex Rodriguez Suarez (eds), *John II Komnenos, Emperor of Byzantium: in the Shadow of Father and Son* (Centre for Hellenic Studies, King’s College London, Publications 17). London and New York, NY: Routledge, pp. 91–109.

Sadek 1983: Mahmoud M. Sadek, *The Arabic Materia Medica of Dioscorides.* St-Jean-Chrysostome: Editions du Sphinx, 1983.

Samberger 1965–1968: Christa Samberger (ed.), *Catalogi codicum graecorum qui in minoribus bibliothecis italicis asservantur in duo volumina collati et novissimis additamentis aucti*, 2 volumes (Catalogi codicum graecorum lucis ope reimpressi). Lipsiae: Zentral Antiquariat der deutschen demokratischen Republik, 1965–1968.

Sangrador Gil 1997: José Sangrador Gil, “La escuela de traductores de Toledo durante la Edad Media”, in Miguel Hernando de Larramendi and Gonzalo Fernández Parrilla (eds), *Pensamiento y circulación de las ideas en el Mediterráneo: el papel de la traducción.* Cuenca: Ediciones de la Universidad de Castilla-La Mancha, 1997, pp. 25–52.

Sarton 1927: George Sarton, *Introduction to the History of Science*, vol. 1, Baltimore MD: Williams and Wilkins, 1927.

Savino 2012: Christina Savino, "Giovanni Battista Rasario and the 1562 – 1563 Edition of Galen. Research, Exchanges and Forgeries", *Early Science and Medicine* 17 (2012), pp. 413 – 445.

Schartau 1994: Bjarne Schartau, *Codices Graeci Haunienses: ein deskriptiver Katalog des griechischen Handschriftenbestandes der Königlichen Bibliothek Kopenhagen* (Danish humanist texts and studies 9). Copenhagen: Museum Tusculanum Press, and The Royal Library, 1994.

Scimone 2021a: Alessandra Scimone, *Galenus Latinus: la traduzione di Burgundio di Pisa del* De causis pulsuum. *Introduzione, testo critico e indici.* PhD. Th., Università degli Studi di Salerno-Université de Reims Champagne-Ardennes, 2021.

Scimone 2021b: Alessandra Scimone, "Burgundio da Pisa tradutore del *De pulsibus ad tirones* e del *De causis pulsuum* di Galeno", *Galenos* 15 (2021), pp. 59 – 92.

Sezgin 1970: Fuat Sezgin, *Geschichte des arabischen Schrifttums*, vol. 3: Medizin, Pharmazie, Zoologie, Tierheilkunde bis ca. 430 H. Leiden: E.J. Brill, 1970.

Steckerl 1958: Fritz Steckerl, *The Fragments of Praxagoras of Cos and his School.* Leiden: E.J. Brill, 1958.

Stedman's Medical Dictionary: *Stedman's Medical Dictionary* 27th edition. Baltimore, MD: Lippincott, Williams and Wilkins, 2000.

Stevenson 1885: Henry Stevenson, *Codices manuscripti Palatini graeci Bibliothecae Vaticanae descripti.* Romae: Ex Typographeo Vaticano, 1885.

Striker 1999: Gisela Striker, "Timon of Phlius", in *Oxford Classical Dictionary* 1996, p. 1529.

Strohmaier 1970: Gothard Strohmaier (ed.), *Galeni De partium homoeomerium differentia* (Corpus Medicorum Graecorum, Supplementum Orientale III). Berlin: Akademie Verlag, 1970.

Studemund and Cohn 1890: Wilhelm Studemund and Leopold Cohn, *Codices ex Bibliotheca Meermanniana Phillippici Graeci nunc Berolinenses* (Verzeichniss der griechischen Handschriften der königlichen Bibliothek zu Berlin 1). Berolini: A. Asher & Co, 1890.

Tassinari 2019: Piero Tassinari, "Galen into the Modern World: from Kühn to the *Corpus Medicorum Graecorum*", in Bouras-Vallinatos and Zipser 2019, pp. 508 – 534.

Touwaide 2008a: Alain Touwaide, "Galēn, pseudo, *Definitiones Medicinales* (100 – 150 CE)", in Keyser and Irby-Massie 2008, pp. 340 – 341.

Touwaide 2008b: Alain Touwaide, "Philaretos", in Keyser and Irby-Massie 2008, pp. 644 – 645.

Touwaide 2016: Alain Touwaide, *A Census of Greek Medical Manuscripts. From Byzantium to the Renaissance* (Medicine in the Medieval Mediterranean 6). London, and New York, NY: Routledge, 2016.

Touwaide 2021: Alain Touwaide, *Greek Medical Manuscripts. Diels' Catalogue*, 5 tomes (Medical Traditions 2). Berlin and Boston, MA: De Gruyter, 2021.

Tunis 1989: Elizabeth Tunis, *Early Western Manuscripts in the National Library of Medicine. A Short-Title List.* Bethesda, MD: History of Medicine Division, National Library of Medicine, 1989.

Ullmann 1970: Manfred Ullmann, *Die Medizin in Islam* (Handbuch der Orientalistik, Erste Abteilund: Der nahe und der mittlere Osten, Ergänzungsband VI, Erster Abschnitt). Leiden, and Köln: E.J. Brill, 1970.

Urmson 1990: James Opie Urmson, *The Greek Philosophical Vocabulary.* London: Duckworth, 1990.

Urso 2019: Anna Maria Urso, "Translating Galen in the Medieval West: the Greek-Latin Translations", in Bouras-Vallianatos and Zipser 2019, pp. 359 – 380.

Vagelpohl 2016: Uwe Vagelpohl (ed.), *Galen, Commentary on Hippocrates' Epidemics Book II, Parts I–VI. Edition of the Arabic Version and English Translation*, volume I: Parts I–III (Corpus Medicorum Graecorum, Supplmentum orientale V 2). Berlin and Boston, MA: De Gruyter, 2016.

Vallance 1990: John T. Vallance, *The Lost Theory of Asclepiades of Bithynia.* Oxford: Clarendon Press, 1990.

Villey 2014: Emilie Villey (ed.), *Les sciences en syriaque* (Etudes syriaques 11). Paris: Geuthner, 2014.

Vladimir 1894: Архимандрит Владимир, *Систематическое описаніе рукописей Московской синодальной (патріаршей) библіотеки: Рукописи греческія.* Москва: Синодальная Типография. 1894.

von Arnim 1903 – 1905: Hans F. A. von Armin, *Stoicorum Veterum Fragmenta.* 4 vols (Biblioteca Teubneriana). Leipzig: Teubner, 1903 – 1905.

von Staden 1989: Heinrich von Staden, *Herophilus: The Art of Medicine in Early Alexandria.* Cambridge: Cambridge University Press, 1989.

Wong and Lien-the 1936: K. Chimin Wong and Wu Lien-the, *History of Chinese Medicine.* Shanghai: National Quarantine Service, 1936.

Wood 1967: Paul Wood, *Diseases of the Heart and* Circulation, 3rd edition. London: Eyre and Spottiswoode, 1967.

Yang 1997: Yang Shou-zheng, *The Pulse Classic: A Translation of the Mai Jing by Wang Shu-he.* Boulder, CO: Blue Poppy Press, 1997.

Indexes

Personal and Place Names

https://doi.org/10.1515/9783110612677-015

Authors and Treatises

1 Self references and references in one of the translated works to other of the translated works in the present volume are not listed in this Index.

General

www.ingramcontent.com/pod-product-compliance
Ingram Content Group UK Ltd.
Pitfield, Milton Keynes, MK11 3LW, UK
UKHW050942290926
15470UKWH00005B/1590